Magazine Markets
for Children's
Writers 2009

**Writer's Institute
Publications**

Acknowledgments

The editors of this directory appreciate the generous contributions of our instructors and students and the cooperation of the magazine editors who made clear their policies and practices.

MARNI McNIFF, Editor

SUSAN TIERNEY, Articles Editor

BARBARA COLE, Assistant Editor

SHERRI KEEFE, Assistant Editor

BECKY FORSTROM, Assistant Editor

AMANDA NORELLI, Research Assistant

Contributing Writers: SUSAN ANDERSON, KRISTEN BISHOP, JENNIFER PONTE CANNING, SUSAN TARRANT

Cover Design: JOANNA HORVATH
Cover illustrations supplied by Getty Images

International Standard Book Number 978-1-889715-43-8

1-800-443-6078. www.writersbookstore.com
email: services@writersbookstore.com

Contents

Step-by-Step through the Submissions Process

This section offers tips and step-by-step instructions for researching the market; writing query letters; preparing manuscript packages and bibliographies; initiating follow-up; and understanding copyrights, permissions, and other common publishing terms.

Gateway to the Markets

Contents (cont.)

Step-by-Step through the Submissions Process

Sales Savvy

"The secret of good writing is to say an old thing in a new way or to say a new thing in an old way." —Richard Harding Davis

The January 2008 issue of *Odyssey* included a nonfiction article called "A Wizard's Vision of a Wireless World," by Mary Beth Cox. The article, which appeared in the magazine's themed issue about "our unwired world," gave a lively summary of nineteenth-century inventor Nikola Tesla and his experiments with wireless power. Cox ended the piece with information about what scientists at MIT are currently doing to make wireless power a reality.

The March 2008 issue of *Highlights for Children* included a fictional story called "The Mystery of the Magic Lunch Bag," by Neal Levin. A young girl named Sheila is at first disappointed to discover that her special new lunch bag doesn't, unfortunately, bring with it a new and exciting lunch. But later, when her tuna-fish sandwich mysteriously turns into a slice of pizza, Sheila believes it's a magic bag. The ending packs a surprise: Sheila's lunch bag has gotten mixed up with another classmate's bag that looks exactly the same. Both girls get a better lunch—and a new friend, as well.

Strategy from the Start

Both of these pieces are standouts, and not because they cover unusual topics or reach new audiences. The reason for their success is simple: Both authors chose interesting, relevant topics and approached them in a fresh way, then pitched their ideas to the right publication. It helps to have a thoughtful sales strategy to accompany your well-written article.

Magazine editors are always looking for the perfect match between a writer's idea and their magazine. Many publications have been around for years and have carefully cultivated their readers, who expect a certain kind of article with a particular style and tone. As a writer, the best way to make that match is to start from the beginning.

Choosing a Topic

Deciding on a subject is the first step toward making a sale. Whether your topic is fiction or nonfiction, you'll increase your publication odds if you know what topics interest children. In the Tesla piece, for example, the writer knew that connecting a historical figure to the contemporary concept of wireless technology would make an article about the inventor much more appealing to middle-graders. It also connects the past to the present, a learning concept that appeals to classroom magazine editors. When you're thinking about ideas, consider how a young person views the world. What is their typical day like? What kinds of things are they interested in? Questions like these can produce a number of interesting subject leads.

Another way to find an appropriate subject is to review current publications. Start by surveying the categories of magazines in *Magazine Markets for Children's Writers*. The Category Index on pages 341–376 is an excellent guide to finding magazines that publish the types of articles or stories you write on the subjects that interest you. You'll find everything from general interest publications, like *Boys' Quest* or *Highlights for Children*, to special interest periodicals on topics such as sports, careers, crafts, current events, parenting, geography, health and fitness, and more.

Continue your research online or at libraries. What magazines are out there, for what ages, and what subjects do they cover? Along with magazines targeted specifically to children, be sure to check parenting, educational, and regional magazines.

You'll find that each magazine covers numerous subjects from month to month or year to year, even special interest publications that cover a niche more deeply than widely. Read several issues of each magazine to find out which subjects a potential target magazine has covered recently and how it has approached particular subjects in the past. Begin to make a list of the magazines that cover subjects of interest to you.

Targeting Your Readers

Subject and audience often go hand in hand, though many subjects can be geared toward a variety of audiences, with the right treatment.

Don't mistake your *subject* for your *slant*: A *subject* is a broad topic, while the *slant* is what gives an article its uniqueness. To help find your slant, narrow down your subject by focusing on its parts, such as key people, places, or events.

Select a subject slant that is age-appropriate for your intended audience and your potential market. For example, if you'd like to take on the subject of architecture you might write a story on the basics of building design for younger readers or the more complex slant of conservationist/green design for high-school readers. Both articles vary in the amount and complexity of information offered, and in tone. Determine each magazine's target age and how the publication speaks to that age, through voice and purpose.

To learn more about the developmental level of your intended age range, go to the Internet and other media, as well as to schools and children's activities. For example, at the Google Directory (www.google.com/dirhp) click on Kids and Teens. Look at the topics under preschool, school time, teen life, and other categories. The arts section in particular has many interesting sites that relate to every age group.

Understanding the Magazine

Once you have a good handle on your subject and your audience, start doing in-depth research for those magazines you intend to target. Create a magazine market file for publications that seem to match your interests. Use index cards, a notebook, or your computer to develop a file for each magazine for your initial list of publications.

Review the Writers' Guidelines

If the listing for a particular magazine indicates that it offers writers' guidelines, either send a letter to the publication along with an SASE (see the sample below) to request the guidelines, or check the magazine's website. (The listing will specify where to look for this information.) Writers' guidelines, editorial calendars, and theme lists may give you specific topics to write about, or they may be more general.

Either way, follow the guidelines carefully, or your submission could be rejected by the editor. They are key to the needs of publications and often new writers give them too little weight.

Some guidelines are more detailed and helpful than others, but virtually all will tell you something about the readership,

Sample Guidelines Request

Name
Address
City, State ZIP

Date

Dear (Name of Current Editor):

I would like to request a copy of your writers' guidelines and editorial calendar. I have provided a self-addressed, stamped envelope for your convenience.

Sincerely,

Your name

philosophy, and voice, as well as word length requirements, submission format, and payment. More than that, some guidelines give writers specific insight into the immediate needs of a magazine. For example, *Adoptive Families* magazine has thorough guidelines that include a listing of its departments, instructions on how to submit your work, subject areas editors are looking to cover, and suggestions for writers, such as "have a clear sense of your central theme," and "focus on choices made and strategies used to deal with a particular situation." *Pack-O-Fun*'s guidelines offer a detailed list of the kinds of crafts its editors are looking for year-round.

Depending on their level of detail, some guidelines may also indicate the rights a publication purchases, payment policies, and many more specifics—factors you'll consider as you get closer to submission. Many experienced writers do not sell all rights, unless the fee is high enough to be worth it; reselling articles or stories for reprint rights can be an additional source of income. (See the discussion of rights on page 27.)

Review Sample Issues

Get sample issues of the magazines and read them, either by requesting them from the publisher; finding copies at the library, the bookstore, or through friends; or by reviewing articles on the magazine's website. The listings in *Magazine Markets for Children's Writers* will tell you if writers' guidelines, an editorial calendar, or a theme list are available, as well as the cost of a sample copy requested directly from the publisher.

Review each of the magazines in more detail for subjects similar to yours. You should also check the *Readers' Guide to Periodical Literature* in your library to see if a

---

Magazine Description Form

Name of Magazine: *Justine Magazine*   **Editor:** Jana Petty
Address: 6263 Poplar Ave., Suite 1154, Memphis, TN 38119

Freelance Percentage: 20%   **Percentage of Authors Who Are New to the Magazine:** 90%

Description
What subjects does this magazine cover? *Justine Magazine* features informational articles, profiles, and personal experience pieces on topics such as room decorating, health, recreation, travel, fashion, and more.

Readership
Who are the magazine's typical readers? 13- to 18-year-olds

Articles and Stories
What particular slants or distinctive characteristics do its articles or stories emphasize? Articles are wholesome and tasteful, and focus on building self-confidence and healthy relationships.

Potential Market
Is this magazine a potential market? Yes. My article idea is aimed at this audience and fits in with the magazine's mission.

Ideas for Articles or Stories
What article, story, or department idea could be submitted? An article about the Rock 'n Roll Camp for Girls, which builds self-esteem through music creation and performance. It would include interviews with campers and a sidebar on how teens can start their own band.

target magazine has covered your topic within the past two years. If so, you may want to find another magazine or, depending on the publication, develop a new slant if you find that your subject is already well covered.

Use the Magazine Description Form (see page 8) to continue your detailed analysis of the publications, especially those you're beginning to hone in on as good matches. Evaluate how you could shape or present your manuscript to improve your chances of getting it published. If a particular idea or target magazine doesn't work out now, it may in the future—or it may lead to other ideas, angles, or possible markets. Review your market files periodically to generate ideas.

Your review of sample magazines and guidelines should include:

- **Editorial objective.** In some cases, the magazine's editorial objective is stated inside the publication, on the same page as the masthead, where the names of the editors are listed; it is also usually stated on the website. For example, the editorial objective of *Highlights for Children* is summarized in its simple subhead, "Fun with a Purpose." Does your story or article fit your targeted magazine's purpose?

- **Audience.** What is the age range of the readers and the characters or children portrayed? For fiction, is your main character at the upper end of that range? Kids want to read about characters their own age or older.

- **Article and story types.** Examine the types of articles and stories in the issue, paying careful attention to article construction. Are the articles informal, or are they filled with facts and statistics? Are they illustrated with personal experience stories? Are sidebars, subheads, and/or other elements included, and are they interactive? Think about the presentation as well: Is the magazine highly visual or does it rely primarily on text? Will photographs or illustrations be a consideration for you?

- **Style.** Become familiar with the magazine's editorial style, and how it is

New Angles, New Audiences

A substantial amount of research goes into writing magazine articles, and it's to your advantage to make the most of this time. In many cases, the research for one article can be used for other articles, as long as they reach different audiences and have fresh angles. To find additional sales opportunities, consider other markets that might be interested in the information—religious publications, arts and crafts magazines, science magazines, etc. Then investigate angles that would be of most interest to them. Make a list of the possibilities, like the ideas below, and keep it handy for future reference.

Subject: Green Energy

Magazine	Audience Age	Slant
ChemMatters	14-18 years	An interview with ethanol expert David Blume, describing a simple technique used to convert vehicles to run on ethanol
Boys' Life	6-18 years	An article on green chemistry in relation to energy production
SuperScience	Grades 3-6	An article about how corn is used to produce ethanol, covering the pros and cons of this technique

impacted by the age of the audience. Does it strive for a conversational, energetic style, or a straightforward, educational one? Are the sentences simple or complex, or a mixture of both? Are there numerous three-syllable words, or mostly simple words? Do the writers speak directly to the readers or is the voice appropriately authoritative?

- **Editor's comments.** Very often the writers' guidelines include insight from the editor about the feel of a magazine. For example, *Instructor* magazine's guidelines state, "Write in your natural voice, as if you were talking to a colleague. We shy away from wordy, academic prose." *Breakaway* magazine's guidelines describe it as "fast-paced," "compelling," and "out of the ordinary." Remarks such as these offer valuable clues for writers who are trying to break in. *Magazine Markets for Children's Writers* includes a section called Editor's Comments in each listing. Study this section carefully for similar tips about what editors want to see, or don't need.

Refine Your Magazine List

After you analyze your selected magazines, rank them by how well they match your idea, subject, style, and target age. Then return to the listings to examine other factors, such as the magazine's freelance potential, its receptivity to new or unpublished writers, rights purchased, and payment.

Not only should your decision to submit be based on how well your idea matches a particular publication, you should also consider which publications match you as a writer. An examination of magazine business policies—not just current editorial needs—can reveal significant details about the magazine that you can use to your advantage as a freelance writer. For example, many published writers prefer magazines that:

- respond in one month as opposed to three or more;
- pay on acceptance rather than on publication;
- do not purchase all rights (see the rights discussion on p. 27);

- publish a high percentage of authors who are new to the magazine.

If you're not yet published, however, writing for a nonpaying market or taking risks in other areas (such as signing a work-for-hire contract or agreeing to payment on publication rather than on acceptance) may be worth the effort to earn the clips needed for future submissions. Once you've acquired credentials in these markets, you can list these published pieces in your queries to other markets.

If you don't yet have published clips to send with your submission, choose writing samples that demonstrate your skills as a writer, such as blog entries, Web content, sales letters, newsletters, and letters to editors. All are considered acceptable forms of writing samples.

Submitting Your Work to an Editor

Submission policies vary across the board; some magazines accept queries only, others accept complete manuscripts or queries, while others want queries accompanied by writing samples, a synopsis, an outline, or other information. A query may be sufficient for some editors; others prefer to get a more complete sense of you and your work before making a determination. To know for sure what to send, check the writers' guidelines for the publication you're interested in. Expect that the editor who accepts a complete, unsolicited manuscript may require even more revisions or rewrites than if you had queried first.

The Right Stuff

So you've looked it up, and the guidelines of your target magazine say to "query with outline, bibliography, and clips or writing samples." In this case, you should send the following:

- One-page query letter
- Brief outline of the topics your article will cover
- Bibliography of research sources
- Selected clips (published writing) or writing samples (unpublished writing)
- SASE

For nonfiction submissions, it's always a good idea to include a bibliography of your research sources, whether you're sending a query letter or a complete manuscript. A well-rounded bibliography with a variety of sources demonstrates a professional approach and what is likely a fresh take on a subject. For complete manuscript submissions of both fiction and nonfiction, include a cover letter that briefly introduces the work (see p. 20). And unless you're submitting via email (see below), an SASE is a necessary part of any submission package.

Email Submissions

Email is an efficient means of communication in business today, but publications still vary widely in their policies regarding email submissions: Some publishers prefer to receive submissions via email, others accept both print and email submissions, and still others avoid email submissions altogether. The only way to know for sure is to check each publisher's guidelines.

Before you submit via email, be sure that your electronic query is as carefully crafted as a print one. Email queries by nature are slightly less formal, but not sloppy, and the content of your query should be as informative and engaging as a traditional one. Beware, however, the conversational, too-familiar tone of many emails, which is inappropriate for queries and submissions. Your query is your first contact with an editor, so write in a professional manner. Pay close attention to grammar and punctuation and avoid using cutesy email addresses and emoticons. And remember—email does not always mean faster response times, so be patient with the editor and respect his or her time.

Before you hit the send button, check that you've complied with any guidelines specific to email submissions, such as:

- *File formats:* Should the submission be included in the body of the email or as an attachment? Should files be sent in Microsoft Word or Rich text format?
- *Subject line:* Should your subject line say "Query" or "Submission," and include your name or the title of the work?
- *Contact information:* Your name, address, email address, and telephone number should be included below your "signature."

Crafting a Query Letter

The query letter is a writer's most important tool: It must sell the editor on you and your ability to write, amongst hundreds of similar letters. The truth is that good query letters are few and far between, so mastering this tool is well worth your while.

There are several advantages to using a query letter. First, it is the preferred submission method of many publications. Editors are deluged with pitches, and queries offer a quick, easy way to identify those with potential. For this reason, editors typically respond faster to queries than manuscripts. Also, many query letters are written before the manuscript. Whether or not this is the case, phrase your letter as if the article is in the planning stage. Editors prefer pieces written specifically for their publication; early involvement also allows them to mold the piece to their specifications.

Essential Elements of a Query

Editors always appreciate a well-written query that catches their interest and gets to the point quickly. The following elements are part of every good query letter, though some elements may be emphasized more than others.

Article/story summary. The purpose of a query is to pique the interest of an editor, not to provide a detailed outline of your non-fiction article or in-depth coverage of your story's plot. More important at this stage is getting across the basic idea of your piece—what it is about and who are the key players—in a lively, professional manner. Start your letter with a lead-in that hooks the editor, and state your article or story's unique point of view upfront. It takes practice to craft such a tightly focused summary, but your skill in doing so will convince the editor that you are, indeed, a qualified writer. Remember to include the approximate word count of the article, as well as any sidebars or other elements, in your letter.

Your qualifications. Pitching yourself—not just your article—is another key component. List your publishing credits if you have them; if not, don't mention it at all. Even if you haven't been published, you may have experience in some other area that proves relevant to the topic you're writing about. For example, if you're pitching an article to a children's science magazine on how bees make honey and you once worked as a researcher in this area,

---

Query Checklist

❑ Direct your query to a specific editor. Verify the spelling of the name and address.

❑ Begin with a lead paragraph that "hooks" the editor and conveys your slant.

❑ Include a brief description of your article that conveys your central idea.

❑ Show how your idea meets the editorial goals of the targeted magazine.

❑ Indicate approximate word length of main article, along with any sidebars.

❑ Provide specific details about the content—anecdotes, case histories, personal experience stories, etc.

❑ Cite sources and planned interviews.

❑ If applicable, indicate number and type of photographs or illustrations available.

❑ List your publishing credits, if you have any. If not, emphasize your relevant or unique experience.

❑ Close by asking if the magazine is interested; mention whether your query is a simultaneous submission.

❑ Include other information if requested, such as an outline, bibliography, or resume.

you are particularly qualified to write such a piece even if you've never been published before. Note any background or experience you have that gives you credibility in writing this piece for this particular audience.

Knowledge of the market. Know your audience, and know the magazine you're pitching to inside and out. Tailor your idea to work specifically for that publication. Know the word limit the magazine prefers and whether or not it requires a bibliography of sources. Know its tone and style, and mold your article to match. For example, if the magazine often includes information from an expert's perspective, find an expert who's willing to participate in an interview for your article and include his or her name in the letter. Or if you're pitching a story to *Pockets*, which prefers articles about real children, find one or several you can talk to and let the editor know. Understanding a magazine's "personality" will help you make a convincing

argument as to why your article will benefit that publication.

Professional presentation. Some editors stop reading a query altogether after finding a mistake, so take a few extra minutes to make sure your letter is ready to send. Proofread for grammar, spelling, and typos throughout. Double-check the spelling of the publication, the address, and the editor's name. Use a letter-quality printout, with crisp, dark type, single-spaced, and a font close to Times Roman 12-point. Make sure your contact information is included, along with a self-addressed, stamped envelope or postcard for the editor's reply.

A good query letter is short and to the point. If you can't get your idea across in one page or less, your article may not be as tightly focused as it should be. Use this list of essentials to determine if your query is getting the point across in the most effective and professional manner possible.

Article Appeal

If you're pitching a nonfiction article, think about how you can make your submission a cut above the rest. One of the ideas below might be just what your manuscript needs to get noticed.

- *Photographs.* While most publications don't require photos with a submission, most welcome them. To add visuals to your article, you can either take pictures yourself and send them with your query, or search online for images. Government sources, PR departments, museums, and historical associations are all good places to search. Let the editor know that you've located good-quality, accessible photos and where. A submission with a visual component is more likely to get an editor's attention.

- *Sidebars.* Magazines specialize in serving up small bits of information, and one way to "chunk" your manuscript is through the use of sidebars. The contents of a sidebar may include fast facts, checklists, instructions, definitions, quizzes, quotes, crafts, or stories to support the article. Sidebars offer readers an additional point of entry to the larger article, and serve as an added visual element. Sidebar titles, concepts, and word counts should be mentioned in your query as an optional addition to the main article.

- *Unusual formatting.* There are many ways to present facts, and the traditional narrative format is just one of them. Some magazines are very visual, and prefer art-driven pieces, relying heavily on charts, photos, games, captions, etc. Other innovative formats include Q & A's, profiles, interviews, quizzes, lists, and crossword puzzles. Lists (such as "10 ways to break the ice") provide a snappy and memorable way to communicate information. Use your imagination to think of alternative formats that might be appropriate for your subject.

Sample Query Letter

Address
Phone Number
Email
Date

Elizabeth Lindstrom, Editor
Odyssey
Carus Publishing
30 Grove Street, Suite C
Peterborough, NH 03458

Dear Ms. Lindstrom:

In math classrooms, students seem to understand the concepts in class only to turn around and fail a test given the next day. What's the deal? The majority of students don't know how or what to study for math tests because studying for math is totally different than studying for other school subjects.

I've prepared a 966-word article tentatively titled "Passing with Flying Colors: How to Ace a Mathematics Test" that outlines specific tips to help students receive an excellent grade on any math test. The article begins with a brief conversation between study partners to grab attention. Following this, I cite math test-taking tips such as active daily learning, using flash cards, studying over time, and taking a practice test.

As a middle school math teacher, I witness the common studying mistakes that students make. By using the tips mentioned in the article, teenagers can learn how to prepare and study for a math test and wow their teachers and parents with their grades.

If you are interested in seeing the finished article, I would be happy to submit it for your consideration. I have enclosed a self-addressed, stamped envelope and I look forward to your reply.

Best regards,

Janae Rosendale

Enc.: SASE

Sample Query Letter

Address
Phone Number
Email
Date

Rachel Buchholz
Boys' Life
1325 West Walnut Hill Lane
P.O. Box 152079
Irving, TX 75015-2079

Dear Ms. Buchholz:

Rumble, rumble. Kaboom! An underwater volcano named Loihi is spitting out hot lava into the waters of the Pacific Ocean. It is creating a new Hawaiian island.

Would your *Boys' Life* readers like to read about Loihi? I propose a 600-word article, "Island in the Womb," for your science column. When will Loihi become an island? How it is being formed? How are scientists monitoring it? Why is it of interest to scientists? This article will answer these questions.

Enclosed is a bibliography. I could complete the article on speculation within three weeks of the go-ahead.

Recently, my stories have been accepted by *Our Little Friend* and *Skipping Stones*. A story of mine was also a finalist in the Half Price Book Bedtime Story contest and was published in *Once Upon A Bedtime*, released by Half Price Books.

Would you like to read my article on Loihi? Enclosed is an SASE for your decision. Thank you for your time and consideration.

Sincerely,

Vijayalakshmi Chary

Inc: Bibliography

Sample Query Letter

Address
Phone Number
Email
Date

Cindy Davis
Characters
P.O. Box 708
Newport, NH 03773

Dear Ms. Davis:

Have you ever finished mopping the kitchen floor only to find it covered with muddy footprints a half hour later? Wouldn't you love it if the person who made those muddy footprints had to clean them up?

I have a 700-word mystery titled "Muddy Footprints," targeted for 8- to 10-year-old readers. In the story a seven-year-old has recently received a detective's kit for his birthday, and has become a sleuth. At the request of his mother, "Inspector Morgan" explores the possibilities of who made the muddy footprints on Mom's newly mopped floor. At the end of the story, the sleuth discovers that he is actually the culprit, and cleans up the mess.

Most children enjoy the intrigue of a mystery and would love to be able to solve a case such as this on their own. If you would like to see a copy of the piece, I can send it to you within ten days.

I have enclosed a self-addressed, stamped enveloped for your reply. Thanks for your consideration.

Sincerely,

Bonita Wilborn

Sample Query Letter

Address
Phone Number
Email
Date

Joelle Dujardin, Associate Editor
Highlights for Children
803 Church Street
Honesdale, PA 18431

Dear Ms. Dujardin:

Imagine having a dog as your doctor! My article, "Doctor Dog," profiles a working dog who is a full time member of the medical staff at Columbus Children's Hospital in Ohio. This two-year-old Labrador/Golden Retriever mix named Ansley is part of the hospital's new Animal Assisted Therapy Program. The program goes beyond using therapy dogs to provide comfort and companionship for patients. Ansley is directly involved in intense rehabilitation sessions with children who need therapy related to cognitive thinking, speech, or mobility.

My 450-word article for children ages 4-8 takes readers through a day at work with Ansley. I show how he brightens the mood and motivates young patients to achieve their goals. As a teacher, parent, and dog owner myself, I think your readers would love to learn about this amazing working dog.

I have enclosed a bibliography and an SASE for your convenience. Please be aware that this is a simultaneous submission. Photos of Ansley in action are available upon your request. Thank you for your time, and I look forward to receiving your response.

Sincerely,

Jennifer Mattox

Preparing a Manuscript Package

The following guide shows how to prepare and mail a professional-looking manuscript package. However, you should always adhere to an individual magazine's submission requirements as detailed in its writers' guidelines and its listing in this directory.

Cover Letter Tips

Always keep your cover letter concise and to the point. Provide essential information only.

If the letter accompanies an unsolicited manuscript submission (see below), indicate that your manuscript is enclosed and mention its title and word length. If you're sending the manuscript after the editor responded favorably to your query letter, indicate that the editor requested to see the enclosed manuscript.

Provide a brief description of the piece and a short explanation of how it fits the editor's needs. List any publishing credits or other pertinent qualifications. If requested in the guidelines or listing, note any material or sources you can provide. Indicate if the manuscript is being sent to other magazines as well (a simultaneous submission). Mention that you have enclosed a self-addressed, stamped envelope for return of the manuscript.

Sample Cover Letter

Address
Phone Number
Email
Date

Mr. Marvin Wengerd, Editor
Nature Friend
2673 Township Road 421
Sugarcreek, OH 44681

Dear Mr. Wengerd:

Most kids would love a chance to go out at night and explore. The enclosed 850-word article, "Sounds in the Night," teaches children about God's mysterious creatures of the night—owls!

The article starts by explaining the hooting calls and some other interesting facts about the four most common owls found in backyards. During the day, kids become "owl detectives," searching for evidence of owls in their backyard. Finally, on a night when the moon is bright, they are invited to ask an adult to take them on an owl search—using the information they have learned to identify the owls they hear.

I hope you'll consider this article for your publication. I have enclosed the manuscript and a self-addressed, stamped envelope for your reply.

Thank you for your time and consideration. I look forward to hearing from you.

Sincerely,

Beth A. Deitzel

Enc: Ms, Bibliography, SASE

Subject/Specifications: A brief description of the topic and its potential interest to the magazine's readers. Word lengths, age range, availability of photos, and other submission details.

Closing: Be formal and direct.

Standard Manuscript Format

The format for preparing manuscripts is fairly standard—an example is shown below. Double-space manuscript text, leaving 1- to 1½-inch margins on the top, bottom, and sides. Indent 5 spaces for paragraphs.

In the upper left corner of the first page (also known as the title page), single space your name, address, phone number, and email address. In the upper right corner of that page, place your word count.

Center the title with your byline below it halfway down the page, approximately 5 inches. Then begin the manuscript text 4 lines below your byline.

In the upper left corner of the following pages, type your last name, the page number, and a word or two of your title. Then, space down 4 lines and continue the text of the manuscript.

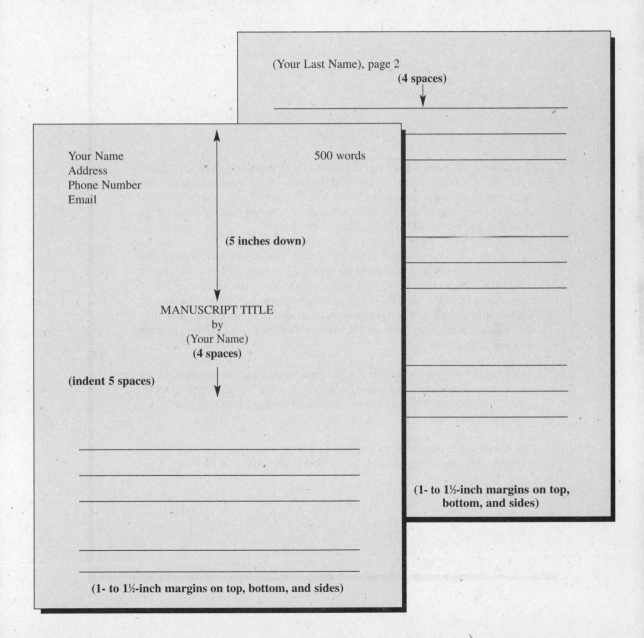

(Your Last Name), page 2

(4 spaces)

Your Name
Address
Phone Number
Email

500 words

(5 inches down)

MANUSCRIPT TITLE
by
(Your Name)

(4 spaces)

(indent 5 spaces)

(1- to 1½-inch margins on top, bottom, and sides)

(1- to 1½-inch margins on top, bottom, and sides)

Sample Cover Letter

Address
Phone Number
Email
Date

Submissions Editor, *Cricket*
Carus Publishing
70 East Lake Street, Suite 300
Chicago, IL 60601

Dear Editor,

Eleven-year-old Aden returns to his favorite fishing spot early one spring afternoon only to find all the fish lying dead upon the shore. What caused this horrible tragedy? How did the fish die? Will Aden's favorite spot ever be the same again? As one clue leads to another, Aden and his friend Sammy scramble to solve this science-based mystery and save the fish.

Touching on the potential loss of the fish and of one child's special place, this 1,130-word story is sure to capture the interest of young readers ages 8 to 10. In addition, this story is based on a true ecologic crisis, the invasion of the zebra mussel that has recently occurred in many parts of the United States. A sidebar with factual information about nonnative zebra mussels and how they are affecting local ecosystems like the one in this story can easily be included for even greater interest.

As a high school science teacher, former park ranger, and environmental educator, ecological mysteries and animal adaptations are a favorite teaching topic of mine. Enclosed you will find a copy of "The Mystery of the Dead Fish" for your consideration.

I look forward to hearing from you and I have included an SASE for your convenience. This manuscript has been sent exclusively to you at this time.

Sincerely,

Eliza Bicknell

Sample Cover Letter

Address
Phone Number
Email
Date

Editor
Boys' Quest
P.O. Box 227
Bluffton, OH 45817-0227

Dear Editor,

Don't think you can enjoy your favorite pastime of fishing in the dead of winter? Well, think again. Unbelieveably, fishing on a frozen lake is possible and fish can be plentiful. Not only is it exciting to learn a new way to fish, you will also enjoy the great outdoors in a very different way.

For many years my father took me on ice fishing trips. I always found it amazing that you could drill holes in the ice and catch live fish out of the frozen lake.

Enclosed for your consideration is a 500-word article on ice fishing entitled, "Fishing Through a Frozen Lake." The article includes information on safety tips, equipment, clothing, and tips for ice fishing. As per your writers' guidelines, I have included two black-and-white photos, which I have the JPEG files for if you are interested. I have also included a bibliography for your review.

I am a graduate of the Institute of Children's Literature and a member of SCBWI. My writing has been published in *Stories for Children Magazine, A Long Story Short, Kid Magazine Writers,* and *The National Writing for Children Center.*

Thank you for your time and consideration. I look forward to your response. An SASE is enclosed for your reply.

Sincerely,

Donna M. McDine

Enc: Ms, bibliography, B&W photos & SASE

Sample Cover Letter

Address
Phone Number
Email
Date

Elizabeth Crooker Carpentiere
Cobblestone Publishing
30 Grove Street, Suite C
Peterborough, NH 03458

Dear Ms. Carpentiere:

Have you ever fallen asleep before midnight on New Year's Eve? I have just the remedy for you and your readers! I lived in China and love writing about Chinese traditions, culture, and legends. My nonfiction article, "Welcome to the Year 4706!" is about Chinese New Year celebrations and would be perfect for your upcoming issue, "Let's Celebrate: Holidays and Festivals around the World."

The length of my article is 600 words, including a favorite recipe for New Year's cake or Nian Gao (465 words). It is aimed at 9- to 14-year-old readers. The article is full of exciting and fun celebration suggestions, and after reading it, I can guarantee you and every reader will choose to celebrate the Chinese New Year.

I would like to send you my article. Will you please reply to me through email concerning this query? Thank you so much for your consideration.

Sincerely,

Jewelene Carter

Sample Outline

Outline for "Welcome to the Year 4706!"

Introduction: This year, 2008, we'll be saying, "Welcome to the Chinese year 4706!" Enjoy twice the fun as we celebrate in a totally new way.

A. Chinese lunar and animal calendar

    1. Brief explanation of Chinese lunar calendar and dates for 2008
    2. Clear, easy-to-understand explanation of Chinese animal calendar

B. Ancient and modern traditions and legends of Chinese New Year

    1. How Chinese New Year celebrations began, told through legends
    2. Specific traditions enjoyed by Chinese families today

C. Celebrating Chinese New Year
    1. Fun festivites for readers to try as they celebrate Chinese New Year
    2. Recipe for Nian Gao—New Year's Cake

Conclusion: These are only a few of the ways you can enjoy the New Year again! There are many cities across America where fabulous Dragon Parades and Lantern Festivals will take place to celebrate the Chinese New Year. Check your local library to see if your community is doing something special to welcome in the year 4706. Whatever you do on February 6th, be sure to stay awake past midnight. Just tell your parents, "I'm adding minutes to your life! Gung hay fat choy!"

Sample Cover Letter

Address
Phone Number
Email
Date

Richard Sagall, M.D.
Pediatrics for Parents
P.O. Box 63716
Philadelphia, PA 19147

Dear Dr. Sagall:

Did you know that thousands of American children and teenagers under age 16 suffer from juvenile idiopathic arthritis? Although this disease is associated with the elderly, it can occur as early as infancy.

My manuscript, "Kids With Arthritis," is 750 words. It describes the different types of juvenile arthritis, its symptoms, and treatment options. This article was thoroughly researched and reviewed by a pediatric rheumatologist. It is a great fit for *Pediatrics for Parents*.

As the mother of a five-year-old child who suffers from juvenile arthritis, I know firsthand the emotional and physical challenges attributed to the disease.

I am a freelance writer living in San Diego with several publishing credits, including *New Moon* and *Skipping Stones* magazines. The manuscript and bibliography are enclosed. I look forward to hearing from you.

Sincerely,

Celia Taghdiri

Bibliography

A thorough bibliography is particularly valuable to editors, who use these source lists as a way to determine how extensively a manuscript has been researched. In fact, many nonfiction editors review the bibliography even before the manuscript. A well-rounded, diverse bibliography not only shows that you've properly acknowledged and credited your sources of information, but that the finished piece is likely to present an original point of view that is supported by credible sources.

Some magazines require a bibliography as part of a submission package for nonfiction articles. (Bibliographies are rarely required for fiction, with the possible exception of historical fiction.) Unique research methods are required for every project but, in general, the bibliography should be made up of resources that are both historical and current, target adults and children, and cover primary and secondary sources. Search museums, libraries, and other organizations for primary documents such as naval reports, maps, bills of lading, court documents, song lyrics, and more. Your bibliography should also show that you plan on culling information from other sources, such as interviews with experts or other relevant individuals (a profile subject and those who know him or her, for example). It's also a good idea to mention your research in your cover letter, to explain your main sources and any other research-related issues.

Citation styles vary greatly, but several references are available and generally accepted for bibliographic format. Among these are: *The Chicago Manual of Style*; *Modern Language Association (MLA) Handbook*; or handbooks by news organizations such as the *New York Times*.

> **To find expert sources to interview, read newspapers, magazines, and journals with articles about your topic. Current articles may include the names of specialists who are doing work in your area of interest, some of whom may be available for interviews.**

Sample Bibliography

Bibliography for "Doctor Dog" by Jennifer Mattox

"Animal Assisted Therapy Brings Patients One Step Closer to Healing," *Columbus Parent Magazine*, March 2007.

"Canine Companions for Independence," www.caninecompanions.org.

"Columbus Children's Hospital Staffs Full-Time 'Medical Dog' In New Animal Assisted Therapy Program," March 2007, www.columbuschildrens.com.

Lundine, Jennifer. Telephone Interview. March 31, 2007. Speech pathologist at Children's Hospital, Columbus, Ohio.

Mehus-Roe, Kristen. *Working Dogs: True Stories of Dogs and Their Handlers.* Irvine, California: BowTie Press, 2003.

Writing Your Résumé

Many publications request a resume along with queries or sample clips. You may have a resume you use for job-hunting, but it isn't likely to address the needs of an editor, who is only interested in your writing experience. By reviewing a resume, an editor can determine if a prospective writer has the necessary experience to research and write material for his or her publication.

If you have few writing credits, format your resume so that your writing-related skills and experience appear first, such as writing a company newsletter or reports, or creating volunteer materials. Then briefly list your work history and educational information, followed by awards and professional memberships such as SCBWI. Pertinent job experience should always be highlighted, both on your resume and in your cover or query letter.

Many writers tailor their resume to be most relevant to the particular job or opportunity they are seeking. For example, if you're querying an educational publication, emphasize your experience as a writing instructor or a teacher; for a nonfiction editor, highlight any nonfiction writing credits first.

No one style or format is preferred, but make sure your name and email address appear at the top of the page. Keep your resume concise—it should not be more than one page.

Sample Résumé

Joanna Coates
Address
Phone
Email

EDUCATION:

University of Missouri, Columbia, MO
1980   M.Ed. Reading
1975   B.A. English Education

Missouri Certified Teacher of English and Reading Specialist

TEACHING EXPERIENCE:
1997–present   Instructor
                 Adult Continuing Education ESL Classes
                 Springfield College, Springfield, MO

1981-1995   Classroom Teacher
                 Middle School English and Reading
                 John Jay Middle School, Thornfield, MO

EDUCATIONAL MATERIAL PUBLISHED:
                 Educational Insights
1995   FUN WITH READING II
                 Story/activity kit
1993   FUN WITH READING I
                 Story/activity kit

MEMBERSHIP:
Society of Children's Book Writers and Illustrators

Copyright and Permissions

Just like the movie you watched last night and the CD you listened to on the way to work this morning, your magazine article is one of many creative works that is afforded the protection of copyright. As one of the nation's "copyright-based industries," publishing relies heavily on the concept of obtaining legal ownership of written works. When you write an article, you own the legal rights to the manuscript, as well as the right to decide how it is reproduced and, for certain works, how it is performed or displayed.

As of 1998, your heirs can also enjoy the fruits of your labor: That's when Congress passed the Copyright Term Extension Act, which offers you copyright protection for your work created during or after 1978 for your lifetime plus 70 years, until you choose to sell all or part of the copyright for this work.

Do You Need to Register Your Work?

Thanks to copyright law, your work is protected from the moment it is recorded in a tangible medium, such as a computer file or on paper, without any need for legal action or counsel. You don't even need to register your work with the United States Copyright Office; in fact, most editors view an author's copyright notice on manuscripts as the sign of an amateur. A copy of the manuscript and a dated record of your submission will provide proof of creation, should the need arise.

If you do decide to register your work, obtain an application form and directions on the correct way to file your copyright application. Write to the Library of Congress, Copyright Office, 101 Independence Ave. S. E., Washington, DC 20559-6000. These forms and directions are also available online at: www.copyright.gov/forms. Copyright registration fees are currently $35 (online) and $45 (print).

If you have registered your unpublished manuscript with the Library of Congress, notify your editor of that fact once it is accepted for publication.

Rights Purchased by Magazines

As a writer and copyright holder, you have the right to decide how your work should be shared with the world. By agreeing to publication in a magazine, you also agree to transfer some of your rights over to the magazine so that your article can be printed and distributed as part of that publication. A publisher is restricted, however, on when, how, and where he or she may publish your manuscript—terms that are set down in a publishing contract. Below is a list of common rights that are purchased by magazines:

All World Rights: The publisher purchases all rights to publish your work anywhere in the world any number of times. This includes all forms of media (both current and those which may be developed later). The publisher also has the right to all future use of the work, including reprints, syndication, creation of derivative works, and use in databases. You no longer have the right to sell or reproduce the work, unless you can negotiate for the return of certain rights (for example, book rights).

All World Serial Rights: The publisher purchases all rights to publish your work in newspapers, magazines, and other serial publications throughout the world any number of times. You retain all other rights, such as the right to use it as a chapter in a book.

First Rights: A publisher acquires the right to publish your work for the first time in any specified media. Electronic and nontraditional markets often seek these rights. All other rights, including reprint rights, belong to you.

Electronic Rights: Publishers use this as a catch-all for inclusion in any type of electronic publication, such as CD-ROM, websites, ezines, or in an electronic database.

First North American Serial Rights: The publisher can publish your work for the first time in a U.S. or Canadian periodical. You retain the book and North American reprint rights, as well as first rights to a foreign market.

Second or Reprint Rights: This allows a publication non-exclusive rights to print the material for the second time. You may not authorize second publication to occur until after the work has appeared in print by the publisher who bought first rights. ***One-time Rights:*** Often bought by regional publications, this means the publication has bought the right to use the material once. You may continue to sell the material elsewhere; however, you should inform the publisher if this work is being simultaneously considered for publication in a competing magazine.

You should be aware that an agreement may limit a publisher to the right to publish your work in certain media (e.g., magazines and other periodicals only) or the agreement may include wider-ranging rights (e.g., the right to publish the manuscript in a book or an audiocassette). The right may be limited to publishing within a specific geographic region or in a specific language. Any rights you retain allow you to resell the manuscript within the parameters of your agreement.

It is becoming increasingly common for magazines to purchase all rights, especially those that host Internet sites and make archives of previously published articles available to readers. Unless you have extensive publishing credentials, you may not want to jeopardize the opportunity to be published by insisting on selling limited rights.

Contracts and Agreements

Typically, when a publisher indicates an interest in your manuscript, he or she specifies what rights the publication will acquire. Then usually a publisher will send you a letter of agreement or a standard written contract spelling out the terms of the agreement.

If a publisher does not send you a written contract or agreement, you need to consider your options. While an oral agreement may be legally binding, it is not as easy to enforce as a written one. To protect your interests, draft a letter outlining the terms as you understand them (e.g., a 500-word article without photos, first North American serial rights, paying on acceptance at $.05 a word). Send two copies of the letter to the editor (with a self-addressed, stamped envelope), asking him or her to sign one and return it to you if the terms are correct.

Work Made for Hire

Another term that is appearing more frequently in contracts is work made for hire. As a freelance writer, most editors treat you as an independent contractor (not an employee) who writes articles for their publication. Magazine editors can assign or commission articles to freelancers as works-made-for-hire, making the finished article property of the publisher.

E-zines typically purchase both exclusive and non-exclusive electronic rights. They may want to publish your article exclusively for a particular period of time (usually one year). After that you may be free to sell it elsewhere or place it on your own website, while the original e-zine continues to display or archive the article.

Under current copyright laws, only certain types of commissioned works are considered works-made-for-hire, and only when both the publisher and the commissioned writer agree in writing. These works typically include items such as contributions to "collective works" such as magazines. A contract or agreement clearly stating that the material is a work-made-for-hire must be signed by both parties and be in place before the material is written. Once a writer agrees to these terms, he or she no longer has any rights to the work.

Note that a pre-existing piece, such as an unsolicited manuscript that is accepted for publication, is not considered a commissioned work.

Guidelines for Permission to Quote

When you want to quote another writer's words in a manuscript you're preparing, you must get that writer's permission. If you don't, you could be sued for copyright infringement. Here are some guidelines:

- Any writing published in the U.S. prior to 1923 is in the public domain, as are works created by the U.S. government. Such material may be quoted without permission, but the source should be cited.

- No specific limits are set as to the length of permitted quotations in your articles: different publishers have various requirements. Generally, if you quote more than a handful of words, you should seek permission. Always remember to credit your sources.
- The doctrine of "fair use" allows quoting portions of a copyrighted work for certain purposes, as in a review, news reporting, nonprofit educational uses, or research. Contrary to popular belief, there is no absolute word limit on fair use. But as a general rule, never quote more than a few successive paragraphs from a book or article and credit the source.
- If you're submitting a manuscript that contains quoted material, you'll need to obtain permission from the source to quote the material before it is published. If you're uncertain about what to do, your editor should be able to advise you.

Before submitting, know what your potential markets pay, when they pay, what rights they routinely acquire, and what's reasonable when it comes to negotiating your contract terms. Writers' groups such as the Authors Guild offer real-world advice on these topics and more.

Resources

Interested in finding out more about writers and their rights under the law? Check these sources for further information:

The Publishing Law Center
www.publaw.com/legal.html
The Copyright Handbook: How to Protect and Use Written Works, 8th Edition by Attorney Stephen Fishman. Nolo, 2005.
The Writer's Legal Guide, 3rd Edition by Tad Crawford. Allworth Press, 2002.

Last Steps and Follow Up

Before mailing your manuscript, check the pages for neatness, readability, and proper page order. Proofread for typographical errors. Redo pages if necessary. Keep a copy of the manuscript for your records.

Mailing Requirements

Assemble the pages (unstapled) and place your cover letter on top of the first page.

Send manuscripts over 5 pages in length in a 9x12 or 10x13 manila envelope. Include a same-size SASE marked "First Class." If submitting to a foreign magazine, enclose the proper amount of International Reply Coupons (IRC) for return postage. Mail manuscripts under 5 pages in a large business-size envelope with a same-size SASE folded inside.

Package your material carefully and address the outer envelope to the magazine editor. Send your submission via first-class or priority mail. Don't use certified or registered mail. (See Postage Information, page 38.)

Follow Up with the Editor

Some writers contend that waiting for an editor to respond is the hardest part of writing. But wait you must. Editors usually respond within the time period specified in the listings.

If you don't receive a response by the stated response time, allow at least three weeks to pass before you contact the editor. At that time, send a letter with a self-addressed, stamped envelope requesting to know the status of your submission.

The exception to this general rule is when you send a return postcard with a manuscript. In that case, look for your postcard about three weeks after mailing the manuscript. If you don't receive it by then, write to the editor requesting confirmation that it was received.

If more than two months pass after the stated response time and you don't receive any response, send a letter withdrawing your work from consideration. At that point, you can send your query or manuscript to the next publication on your list.

What You Can Expect

The most common responses to a submission are an impersonal rejection letter, a personalized rejection letter, an offer to look at your material "on speculation," or an assignment.

If you receive an impersonal rejection note, revise your manuscript if necessary, and send your work to the next editor on your list. If you receive a personal note, send a thank-you note. If you receive either of the last two responses, you know what to do!

Set Up a Tracking System

To help you keep track of the status of your submissions, you may want to establish a system in a notebook, in a computer file, or on file cards (see below).

This will keep you organized and up-to-date on the status of your queries and manuscripts and on the need to follow up with certain editors.

SENT QUERIES TO THE FOLLOWING PUBLICATIONS

Editor	Publication	Topic	Date Sent	Postage	Accepted/ Rejected	Rights Offered

SENT MANUSCRIPTS TO THE FOLLOWING PUBLICATIONS

Editor	Publication	Title	Date Sent	Postage	Accepted/ Rejected	Rights Offered

Frequently Asked Questions

How do I request a sample copy and writers' guidelines?

Write a brief note to the magazine: "Please send me a recent sample copy and writers' guidelines. If there is any charge, please enclose an invoice and I will pay upon receipt." The magazine's website, if it has one, offers a faster and less expensive alternative. Many companies put a part of the magazine, writers' guidelines, and sometimes a theme list or editorial calendar on the Internet.

How do I calculate the amount of postage for a sample copy?

Check the listing in this directory. In some cases the amount of postage will be listed. If the number of pages is given, use that to estimate the amount of postage by using the postage chart at the end of this section. For more information on postage and how to obtain stamps, see page 38.

Should my email submission 'package' be different than a submission via snail mail?

In general, an email submission should contain the same elements as a mailed one—i.e. a solid article description, sources, etc. In all cases, writers' guidelines should be followed to the letter when it comes to sending writing samples, bibliographies, and other requirements, either as separate file attachments or embedded in the email text.

What do I put in a cover letter if I have no publishing credits or relevant personal experience?

In this case, you may want to forego a formal cover letter and send your manuscript with a brief letter stating: "Enclosed is my manuscript, (Insert Title), for your review." For more information on cover letters, see pages 20–26.

How long should I wait before contacting an editor after I have submitted my manuscript?

The response time given in the listings can vary, and it's a good idea to wait three to four weeks after the stated response time before sending a brief note to the editor asking about the status of your manuscript. You might use this opportunity to add a new sales pitch or include additional material to show that the topic is continuing to generate interest. If you do not get a satisfactory response or you want to send your manuscript elsewhere, send a certified letter to the editor withdrawing the work from consideration and requesting its return. You are then free to submit the work to another magazine.

I don't need my manuscript returned. How do I indicate that to an editor?

With the capability to store manuscripts electronically and print out additional copies easily, some writers keep postage costs down by enclosing a self-addressed, stamped postcard (SASP) saying, "No need to return my manuscript. Please use this postcard to advise me of the status of my manuscript. Thank you."

Common Publishing Terms

All rights: Contractual agreement by which a publisher acquires the copyright and all use of author's material (see page 24).

Anthology: A collection of selected literary pieces.

Anthropomorphization: Attributing human form or personality to things not human (i.e., animals).

Assignment: Manuscript commissioned by an editor for a stated fee.

Bimonthly: A publication that appears every two months.

Biweekly: A publication issued every two weeks.

Byline: Author's name credited at the heading of an article.

Caption: Description or text accompanying an illustration or photograph.

CD-ROM (compact disc read-only-memory)**:** Non-erasable compact disc containing data that can be read by a computer.

Clip: Sample of a published work.

Contributor's copies: Copies of the publication issue in which the writer's work appears.

Copyedit: To edit with close attention to style and mechanics.

Copyright: Legal rights that protect an author's work (see page 27).

Cover letter: Brief letter sent with a manuscript introducing the writer and presenting the materials enclosed (see page 18).

Disk submission: Manuscript that is submitted on a computer disk.

Early readers: Children 4 to 7 years.

Editorial calendar: List of topics, themes, or special sections that are planned for upcoming issues for a specific time period.

Electronic submission: Manuscript transmitted to an editor from one computer to another through a modem.

Email (electronic mail)**:** Messages sent from one computer to another via computer network or modem.

English-language rights: The right to publish a manuscript in any English-speaking country.

Filler: Short item that fills out a page (e.g., joke, light verse, or fun fact).

First serial rights: The right to publish a work for the first time in a periodical; often limited to a specific geographical region (e.g., North America or Canada) (see page 27).

Genre: Category of fiction characterized by a particular style, form, or content, such as mystery or fantasy.

Glossy: Photo printed on shiny rather than matte-finish paper.

Guidelines: See **Writers' guidelines.**

In-house: See **Staff written.**

International Reply Coupon (IRC): Coupon exchangeable in any foreign country for postage on a single-rate, surface-mailed letter.

Kill fee: Percentage of the agreed-upon fee paid to a writer if an editor decides not to use a purchased manuscript.

Layout: Plan for the arrangement of text and artwork on a printed page.

Lead: Beginning of an article.

Lead time: Length of time between assembling and printing an issue.

Libel: Any false published statement intended to expose another to public ridicule or personal loss.

Manuscript: A typewritten or computer-printed version of a document (as opposed to a published version).

Masthead: The printed matter in a newspaper or periodical that gives the title and pertinent details of ownership, advertising rates, and subscription rates.

Middle-grade readers: Children 8 to 12 years.

Modem: An internal device or a small electrical box that plugs into a computer; used to transmit data between computers, often via telephone lines.

Ms/mss: Manuscript/manuscripts.

One-time rights: The right to publish a piece once, often not the first time (see page 28).

On spec: Refers to writing "on speculation," without an editor's commitment to purchase the manuscript.

Outline: Summary of a manuscript's contents, usually nonfiction, organized under subheadings with descriptive sentences under each.

Payment on acceptance: Author is paid following an editor's decision to accept a manuscript.

Payment on publication: Author is paid following the publication of the manuscript.

Pen name/pseudonym: Fictitious name used by an author.

Pre-K: Children under 5 years of age; also known as *pre-school*.

Proofread: To read and mark errors, usually in printed text.

Query: Letter to an editor to promote interest in a manuscript or an idea.

Rebus story: A "see and say" story form, using pictures followed by the written words; often written for pre-readers.

Refereed journal: Publication that requires all manuscripts be reviewed by an editorial or advisory board.

Reprint: Another printing of an article or story; can be in a different magazine format, such as an anthology.

Reprint rights: See **Second serial rights.**

Response time: Average length of time for an editor to accept or reject a submission and contact the writer with his or her decision.

Résumé: Account of one's qualifications, including educational and professional background, as well as publishing credits.

SAE: Self-addressed envelope (no postage).

SASE: Self-addressed, stamped envelope.

SASP: Self-addressed stamped postcard.

Second serial rights: The right to publish a manuscript that has appeared in another publication; also known as *Reprint rights* (see page 27).

Semiannual: Occurring every six months or twice a year.

Semimonthly: Occurring twice a month.

Semiweekly: Occurring twice a week.

Serial: A publication issued as one of a consecutively numbered and indefinitely continued series.

Serial rights: See **First serial rights.**

Sidebar: A short article that accompanies a feature article and highlights one aspect of the feature's subject.

Simultaneous submission: Manuscript submitted to more than one publisher at the same time; also known as multiple submission.

Slant: Specific approach to a subject to appeal to a certain readership.

Slush pile: Term used within the publishing industry to describe unsolicited manuscripts.

Solicited manuscript: Manuscript that an editor has requested or agreed to consider.

Staff written: Prepared by members of the magazine's staff; also known as *in-house*.

Syndication rights: The right to distribute serial rights to a given work through a syndicate of periodicals.

Synopsis: Condensed description or summary of a manuscript.

Tabloid: Publication printed on an ordinary newspaper page, turned sideways and folded in half.

Tearsheet: A page from a newspaper or magazine (periodical) containing a printed story or article.

Theme list: See **Editorial calendar.**

Transparencies: Color slides, not color prints.

Unsolicited manuscript: Any manuscript not specifically requested by an editor.

Work-made-for-hire: Work specifically ordered, commissioned, and owned by a publisher for its exclusive use (see page 28).

World rights: Contractual agreement whereby the publisher acquires the right to reproduce the work throughout the world (see page 27); also known as *all rights*.

Writers' guidelines: Publisher's editorial objectives or specifications, which usually include word lengths, readership level, and subject matter.

Writing sample: Example of your writing style, tone, and skills; may be a published or unpublished piece.

Young adult: Readers 12 to 18 years.

Postage Information

How Much Postage?

When you're sending a manuscript to a magazine, enclose a self-addressed, stamped envelope with sufficient postage; this way, if the editor does not want to use your manuscript, it can be returned to you. To help you calculate the proper amount of postage for your SASE, here are the U.S. postal rates for first-class mailings in the U.S. and from the U.S. to Canada based on the latest increase (2008). Rates are expected to increase again, so please check with your local Post Office, or check the U.S. Postal Service website at usps.com.

Ounces	9x12 Envelope (Approx. no. of pages)	U.S. First-Class Postage Rate	Rate from U.S. to Canada
1	1–5	$0.42	$0.72
2	6–10	0.59	0.96
3	11–15	0.76	1.20
4	16–20	1.34	1.70
5	21–25	1.51	1.94
6	26–30	1.68	2.18
7	31–35	1.85	2.42
8	36–40	2.02	2.66

The amount of postage and size of envelope necessary to receive a sample copy and writers' guidelines are usually stated in the magazine listing. If this information is not provided, use the chart above to help gauge the proper amount of postage.

How to Obtain Stamps

People living in the U.S., Canada, or overseas can acquire U.S. stamps through the mail from the Philately Fulfillment Service Center. Call 800-STAMP-24 (800-782-6724) to request a catalogue or place an order. For overseas, the telephone number is 816-545-1100. You pay the cost of the stamps plus a postage and handling fee based on the value of the stamps ordered, and the stamps are shipped to you. Credit card information (MasterCard, Visa, and Discover cards only) is required for fax orders. The fax number is 816-545-1212. If you order through the catalogue, you can pay with a U.S. check or an American Money Order. Allow 3–4 weeks for delivery.

Gateway to
the Markets

A Pretty Picture of Preschool Vitality

The preschool audience can be a paradox. It is a highly appealing, exciting readership that lends itself to simple but fun concepts and ideas for writers to pursue. But under the simple forms, good preschool writing is much more sophisticated than many recognize. In the marketplace, picture books for the youngest readers are everywhere, gorgeous and fun. But they have become a hard sell in a glutted market. Yet preK magazines have seen the successful recent launches of two powerhouse publications—*Highlights High Five* and *National Geographic Little Kids.* There's a pretty picture to paint for preK magazine writers.

"These new titles joined several established magazines for this age group, so I would say that the category of preschool magazines is indeed strong and growing," says *Highlights for Children* Editor in Chief Christine French Clark. "As parents become increasingly aware of the importance of early childhood and the benefits of exposing young children to good literature early, this market should continue to grow. More and more parents seem to appreciate that maga-zines in particular that are addressed to and delivered directly to the young child are a wonderful product to help them foster in their children a lifelong love of reading."

Other preK magazine editors agree. "I see the preK magazine market continuing to flourish and staying stable," says Jenny Gillespie, Associate Editor at the Cricket Magazine Group. "Beyond our own magazines, *Babybug, Click,* and *Ladybug,* which are some of our strongest sellers, preK magazines such as *High Five,*  *Wild Animal Babies,* and *National Geographic Little Kids* are all doing well. I believe there will always be a need for preK magazines. Young children love receiving their own mail and flipping through the pages. Even with the advent of interactive DVDs and computer games for preschoolers, parents still want their children to have the tangible experience of a magazine or book."

At the Children's Better Health Institute (CBHI), *Turtle* and *Children's Playmate* Editor Terry Webb Harshman says, "*Turtle*'s readership is very strong and always has been. I think preK will always do well."

Set Apart

No writer should underestimate the full compass of interests of the littlest readers. *Turtle*'s content and needs illustrate. For her two- to five-year-old readers, Harshman says, "*Turtle* has a good mix, which includes stories, poems, simple science, recipes, activities, art contests, rebuses, finger plays, and our new PokeyToes' Pet Show for kids and their pets. We also have a health column in the back for parents and teachers." Of greatest interest to her now are short, clever stories about things little children can relate to, such as camping with their family, sleeping over at grandma's house, or caring for a new pet. "I'm also looking for finger plays, easy crafts, and simple science experiments."

Turtle and the other CBHI magazines are distinguished by their commitment to the physical, educational, creative, social and emotional growth of children at each stage of development.

Cricket Magazine Group distinguishes its titles as literary magazines. "Our magazines, *Babybug* (ages birth to three) and *Ladybug* (ages three to six) are different in that they are mainly literary magazines, emphasizing language and imagination through stories and poems," says Gillespie. "We work with the top chil-

dren's illustrators to feature rich and varied artwork and few photographs. *Babybug* is a board book-style magazine, and seems to be the only one of its kind out there today. *Ladybug* is a unique mix of fiction and nonfiction pieces, as well as songs and crafts. Another feature that sets us apart is a song each month in *Ladybug*; readers can then find a recording of the song on the Web."

Cricket Group's nonfiction magazine *Click*, says Editor Amy Tao, "encourages children to look closely at the world around them and to ask questions about that world. Unlike other nonfiction children's magazines, *Click* doesn't just pres-

> ## "Writing that addresses and honors the lived experiences of young children is most successful."

ent random factoids of information, but tries to interest children in the bigger picture, to stimulate their intellectual curiosity, and to cultivate a spirit of inquiry. Our goal is to offer stories and articles that will hold the interest of not only preschoolers but also their parents." *Click* readers are ages three to seven.

Highlights has always targeted children ages 2 to 12 each month, but the editors also determined that the preschool audience could be even better served by its own magazine. "Like its big sister *Highlights*, *High Five* is a general interest magazine that addresses the whole child. Each issue encourages a child's cognitive, social/emotional, motor, and language development," says Clark. "It reflects the belief that child development is very individual and that kids have different learning styles and grow unevenly and at their own unique paces. With that in mind, *High Five* offers a wide variety of content—rather than a narrow selection of themed material—that is intended to be enjoyed by the kids alongside a parent or older sibling,

but some of which may also be read alone by the very beginning reader. It's our intent to deliver magazines that will be saved in a child's home library alongside his or her books."

Magical

Another of those preschool paradoxes is that beginning writers tend both to overwrite and to underestimate their young audience. The hallmarks of a good preK author's work are intelligent simplicity and respect for children's curiosity.

High Five Editor Kathleen Hayes says, "Writing that addresses and honors the lived experiences of young children is most successful. If children cannot bring their own prior knowledge to bear, they may not be able to comprehend or make sense of the material, no matter how well written it may be. Writing for this age requires the ability to create pieces that resonate with the young child's emotional and social development. Some of the best pieces deal with moving out into the larger world and discovering how to get along with others. Others may provide information and facts to satisfy and extend a preschooler's natural curiosity. We also like stories with some tension that is resolved, and, finally, stories with humor and a playful use of language that helps readers learn to enjoy the sounds of language."

Clarity and, again, respect are at the heart of preK stories, according to Gillespie. "The best writing is the kind that is very clear, yet isn't didactic and doesn't patronize the young reader. Showing over telling is always an asset, as are humor and pathos. Most important, we look for stories that hold the young reader's attention the whole way through, with an engaging story arc and identifiable characters. Rhythmic, pleasing language is a plus as well, as these are read-aloud stories that the parent shares with the child."

At *Ladybug*, current needs include "non-didactic, interesting World Around You pieces—short, clear pieces on nature and world cultures made understandable for preschoolers. These can take the form of nonfiction or narrative," says Gillespie. "We also need fun, simple activities including visual games (matching, shapes, etc.) and Can You Do This action rhymes or finger plays. Activities don't come our way enough because I don't think writers know that we publish unsolicited activities as well. However, recently we have received great activities from preschool teachers who really understand what stimulates a young child."

Tao cautions, "It's not easy to write for this age group. Word counts are so low, but you can't assume much prior knowledge. Sentences need to be simple and straightforward but still lively and charming. Every word has to count. Furthermore, because *Click* is a nonfiction magazine, all content must be thoroughly researched and current." Because *Click* covers a designated theme each month, "our story needs are always changing. We prefer articles in which children are active participants, investigating and making connections on their own rather than learning from an adult."

When asked what makes one submission to *Turtle* stand out over another, Harshman comments, "I want to say *magic*. It's so rewarding to be on the receiving end of a well-spun story or poem. Skillful writers make the whole process more exciting, because their words conjure pictures, too. Then you see the illustration possibilities. Humor plays a big role as well. Stories may have wacky characters or wacky situations, as long as they have an element a child can identify with."

Submission Preferences

"At present, *Turtle* is published six times a year," says Harshman. "Therefore, I purchase about six fiction stories (350 words or less), nonfiction articles, recipes, rebuses, science experiments (200 words or less), and poems (4 to 12 lines) each year. I purchase three to four finger plays and a variety of activities." She prefers to see complete manuscripts. "No queries or outlines, please. Submissions should look professional, but don't put too much energy into your cover letter. Just the basics." She cites "Measuring Fun," by Jill Burrows (July/August 2008), as a model article. "This was an exercise in volume measurement for small children. It was fun and educational and the art meshed perfectly. I would love to receive more educational activities like this." Another good example was Jean Shannon

Gomez's "My Playtime Garden" (May/June 2008). "It was a how-to article on helping children plan their own special garden in their backyard. The author included wonderful kid-friendly ideas, like stepping stones, garden flags, animal figurines, a rock village, and a list of garden games to play with friends."

While *High Five* is "not actively encouraging unsolicited manuscripts," says Clark, *Highlights* continues to look for submissions for younger as well as older children. Current needs for the four to eight age range include 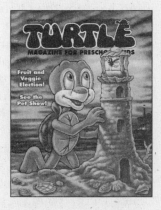 humorous stories, holiday stories, and nonfiction pieces on animals and nature, as well as photo-essays.

Gillespie says that the Cricket Group magazines "are always open to acquiring new material. We purchase about 10 to 15 activities per year, 5 to 10 nonfiction pieces, and 5 to 10 stories. Our length requirement is 800 words maximum for stories. For poems, think simple; toddlers and preschoolers like simple rhymes. If you are going the narrative route with a poem, try not to go beyond two typed pages." Do not send queries or outlines, only full manuscripts. "A successful submission is one that understands our magazines and is aimed toward our readership, is clearly typed and free of grammatical errors, and includes a clear, no-nonsense cover letter, and an SASE that fits the manuscript."

Among "the stories I really admire," Gillespie continues, "is 'Golden Eggs and a Silvery Voice,' by Elizabeth Varadan (*Ladybug*, April 2008). This is an original folktale about a hen and a rooster who are unappreciated by their owners; they run away to try to find out how to make golden eggs and sing with a silvery voice to win the owners' approval. The story has a clear, engaging plot, a tinge of pathos that makes the reader sympathize and care for the characters, and humor dashed throughout. It's a very well-crafted story, perfectly pitched for this age group. An example of a World Around You piece in *Ladybug* that I thought was quite fun was 'Dim Sum with Daddy,' by Lisa Franich Lee (February 2008). It's a narrative poem that depicts a little boy going to eat dim sum with his daddy, and I like that it introduces an interesting cultural experience to the young reader through the rhythms and colorful language of a poem."

At the time of the writing of this article, Tao reported, "*Click* is currently closed to queries and submissions; changes to this policy will be posted at www.cricketmag.com." But she joins with other editors in seeing the overall vitality of the preschool market. "The number of magazines in the preK category is certainly growing. Let's hope the market is growing as well." The evidence all points to the fact that it is.

Looking to the Future of College and Career

College and career publishing is a niche with a mission. Whether geared to teens or parents, to funds or faith, to regional or general interest, publications in this category educate about education. They look to the future of young people.

Today that may mean considering the impact of current economic struggles and how to serve readerships more uncertain about the future. Yet editors also remain committed to their distinct missions in informing and inspiring teens who are beginning to shape their lives.

most affecting Gen Y and Z. But this year, we're focusing on college and interactive content, having kids contribute articles and divulge what they want to read about and develop their sense of self. With the Internet and contemporary consumer culture, kids today have access to a wealth of information. We strive to help kids sort through the information in order to

> ## "What's an excellent student without confidence? Self-empowerment plays a role in every article we run."

Self-Empowerment

The magazine *t&T News* (for *Tweens & Teens*) covers all the issues of adolescence, including education. Despite all the other young adult subjects it treats, the monthly—newly relaunched in September 2008—has been making college coverage a priority. Editor Jenna Greditor says, "We'll continue to feature education, health, fitness, social and family life, teen trends, and other topics

positively express themselves and find a college that best suits their needs and interests."

The nation's economic difficulties decidedly impact content for many magazines. "While finances have always played a role in choosing a suitable college, in today's struggling economy, the subject of finance is of paramount importance," Greditor says. "We give teens and their parents pointers on applying for financial aid and scholarships, along with the tools to excel in

school, write compelling college essays and applications, and ace standardized tests or job interviews in order to become superior college and job applicants."

Nonetheless, a recession or slowdown doesn't change the overall mission of *t&T*. "What's an excellent student or applicant without confidence? Not a happy person, meaning that attaining self-empowerment plays a role in pretty much every article we run," says Greditor, who also edits New York's *PARENTGUIDE News*. "I aim to help teens navigate the high school and college experience, and prepare for their futures as confident and competent people. Reading *t&T* in print and online, kids enjoy some of the fun stuff of being a teenager, with light articles on social life and fashion, as well as have access to articles that intend to take the turbulence out of adolescence and the college search process."

Outlook

American Careers is nothing if not serious about providing students, parents, teachers, and counselors with all they need to make knowledgeable decisions to brighten the outlook for young people.

"Through our American Careers Publications, we provide students with insight into a variety of occupations by publishing interviews with professionals and related career information. In fact, we cover careers in 16 career clusters," says Editor in Chief Mary Pitchford. "We also produce companion teaching guides with standards-based classroom projects that integrate careers with academic and career-technical education." The company's annual *American Careers, High School Edition* "features individuals at work in their occupations." Articles are based on interviews with professionals and on research.

Well before now and whatever the state of the economy, the company's publications and programs "strive to include career and educational data from the Bureau of Labor Statistics related to fast-growing and/or high-growth occupations. Current data are based on 2006 to 2016 projections," says Pitchford. The career cluster sections cover education and salary data, which are meant to

be used as planning tools by students and parents. Readers of the materials published by the American Careers programs, which include teachers' guides, are also school counselors, teachers of English, social studies, career and technical education, business classes, and consumer science classes.

"We approach our stories in a relatively conversational style," says Pitchford, "but we don't use current teen slang. Writers who are new to us may submit a query letter (on relevant career and college topics), along with a résumé and a couple of samples. We do not accept manuscripts. We don't need how-to-interview or how-to-write-a-résumé types of stories."

The publisher also offers *ACK! American Careers for Kids,* targeting fourth and fifth graders and their teachers. The magazine is undergoing changes at the time of this writing, so specifications are not available. Keep an eye out for it in the future.

Starting Young

Imagination-Café was begun by Editor Rosanne Tolin to give children a safe interactive educational and entertainment website. One of its highlights is the Career-O-Rama section. "Our mission is to inspire young people to explore a variety of careers, just by surfing the website," says Tolin. "We want kids to be able to hear both sides of the story: both the good and the bad, so they know *dream* jobs don't always come easy." She is looking for pieces on occupations and professions. "It might be fun to see more quizzes related to these areas. Interested writers should first look to see which jobs are already featured in our Career-O-Rama section. Among them, for example, are articles on being a graphic designer and a professional athlete.

"The articles we use are firsthand interviews, with a journalistic feel," says Tolin. "We also break out the types of specialties within a particular profession. For instance, under *doctor* we might include *surgeon.* Then, there are suggestions on how to get started on the path to the specific career, along with information on how well it pays, etc." The interview is key, and the article should also suggest books and websites readers can look to.

Tolin accepts either queries or complete manuscripts and purchases

Career Preparation

Student Leader, the official publication of the American Student Government Association, isn't a career magazine per se. "It provides college student governments with advice, examples of best practices, and research that helps them become more efficient and effective on their own campuses," says Editor Anna Campitelli. "Students involved in student government gain valuable management and life skills (such as multitasking) that can make them stronger job candidates in almost any field. Many of the nation's presidents, CEOs, congressmen, etc. gained leadership experience in college student government."

Campitelli explains that *Student Leader* "uses profiles of successful student governments and programs, journalistic articles about important student government issues, and how-to's (i.e., how to improve your funding/reimbursement system). Each article should have at least five sources, such as members of the government, or the advisor. If it's an article about an overall SG issue—ways to make textbooks cheaper for students, such as starting a bookswap, for example, the sources can come from different colleges and universities."

Submit article ideas and outlines via the website (www.asgaonline.com/studentleader), after reviewing the style guide and writer's guidelines. "We typically do *not* pay for submissions from writers unless they have already written something for us in the past and we know they will meet deadlines, and that the quality will be there. More often, we give college journalists the opportunity to acquire additional clips," says Campitelli.

about ten career pieces annually, usually 350 to 500 words in length. "Since these career pieces can be longer than some other articles on the website, it might be preferable for a writer to query an idea first. That way, he or she knows whether we've already purchased something similar. And of course, keep up to date on what's currently online at *Imagination-Café*."

Positive Light

Relate has a calling of faith, and a dedication to girls, in particular helping them with education and satisfying career goals.

"Our mission is to inspire teen girls to pursue their dreams with confidence and to teach them to be an example for each other in their speech, life, love, faith, and purity," says Editor in Chief Mary Bowman. "We feel as though teens are bombarded by media that constantly makes them feel less than perfect, and we want to be a positive light for them."

Bowman looks "for several types of articles. The most important thing is that our articles maintain a fun, teen-friendly tone, and that they build teens up. Our magazine is for entertainment and, although informative, we don't want it to read like a textbook." She continues, "We are always looking for college advice relevant to the time of year (since we publish quarterly) and we like to highlight careers that teens might not have heard of or that could seem too good to be true but in fact aren't. We also regularly feature teen entrepreneurs, designers (or talent in any of the fine arts), and teen volunteers to inspire our readers to reach for their own dreams and fulfill their potentials."

Despite the current tough financial times, *Relate*'s approach remains the same. "The economy has not changed the way we address the topics of college and careers. We still encourage our readers to go to college, try hard in school, and look for careers that will be fulfilling and enjoyable," says Bowman. Writers for *Relate* "must understand teens. It is important to know what issues they face, what is trendy, and how they like to be talked to."

Bowman prefers to receive e-mail queries, and may then request an outline with more details. "Any submission that shows a well-thought out idea, with a real flavor for the article, stands out. I get too many generic ideas. I also like it when a query includes an opening paragraph that helps give a feel for the tone of the proposed article. The other important thing is that submissions show an understanding of *Relate* and our mission. We want our readers to know our niche and reflect that in their work."

About 85 percent of each issue is written by freelancers. Typical articles are 600 or 1,200 words, but that may vary.

Model Articles

Also a strong freelance market in this niche, *t&T* publishes about ten 1,000-word articles in every monthly issue. The website also needs freelance pieces, from 250 to 750 words.

In both forms, *t&T* "runs profiles, service pieces or how-to articles, features, and celebrity/expert question-and-answer articles. All facts must be properly sourced and geared for the teen audience. Articles should inform and advise in a friendly yet factual manner, rather than preach or talk down to teens," says Greditor. "The best pitches present the idea in a captivating yet concise way, and show the prospective writer has expertise, experience, and a knack for journalism or conveying a story that teens will embrace. Actual manuscripts are fine but I prefer queries."

Greditor is proud that the magazine "consistently publishes strong articles relating to teen health, fitness, and education." Articles that stand out, she says, incorporate first-person accounts. Among them was "Surviving Meningococcal Meningitis: What You Need to Know," by Lynn Bozof. It profiled "teen survivors of meningitis who have become champions of the cause, informing other teens how to pinpoint the symptoms of this preventable disease."

Other articles included the "insightful" pieces, "Work It! Nailing the Job Interview, from Résumé Writing to Sealing the Deal," by Cindy Haygood, and "Studying for the SAT: Using the School Vacation for Test Prep," by Rachim Baskin (in a combined May/June 2008 issue).

Writers can follow the successful career model and prep for the freelance college and career market, too: Study up, do the research, and nail the sale.

Keep It Real:
Writing to Inspire

"Fear of failure, depression, pre-marital sex (how to avoid it, how to deal with guilt related to it, how to talk with friends about the importance of sexual purity), bad language, anger, homosexuality, cliques, judgmentalism, parental conflicts, pornography, peer pressure, suicide, eating disorders, hypocrisy." These, says Editor Christopher Lutes, are the topics *Ignite Your Faith* teen readers revealed as their concerns in a recent survey.

Fiction and nonfiction—whether faith-based or not—help young readers to overcome difficulties; to grow in beliefs, hope, or self-confidence; examine their place in the world; and carve their futures or build their morality: That is inspirational. Such writing can be fraught with preachy pitfalls. But when done well, it can also have an impact perhaps larger than that of any other genre for young people.

Make a Change

"The most important thing I look for in inspirational fiction is this: Does it make me think about my own life and consider making a change in my attitudes, actions, and/or relationship with God? And it doesn't tell me what I should do, but leads me to make those conclusions for myself," says Richard Bennett, Editor at Radiant Life

"Does the story make me think about my own life and consider making a change? Does it lead me to make conclusions for myself?"

Resources. Among its publications is *LIVE,* a weekly take-home story paper distributed in young adult and adult Sunday School classes.

An important part of what makes fiction inspirational, says VS Grenier, Founder & Editor in Chief of *Stories for Children Magazine,* is leaving readers "knowing that everyone isn't perfect and that taking off their masks is okay. It's really refreshing for children to know we all have struggles and challenges in our lives."

For older readers, Lutes says, "Inspirational fiction that connects with teens must connect with their world and it must feel real and genuine. It doesn't have to be overly dramatic. In fact, it often works best when it's about an everyday issue—feelings of inadequacy, fear of failure, worries about a friend's spiritual state, those sorts of issues. It simply must connect with the internal emotional drama that's inside every teen."

At *InTeen*, Editor LaTonya Taylor also looks for inspirational fiction that is "clued into the teen world—the way teens act and speak. Stories for *InTeen* should serve an educational purpose without being didactic." The magazine is published by Urban Ministries for urban and African American teens and is distributed through Christian education classes.

For any age, strong inspirational fiction "has a long-lasting impact with a fresh or new approach," says Grenier. It touches "on the moral without being preachy or heavyhanded, and uses humor to lighten up the plot line. Another great way to have a good inspirational, fictional story is to have an engaging hook and narrative flow that keeps the reader engaged. This leaves readers with the *wow* factor. They'll come away from the story feeling inspired."

Editors all agree that writing from the child's or teen's perspective is key, as is a naturalness in a story's depiction of conflict and resolution. "Showing the reader what led the character to that decision and the result of the choice is important. Let the moral lessons flow naturally into the story" and draw the reader in, says Grenier.

Bennett agrees about natural flow and advises writers, "Good, believable dialogue is important while avoiding too much *thought* dialogue. While the miraculous does happen and can be a part of the story, it should neither be so momentous (at the level of the sun and moon standing still) that it is almost unbelievable or so coincidental that it appears staged. It should be part of the natural flow of the story without being the climax of the story. In my mind, good fiction means I should wonder if it really happened or if the story could happen in real life."

Tune In

Potential writers should study good stories published in this genre to avoid being too heavy handed or didactic, says Taylor. "A writer might have a group of teen readers or a youth group read their piece to see if it conveys the point and reads true."

Bennett also strongly urges writers not to be moralistic. "If you have to spell out the point of the story for the reader, it probably isn't well written and likely wouldn't be accepted for publication. Preachy, over-the-top characters generally are unbelievable and annoying. They don't make your point; they cause people to tune out (and stop reading)." Because *LIVE* is a Christian publication, he adds, "If you have a character doing a lot of prooftexting (using biblical texts to support an argument), your story will likely end up being didactic and preachy." Bennett suggests attending writing groups and peer readings, and being very open to honest criticism.

"Know your audience very, very well. Don't write for teens if you don't know teens," Lutes recommends. "I'd really encourage younger writers not to even think about writing fiction for teens until they've successfully crafted several as-told-to first-person stories based on interviews with teens or with young adults sharing an experience from their teen years. A real story about a real person (that digs below the surface) will most likely avoid those typical pitfalls."

Ignite Your Faith publishes inspirational fiction, but Lutes says it's hard to find good-quality stories in the genre. He points alternatively to the true stories told in departments such as My Life as a Student, My Life in Youth Group, and How I Share Christ. Helping readers with any of those topics the readers listed in the survey is important to *Ignite Your Faith,* Lutes says. "Now if those issues don't sound particularly inspirational, then go for the how I managed to avoid/survive/find-hope-in-the-midst-of approach. Speaking of approach: The as-told-to personal experience story is always the way to go for us. Well-crafted personal experience stories don't come across my desk enough."

Taylor is "interested in stories that

teach discipleship and faith in Christian life, especially through relationships with family and friends. I am even more interested in inspirational nonfiction stories that use the elements of fiction and story-telling." Writers for *InTeen* "should have an understanding of and respect for teens. They need to be able to enter into the world of today's teens." Taylor prefers query letters, but will consider unsolicited manuscripts no longer than 1,200 words. *InTeen* publishes seven or eight inspirational fiction pieces each year.

Stories for Children Magazine looks for fiction that is free of stereotypes but "pulls from different cultures and promotes and encourages cooperation and creativity in solving problems," says Grenier. "Fiction for our readership (ages 3 to 12) must delight and educate while still bringing an air of imagination or humor to the story. The writing needs to be of high quality and reflect the truth of childhood experiences. I get excited when I see a writer has a new or fresh idea or when a writer knows what a child of a certain age is like."

Submissions to *Stories for Children Magazine* should be emailed to Angelika Lochner, Assistant Editor (angelikalochner @storiesforchildrenmagazine.org). The editors prefer complete manuscripts. "This allows us to best judge the quality of the overall storyline. However, writers are welcome to query if they're not sure their plot is something we're interested in," says Grenier. "We're not a themed magazine and we haven't touched on all the topics we would like to, so query us if you have an idea and want to see if we'd be willing to look it over."

Fiction should have a "strong story-line with a beginning, middle, and end that wraps up the conflict or at least shows it's going to be fixed over time." Grenier also wants to see "tight writing. I see so many submissions with words like *as, and, so, just,* and *that* overused in each sentence or paragraph. If you can break the sentence up into two sentences then do it. Kids don't like really long sentences, especially beginning and young readers."

Stories for Children Magazine publishes 144 stories in the course of a year; about 10 percent of them are inspirational fiction. Grenier explains, "Even though we're a nonpaying magazine, 12 stories from each publishing year are purchased for our *Best Of* anthology." Among such potential best stories are "Peter the Planner," by Bish Denham, for ages three to six (January 2008) and "The Friendship Club," by Richard L. Provencher, for ages seven to nine (February 2008).

No Cheese, Please

"*LIVE* is intended to be a journal of practical Christian living," says Bennett. "One topic we don't see much of is Spirit-filled living. These can be tougher to write, but I am always looking for good stories on this theme." As for other material, he says, "While we use how-to stories, we limit how many we use each quarter (two or three). The majority of our stories are true stories. Of the stories accepted up to this point for 2009, 62 percent are true stories, 28 percent are nonfiction, and 10 percent are fiction. We label how-to stories and stories in which the writer advises readers what to do as nonfiction. I think of the latter as life application nonfiction, usually when the writer tells about an experience, how he or she dealt with it, and makes sug-

gestions to readers about how they might deal with similar experiences. Of the nonfiction stories we use, most would fall into the life application category." The potential topics are wide open.

The publication does not accept science fiction or Bible fiction. Bennett looks for well-written pieces without spelling or punctuation errors. "A good introduction and a good conclusion (ending the story where the story ends; not trying to tie up every loose end; not summarizing the story) are vital." Among Bennett's needs are "stories and poems that relate to New Year's, Sanctity of Life Sunday, Valentine's Day, Easter, Mother's Day, Memorial Day, Father's Day, Independence Day, Pastor Appreciation Sunday, Veteran's Day, Halloween, Thanksgiving, and Christmas. I am more open to nonfiction stories for holidays and special days," says Bennett. E-mailed manuscript submissions are preferred.

Recent stories in *LIVE* include "God, Where Are You?" a true story by Linda D. Roth (September 14, 2008). Bennett describes it as the "poignant story of how God reveals to a widow with younger children that He is there for her and is helping her. It is easy to empathize with the woman's situation and to be reminded of God's goodness to us, often when we are not aware of it." Another was Linda B. O'Connor's "Thanksgiving Guest" (November 16, 2008), "an interesting story that brings out how we tend to prejudge others, particularly based on outward appearances. The story is believable and interesting, has tension, and a good conclusion."

Ignite Your Faith accepts queries only, says Lutes. "We want well-crafted personal experience stories that offer hope through Christ and through

Christian community." Among the recent inspirational stories that worked well for *Ignite Your Faith* were "Torn by Divorce," by Amy Adair (March/April 2008); and "My Secret Struggle," by Scott Kelly (June/July 2008).

Lutes sums up the characteristics of fiction that truly inspires: "What makes one idea stand out from the others: the ability to tell a compelling, descriptive, scene-driven story (using fictional techniques) that's honest and hopeful, and not overwritten, overly religious, or cheesy."

Listings

How to Use the Listings

The pages that follow feature profiles of 636 magazines that publish articles and stories for, about, or of interest to children and young adults. Throughout the year, we stay on top of the latest happenings in children's magazines to bring you new and different publishing outlets. This year, our research yielded over 60 additional markets for your writing. They are easy to find; look for the listings with an exclamation point in the upper right corner.

A Variety of Freelance Opportunities

This year's new listings reflect the interests of today's magazine audience. You'll find magazines targeted to readers interested in nature, the environment, computers, mysteries, child care, careers, different cultures, family activities, and many other topics.

Along with many entertaining and educational magazines aimed at young readers, we list related publications such as national and regional magazines for parents and teachers. Hobby and special interest magazines generally thought of as adult fare but read by many teenagers are listed too.

In the market listings, the Freelance Potential section helps you judge each magazine's receptivity to freelance writers. This section offers information about the number of freelance submissions published each year.

Further opportunities for selling your writing appear in the Additional Listings section on page 275. This section profiles a range of magazines that publish a limited amount of material targeted to children, young adults, parents, or teachers. Other outlets for your writing can be found in the Selected Contests and Awards section, beginning on page 325.

Using Other Sections of the Directory

If you are planning to write for a specific publication, turn to the Magazine and Contest Index beginning on page 377 to locate the listing page. The Category Index, beginning on page 341, will guide you to magazines that publish in your areas of interest. This year, the Category Index also gives the age range of each publication's readership. To find the magazines most open to freelance submissions, turn to the Fifty+ Freelance index on page 340, which lists magazines that rely on freelance writers for over 50% of the material they publish.

Check the Market News, beginning on page 338, to find out what's newly listed, what's not listed and why, and to identify changes in the market that have occurred during the past year.

About the Listings

We revisited last year's listings and, through a series of mailed surveys and phone interviews, verified editors' names, mailing addresses, submissions and payment policies, and current editorial needs. All entries are accurate and up-to-date when we send this market directory to press. Magazine publishing is a fast-moving industry, though, and it is not unusual for facts to change before or shortly after this guide reaches your hands. Magazines close, are sold to new owners, or move; they hire new editors or change their editorial focus. Keep up to date by requesting sample copies and writers' guidelines.

Note that we do *not* list:

- Magazines that did not respond to our questionnaires or phone queries. Know that we make every effort to contact each editor before press date.

- Magazines that *never* accept freelance submissions or work with freelance writers.

To get a real sense of a magazine and its editorial slant, we recommend that you read several recent sample issues cover to cover. This is the best way to be certain a magazine is right for you.

The Kids' Ark

 — New listing

P.O. Box 3160
Victoria, TX 77903

Who to contact —

Editor: Joy Mygrants

DESCRIPTION AND INTERESTS

Profiles the publication, its interests, and readers —

The mission of this Christian magazine is to provide children with a biblical foundation on which to base their life choices. Each issue is theme-based and features three fun and adventurous stories along with games, puzzles, and comics. There is also an Internet version of the magazine. Circ: 8,000+.

Audience: 6–12 years
Frequency: Quarterly
Website: www.thekidsark.com

FREELANCE POTENTIAL

90% written by nonstaff writers. Publishes 12 freelance submissions yearly; 80–100% by authors who are new to the magazine.

Designates the amount and type of freelance submissions published each year; highlights the publication's receptivity to unpublished writers

SUBMISSIONS

Provides guidelines for submitting material; lists word lengths and types of material accepted from freelance writers —

Send complete ms. Accepts email submissions to thekidsarksubmissions@yahoo.com (Microsoft Word attachments). Responds in 2 months.

- Fiction: 600 words. Genres include contemporary, historical, and science fiction, all with a Christian base.
- Other: Puzzles, games, and comics.

SAMPLE ISSUE

24 pages: 3 stories; 5 activities; 1 comic. Guidelines and theme list available at website.

- "Randy's Victory." Story tells of a young boy who learns to deal with his anger.
- "Responding to Anger." Story is about two friends who reconcile their differences after an argument.
- "Remembering Yesterday." Story follows a mother and her son as they learn a valuable lesson about unkind people.

Analyzes a recent sample copy of the publication; briefly describes selected articles, stories, departments, etc.

RIGHTS AND PAYMENT

Lists types of rights acquired, payment rate, and number of copies provided to freelance writers —

First North American, worldwide, and electronic rights, with reprint rights. Fiction, $100. Reprints, $25. Pays on publication.

EDITOR'S COMMENTS

All material must match one of our quarterly themes. The first page of your submission should include the theme for which you are submitting as well as your name, phone number, and email address. We do not accept unsolicited poems, games, or comics. Please be sure stories are age-appropriate and include children of all races and ethnic backgrounds. Please refer to our website for a list of upcoming themes.

Offers advice from the editor about the publication's writing style, freelance needs, audience, etc.

Icon Key

 New Listing   E-publisher   Overseas Publisher

 Accepts agented submissions only   Not currently accepting submissions

Abilities

401-340 College Street
Toronto, Ontario M5T 3A9
Canada

Managing Editor: Jaclyn Law

DESCRIPTION AND INTERESTS
The purpose of this cross-disability lifestyle magazine is to provide information and inspiration to people with disabilities, their families, and the professionals who serve them. It contains articles on travel, health, careers, relationships, social policy, and more. Circ: 20,000.
Audience: Adults with disabilities
Frequency: Quarterly
Website: www.abilities.ca

FREELANCE POTENTIAL
50% written by nonstaff writers. Publishes 25 freelance submissions yearly; 15% by unpublished writers, 25% by authors who are new to the magazine. Receives 180 queries, 120 unsolicited mss yearly.

SUBMISSIONS
Prefers query with writing samples; will accept complete ms. Prefers email submissions to jaclyn@abilities.ca; will accept hard copy. SASE. No simultaneous submissions. Responds in 2–3 months.

- Articles: 800–2,000 words. Informational, self-help, and how-to articles; profiles; and interviews. Topics include health, technology, careers, and social issues.
- Depts/columns: 500–1,500 words. News, media and product reviews, profiles, family life, parenting, education, sports, travel, health, relationships, employment, housing, sexuality, and humor.

SAMPLE ISSUE
50 pages (50% advertising): 3 articles; 9 depts/columns. Guidelines available.

- "Rock 'n' Roll Soul." Article presents an interview with musician Martin Deschamps, whose disability does not stop his rock.
- Sample dept/column: "Family Life" explains why children of all abilities benefit from learning sign language.

RIGHTS AND PAYMENT
First-time serial and non-exclusive electronic rights. Written material, $25–$300. Kill fee, 50%. Pays 2 months after publication. Provides 2 contributor's copies.

➤➤EDITOR'S COMMENTS
We employ a conversational style, and we strive to give readers practical information that they can apply to their daily lives.

Ad Astra

1620 I Street NW, Suite 615
Washington, DC 20006

Editor: Katherine Brick

DESCRIPTION AND INTERESTS
Space enthusiasts, scientists, and technologists alike read this magazine published by the National Space Society. Its editorial content chronicles the most important developments in space exploration and aerospace science. Circ: 20,000.
Audience: YA–Adult
Frequency: Quarterly
Website: www.nss.org

FREELANCE POTENTIAL
95% written by nonstaff writers. Publishes 50 freelance submissions yearly. Receives 65 queries and unsolicited mss yearly.

SUBMISSIONS
Query or send complete ms with résumé. Accepts disk submissions and email submissions to adastra@nss.org. SASE. Response time varies.

- Articles: Word lengths vary. Informational and factual articles; profiles; and interviews. Topics include science and technology related to space exploration and issues related to the aerospace industry.
- Depts/columns: Word lengths vary. Reviews and opinion pieces.

SAMPLE ISSUE
64 pages (14% advertising): 6 articles; 11 depts/columns. Sample copy, $11.25 ($4 postage). Writers' guidelines and editorial calendar available.

- "Return to the Moon." Article takes an in-depth look at NASA's next generation program, Constellation, and notes its similarities and differences from Apollo.
- "Lunar Fly-In." Article reports on the two spacecraft now orbiting the Moon and the two more that are nearing launch.
- Sample dept/column: "Remembrance" pays tribute to a former member of the National Space Society's Board of Governors.

RIGHTS AND PAYMENT
First North American serial rights. Written material, payment rates vary. Payment policy varies.

➤➤EDITOR'S COMMENTS
If you have an expertise in the field, we welcome your submissions. With the exception of "Spotlight" and "Community," all sections are open to new writers.

ADDitude

39 West 37th Street, 15th Floor
New York, NY 10018

Editorial Assistant: Caitlin Ford

DESCRIPTION AND INTERESTS
A trusted resource for families touched by attention deficit disorder, the pages of *ADDitude* contain medical, behavioral, and educational information related to the condition. Articles address issues faced by both children and adults with ADD. Circ: 40,000.

Audience: Parents/Adults
Frequency: 5 times each year
Website: www.additudemag.com

FREELANCE POTENTIAL
80% written by nonstaff writers. Publishes 15–20 freelance submissions yearly; 30% by unpublished writers, 30% by authors who are new to the magazine. Receives 96 queries each year.

SUBMISSIONS
Query. Accepts email queries to submissions@additudemag.com. Responds in 6–8 weeks.

- Articles: To 2,000 words. Informational articles and personal experience pieces. Topics include ADD and ADHD, education, medication, recreation, organization, parenting, and child development.
- Depts/columns: Word lengths vary. Profiles of students, teachers, and schools; first-person essays; healthy living; organization; product reviews; and ADD/ADHD news.

SAMPLE ISSUE
62 pages (33% advertising): 2 articles; 22 depts/columns. Sample copy, $6.95. Guidelines available at website.

- "Calm the Chaos." Article provides nine strategies to help weeknights go more smoothly.
- Sample dept/column: "Lives" includes an essay by a single dad with ADHD about his worries and hopes for his children.

RIGHTS AND PAYMENT
First rights. Written material, payment rates vary. Kill fee, $75. Pays on publication. Provides a 1-year subscription.

☛EDITOR'S COMMENTS
Most of the articles we publish are written by journalists and mental-health professionals. However, we are willing to consider first-person articles by parents, employers, and teachers who have personal experience with ADD/ADHD.

Administrator

Scholastic Inc.
557 Broadway
New York, NY 10012

Editorial Director: Dana Truby

DESCRIPTION AND INTERESTS
This magazine from Scholastic offers "smart strategies" and "proven solutions" for school administrators working at all grade levels. It also covers education news, trends, and technology. Circ: 85,000.

Audience: School administrators, K–12
Frequency: 8 times each year
Website: www.scholastic.com/administrator

FREELANCE POTENTIAL
80% written by nonstaff writers. Publishes 25 freelance submissions yearly; 10% by unpublished writers.

SUBMISSIONS
Query with résumé. Accepts hard copy and email queries to dtruby@scholastic.com. SASE. Responds in 3–4 months.

- Articles: 1,200 words. Informational and how-to articles; profiles; and interviews. Topics include teacher recruitment and retention; salary negotiation; intervention programs; working with school boards; gifted education; school menus; grants and spending; crisis management; social issues; education trends and legislation; and technology.
- Depts/columns: Word lengths vary. News, opinions, technology, and book reviews.

SAMPLE ISSUE
64 pages: 4 articles; 7 depts/columns. Sample copy available at website. Guidelines available.

- "Retain and Deliver." Article tells how to stem the new-teacher exodus.
- "What's In Your Package?" Article presents tips for negotiating school superintendent salaries and benefits.
- Sample dept/column: "Plugged In" includes information on leasing school computers.

RIGHTS AND PAYMENT
All rights. All material, payment rates vary. Pays on publication.

☛EDITOR'S COMMENTS
We encourage article submissions from school administrators and other educators, as well as from freelance writers and journalists. Articles written by vendors are not accepted. If you are interested in submitting an article, please email Dana Truby or call her at 212-343-4652.

Adoptalk

North American Council on
Adoptable Children
970 Raymond Street, Suite 106
St. Paul, MN 55114-1149

Editor: Diane Riggs

DESCRIPTION AND INTERESTS
Adoptalk is a newsletter dedicated to the issues and people involved in foster care and adoption, as well as news and updates from the NACAC. Circ: 3,700.

Audience: Adults
Frequency: Quarterly
Website: www.nacac.org

FREELANCE POTENTIAL
25% written by nonstaff writers. Publishes 8 freelance submissions yearly; 5% by unpublished writers, 10% by authors who are new to the magazine. Receives 6 queries, 2 unsolicited mss yearly.

SUBMISSIONS
Query or send complete ms with bibliography. Accepts hard copy and email to dianeriggs@nacac.org. SASE. Responds in 2–3 weeks.

- Articles: To 2,000 words. Informational articles; personal experience pieces; profiles; and interviews. Topics include adoptive and foster care, parenting, recruitment, adoption news, conference updates, and NACAC membership news and updates.
- Depts/columns: Word lengths vary. Book reviews and first-person essays.

SAMPLE ISSUE
20 pages (no advertising): 7 articles; 6 depts/columns. Sample copy, free with 9x12 SASE ($1.14 postage).

- "GLBT Communities & Adoption: Courting an Untapped Resource." Article discusses the lack of recruitment by adoption agencies of gay, bisexual, or transgendered adults, and offers tips on increasing those numbers.
- Sample dept/column: "Food for Thought: Pain" is a first-person essay on the many types of pain that foster children endure.

RIGHTS AND PAYMENT
Rights vary. No payment. Provides 5 contributor's copies.

➤EDITOR'S COMMENTS
We would like to see more articles on parenting children with specific behavioral or educational issues, such as PTSD, FASD, ADHD, and drug exposure. Please be aware that most of our material comes from in-house staff or volunteers, but we certainly welcome well-written submissions.

Adoptive Families

39 West 37th Street, 15th Floor
New York, NY 10018

Editorial Assistant: Caitlin Ford

DESCRIPTION AND INTERESTS
A leading informational resource for families before, during, and after the adoption process, this magazine provides independent and authoritative information in a reader-friendly format. Circ: 40,000.

Audience: Adoptive parents
Frequency: 6 times each year
Website: www.adoptivefamilies.com

FREELANCE POTENTIAL
75% written by nonstaff writers. Publishes 100 freelance submissions yearly; 20% by unpublished writers, 50% by authors who are new to the magazine. Receives 500–600 queries and unsolicited mss yearly.

SUBMISSIONS
Query with clips for articles. Send complete ms for personal essays. Prefers email to submissions@adoptivefamilies.com (Microsoft Word attachments); will accept hard copy. SASE. Responds in 6–8 weeks.

- Articles: 500–1,800 words. Informational, self-help, and how-to articles; and personal experience pieces. Topics include preparing for adoption; health; education; birth families; and parenting tips.
- Depts/columns: To 1,200 words. Pre-adoption issues, birthparents, advice, opinion pieces, personal essays, single parenting, cultural diversity, child development, legal issues, and book reviews.

SAMPLE ISSUE
66 pages (40% advertising): 4 articles; 10 depts/columns. Sample copy, $6.95. Guidelines available at website.

- "Split Decision." Article explains ways to work through disagreements when one spouse is undecided about adoption.
- Sample dept/column: "Waiting Game" recounts how a family survived a 10-month wait with a series of updated photos.

RIGHTS AND PAYMENT
All rights. Written material, payment rates vary. Payment policy varies. Provides 2 copies.

➤EDITOR'S COMMENTS
We are currently looking for articles on adoption and school, and personal essays for "The Waiting Game" and "Been There" columns.

Advocate, PKA's Publication

1881 Little Westkill Road
Prattsville, NY 12468
Publisher: Patricia Keller

DESCRIPTION AND INTERESTS
This magazine provides a forum in which new writers can see their short stories, essays, and poems published. Most of its editorial relates to horses and other animals. Circ: 10,000.
Audience: YA–Adult
Frequency: 6 times each year

FREELANCE POTENTIAL
90% written by nonstaff writers. Publishes 150 freelance submissions yearly; 65% by unpublished writers, 35% by authors who are new to the magazine. Receives 1,500 unsolicited mss each year.

SUBMISSIONS
Send complete ms. Accepts hard copy. No simultaneous submissions. SASE. Responds in 6–10 weeks.

- Articles: To 1,500 words. Topics include horses, animals, the arts, humor, nature, and recreation.
- Fiction: To 1,500 words. Genres include adventure; fantasy; folktales; folklore; romance; historical, contemporary, realistic, and science fiction; humor; mystery; suspense; and stories about animals, nature, and the environment.
- Artwork: All forms of prints and illustrations.
- Other: *Gaited Horse Association Newsletter*, published within. Poetry, no line limit.

SAMPLE ISSUE
20 pages (50% advertising): 1 article; 1 story; 13 poems. Sample copy, $4. Writers' guidelines available.

- "Aruba: Fun in the Sun on Horse." Article provides a personal experience piece on the author's trip to Aruba for an island horseriding adventure.
- "In the Eye of the Stallion." Serialized story revolves around a man's quest to find, and ride, horses from a wild herd.

RIGHTS AND PAYMENT
First rights. No payment. Provides 2 contributor's copies.

➹EDITOR'S COMMENTS
You are welcome to submit well-written stories on any subject, however, pieces about horses, animals, and nature really get our attention. No previously published or simultaneous submissions, please.

African American Family

22041 Woodward Avenue
Ferndale, MI 48220
Editor: Lori Robinson

DESCRIPTION AND INTERESTS
African American Family covers parenting, family health, education, travel, and other subjects of particular interest to this demographic. Geared to families living in the Detroit area, it also features community projects, profiles of local leaders, and current events. Circ: Unavailable.
Audience: Parents
Frequency: Monthly
Website: www.africanamericanfamily-magazine.com

FREELANCE POTENTIAL
80% written by nonstaff writers. Publishes 20 freelance submissions yearly.

SUBMISSIONS
Query. Accepts email queries to lrobinson@metroparent.com. Response time varies.

- Articles: Word lengths vary. Informational and self-help articles; profiles; interviews; personal experience pieces; and reviews. Topics include current events, education, history, nature, the environment, recreation, social issues, sports, and travel.
- Depts/columns: Word lengths vary. Community issues, regional resources and events, and health and fitness.

SAMPLE ISSUE
46 pages: 2 articles; 7 depts/columns. Guidelines available at website.

- "The Cosby Parenting Prescription." Article profiles the famous actor and discusses his sometimes controversial views on parenting within the African American community.
- "Fired Up for Obama." Article describes the excitement from Detroit-area residents for the Barack Obama campaign.
- Sample dept/column: "Arts & Trends" provides information on a hip hop art exhibit opening at a local museum.

RIGHTS AND PAYMENT
Rights vary. Written material, payment rates vary. Pays on publication.

➹EDITOR'S COMMENTS
We prefer material that reflects sources and ideas from the greater Detroit metropolitan area. Articles should be professional with solid reporting and writing skills. Please review our guidelines for more information.

AIM Magazine

P.O. Box 856
Forest Grove, OR 97116

Editor: Kathleen Leatham

DESCRIPTION AND INTERESTS
This magazine supports the eradication of racism and promotes intercultural awareness, respect, heightened understanding, and unconditional acceptance of all people and ideas. It publishes articles, fiction, and poetry with the same mission. Circ: 1,000.

Audience: Educators
Frequency: Quarterly
Website: www.aimmagazine.org

FREELANCE POTENTIAL
95% written by nonstaff writers. Publishes 50 freelance submissions yearly; 50% by unpublished writers, 80% by authors who are new to the magazine. Receives 98 unsolicited mss each year.

SUBMISSIONS
Send complete ms. Accepts email submissions to submissions@aimmagazine.org (Microsoft Word attachments). No simultaneous submissions. Responds in 3–4 months.

- Articles: To 4,000 words. Informational articles; profiles; interviews; essays; and personal experience and opinion pieces. Topics include race, social issues, bigotry, culture, ethnic holidays, politics, and humor.
- Fiction: To 4,000 words. Genres include inspirational, historical, contemporary, ethnic, multicultural, humorous, and literary fiction.
- Artwork: B/W or color prints; JPEG, TIFF, or BMP images at 300 dpi. Line art.
- Other: Poetry, to 100 lines; to 5 poems per submission.

SAMPLE ISSUE
50 pages: 1 article; 9 stories; 4 poems. Sample copy, $5 with 9x12 SASE (5 first-class stamps). Writers' guidelines and editorial calendar available.

- "A Mother's Heart." Story tells the grief of a mother whose child was killed during an ambush in Iraq.
- "Drowning in Los Angeles." Story portrays the seedier side of life in Hollywood.

RIGHTS AND PAYMENT
First rights. Articles, $15–$25. Fiction, $100. Pays on publication. Provides 1 author's copy.

➥EDITOR'S COMMENTS
We seek only the most compelling works of nonfiction and fiction relevant to our mission.

Alateen Talk

Al-Anon Family Group Headquarters
1600 Corporate Landing Parkway
Virginia Beach, VA 23454-5617

Associate Director: Mary Lou Mahlman

DESCRIPTION AND INTERESTS
Written for children and teens who have been affected by a family member's alcoholism, *Alateen Talk* is a newsletter that allows those children to share their experiences and thoughts through their writing. Circ: 4,000.

Audience: 6–18 years
Frequency: Quarterly
Website: www.al-anon.alateen.org

FREELANCE POTENTIAL
90% written by nonstaff writers. Publishes 85–120 freelance submissions yearly; 80% by unpublished writers, 75% by authors who are new to the magazine. Receives 100–150 unsolicited mss yearly.

SUBMISSIONS
Accepts complete mss from Alateen members only. Accepts hard copy. SASE. Responds in 2 weeks.

- Articles: Word lengths vary. Self-help and personal experience pieces. Topics include alcoholism and its effects on relationships; social issues; and family life.
- Depts/columns: Staff written.
- Artwork: B/W line art.
- Other: Poetry.

SAMPLE ISSUE
8 pages (no advertising): 2 articles; 14 letters; 1 poem; 1 activity. Sample copy, free with 9x12 SASE ($.87 postage). Guidelines available to Alateen members only.

- "From the WSO." Article reports on two Regional Service Seminars, in Canada and West Virginia, and begins laying the groundwork for Alateen's 50th anniversary year.
- "Sharings on the Serenity Prayer" presents writings from children whose lives have been turned around through Alateen, its meetings, and the Serenity Prayer.

RIGHTS AND PAYMENT
All rights. No payment.

➥EDITOR'S COMMENTS
Because we are a membership-service organization, all of our material is written by either staff or by our members—children and teens whose lives have been affected by the alcoholism of family members. We welcome submissions from our young members all over the world.

Alfred Hitchcock Mystery Magazine

475 Park Avenue South
New York, NY 10016

Editor: Linda Landrigan

DESCRIPTION AND INTERESTS
Alfred Hitchcock Mystery Magazine offers a selection of the finest crime story writing. This includes tales of mystery, espionage, and suspense that feature crime themes. Puzzles and movie and book reviews also appear regularly. Circ: 100,000.
Audience: YA–Adult
Frequency: 10 times each year
Website: www.themysteryplace.com

FREELANCE POTENTIAL
97% written by nonstaff writers. Publishes 90–100 freelance submissions yearly; 5% by unpublished writers, 5–10% by authors who are new to the magazine. Receives 2,400 unsolicited mss yearly.

SUBMISSIONS
Send complete ms. Accepts hard copy. No simultaneous submissions. SASE. Responds in 3–5 months.

- Fiction: 12,000 words. Genres include classic crime mysteries, detective stories, suspense, private-investigator tales, courtroom drama, and espionage.
- Depts/columns: Word lengths vary. Reviews, puzzles, and profiles of bookstores.

SAMPLE ISSUE
144 pages (2% advertising): 8 stories; 8 depts/columns. Sample copy, $5. Writers' guidelines available.

- "Horse Friends." Story of a murder features a twist of revenge for the insurance company representative sent to investigate.
- "The Length of a Straw." Story set in Tsarist Russia tells of a hunting party's confrontation with the army.
- Sample dept/column: "Reel Crime" reviews the AHMM Winter Film Fest picks by top crime writers.

RIGHTS AND PAYMENT
First serial, anthology, and foreign rights. Written material, payment rates vary. Pays on acceptance. Provides 3 contributor's copies.

⟡EDITOR'S COMMENTS
Crime must be the central theme of every submission. Tales of the supernatural will be considered, as long as a crime is part of the plot. We accept fiction only, so please do not send stories based on actual real-life events.

American Baby

375 Lexington Avenue, 9th Floor
New York, NY 10017

Editorial Assistant: Katie Rockman

DESCRIPTION AND INTERESTS
This magazine is filled with articles on pregnancy, childbirth, and the ups and downs of caring for little ones. It covers a wide spectrum of topics related to parenting babies and toddlers. Circ: 2.1 million.
Audience: Parents
Frequency: Monthly
Website: www.americanbaby.com

FREELANCE POTENTIAL
55% written by nonstaff writers. Publishes 24 freelance submissions yearly; 20% by unpublished writers. Receives 996 queries and unsolicited mss yearly.

SUBMISSIONS
Query with clips and writing samples; or send complete ms. Accepts hard copy and simultaneous submissions if identified. SASE. Responds in 2 months.

- Articles: 1,000–2,000 words. Informational and how-to articles; profiles; interviews; humor; and personal experience pieces. Topics include pregnancy, childbirth, parenting, health, fitness, nutrition, child care, child development, religion, and travel.
- Depts/columns: 1,000 words. Health briefs, fitness tips, new products, and fashion.
- Other: Submit seasonal material 3 months in advance.

SAMPLE ISSUE
80 pages (50% advertising): 13 articles; 6 depts/columns. Sample copy available at newsstands. Guidelines available.

- "Everyone's an Expert." Article offers ways parents can deal with well-meaning, but unwanted, advice from others.
- Sample dept/column: "Preschooldays" offers ideas for packing the most nutrition into a quick breakfast.

RIGHTS AND PAYMENT
First serial rights. Articles, to $2,000. Depts/columns, to $1,000. Pays on acceptance. Provides 5 contributor's copies.

⟡EDITOR'S COMMENTS
If it's a topic that new parents are talking about—or even wondering about—we'd be interested in it. We don't forget the fact that new fathers want information, too, and welcome dad-centric articles as well.

American Careers

6701 West 64th Street
Overland Park, KS 66202

Editor: Mary Pitchford

DESCRIPTION AND INTERESTS

American Careers is published in three annual editions: elementary, for grades four through six; middle school, for grades seven through nine; and high school, for grades nine through twelve. Each edition focuses on career development for students. Circ: 400,000.

Audience: 12–18 years
Frequency: Annually
Website: www.carcom.com

FREELANCE POTENTIAL

25% written by nonstaff writers. Publishes 5 freelance submissions yearly; 20% by authors who are new to the magazine. Receives 240+ queries yearly.

SUBMISSIONS

Query with résumé and clips. Accepts hard copy. SASE. Responds in 2 months.

- Articles: 300–750 words. Informational and how-to articles; profiles; and personal experience pieces. Topics include specific and general careers, career planning and education, self-evaluation, and entrepreneurship.
- Depts/columns: 300–750 words. Educational opportunities, employability, and life skills.
- Artwork: Color prints or digital images.
- Other: Quizzes and self-assessments.

SAMPLE ISSUE

66 pages (no advertising): 16 articles; 7 depts/columns. Sample copy, $4 with 9x12 SASE (5 first-class stamps). Guidelines available.

- "Wind Farming." Article describes the various job opportunities in the new and growing field of wind energy production.
- "Making a Difference." Article profiles a young chiropractor, who explains her reasons for choosing this career.
- Sample dept/column: "Employability and Life Skills" offers advice and tips for first-time job seekers.

RIGHTS AND PAYMENT

All rights. All material, payment rates vary. Payment policy varies. Provides 2 copies.

➥EDITOR'S COMMENTS

Prospective authors should keep in mind that articles must be age- and grade-appropriate. We do not need queries related to employability skills such as résumé and cover letter writing, or interviewing.

American Cheerleader

110 William Street, 23rd Floor
New York, NY 10038

Editor-in-Chief: Marisa Walker

DESCRIPTION AND INTERESTS

The only magazine of its kind, *American Cheerleader* offers comprehensive coverage of the sport. Profiles, training advice, competition guides, and general beauty and style tips fill its pages. Circ: 150,000.

Audience: 13–18 years
Frequency: 6 times each year
Website: www.americancheerleader.com

FREELANCE POTENTIAL

40% written by nonstaff writers. Publishes 15 freelance submissions yearly; 20% by unpublished writers, 10% by authors who are new to the magazine. Receives 12–24 queries and unsolicited mss yearly.

SUBMISSIONS

Query with clips; or send complete ms. Prefers email submissions to mwalker@americancheerleader.com; will accept hard copy. SASE. Responds in 3 months.

- Articles: To 1,000 words. Informational and how-to articles; profiles; personal experience pieces; and photo-essays. Topics include cheerleading, workouts, competitions, scholarships, fitness, college, careers, and popular culture.
- Depts/columns: Word lengths vary. Safety issues, health, nutrition, beauty, fashion, fundraising, and new product information.
- Artwork: High-resolution digital images; 35mm color slides.

SAMPLE ISSUE

112 pages (40% advertising): 19 articles; 6 depts/columns; 1 quiz. Sample copy, $3.99 with 9x12 SASE ($1.70 postage). Editorial calendar available.

- "My Worlds Diary." Personal experience piece reveals a cheerleader's views while accompanying her team to a major competition.
- "Camp Bond-a-Thon." Article highlights 10 ways to bond with teammates during camp.

RIGHTS AND PAYMENT

All rights. All material, payment rates vary. Pays 2 months after acceptance. Provides 1 contributor's copy.

➥EDITOR'S COMMENTS

Our writing style is upbeat and "sporty" to catch the attention of our readers. We invite proposals from those involved in the sport.

American Educator

555 New Jersey Avenue NW
Washington, DC 20001

Assistant Editor: Jennifer Jacobson

DESCRIPTION AND INTERESTS
This professional journal for the American Federation of Teachers seeks to improve public education by exploring the issues, trends, and policies that affect the country's school system. It is read by teachers, professors, and policymakers. Circ: 850,000.
Audience: Teachers, professors, policymakers
Frequency: Quarterly
**Website: www.aft.org/pubs-reports/
    american-educator/index.htm**

FREELANCE POTENTIAL
25% written by nonstaff writers. Publishes 5–6 freelance submissions yearly; 10% by unpublished writers, 20% by authors who are new to the magazine. Receives 360 queries each year.

SUBMISSIONS
Query with contact information. Accepts hard copy and email submissions to amered@aft.org. SASE. Responds in 2 months.

- Articles: 1,500–3,000 words. Informational articles; profiles; and opinion pieces. Topics include trends in education, politics, education law, professional ethics, social and labor issues, and international affairs.

SAMPLE ISSUE
44 pages (9% advertising): 8 articles; 1 dept/column. Guidelines available at website.

- "What's Missing from Math Standards." Article examines the reasons behind lagging math scores in the U.S.
- "Before Their Time." Photo-essay takes a look at child labor around the world.
- "Plugging the Holes in State Standards." Opinion piece by a college professor discusses what is needed to improve state standards in education.

RIGHTS AND PAYMENT
All rights. Written material, payment rates vary. Pays on publication. Provides 10 copies.

☛EDITOR'S COMMENTS
We would like to see more articles on school improvements and college readiness. Articles should be written in a journalistic style that avoids education and research jargon. We do not publish research papers or doctoral theses. Please be sure to include your contact information with your query.

American Girl

American Girl Publishing
8400 Fairway Place
Middleton, WI 53562

Editorial Assistant

DESCRIPTION AND INTERESTS
Dedicated to girls ages eight and up, this magazine encourages big dreams, self-confidence, and curiosity through its articles profiling interesting girls and presenting interesting topics. Circ: 700,000.
Audience: 8–12 years
Frequency: 6 times each year
Website: www.americangirl.com

FREELANCE POTENTIAL
4% written by nonstaff writers. Publishes 5 freelance submissions yearly; 2% by unpublished writers, 2% by authors who are new to the magazine. Receives 780 queries each year.

SUBMISSIONS
Query. Accepts hard copy and simultaneous submissions if identified. SASE. Responds in 4 months.

- Articles: 500–1,000 words. Informational articles; profiles; and interviews. Topics include history, nature, food, hobbies, crafts, sports, and culture.
- Depts/columns: 175 words. Profiles, how-to pieces, and craft ideas.

SAMPLE ISSUE
48 pages (no advertising): 2 articles; 1 story; 7 depts/columns. Sample copy, $4.50 at newsstands. Guidelines available.

- "Alex's Story." Article relates a young girl's experiences dealing with childhood cancer, and reports on her inspirational fundraising efforts now that she is healthy.
- Sample dept/column: "Looking Good" offers advice on girls' biggest worries about glasses.

RIGHTS AND PAYMENT
Rights vary; all material purchased under work-for-hire agreement. Written material, payment rates vary. Pays on acceptance. Provides 3 contributor's copies.

☛EDITOR'S COMMENTS
We are looking for articles about individual girls or groups who are doing something other girls would love to read and learn about. Find a topic about which girls are passionate. Our Girls Express section, featuring short profiles, is the best place for new writers to get published here. We are no longer accepting fiction submissions.

American Libraries

American Library Association
50 East Huron Street
Chicago, IL 60611

Acquisitions Editor: Pamela Goodes

DESCRIPTION AND INTERESTS

As the official publication of the American Library Association, *American Libraries* publishes articles on the professional concerns of librarians, along with news of the association, library-related legislation, and libraries around the world. Circ: 56,000.

Audience: Librarians
Frequency: 10 times each year
Website: www.ala.org/alonline

FREELANCE POTENTIAL

60% written by nonstaff writers. Publishes 10 freelance submissions yearly; 50% by authors who are new to the magazine, 100% by experts. Receives 492 unsolicited mss yearly.

SUBMISSIONS

Send complete ms. Accepts hard copy, IBM disk submissions (Microsoft Word), and email submissions to americanlibraries@ala.org. No simultaneous submissions. SASE. Responds in 8–10 weeks.

- Articles: 600–2,000 words. Informational articles; profiles; interviews; and personal experience pieces. Topics include modern libraries, library and ALA history, technology, leadership, advocacy, funding, and privacy.
- Depts/columns: Word lengths vary. News, opinions, profiles, media reviews, information technology, ALA events, career leads, and professional development.

SAMPLE ISSUE

120 pages (27% advertising): 5 articles; 22 depts/columns. Sample copy, $6. Guidelines available.

- "Be Outstanding in Your Fieldwork." Article recommends that graduate students become worksite supervisors.
- Sample dept/column: "Technology News" describes the "Library of Congress Experience," which uses interactive technology.

RIGHTS AND PAYMENT

First North American serial rights. Written material, $50–$400. Pays on acceptance. Provides 1+ contributor's copies.

➭EDITOR'S COMMENTS

We must stress that we accept submissions only from professional librarians. Our intention is to publish experience-based articles that aid our ALA colleagues in performing their jobs.

American School & University

9800 Metcalf Avenue
Overland Park, KS 66212

Executive Editor: Susan Lustig

DESCRIPTION AND INTERESTS

Leaving academics to the educators, this magazine covers all things related to school infrastructure: the design, construction, retrofit, operations, maintenance, security, and management of the nation's educational facilities. Circ: 65,000.

Audience: School facilities managers
Frequency: Monthly
Website: www.asumag.com

FREELANCE POTENTIAL

35% written by nonstaff writers. Publishes 40 freelance submissions yearly; 30% by authors who are new to the magazine. Receives 180 queries yearly.

SUBMISSIONS

Query with outline. Prefers email queries to slustig@asumag.com; will accept hard copy. SASE. Responds in 2 weeks.

- Articles: 1,200 words. Informational and how-to articles. Topics include educational facilities management, maintenance, security, planning, design, construction, operations, and furnishings.
- Depts/columns: 250–350 words. Opinions, environmental practices, technology, planning issues, and new products.

SAMPLE ISSUE

58 pages (55% advertising): 4 articles; 5 depts/columns, 1 special section. Sample copy, $10. Guidelines and editorial calendar available.

- "Balancing Act." Article offers 10 key concepts that administrators and planners must consider as they build, maintain, manage, and improve education facilities.
- Sample dept/column: "Viewpoint" includes an article on using "green" cleaning and maintenance products.

RIGHTS AND PAYMENT

All rights. Written material, payment rates and payment policy vary. Provides 2 author's copies.

➭EDITOR'S COMMENTS

In general, we work about three months in advance for feature stories, and often farther in advance. Please keep in mind that we get hundreds of press releases, phone calls, and emails per day. We will contact you regarding your submission if it applies to our audience, but sometimes the process takes a few weeks.

American Secondary Education

Ashland University
Dwight Schar COE, Room 231
401 College Avenue
Ashland, OH 44805

Editorial Assistant: Gay Vanderzyden

DESCRIPTION AND INTERESTS
American Secondary Education is a refereed journal that focuses specifically on secondary education theory and practice. Articles cover a broad range of topics in curriculum, instruction, and assessment. Circ: 450.
Audience: Secondary school educators
Frequency: 3 times each year
Website: www3.ashland.edu/ase

FREELANCE POTENTIAL
99% written by nonstaff writers. Publishes 20 freelance submissions yearly; 75% by authors who are new to the magazine. Receives 40–50 unsolicited mss yearly.

SUBMISSIONS
Send 3 copies of complete ms with 100-word abstract. Accepts hard copy. No simultaneous submissions. SASE. Response time varies.

- Articles: 10–30 double-spaced ms pages. Informational articles. Topics include secondary and middle school education research and practice.
- Depts/columns: Word lengths vary. Book reviews.

SAMPLE ISSUE
104 pages (no advertising): 7 articles; 1 book review. Sample copy, free. Guidelines available in each issue.

- "The Effect of Teacher Communication with Parents on Students' Mathematics Achievement." Article tells how a controlled study of four math classes taught by the same teacher showed that those with regular parental involvement performed better.
- "Reassigning the Identity of the Pregnant and Parenting Student." Article describes how separate schools for teen mothers have the potential to provide unique, alternative learning spaces.
- Sample dept/column: "Book Review" examines a recent publication about collaborative planning and teaching for universally designed learning.

RIGHTS AND PAYMENT
All rights. No payment. Provides 1 author's copy.

➜EDITOR'S COMMENTS
Please note that it takes approximately eight months from the time we receive a submission to publish it, if accepted.

American String Teacher

4153 Chain Bridge Road
Fairfax, VA 22030

Editor: Mary Jane Dye

DESCRIPTION AND INTERESTS
This award-winning journal features well-researched information used to enhance the professional growth of teachers of violin, viola, cello, bass, guitar, and harp. Teaching, methodology, techniques, and performing are all covered. Circ: 11,500.
Audience: String instrument teachers
Frequency: Quarterly
Website: www.astaweb.com

FREELANCE POTENTIAL
100% written by nonstaff writers. Publishes 30 freelance submissions yearly; 5% by unpublished writers, 50% by authors who are new to the magazine. Receives 50 queries yearly.

SUBMISSIONS
Prefers query; will accept 5 copies of complete ms. Prefers email to maryjane@astaweb.com; will accept hard copy. No simultaneous submissions. SASE. Responds in 3 months.

- Articles: 1,000–3,000 words. Informational and factual articles; profiles; and association news. Topics include teaching, methodology, techniques, competitions, and auditions.
- Depts/columns: Word lengths vary. Teaching tips, opinion pieces, and industry news.

SAMPLE ISSUE
80 pages (45% advertising): 5 articles; 9 depts/columns. Sample copy, free with 9x12 SASE ($3.25 postage). Writers' guidelines available at website.

- "Brahms and Joachim: An 1850s Comedy?" Article researches the unusual friendship between two violinists in the mid-1800s.
- "Therapy for Sight-Reading Woes." Article discloses attributes necessary to succeed in sight reading as well as strategies to enhance the skill.
- Sample dept/column: "Teaching Tips" provides information on violin practice patterns.

RIGHTS AND PAYMENT
All rights. No payment. Provides 5 copies.

➜EDITOR'S COMMENTS
We accept material from members of the American String Teacher Association only. Articles must be based on research, have a national perspective, and avoid promoting products or commercial programs. Please review our guidelines.

Analog Science Fiction and Fact

Dell Magazine Fiction Group
475 Park Avenue South
New York, NY 10016

Editor: Stanley Schmidt

DESCRIPTION AND INTERESTS

Science and technology are the key themes in the fiction and nonfiction contents of this magazine. Circ: 40,000.

Audience: YA–Adult
Frequency: 10 times each year
Website: www.analogsf.com

FREELANCE POTENTIAL

100% written by nonstaff writers. Publishes 80–90 freelance submissions yearly; 10% by unpublished writers, 10% by authors who are new to the magazine. Receives 6,000 unsolicited mss yearly.

SUBMISSIONS

Query for serials. Send complete ms for shorter works. Accepts hard copy. SASE. Responds in 6 weeks.

- Articles: To 6,000 words. Informational articles. Topics include science and technology.
- Fiction: Serials, 40,000–80,000 words. Novellas and novelettes, 10,000–20,000 words. Short stories, 2,000–7,000 words. Physical, sociological, psychological, and technological science fiction.
- Depts/columns: Staff written.

SAMPLE ISSUE

144 pages (7% advertising): 2 articles; 1 serial; 3 novelettes; 3 stories; 1 poem; 6 depts/columns. Guidelines available.

- "Nuclear Autumn: The Global Consequences of a Small Nuclear War." Article examines the current political and technological conditions that could precipitate a nuclear war.
- "Into that Good Night." Story tells of a man who makes the ultimate sacrifice to save others in a nuclear fusion accident.
- "Righteous Bite." Story follows a military dog in the war against terror.

RIGHTS AND PAYMENT

First North American serial and non-exclusive rights. Serials, $.04 per word; other written material, $.05–$.08 per word. Pays on acceptance. Provides 2 contributor's copies.

➾EDITOR'S COMMENTS

We consider quality of writing before anything else; so new, unpublished writers are at no disadvantage. We look for realistic plot lines, believable characters (doing believable things), and a strong scientific foundation.

AppleSeeds

Cobblestone Publishing Company
30 Grove Street
Peterborough, NH 03458

Editor: Susan Buckley

DESCRIPTION AND INTERESTS

AppleSeeds offers lively, age-appropriate articles for curious children in grades three and four. It explores a different theme in each issue, from the history of presidential elections to the geological wonders of the Grand Canyon. Circ: 10,000.

Audience: 8–10 years
Frequency: 9 times each year
Website: www.cobblestonepub.com

FREELANCE POTENTIAL

80% written by nonstaff writers. Publishes 90–100 freelance submissions yearly; 33% by authors who are new to the magazine. Receives 600+ queries yearly.

SUBMISSIONS

Query with clips, source list, and proposed word count. Accepts email queries to swbuc@aol.com (no attachments). Responds in 2–3 months only if interested.

- Articles: 150–600 words. Informational and how-to articles; profiles; and interviews. Topics include history, biography, biology, geology, technology, geography, literature, and the environment.
- Depts/columns: "Reading Corner," 300–600 words. "By the Numbers," "Your Turn," and "Experts in Action," 150 words. "From the Source," 150–300 words.
- Artwork: B/W and color prints.
- Other: Theme-related games and activities.

SAMPLE ISSUE

34 pages (no advertising): 7 articles; 5 depts/columns. Sample copy available. Guidelines and theme list available at website.

- "People of the Blue-Green Waters." Article describes life in Supai Village, the most remote settlement in the U.S.
- Sample dept/column: "Your Turn" provides tips for safe hiking.

RIGHTS AND PAYMENT

All rights. Written material, $50–$200. Artwork, $15–$100. Pays on publication. Provides 2 contributor's copies.

➾EDITOR'S COMMENTS

We seek articles that exhibit an original approach to the issue's theme. Scientific and historical accuracy is extremely important, and authors should use current sources.

Arizona Parenting

2432 West Peoria Avenue, Suite 1206
Phoenix, AZ 85029

Editor: Lyn Wolford

DESCRIPTION AND INTERESTS
This monthly magazine offers Arizona parents a wealth of helpful advice on many family- and child-related issues, all in a colorful, easy-to-navigate format. Circ: 80,000.
Audience: Parents
Frequency: Monthly
Website: www.azparenting.com

FREELANCE POTENTIAL
50% written by nonstaff writers. Publishes 20 freelance submissions yearly; 5% by unpublished writers, 25% by authors who are new to the magazine. Receives 520 queries, 300 unsolicited mss yearly.

SUBMISSIONS
Query or send complete ms. Prefers email to lynwolford@azparenting.com; will accept hard copy. SASE. Responds in 2–3 months.

- Articles: 850–2,400 words. Informational articles; profiles; interviews; and humorous pieces. Topics include parenting and family issues, education, personal finances, health, fitness, recreation, travel, computers, electronics, and sports.
- Depts/columns: 400–850 words. Child development, cooking and nutrition, new product reviews, and short news items.
- Artwork: B/W prints and transparencies.

SAMPLE ISSUE
66 pages (50% advertising): 4 articles; 6 depts/columns. Sample copy and editorial calendar, free with 9x12 SASE ($2 postage).

- "Listen Up!" Article offers first-hand advice from a teen author on how parents can get their teenagers to share their thoughts.
- Sample dept/column: "Family Table" describes how the nation's largest organic farmers' cooperative raises awareness about local, organic foods.

RIGHTS AND PAYMENT
First North American serial and electronic rights. All material, payment rates vary. Pays on publication. Provides 2–3 author's copies.

✎EDITOR'S COMMENTS
We are always looking for submissions from experienced and knowledgeable writers who can give our readers the best parenting advice and information. Keep the focus on Arizona parents and families.

Art Education

University of Cincinnati
Department of Art Education
Aronoff 6431C, P.O. Box 210016
Cincinnati, OH 45221-0016

Editor: Dr. Flavia Bastos

DESCRIPTION AND INTERESTS
Art teachers of all grade levels turn to this magazine for the latest developments in the art field as well as for instructional resources. It is the official publication of the National Art Education Association. Circ: 20,000.
Audience: Art educators
Frequency: 6 times each year
Website: www.naca-reston.org/ publications.html

FREELANCE POTENTIAL
88% written by nonstaff writers. Publishes 36 freelance submissions yearly; 40% by unpublished writers, 40% by authors who are new to the magazine. Receives 120 unsolicited mss each year.

SUBMISSIONS
Send 3 copies of complete ms. Accepts hard copy, disk submissions, and simultaneous submissions if identified. SASE. Responds in 8–10 weeks.

- Articles: To 3,000 words. Research-based informational articles; personal experience pieces; interviews; and profiles. Topics include the visual arts, curriculum planning, art history, and art criticism.
- Depts/columns: To 2,750 words. "Instructional Resources" features lesson plan ideas.
- Artwork: 8x10 or 5x7 B/W prints or slides. Digital images.

SAMPLE ISSUE
54 pages (3% advertising): 6 articles; 2 depts/columns. Sample copy, $1.25 with 9x12 SASE ($.87 postage). Guidelines available.

- "The Individual Video Experience (iVE)." Article discusses how the iPod can be used as an educational tool in art museums.
- Sample dept/column: "Instructional Resources" features a lesson plan on landscape photographer Eliot Porter's influence.

RIGHTS AND PAYMENT
All rights. No payment. Provides 2 copies.

✎EDITOR'S COMMENTS
We are currently seeking more research-based articles written by art teachers of kindergarten through grade 12. Manuscripts on museum objects, artwork, and sample curricula are welcome for our "Instructional Resources" department.

Arts & Activities

12345 World Trade Drive
San Diego, CA 92128

Editor: Maryellen Bridge

DESCRIPTION AND INTERESTS
Written exclusively for and by art teachers, this magazine presents a plethora of new ideas for projects for all grade levels. *Arts & Activities* is also a place where art teachers can share experiences and expertise. Circ: 20,000.

Audience: Art educators, grades K–12
Frequency: 10 times each year
Website: www.artsandactivities.com

FREELANCE POTENTIAL
95% written by nonstaff writers. Publishes 100 freelance submissions yearly; 50% by unpublished writers, 50% by authors who are new to the magazine. Receives 240 unsolicited mss each year.

SUBMISSIONS
Send complete ms. Accepts disk submissions with hard copy. No simultaneous submissions. SASE. Responds in 4–8 months.

- Articles: Word lengths vary. Informational, how-to, and practical application articles; and personal experience pieces. Topics include art education, program development, collage, printmaking, art appreciation, and composition.
- Depts/columns: Word lengths vary. New product information, news items, and reviews.
- Artwork: Color photos; digital images, minimum pixel resolution of 2550 x 3400.
- Other: Lesson plans for classroom projects.

SAMPLE ISSUE
50 pages (29% advertising): 15 articles; 3 depts/columns. Sample copy, $3 with 9x12 SASE ($2 postage). Writers' guidelines and theme list available.

- "Hip Like Haring." Article outlines a lesson plan using information about artist Keith Haring to launch into an art project that studies the use of color to portray a mood.
- Sample dept/column: "Media Reviews" offers reviews of art education books.

RIGHTS AND PAYMENT
First North American serial rights. All material, payment rates vary. Pays on publication. Provides 2 contributor's copies.

➥EDITOR'S COMMENTS
Our readers are art educators; we are not interested in home craft projects. We are very interested in high school-level projects.

Asimov's Science Fiction

Dell Magazine Group
475 Park Avenue South, 11th Floor
New York, NY 10016

Editor: Sheila Williams

DESCRIPTION AND INTERESTS
Fans of science fiction and fantasy turn to this award-winning journal for a mix of novelettes, short stories, and poetry. The character-driven stories found on its pages seek to examine or illuminate some aspect of human existence. Book and website reviews are also included regularly. Circ: 60,000.

Audience: YA–Adult
Frequency: 10 times each year
Website: www.asimovs.com

FREELANCE POTENTIAL
97% written by nonstaff writers. Publishes 85 freelance submissions yearly; 10% by unpublished writers, 30% by authors who are new to the magazine. Receives 8,400 unsolicited mss each year.

SUBMISSIONS
Send complete ms. Accepts hard copy. No simultaneous submissions. SASE. Responds in 6–8 weeks.

- Fiction: To 20,000 words. Genres include science fiction and "borderline" fantasy.
- Depts/columns: Word lengths vary. Book and website reviews.
- Other: Poetry, to 40 lines.

SAMPLE ISSUE
144 pages (10% advertising): 3 novelettes; 5 stories; 3 poems; 4 depts/columns. Sample copy, $5 with 9x12 SASE ($.77 postage). Guidelines available.

- "Soldier of the Singularity." Story tells of a doctor's encounter with a malfunctioning robot with an interesting history.
- "Midnight Blue." Novelette set in a suburban neighborhood describes the discovery of charms with mystical powers.
- Sample dept/column: "Reflections" looks back on 30 years of science fiction.

RIGHTS AND PAYMENT
First world-wide English-language serial rights. Fiction, $.06–$.08 per word. Poetry, $1 per line. Depts/columns, payment rates vary. Pays on acceptance. Provides 2 author's copies.

➥EDITOR'S COMMENTS
We're looking for character-oriented stories. Most of the stories we publish are serious and thoughtful; however, we do at times purchase humorous pieces.

ASK

Carus Publishing
70 East Lake Street, Suite 300
Chicago, IL 60601

Submissions Editor

DESCRIPTION AND INTERESTS
ASK is read by curious children who have questions about the real world. From ancient civilizations to space travel, topics are covered in enlightening, entertaining, age-appropriate articles. Circ: 42,000.

Audience: 7–10 years
Frequency: 9 times each year
Website: www.cricketmag.com

FREELANCE POTENTIAL
80% written by nonstaff writers.

SUBMISSIONS
Send résumé and clips. All material is commissioned from experienced authors.

- Articles: To 1,500 words. Informational articles; interviews; and photo-essays. Topics include nature, animals, the environment, computers, science and technology, history, math, and the arts.
- Depts/columns: Word lengths vary. News items and contests.

SAMPLE ISSUE
34 pages (no advertising): 5 articles; 5 depts/columns. Sample copy available at website.

- "Why Is Everyone Worried About Global Warming?" Article explains how human activities cause global warming, how global warming causes ice caps to melt, and how changes in human behavior can help to turn the tide.
- "Climate Clues." Article shows how examining tree rings, ice sheets, and fossilized pollen helps scientists prove that global warming is occurring.
- Sample dept/column: "Ask Jimmy the Bug" answers the question "If you eat too much, will you explode?"

RIGHTS AND PAYMENT
Rights vary. Written material, payment rates vary. Payment policy varies.

➥EDITOR'S COMMENTS
We will assign material to experienced writers only. We work with writers who have demonstrated the ability to accurately explain how the world works in a way that is interesting and exciting to young children. We urge prospective authors to read our magazine, and see if their style and expertise matches our needs.

Atlanta Baby

2346 Perimeter Park Drive, Suite 101
Atlanta, GA 30341

Editor: Liz White

DESCRIPTION AND INTERESTS
This younger sibling of *Atlanta Parent Magazine* offers new and expectant parents the scoop on pregnancy, labor, delivery, and life with a little one. Of course, as a regional publication, it also includes invaluable local resources, from childbirth classes to pediatricians and beyond. Circ: 25,000.

Audience: Parents
Frequency: Quarterly
Website: www.atlantaparent.com

FREELANCE POTENTIAL
25% written by nonstaff writers. Publishes 50 freelance submissions yearly; 5% by unpublished writers, 30% by authors who are new to the magazine. Receives 2,400 unsolicited mss each year.

SUBMISSIONS
Send complete ms. Prefers email submissions to lwhite@atlantaparent.com; will accept hard copy. SASE. Responds in 2 months.

- Articles: 600–1,200 words. Informational and how-to articles; and humor. Topics include pregnancy, childbirth, child development, early education, health, fitness, parenting, and disability resources.
- Depts/columns: Word lengths vary. Short essays and resource guides.
- Other: Submit seasonal material at least 6 months in advance.

SAMPLE ISSUE
36 pages (50% advertising): 6 articles; 3 depts/columns. Sample copy, $2 with 9x12 SASE. Guidelines available.

- "Mama Massages." Article tells how prenatal massage can alleviate pain and stress during pregnancy, and suggests questions to ask the massage therapist.
- Sample dept/column: "Nursery Giggles" offers a husband's guide to nesting.

RIGHTS AND PAYMENT
One-time rights. Written material, $35–$50. Pays on publication. Provides 1 tearsheet.

➥EDITOR'S COMMENTS
All of our articles are very down-to-earth—we print no philosophical or theoretical articles. Our readers also want to read articles that are activity-based. Keep the instructions short and to the point.

Atlanta Parent

2346 Perimeter Park Drive, Suite 101
Atlanta, GA 30341

Managing Editor: Tali Toland

DESCRIPTION AND INTERESTS
Parents in the greater Atlanta area turn to this magazine for information about parenting, education, and health issues, as well as regional recreation and educational opportunities for children of all ages. Circ: 120,000.

Audience: Parents
Frequency: Monthly
Website: www.atlantaparent.com

FREELANCE POTENTIAL
40% written by nonstaff writers.

SUBMISSIONS
Send complete ms. Prefers email to editor@atlantaparent.com (Microsoft Word attachment); will accept hard copy and disk submissions. SASE. Response time varies.

- Articles: 800–1,200 words. Informational and how-to articles; and humor. Topics include education, child care, child development, health, and parenting issues.
- Depts/columns: Word lengths vary. Health, trends, medical advice, parenting tips, and humor pieces.
- Other: Submit secular holiday material 6 months in advance.

SAMPLE ISSUE
114 pages (55% advertising): 6 articles; 8 depts/columns. Sample copy, $3 with 9x12 SASE. Guidelines available.

- "Worry-Proof Your Child." Article provides ideas to ensure your children are enjoying their childhood years.
- "Teaching Outside the Box." Article profiles the Montessori tradition of education, as well as Atlanta schools with similar philosophies.
- Sample dept/column: "The Medicine Mom" offers ideas for poison-proofing the home.

RIGHTS AND PAYMENT
One-time print and Internet rights. Written material, $35–$50. Pays on publication. Provides 1 tearsheet.

➥EDITOR'S COMMENTS
We are a service journalism publication. When writing about a problem, give parents the symptoms or signs of the problem and then the solution or where they can turn to for help. We also want articles that are activity-based, but please keep the instructions short and to the point.

Autism Asperger's Digest

P.O. Box 1519
Waynesville, NC 28786

Managing Editor: Veronica Zysk

DESCRIPTION AND INTERESTS
The mission of this magazine is to share information that can assist parents and professionals in better understanding autism spectrum disorders. It looks for articles that take a strong hands-on approach, rather than a theoretical exploration. Circ: 12,000.

Audience: Parents and professionals
Frequency: 6 times each year
Website: www.autismdigest.com

FREELANCE POTENTIAL
90% written by nonstaff writers. Publishes 45 freelance submissions yearly; 60% by unpublished writers, 60% by authors who are new to the magazine. Receives 48 queries each year.

SUBMISSIONS
Query with 25- to 30-word author bio. Accepts hard copy and email queries to editor@autismdigest.com. SASE. Responds in 2–4 weeks.

- Articles: Personal experience pieces, 1,000–1,200 words. Informational and how-to articles, 1,200–2,000 words. Topics include living with autism; strategies for parents and educators; and current research.
- Depts/columns: Newsbites, 100–300 words.
- Artwork: Color prints; JPEG images at 300 dpi.

SAMPLE ISSUE
50 pages: 7 articles; 4 depts/columns. Guidelines available.

- "Homework and Beyond!" Article provides 10 steps to teach organizational skills to children with autism disorders.
- "Developing Dating and Relationship Skills." Personal essay tells how a woman with Asperger's Syndrome became happily wed.
- Sample dept/column: "Newsbites" includes a brief about a social networking website for parents of nonverbal children.

RIGHTS AND PAYMENT
First rights. No payment. Provides contributor's copies and a 1-year subscription.

➥EDITOR'S COMMENTS
We're especially looking for articles on effective strategies or treatment programs related to behavior, communication, sensory processing, social skills, recreation, teaching, self-help skills, and social connection.

BabagaNewz

11141 Georgia Avenue, Suite 406
Silver Spring, MD 20902

Editor: Mark Levine

DESCRIPTION AND INTERESTS
This classroom magazine for Jewish preteens and teenagers analyzes major news stories, religious holidays, cultural events, and youth trends. Each issue is organized around a specific Jewish value. Circ: 41,029.

Audience: 9–13 years
Frequency: 8 times each year
Website: www.babaganewz.com

FREELANCE POTENTIAL
30% written by nonstaff writers. Publishes 20 freelance submissions yearly; 10% by authors who are new to the magazine.

SUBMISSIONS
Query. All material is written on assignment. Accepts hard copy. SASE. Response time varies.

- Articles: Word lengths vary. Informational and how-to articles; profiles; and interviews. Topics include renewal, friendship, personal satisfaction, peace, the environment, truth, responsibility, heroism, health, the Torah, Jewish holidays, history, political science, social studies, geography, and sports.
- Depts/columns: Word lengths vary. World news, short profiles, and science news.

SAMPLE ISSUE
22 pages (no advertising): 6 articles; 3 depts/columns; 4 activities. Sample copy and guidelines available by email request to aviva@babaganewz.com.

- "Speaking Out for Those Who Can't." Article interviews Sen. Barbara Boxer about her leadership and role in environmental causes.
- "High Court Limits Students' Speech." Article details a First Amendment rights case.
- Sample dept/column: "W.O.W. (World of Wonders)" explains how cell phones work.

RIGHTS AND PAYMENT
All rights. Written material, payment rates vary. Pays on acceptance. Provides contributor's copies upon request.

✏️EDITOR'S COMMENTS
Because we are often used as a classroom supplement, each issue includes a teacher's guide. All articles are assigned, but if you are a writer who understands Jewish culture and have an interesting idea that fits with our editorial themes, we welcome you to query us.

Babybug

70 East Lake Street, Suite 300
Chicago, IL 60601

Submissions Editor: Jenny Gillespie

DESCRIPTION AND INTERESTS
This cleverly designed board book-style magazine is ideal for small hands, and is written so that parents and toddlers can enjoy it together. Each colorfully illustrated issue features simple read-aloud stories and poems that convey the everyday wonders of a very young child's world. Circ: 50,000.

Audience: 6 months–2 years
Frequency: 10 times each year
Website: www.cricketmag.com

FREELANCE POTENTIAL
100% written by nonstaff writers. Publishes 30–40 freelance submissions yearly; 50% by authors who are new to the magazine. Receives 2,400 unsolicited mss yearly.

SUBMISSIONS
Send complete ms. Accepts hard copy and simultaneous submissions if identified. SASE. Responds in 6 months.

- Articles: 10 words. Features material that conveys simple concepts and ideas.
- Fiction: 3–6 short sentences. Age-appropriate humor and short stories.
- Other: Parent/child activities, to 8 lines. Rhyming and rhythmic poetry, to 8 lines.

SAMPLE ISSUE
24 pages (no advertising): 1 story; 6 poems; 1 activity. Sample copy, $5. Guidelines available.

- "Kim and Carrots." Story follows a little girl and stuffed rabbit, characters familiar to *Babybug* readers, as they happily make music wherever they go.
- "Crossing the Street." Poem incorporates an important safety tip for a family crossing a busy street.
- "Gingerbread Cookies." Poem tells of a little boy and his mother baking and then decorating gingerbread cookies.

RIGHTS AND PAYMENT
Rights vary. Written material, $25 minimum. Pays on publication. Provides 6 copies.

✏️EDITOR'S COMMENTS
Writers who are thoroughly familiar with our magazine will know that *Babybug* selections typically include strong rhymes and rhythmic patterns. We particularly appreciate age-appropriate humor and work that features a light-hearted approach.

Baby Dallas

Lauren Publications
4275 Kellway Circle, Suite 146
Addison, TX 75001

Editorial Director: Shelley Hawes Pate

DESCRIPTION AND INTERESTS
Read by expectant and new parents, this regional magazine offers guidelines, advice, comfort, and a sense of humor, as well as accurate, up-to-date information on health and safety issues. Circ: 120,000.

Audience: Parents
Frequency: Twice each year
Website: www.babydallas.com

FREELANCE POTENTIAL
25% written by nonstaff writers. Publishes 12–15 freelance submissions yearly; 20% by authors who are new to the magazine. Receives 240 queries yearly.

SUBMISSIONS
Query with résumé. Accepts hard copy, simultaneous submissions if identified, and email queries to editorial@dallaschild.com. SASE. Responds in 2–3 months.

- Articles: 1,000–2,500 words. Informational, self-help, and how-to articles; profiles; interviews; humor; and personal experience pieces. Topics include pregnancy, parenting, education, current events, social issues, multicultural and ethnic issues, health, fitness, crafts, and computers.
- Depts/columns: 800 words. Parenting resources, health information.

SAMPLE ISSUE
42 pages (14% advertising): 1 article; 6 depts/columns. Sample copy, free with 9x12 SASE. Guidelines available.

- "Smart from the Start." Article discusses whether a child's brain development is set during the 0-to-3-year period.
- Sample dept/column: "Health" offers eco-friendly choices for baby items.
- Sample dept/column: "Real Moms" profiles a local mother and her new life with her 5-month-old son.

RIGHTS AND PAYMENT
First serial rights. Written material, payment rates vary. Pays on publication. Provides contributor's copies upon request.

➦EDITOR'S COMMENTS
We strive to include fresh voices, ideas, and perspectives in our magazine. We prefer to work with writers from the Metroplex area, and articles should have a local focus.

Baby Talk

The Parenting Group
135 West 50th Street, 3rd Floor
New York, NY 10020

Senior Editor: Patty Onderko

DESCRIPTION AND INTERESTS
Everything new moms need to know about their babies is located here. The articles in *Baby Talk* are designed to inform and support women who are trying to become pregnant, who are pregnant, or who have children under the age of 18 months. Circ: 2 million.

Audience: Parents
Frequency: 10 times each year
Website: www.babytalk.com

FREELANCE POTENTIAL
50% written by nonstaff writers. Publishes 40 freelance submissions yearly; 20% by authors who are new to the magazine. Receives 504 queries yearly.

SUBMISSIONS
Query with clips or writing samples. No simultaneous submissions. Accepts hard copy. SASE. Responds in 2 months.

- Articles: 1,500–2,000 words. Informational and how-to articles; and personal experience pieces. Topics include fertility, pregnancy, baby care, infant health and development, juvenile equipment and toys, day care, marriage, and relationships.
- Depts/columns: 500–1,200 words. Humor, finances, women's and infant health, advice, and personal experiences from new parents.

SAMPLE ISSUE
80 pages (50% advertising): 6 articles; 15 depts/columns. Sample copy, free with 9x12 SASE ($1.60 postage). Guidelines and theme list available.

- "The Truth About Boys and Girls." Article explains research about gender stereotypes.
- "What's Your Newborn's Nursing Style?" Article explains how to make the most of your baby's breastfeeding patterns.
- Sample dept/column: "Baby Steps" explains developmental milestones at various ages.

RIGHTS AND PAYMENT
First rights. Articles, $1,000–$2,000. Depts/columns, $300–$1,200. Pays on acceptance. Provides 2–4 contributor's copies.

➦EDITOR'S COMMENTS
There are a lot of baby magazines out there, and a lot of articles on the same, re-hashed topics. Show us you have something new to bring to the party, and we'll be delighted.

Baltimore's Child

11 Dutton Court
Baltimore, MD 21228

Editor: Dianne R. McCann

DESCRIPTION AND INTERESTS

Helping Baltimore area families find activities and resources in the metropolitan area, this tabloid covers parenting, child development, and health and fitness issues. Events calendars are included in each issue. Circ: 50,000.

Audience: Parents
Frequency: Monthly
Website: www.baltimoreschild.com

FREELANCE POTENTIAL

90% written by nonstaff writers. Publishes 250 freelance submissions yearly; 5% by unpublished writers, 10% by authors who are new to the magazine.

SUBMISSIONS

Prefers query; will accept complete ms. Accepts email submissions to dianne@ baltimoreschild.com. Response time varies.

- Articles: 1,000–1,500 words. Informational articles. Topics include parenting issues, education, health, fitness, child care, social issues, and regional news.
- Depts/columns: Word lengths vary. Music, family cooking, pet care, baby and toddler issues, parenting children with special needs, parenting teens, and family finances.

SAMPLE ISSUE

92 pages: 4 articles; 14 depts/columns; 4 calendars; 1 party directory. Sample copy, $4 with 9x12 SASE ($.77 postage); also available at website. Guidelines and theme list available at website.

- "Wholesome, Healthy Alternatives: There's a Growing Interest in Community Supported Agriculture." Article explains how to become involved in these programs and why they are important.
- "Charitable Parties: A New Way to Celebrate Your Child's Birthday." Article discusses the growing popularity of these parties.
- Sample dept/column: "Baby and Toddler" discusses the benefits of infant massage.

RIGHTS AND PAYMENT

One-time rights. Written material, payment rates vary. Pays on publication.

☛EDITOR'S COMMENTS

We will consider new writers with really good ideas that show they know our magazine. We also accept previously published articles.

Baseball Youth

Dugout Media
P.O. Box 983
Morehead, KY 40351

Managing Editor: Nathan Clinkenbeard

DESCRIPTION AND INTERESTS

This four-color glossy magazine celebrates young baseball players as well as their professional heroes with profiles, interviews, training tips, and information about card collecting. Circ: 100,000.

Audience: Boys, 7–14 years; coaches; parents
Frequency: 6 times each year
Website: www.baseballyouth.com

FREELANCE POTENTIAL

50% written by nonstaff writers. Publishes 10–20 freelance submissions yearly.

SUBMISSIONS

Query with word length and availability of artwork. Prefers email queries to mailbox@ baseballyouth.com (Microsoft Word attachments); will accept hard copy. Availability of artwork improves chance of acceptance. SASE. Response time varies.

- Articles: Word lengths vary. Informational and how-to articles; profiles; interviews; photo-essays; and personal experience pieces. Topics include youth baseball, major and minor league baseball, players, coaches, training, ballparks, baseball equipment, and baseball cards.
- Depts/columns: Word lengths vary. Baseball news; card collections; mascot interviews; fan club information; first-person essays; and video game reviews.
- Artwork: Color digital images and prints.
- Other: Puzzles, quizzes, and comics.

SAMPLE ISSUE

48 pages: 4 articles; 10 depts/columns. Sample copy available at website. Guidelines available.

- "Own a Piece of the Game." Article describes newly introduced autograph and relic cards.
- Sample dept/column: "The Clubhouse" includes a brief summary of the latest Boston Red Sox World Series victory.

RIGHTS AND PAYMENT

All rights. Written material, payment rates vary. Payment policy varies.

☛EDITOR'S COMMENTS

We are most interested in profiles of Major League Baseball players, as well as human-interest stories featuring Major League Baseball players or youth players who have overcome adversity.

Bay State Parent

124 Fay Road
Framingham, MA 01702

Editor: Susan S. Petroni

DESCRIPTION AND INTERESTS
This parenting magazine covers education, health, travel, child development, pregnancy, and family entertainment for Massachusetts parents with children up to middle school age. Circ: 42,000.

Audience: Parents
Frequency: Monthly
Website: www.baystateparent.com

FREELANCE POTENTIAL
95% written by nonstaff writers. Publishes 72–144 freelance submissions yearly; 5% by unpublished writers, 30% by authors who are new to the magazine. Receives 120 queries each year.

SUBMISSIONS
Query. Accepts email queries to editor@ baystateparent.com (Microsoft Word attachments). Availability of artwork improves chance of acceptance. Responds in 1 month.

- Articles: To 2,000 words. Informational and how-to articles; and humor. Topics include parenting and family issues, regional and local events, health, travel, books, arts and crafts, family finance, and computers.
- Depts/columns: To 1,500 words. Family health, working mothers, adoption, entertainment, relationships.
- Artwork: B/W and color prints; JPEG images at 200 dpi.
- Other: Submit seasonal material 4 months in advance.

SAMPLE ISSUE
72 pages (15% advertising): 10 articles; 10 depts/columns. Sample copy, free. Writers' guidelines available.

- "When Getting Pregnant Isn't Easy." Article explains various fertility treatments and methods of paying for them.
- Sample dept/column: "Parenting 1-2-3" provides five keys to raising a healthy child.

RIGHTS AND PAYMENT
First Massachusetts exclusive and electronic rights. Articles, $50–$85. Depts/columns, no payment. Kill fee varies. Pays on publication.

⇨EDITOR'S COMMENTS
While we are open to article ideas on a variety of parenting and family topics, each article should have a Massachusetts perspective.

BC Parent News Magazine

Sasamat RPO 72086
Vancouver, British Columbia V6R 4P2
Canada

Editor: Elizabeth Shaffer

DESCRIPTION AND INTERESTS
For more than 13 years, parents and caregivers living in British Columbia have come to rely on *BC Parent News* for relevant, up-to-date information. It provides articles on health care, education, birthing, pregnancy, the arts, entertainment, and other topics related to raising children. Circ: 45,000.

Audience: Parents
Frequency: 9 times each year
Website: www.bcparent.ca

FREELANCE POTENTIAL
80% written by nonstaff writers. Publishes 25 freelance submissions yearly; 10–30% by authors who are new to the magazine.

SUBMISSIONS
Send complete ms. Accepts email submissions to eshaffer@telus.net (RTF attachments) and IBM disk submissions. SAE/IRC. No simultaneous submissions. Responds in 2 months.

- Articles: 500–1,000 words. Informational articles. Topics include health care, education, pregnancy and childbirth, adoption, computers, sports, money matters, the arts, community events, teen issues, baby and child care, and family issues.
- Depts/columns: Word lengths vary. Parent health, family news, and media reviews.

SAMPLE ISSUE
30 pages: 4 articles; 3 depts/columns; 1 camp guide; 1 activity guide. Guidelines and editorial calendar available.

- "New Millennium Dads." Article compares the 'culture of fatherhood' with the 'conduct of fatherhood.'
- Sample dept/column: "Family Health" takes a look at the health care needs of a newborn.

RIGHTS AND PAYMENT
First rights. Articles, $85; reprints, $50. Depts/columns, payment rates vary. Pays on acceptance.

⇨EDITOR'S COMMENTS
We believe in going beyond the crafts and the recipes and offering parents factual articles on meaty subjects. Parenting isn't easy, and parents turn to us for information and advice. We welcome articles from writers who can relate to the uneasy feelings experienced by all new parents.

Better Homes and Gardens

1716 Locust Street
Des Moines, IA 50309-3023

Department Editor

DESCRIPTION AND INTERESTS

In addition to information on interior design and landscaping ideas, this popular magazine provides articles on topics of interest to parents and families. Circ: 7.6 million.

Audience: Adults
Frequency: Monthly
Website: www.bhg.com

FREELANCE POTENTIAL

10% written by nonstaff writers. Publishes 25–30 freelance submissions yearly; 25% by authors who are new to the magazine. Receives 240 queries yearly.

SUBMISSIONS

Query with résumé and clips or writing samples. No unsolicited mss. Accepts hard copy. SASE. Responds in 1 month.

- Articles: Word lengths vary. Informational and how-to articles; personal experience pieces; and profiles. Topics include food and nutrition, home design, gardening, outdoor living, travel, the environment, health and fitness, holidays, education, parenting, and child development.
- Depts/columns: Staff written.

SAMPLE ISSUE

236 pages (48% advertising): 37 articles; 5 depts/columns. Sample copy, $3.49 at newsstands. Guidelines available.

- "It Pays to Go Green." Article explains how a Texas family saves money and energy with their new, environmentally-friendly house.
- "On the Same Page." Article discusses ways in which parents can understand the cyber world of their children, and at the same time grow closer through Internet-based activities.
- "Daring Do." Article offers an interview with the authors of a book geared toward inspiring women to reclaim their daring spirit.

RIGHTS AND PAYMENT

All rights. Written material, payment rates vary. Pays on acceptance. Provides 1 author's copy.

✒EDITOR'S COMMENTS

Writers with professional backgrounds or other hands-on experience in design are usually picked for our popular design and style features. However, we are always looking for good writers to provide family, parenting, and education-based articles.

Beyond Centauri

P.O. Box 782
Cedar Rapids, IA 52406

Managing Editor: Tyree Campbell

DESCRIPTION AND INTERESTS

This magazine is by and for younger readers of science fiction, fantasy, and adventure. It publishes short stories, reviews, and informational articles, as well as poetry. Circ: 150.

Audience: 9–18+ years
Frequency: Quarterly
Website: www.samsdotpublishing.com

FREELANCE POTENTIAL

98% written by nonstaff writers. Publishes 100–125 freelance submissions yearly; 25% by unpublished writers, 40% by authors who are new to the magazine. Receives 120 queries, 600 unsolicited mss yearly.

SUBMISSIONS

Query or send complete ms. Accepts hard copy and email submissions to beyondcentauri@ yahoo.com (RTF attachments). SASE. Responds to queries in 2 weeks, to mss in 2–3 months.

- Articles: To 500 words. Informational articles; opinion pieces; and book and movie reviews. Topics include space exploration, science, and technology.
- Fiction: To 2,500 words. Science fiction, fantasy, and stories about adventure and exploration.
- Other: Poetry, to 50 lines. Science fiction, fantasy, and insect themes.

SAMPLE ISSUE

38 pages (no advertising): 12 stories; 17 poems. Sample copy, $7 with 9x12 SASE. Guidelines available.

- "Sir Berenger of the Long Toe." Story details the deceitful behavior of a man who is eventually found out by his clever wife.
- "Frost Your Face: A Spider's Boutique." Story tells of a community of artistic spiders who create beautiful jewelry out of frost.

RIGHTS AND PAYMENT

First North American serial rights. Articles, $3. Fiction, $5. Poetry, $2. B/W illustrations, $5. Pays on publication. Provides 1 author's copy.

✒EDITOR'S COMMENTS

We seek submissions that are kid- and family-friendly, so do not send us anything with foul language or disturbing imagery. We are interested in receiving more science fiction and fantasy stories, especially those set in outer space concerning the dangers of new worlds.

Big Apple Parent

1040 Avenue of the Americas, 4th Floor
New York, NY 10018

Editor-in-Chief: Helen Freedman

DESCRIPTION AND INTERESTS
New York City parents turn to this regional magazine for the latest information on family health and nutrition, education, and raising children. News on sports and entertainment, and a calendar of events are also included. Its parent company also publishes parenting magazines for Queens, Brooklyn, and Westchester County, New York. Circ: 285,000.

Audience: Parents
Frequency: Monthly
Website: www.nymetroparents.com

FREELANCE POTENTIAL
50% written by nonstaff writers. Publishes 300 freelance submissions yearly; 20% by unpublished writers, 10% by authors who are new to the magazine. Receives 12 queries, 72 unsolicited mss yearly.

SUBMISSIONS
Query or send complete ms. Accepts hard copy and email submissions to hfreedman@davlermedia.com. SASE. Responds in 1 week.

- Articles: 800–1,000 words. Informational articles; profiles; interviews; and personal experience pieces. Topics include family issues, health, nutrition, fitness, current events, and regional news.
- Depts/columns: Staff written.
- Other: Submit seasonal material 4 months in advance.

SAMPLE ISSUE
66 pages: 11 articles; 10 depts/columns. Sample copy, free with 10x13 SASE.

- "2008 Resolution: Stop the Nagging." Article offers a guide to getting results from your children without nagging them.
- "Slice Up the Ice." Article lists local places to ice skate.
- "Jen's New Role." Article talks with actress Jennifer Garner about the importance of flu shots for children.

RIGHTS AND PAYMENT
First New York area rights. No payment.

✎EDITOR'S COMMENTS
We only pay for a few, assigned articles. However, we are a great market for writers trying to break in or build up clips. We are looking for pieces on parenting trends and current affairs with a local angle.

Birmingham Christian Family

P.O. Box 383203
Birmingham, AL 35238

Editor: Dee Branch Park

DESCRIPTION AND INTERESTS
This free tabloid publication centers on Christian family values and celebrates those who embody them. Circ: 35,000.

Audience: Families
Frequency: Monthly
Website: www.birminghamchristian.com

FREELANCE POTENTIAL
72% written by nonstaff writers. Publishes 15 freelance submissions yearly; 5% by unpublished writers, 3% by authors who are new to the magazine. Receives 240 queries yearly.

SUBMISSIONS
Query with artwork if applicable. Accepts email queries to editor@birminghamchristian.com. Availability of artwork improves chance of acceptance. Responds in 1 month.

- Articles: To 500 words. Informational, self-help, and how-to articles; profiles; and personal experience pieces. Topics include love, family life, parenting, Christianity, churches, philanthropy, education, recreation, the arts, travel, and sports.
- Fiction: To 500 words. Inspirational and humorous stories.
- Depts/columns: To 500 words. Media and restaurant reviews; faith in the workplace; financial, health, home improvement, and business tips; recipes; and family travel.
- Artwork: B/W or color prints.

SAMPLE ISSUE
32 pages (25% advertising): 5 articles; 21 depts/columns. Sample copy, free with 9x12 SASE ($3 postage). Editorial calendar available.

- "The Croyles: A Legacy of Excellence, Integrity and Generosity." Article profiles a couple who run a ranch for troubled youth.
- Sample dept/column: "Cool Stuff with a Mighty Message" provides creative Valentine's Day gift ideas.

RIGHTS AND PAYMENT
Rights vary. No payment.

✎EDITOR'S COMMENTS
As a Christian community publication, we look for positive, locally-focused articles on entertainment, health, parenting, inspirational literature, and community role models. All submissions must have a tie-in to the Birmingham area.

The Black Collegian Magazine

140 Carondelet Street
New Orleans, LA 70130

Chief Executive Officer: Preston J. Edwards, Sr.

DESCRIPTION AND INTERESTS
Launched in 1970, *The Black Collegian* is a career and self-development magazine targeted to African American students and other students of color seeking information on careers, graduate schools, internships, and other opportunities. Circ: 121,000.

Audience: African Americans, 18–30 years
Frequency: Twice each year
Website: www.blackcollegian.com

FREELANCE POTENTIAL
95% written by nonstaff writers. Publishes 20 freelance submissions yearly; 33% by authors who are new to the magazine. Receives 24 queries yearly.

SUBMISSIONS
Query. Prefers email queries to pres@ imdiversity.com; will accept hard copy. SASE. Responds in 3 months.

- Articles: 1,500–2,000 words. Informational, self-help, and how-to articles; profiles; and personal experience pieces. Topics include careers, job opportunities, graduate and professional school, internships, study-abroad programs, personal development, financial aid, history, technology, and multicultural and ethnic issues.
- Depts/columns: Word lengths vary. Health issues and media reviews.
- Artwork: 5x7 and 11x14 B/W and color transparencies. B/W and color line art.

SAMPLE ISSUE
88 pages: 18 articles; 1 dept/column. Sample copy and guidelines available at website.

- "Nikki Giovanni Kicks Off Black History Month at Southern University." Article recounts the insights shared by the world-renowned poet and activist in her speech.
- "In Honor of Juneteenth: African American Inventions We Take for Granted." Article ponders what the world would be like without African American inventions.

RIGHTS AND PAYMENT
One-time rights. All material, payment rates vary. Pays after publication. Provides 1 contributor's copy.

☛EDITOR'S COMMENTS
We'd like to see articles about making a smooth transition from college to career.

Blaze Magazine

P.O. Box 2660
Niagara Falls, NY 14302

Editor: Brenda McCarthy

DESCRIPTION AND INTERESTS
Kids who love horses read *Blaze Magazine* for its mix of entertaining, educational, and interactive articles. Information on nature, history, and creative arts, all with an equine slant, is found in its pages. Circ: 4,000.

Audience: 8–14 years
Frequency: Quarterly
Website: www.blazekids.com

FREELANCE POTENTIAL
50% written by nonstaff writers. Publishes 25–30 freelance submissions yearly.

SUBMISSIONS
Query. Accepts email queries to brenda@ blazekids.com. Availability of artwork improves chance of acceptance. Response time varies.

- Articles: 200–500 words. Informational and how-to articles; and profiles. Topics include horseback riding, training, and breeds.
- Fiction: Word lengths vary. Stories about horses.
- Depts/columns: Word lengths vary. Short news items, arts and crafts.
- Artwork: B/W and color prints and transparencies.
- Other: Puzzles and games.

SAMPLE ISSUE
62 pages (15% advertising): 4 articles; 1 story; 7 depts/columns; 1 comic. Sample copy, $3.75.

- "Happy Campers." Article discusses ways to make the most of horse camp.
- "Five Fabulous Years." Article reviews editors' favorite memories from the first five years of publication.
- Sample dept/column: "The Best of Legend and Lore" tells about the different legends involving horses.

RIGHTS AND PAYMENT
Rights vary. Written material, $.25 per word. Artwork, payment rates vary. Payment policy varies.

☛EDITOR'S COMMENTS
It is best for writers to first send us a query and then we will send you a copy of the magazine. We are always interested in profiles and programs involving horses. Because our magazine is very visually focused, photography is key with submissions, and quality of photos matters greatly.

bNetS@vvy

1201 16th Street NW, Suite 216
Washington, DC 20036

Editor: Mary Esselman

DESCRIPTION AND INTERESTS
Readers of this e-newsletter—published by the
National Education Association Health Information Network, the National Center for Missing
and Exploited Children, and Sprint—find the
tools needed to help kids stay safe online.
Articles focus on positive, solutions-based
approaches. Hits per month: Unavailable.

Audience: Parents and teachers
Frequency: 6 times each year
Website: www.bnetsavvy.org

FREELANCE POTENTIAL
90% written by nonstaff writers. Publishes 20
freelance submissions yearly; 80% by unpublished writers, 90% by authors who are new to
the magazine.

SUBMISSIONS
Query. Accepts email queries to internetsafety@
nea.org; include clips or links. Responds in
2 weeks.

- Articles: 600–950 words. Informational and
how-to articles; and reviews. Topics include
the Internet and Internet safety, and social
networking sites.
- Depts/columns: 600–950 words. Expert
advice and ideas from parents and teens.

SAMPLE ISSUE
8 pages (no advertising): 1 article; 3 depts/
columns. Sample copy and guidelines available
at website.

- "Balancing Trust and Fear: A Parent's Perspective." Article includes tips for establishing trust with children and their Internet use.
- Sample dept/column: "Ask the Experts" discusses the popularity of social networking
sites among kids and offers safety tips.
- Sample dept/column: "Youth Voices" provides a 13-year-old girl's safety tips based on
personal experience.

RIGHTS AND PAYMENT
All rights. Written material, payment rates vary.
Payment policy varies.

➽EDITOR'S COMMENTS
We would like to see more material related to
interactive gaming, "Web 2.0" safeguards,
and school policies and their creation. We
would love to hear from parents, teachers,
school staff, and tweens who have stories,
ideas, and lessons to share.

Bonbon

2123 Preston Square Court, Suite 300
Falls Church, VA 22043

Publisher: Sitki Kazanci

DESCRIPTION AND INTERESTS
This educational, bilingual magazine is directed
toward Turkish American children. Its goals are
to teach Turkish language skills to those who
do not live in Turkey, and to help Turkish
children improve their English. A nonprofit
magazine, *Bonbon* features Turkish children's
stories, articles, and readers' words and artwork. Circ: 6,000.

Audience: 6–14 years
Frequency: 6 times each year
Website: www.bonbonkids.com

FREELANCE POTENTIAL
60% written by nonstaff writers. Publishes 30
freelance submissions yearly; 25% by unpublished writers, 10% by authors who are new to
the magazine.

SUBMISSIONS
Query with brief author biography. Accepts
email queries to info@bonbonkids.com
(include "Bonbon queries" in subject field).
Responds in 2 weeks.

- Articles: 250 words. Informational articles;
profiles; and interviews. Topics include
Turkish language, culture, and traditions;
news; animals; nature; and history.
- Fiction: 500 words. Stories that feature
Turkish traditions, culture, and values.
- Other: Comics, puzzles, games, jokes, and
activities. Poetry.

SAMPLE ISSUE
28 pages: 6 articles; 2 stories; 1 comic; 18
puzzles, quizzes, and activities.

- "Sunay Akin." Article provides an interview
with the Turkish artist who recently opened
the first toy museum in Turkey.
- "Giraffes." Article provides informative tidbits and interesting facts about the world's
tallest mammal.
- "Little Red Riding Hood." Story is a retelling
of the classic Brothers Grimm tale, but told
in Turkish version.

RIGHTS AND PAYMENT
One-time rights. No payment.

➽EDITOR'S COMMENTS
While many of our articles are published in
English and Turkish, some are published in
just one language to help improve the language skills of our young readers.

Book Links

American Library Association
50 East Huron Street
Chicago, IL 60611

Editor: Laura Tillotson

DESCRIPTION AND INTERESTS
This relative of *Booklist* focuses on children's literature, providing thematic bibliographies and activities, author and illustrator interviews, essays, and articles, all designed to help teachers and librarians use books in the classroom. Booksellers and parents also read *Book Links*. Circ: 20,577.

Audience: Teachers and librarians
Frequency: 6 times each year
Website: www.ala.org/booklinks

FREELANCE POTENTIAL
90% written by nonstaff writers. Publishes 60 freelance submissions yearly; 20% by unpublished writers, 30% by authors who are new to the magazine. Receives 96 queries and unsolicited mss yearly.

SUBMISSIONS
Query or send complete ms. Accepts email to ltillotson@ala.org. Response time varies.

- Articles: Word lengths vary. Informational and how-to articles; profiles; interviews; and annotated bibliographies. Topics include multicultural literature, literacy, language arts, core curriculum areas, education, the arts, authors, illustrators, and teaching techniques.
- Depts/columns: Word lengths vary. Curriculum ideas, personal essays, themed book lists.

SAMPLE ISSUE
64 pages (28% advertising): 12 articles; 10 depts/columns. Sample copy, $6. Guidelines and editorial calendar available at website.

- "Alpha-Best Bets for the Preschool Set." Article explains what makes an ABC book appropriate for preschoolers.
- Sample dept/column: "Thinking Outside the Book" suggests engaging students with online research activities, called WebQuests.

RIGHTS AND PAYMENT
All rights. Articles, $100. Pays on publication. Provides 2 contributor's copies.

☛EDITOR'S COMMENTS
Each issue of *Book Links* explores a curricular area; please consult our editorial calendar before submitting a completed manuscript. We are always seeking submissions for "Classroom Connections" articles, which are annotated bibliographies of 20 to 30 books that advance a subject.

Bop

330 North Brand, Suite 1150
Glendale, CA 91203

Editor-in-Chief: Leesa Coble

DESCRIPTION AND INTERESTS
Tweens get the inside scoop on their favorite young stars in this colorful magazine packed with nitty-gritty details and up-close photos of the stars, culled from interviews and set visits. Objects of affection include the Jonas Brothers, Miley Cyrus, and the cast of *High School Musical*. Circ: 200,000+.

Audience: 10–16 years
Frequency: Monthly
Website: www.bopmag.com

FREELANCE POTENTIAL
1% written by nonstaff writers. Receives 20 queries yearly.

SUBMISSIONS
Query with résumé and clips for celebrity angles only. Prefers email queries to leesa@laufermedia.com; will accept hard copy and simultaneous submissions if identified. SASE. Responds in 2 months.

- Articles: To 700 words. Interviews; profiles; and behind-the-scenes reports. Topics include young celebrities in the film, television, and recording industries.
- Depts/columns: Staff written.
- Artwork: Color digital images.

SAMPLE ISSUE
108 pages (5% advertising): 20 articles; 10 quizzes; 11 depts/columns; 3 contests. Sample copy, $3.99 at newsstands.

- "What's Up With Miley?" Article analyzes the behavior of Miley Cyrus in light of her recent breakup with Nick Jonas.
- "Will Cody Quit Acting?" Interview with actor Cody Linley discusses his stint as a Disney Channel Games reporter.
- "High School Musical 3 Gossip!" Article offers a sneak peek at the stars, thanks to on-set "spies."

RIGHTS AND PAYMENT
All rights. Written material, payment rates vary. Pays on publication. Provides 2 author's copies.

☛EDITOR'S COMMENTS
As always, we welcome queries from freelance writers who have access to teen celebrities, Hollywood events, and production sets. Please note that we will consider interview transcripts with relevant celebrities rather than narrative pieces from freelancers.

The Boston Parents' Paper

670 Centre Street
Jamaica Plain, MA 02130

Associate Editor: Georgia Orcutt

DESCRIPTION AND INTERESTS

Practical and up-to-date information on child rearing, family matters, and health issues can be found in this regional magazine for Boston-area parents. Local resources and events are also included. Circ: 75,000.

Audience: Parents
Frequency: Monthly
Website: www.boston.parenthood.com

FREELANCE POTENTIAL

50% written by nonstaff writers. Publishes 36 freelance submissions yearly; 10% by unpublished writers, 50% by authors new to the magazine. Receives hundreds of queries yearly.

SUBMISSIONS

Query with clips or writing samples. Accepts email queries to georgia.orcutt@parenthood.com. Availability of artwork improves chance of acceptance. SASE. Response time varies.

- Articles: Word lengths vary. Informational articles; profiles; and interviews. Topics include child development, education, parenting, family issues, and health.
- Depts/columns: To 1,800 words. Short news items, parenting tips, and profiles.
- Artwork: B/W prints. Line art.
- Other: Submit seasonal material 6 months in advance.

SAMPLE ISSUE

70 pages (45% advertising): 3 articles; 5 depts/columns; 1 calendar. Guidelines and theme list available.

- "Braking for Baseball." Article features a family road trip to Cooperstown, New York.
- Sample dept/column: "Ages & Stages" discusses better ways to communicate with your children.

RIGHTS AND PAYMENT

All rights. All material, payment rates vary. Pays within 30 days of publication. Provides 5 contributor's copies.

➡️EDITOR'S COMMENTS

We prefer more formal pitch letters. Please review our magazine and understand what we are about. When querying, please explain what you can bring to an article that would make it appealing. Material that offers fresh insight on raising children is always needed.

Boys' Life

Boy Scouts of America
1325 West Walnut Hill Lane
P.O. Box 152079
Irving, TX 75015-2079

Senior Writer: Aaron Derr

DESCRIPTION AND INTERESTS

Published by the Boy Scouts of America, this general-interest magazine features a blend of stories and articles that are of interest to boys. Coverage ranges from professional sports and outdoor activities to American history, animals, and science. It has been in print since 1911. Circ: 1.3 million.

Audience: 6–18 years
Frequency: Monthly
Website: www.boyslife.org

FREELANCE POTENTIAL

80% written by nonstaff writers. Publishes 50 freelance submissions yearly; 1% by unpublished writers, 2% by authors who are new to the magazine.

SUBMISSIONS

Query for articles and depts/columns. Query or send complete ms for fiction. Accepts hard copy. SASE. Responds to queries in 4–6 weeks, to mss in 6–8 weeks.

- Articles: 500–1,500 words. Informational and how-to articles; profiles; and humor. Topics include sports, science, American history, geography, animals, nature, and the environment.
- Fiction: 1,000–1,500 words. Genres include science fiction, humor, mystery, adventure.
- Depts/columns: 300–750 words. Advice, humor, collecting, computers, and pets.
- Other: Puzzles and cartoons.

SAMPLE ISSUE

52 pages (18% advertising): 5 articles; 7 depts/columns; 6 comics. Sample copy, $3.60 with 9x12 SASE. Guidelines available.

- "Speed Racer." Article profiles sprinter Jeremy Wariner and his hopes for the 2008 Summer Olympics in Beijing.
- Sample dept/column: "About Books" offers suggestions for summer reading.

RIGHTS AND PAYMENT

First rights. Articles, $400–$1,500. Fiction, $750+. Depts/columns, $150–$400. Pays on acceptance. Provides 2 contributor's copies.

➡️EDITOR'S COMMENTS

All fiction submissions must feature a boy or boys. Articles should be well-reported with the writing crisp and punchy and sentences relatively short and straightforward.

Boys' Quest

P.O. Box 227
Bluffton, OH 45817-0227

Editor: Marilyn Edwards

DESCRIPTION AND INTERESTS
This magazine's goal is to instill traditional family values while educating and inspiring its young readers through theme-related stories, poems, and activities. Circ: 12,000.

Audience: Boys, 5–14 years
Frequency: 6 times each year
Website: www.boysquest.com

FREELANCE POTENTIAL
80% written by nonstaff writers. Publishes 100–150 freelance submissions yearly; 30% by unpublished writers, 40% by authors who are new to the magazine. Receives 240 queries, 3,000 unsolicited mss yearly.

SUBMISSIONS
Prefers complete ms; will accept query. Accepts hard copy and simultaneous submissions if identified. SASE. Responds to queries in 1–2 weeks, to mss in 2–3 months.

- Articles: 500 words. Informational and how-to articles; profiles; personal experience pieces; and humor. Topics include pets, nature, hobbies, science, sports, careers, family, and cars.
- Fiction: 500 words. Genres include adventure, mystery, and multicultural fiction.
- Depts/columns: 300–500 words. Science projects and experiments.
- Artwork: Prefers B/W prints; will accept color prints.
- Other: Puzzles, activities, and riddles. Poetry.

SAMPLE ISSUE
48 pages (no advertising): 4 articles; 1 story; 8 depts/columns; 14 activities; 2 poems; 4 comics. Sample copy, $6 with 9x12 SASE. Guidelines and theme list available.

- "A Winning Heart." Story tells of a wheelchair-bound boy's chance to play basketball.
- Sample dept/column: "Science" explains how to test manual dexterity.

RIGHTS AND PAYMENT
First and second rights. Articles and fiction, $.05 per word. Depts/columns, $35. Poems and activities, $10+. Artwork, $5–$10. Pays on publication. Provides 1 contributor's copy.

➤EDITOR'S COMMENT
We would like to receive more nonfiction pieces with supporting photos. We always need items for "Ticklers & Teasers."

Bread for God's Children

P.O. Box 1017
Arcadia, FL 34265-1017

Editorial Secretary: Donna Wade

DESCRIPTION AND INTERESTS
The purpose of this magazine is to help Christian families learn to apply the Word of God in their everyday living. Stories must be told from a child's point of view. Circ: 5,000.

Audience: Families
Frequency: 6–8 times each year
Website: www.breadministries.org

FREELANCE POTENTIAL
10% written by nonstaff writers. Publishes 15–20 freelance submissions yearly; 70% by unpublished writers, 80% by authors who are new to the magazine. Receives 1,200 unsolicited mss yearly.

SUBMISSIONS
Send complete ms. Accepts hard copy and simultaneous submissions if identified. SASE. Responds in 2–3 months.

- Articles: To 800 words. Informational, self-help, and personal experience pieces. Topics include religion and spirituality.
- Fiction: Stories for younger children, to 800 words. Stories for middle-grade and young adult readers, to 1,800 words. Features stories that demonstrate Christian values.
- Depts/columns: To 800 words. Bible study, ministry highlights, family activities, living memorials, and book recommendations.
- Other: Filler and crafts.

SAMPLE ISSUE
28 pages (no advertising): 4 stories; 5 depts/columns. Guidelines available.

- "God Speaks to Janet." Story features a young girl who discovers how God speaks to her in everyday life.
- Sample dept/column: "The Foundation" discusses ways to build a strong spiritual base within the family.

RIGHTS AND PAYMENT
First rights. Fiction, $40–$50. Articles, $25. Filler, $10. Pays on publication. Provides 3 contributor's copies.

➤EDITOR'S COMMENTS
We are looking for writers with a solid knowledge of biblical principles. Plot and characters must be realistic and have a clear message with no preaching, moralizing, or tag endings. We appreciate stories that teach both the child and the parents.

Breakaway

Life. God. Truth. For Guys.

Focus on the Family
8605 Explorer Drive
Colorado Springs, CO 80920

Editor: Mike Ross

DESCRIPTION AND INTERESTS
This magazine is designed to creatively teach, entertain, inspire, and challenge the emerging teen boy. Its articles strive to strengthen self-esteem, provide good role models, make the Bible relevant, and deepen love for family, friends, church, and Jesus Christ. Circ: 95,000.

Audience: Boys, 12–18 years
Frequency: Monthly
Website: www.breakawaymag.com

FREELANCE POTENTIAL
20% written by nonstaff writers. Publishes 5 freelance submissions yearly; 1% by unpublished writers, 1% by authors who are new to the magazine. Receives 600 queries and unsolicited mss yearly.

SUBMISSIONS
Prefers query; will accept complete ms. Accepts hard copy and email submissions via website. SASE. Responds in 4–6 weeks.

- Articles: 600–1,200 words. How-to and self-help articles; profiles; personal experience pieces; interviews; and humor. Topics include religion, sports, girls, and multicultural issues.
- Fiction: To 2,000 words. Adventure, inspirational, and contemporary stories.
- Depts/columns: Word lengths vary. Scripture readings, Bible facts, and advice.
- Other: Humorous pieces, 600–1,000 words. Submit seasonal material about religious holidays 6–8 months in advance.

SAMPLE ISSUE
32 pages (9% advertising): 5 articles; 5 depts/columns. Sample copy, $2 with 9x12 SASE (2 first-class stamps). Guidelines available.

- "Chris Sligh: Beyond *American Idol*." Article features an interview with the *American Idol* finalist regarding his relationship with God.
- "My Dad and Yours." Article relates the author's experience of losing his father.

RIGHTS AND PAYMENT
First or one-time rights. Written material, $.12–$.15 per word. Pays on acceptance. Provides 5 contributor's copies.

☛EDITOR'S COMMENTS
Articles must be compelling, bright, and out of the ordinary. They should be spiritually challenging, yet inviting.

Brilliant Star

1233 Central Street
Evanston, IL 60201

Associate Editor: Susan Engle

DESCRIPTION AND INTERESTS
This children's magazine explores the teachings of the Bahá'í faith through fiction, non-fiction, puzzles, music, and art. Circ: 7,000.

Audience: 8–12 years
Frequency: 6 times each year
Website: www.brilliantstarmagazine.org

FREELANCE POTENTIAL
35% written by nonstaff writers. Publishes 5 freelance submissions yearly; 5% by unpublished writers, 80% by authors who are new to the magazine. Receives 24 queries, 120 unsolicited mss yearly.

SUBMISSIONS
Query with clips for nonfiction. Send complete ms for fiction. Accepts hard copy, email queries to brilliant@usbnc.org, and simultaneous submissions if identified. SASE. Responds in 6–8 months.

- Articles: To 700 words. Informational and how-to articles; personal experience pieces; profiles; and biographies. Topics include historical Bahá'í figures, religion, history, ethnic and social issues, travel, music, and nature.
- Fiction: To 700 words. Early reader fiction. Genres include ethnic, historical, contemporary, and problem-solving fiction.
- Depts/columns: To 600 words. Profiles of kids; religion; and ethics.
- Other: Puzzles, activities, and games.

SAMPLE ISSUE
30 pages: 10 articles; 1 story; 12 depts/columns; 14 activities; 1 comic; 1 song; 1 calendar. Sample copy, $3 with 9x12 SASE (5 first-class stamps). Guidelines and theme list/editorial calendar available.

- "Getting Bullies to Back Off." Article suggests verbal responses to bullying to defuse tense situations.
- Sample dept/column: "Shining Lamp" profiles the late Bahá'í author and actor, William Sears.

RIGHTS AND PAYMENT
All or one-time rights. No payment. Provides 2 contributor's copies.

☛EDITOR'S COMMENTS
We always need strong fiction depicting protagonists who work out problems for themselves. Articles must show knowledge of the Bahá'í faith.

Brio

Focus on the Family
8605 Explorer Drive
Colorado Springs, CO 80920

Editor: Susie Shellenberger

DESCRIPTION AND INTERESTS
The goal of *Brio* is to help teen girls develop a positive self-image while fostering a closer relationship with God. Articles teach, entertain, and challenge girls to live a faith-filled life. Circ: 150,000.
Audience: Girls, 12–15 years
Frequency: Monthly
Website: www.briomag.com

FREELANCE POTENTIAL
30% written by nonstaff writers. Publishes 100 freelance submissions yearly; 85% by unpublished writers, 15% by authors who are new to the magazine. Receives 300 unsolicited mss each year.

SUBMISSIONS
Send complete ms with cover letter. Accepts email submissions to freelance@fotf.org. Responds in 4–6 weeks.

- Articles: To 2,000 words. Informational and how-to articles; profiles; interviews; and personal experience pieces. Topics include Christian living, peer relationships, family life, and contemporary issues.
- Fiction: To 2,000 words. Genres include contemporary fiction, romance, and humor—all with Christian themes.
- Depts/columns: Staff written.
- Other: Cartoons, anecdotes, and quizzes.

SAMPLE ISSUE
38 pages (6% advertising): 6 articles; 1 story; 11 depts/columns; 1 quiz. Sample copy, $2 with 9x12 SASE (2 first-class stamps). Guidelines available.

- "Trotting Her Way to a World Championship." Article follows a girl's path to a world champion equestrian title.
- "Brotherly Love." Article explains how to foster friendships with boys by viewing them as siblings.

RIGHTS AND PAYMENT
First or second rights. Written material, $.15–$.25 per word. Pays on acceptance. Provides 3 contributor's copies.

❧EDITOR'S COMMENTS
We are more likely to purchase quizzes and fiction. We are interested in articles written from a godly perspective on relationships, faith, school, makeup, fashion, and sports.

Brio and Beyond

Focus on the Family
8605 Explorer Drive
Colorado Springs, CO 80920

Editor: Susie Shellenberger

DESCRIPTION AND INTERESTS
Targeting older teen girls, this magazine encourages a closer relationship with God while addressing topics most important to its readers, such as relationships, faith, beauty, sports, and health. It is the sister publication of *Brio*. Circ: 50,000.
Audience: Girls, 16–19 years
Frequency: Monthly
Website: www.briomag.com

FREELANCE POTENTIAL
30% written by nonstaff writers. Publishes 100 freelance submissions yearly; 85% by unpublished writers, 15% by new authors.

SUBMISSIONS
Send complete ms with cover letter. Accepts email submissions to freelance@fotf.org. Availability of artwork improves chance of acceptance. Responds in 1 month.

- Articles: Word lengths vary. Informational and how-to articles; profiles; interviews; reviews; personal experience pieces; and humor. Topics include religion, multicultural and social issues, college, careers, crafts, hobbies, health, fitness, music, popular culture, sports, and travel.
- Fiction: Word lengths vary.
- Depts/columns: Staff written.
- Artwork: Color prints or transparencies. Line art.
- Other: Submit seasonal material 6 months in advance.

SAMPLE ISSUE
38 pages (no advertising): 6 articles; 1 story; 9 depts/columns. Sample copy, free. Guidelines available.

- "Praise Him with Dancing." Article highlights a Christian performing arts program.
- "Persecuted for Christ." Article profiles two young Christians attacked because of their faith and tells how they overcame the ordeal.

RIGHTS AND PAYMENT
First or second rights. Written material, $.15–$.25 per word. Pays on acceptance. Provides 2 contributor's copies.

❧EDITOR'S COMMENTS
We look for articles that provide our readers with faith-building tools they can use in their daily lives.

Brooklyn Parent

1040 Avenue of the Americas, 4th Floor
New York, NY 10018

Editor-in-Chief: Helen Freedman

DESCRIPTION AND INTERESTS
Parents living in Brooklyn, New York, turn to *Brooklyn Parent* for articles on issues of importance to them, such as parenting trends, child-care issues, and education. The magazine also provides a localized calendar of events for family recreation and entertainment opportunities. Circ: 285,000.
Audience: Parents
Frequency: Monthly
Website: www.nymetroparents.com

FREELANCE POTENTIAL
50% written by nonstaff writers. Publishes 300 freelance submissions yearly. Receives 12 queries, 72 unsolicited mss yearly.

SUBMISSIONS
Query or send complete ms. Accepts hard copy and email submissions to hfreedman@davlermedia.com. SASE. Responds in 1 week.

- Articles: 800–1,000 words. Informational articles; profiles; interviews; and personal experience pieces. Topics include family issues, health, nutrition, fitness, current events, and regional news.
- Depts/columns: Staff written.
- Other: Submit seasonal material 4 months in advance.

SAMPLE ISSUE
66 pages: 12 articles; 4 depts/columns. Sample copy, free with 10x13 SASE.

- "Park Slope Happening House." Article profiles a new business that teaches music, art, and creative expression to young children and, at times, their parents.
- "20/20 Vision?" Article examines the need for thorough vision testing for children, as many learning problems can be traced to vision problems.

RIGHTS AND PAYMENT
First New York area rights. No payment.

➤EDITOR'S COMMENTS
If you have an idea for a story or a subject that would be of interest to parents living in Brooklyn, we want to hear about it. Please note that although we generally do not pay our freelancers, our magazine provides a great place for fledgling writers to get their material published and to build a clip file. Local writers are preferred.

Broomstix for Kids

P.O. Box 8139
Bridgewater, NJ 08807

Editor: Natalie Zaman

DESCRIPTION AND INTERESTS
This New Age e-zine for children focuses on nature-based spirituality through stories, articles, poetry, arts, crafts, activities, and journaling exercises. Hits per month: 6,000.
Audience: Children
Frequency: 8 times each year
Website: www.broomstix.com

FREELANCE POTENTIAL
10% written by nonstaff writers. Of the freelance submissions published yearly, 75% is by unpublished writers, 50% is by authors who are new to the magazine. Receives several unsolicited mss yearly.

SUBMISSIONS
Send complete ms. Accepts hard copy. SASE for artwork only. Does not return mss. Responds in 2 weeks.

- Articles: To 750 words. Informational articles. Topics include nature-based spirituality, crafts and hobbies, arts, nature, and keeping a journal.
- Fiction: Word lengths vary. Genres include folktales, folklore, and myth.
- Depts/columns: 500 words. Crafts; cooking.
- Artwork: B/W and color drawings and prints.
- Other: Poetry. Theme- and season-related projects, rituals, and activities.

SAMPLE ISSUE
Sample copy, guidelines, and theme list available at website.

- "Worts and All." Article explains why marigolds are the perfect flower for mid-summer, and relates their history.
- "Wheel . . . of . . . Fortune!" Article discusses the Wheel of Fortune tarot card, and the role the wheel, or circle of life, plays in our lives.

RIGHTS AND PAYMENT
One-time electronic rights. Written material, $.01 per word; $10 maximum. Artwork, payment rates vary. Pays 1 month after publication.

➤EDITOR'S COMMENTS
We are always looking for writing and art for each issue, and we heartily encourage writers and artists to submit their work. Submissions should be suitable for viewing by children and contain no violence or profanity. Please check out our theme list and guidelines before submitting.

ByLine

P.O. Box 111
Albion, NY 14411-0111

Editor: Robbi Hess

DESCRIPTION AND INTERESTS
ByLine offers information and inspiration for experienced and aspiring authors. Circ: 3,500.
Audience: Writers
Frequency: 11 times each year
Website: www.bylinemag.com

FREELANCE POTENTIAL
95% written by nonstaff writers. Publishes 198 freelance submissions yearly; 80% by unpublished writers, 10% by authors who are new to the magazine. Receives 360 queries, 720–900 unsolicited mss yearly.

SUBMISSIONS
Query or send complete ms. Accepts hard copy and email queries to robbi@ bylinemag.com. Do not email mss. SASE. Responds in 3–6 months.

- Articles: 1,500–1,800 words. Informational and how-to articles; personal experience pieces; and interviews with editors. Topics include writing and marketing fiction, nonfiction, poetry, and humor; finding an agent; research; grammar; and writers' conferences.
- Fiction: 2,000–3,000 words. Genres include mainstream and literary fiction.
- Depts/columns: "End Piece," 550 words. "First Sale," 250–300 words. "Only When I Laugh," 50–400 words. "Great American Bookstores," 500–600 words.
- Other: Poetry.

SAMPLE ISSUE
36 pages (8% advertising): 3 articles; 1 story; 10 depts/columns; 6 poems. Sample copy, $5 with 9x12 SASE. Guidelines available.

- "Helping the Non-writer Avoid Desperation." Article explains that clarifying what it is the writer wants to say is the best way to begin.
- "Airport Christmas." Story tells of a heart-broken man, disguised as Santa Claus, who reunites with his lost love.
- Sample dept/column: "End Piece" concludes that setting and meeting deadlines is crucial.

RIGHTS AND PAYMENT
First North American rights. Articles, $75. Depts/columns, $15–$40. Poetry, $10. Pays on publication. Provides contributor's copies.

➔ EDITOR'S COMMENTS
We need more how-to articles on various aspects of writing fiction.

BYU Magazine

218 UPB
Provo, UT 84602

Editor: Jeff McClellan

DESCRIPTION AND INTERESTS
Brigham Young University alumni, students, and friends read this magazine to keep up with university news and absorb its inspirational essays on the Mormon faith. Circ: 200,000.
Audience: YA–Adult
Frequency: Quarterly
Website: www.magazine.byu.edu

FREELANCE POTENTIAL
45% written by nonstaff writers. Publishes 10–15 freelance submissions yearly; 5% by authors who are new to the magazine. Receives 120 queries yearly.

SUBMISSIONS
Query with writing samples. Accepts hard copy. SASE. Responds in 6–12 months.

- Articles: 2,000–4,000 words. Informational, factual, inspirational, self-help, and how-to articles; profiles; and personal experience pieces. Topics include college life, careers, computers, current events, health, fitness, religions, science, technology, sports, and family life.
- Depts/columns: To 1,500 words. Commentary, campus news, book reviews, and alumni updates.
- Artwork: 35mm color prints or transparencies.
- Other: BYU trivia.

SAMPLE ISSUE
80 pages (15% advertising): 4 articles; 12 depts/columns. Sample copy, free. Writers' guidelines available.

- "To the Ends of the Earth." Article details the expansion of BYU Television into nearly 200 countries and territories.
- "One on One." Article includes the personal accounts of those involved in mentoring.
- Sample dept/column: "Family Focus" offers ways to increase family harmony and foster spiritual growth.

RIGHTS AND PAYMENT
First North American serial rights. Articles, $.35 per word. Pays on publication. Provides 10 contributor's copies.

➔ EDITOR'S COMMENTS
We seek engaging features, lively news, and useful columns, in order to extend the learning, events, and spirit of Brigham Young University far and wide.

Cadet Quest

Calvinist Cadet Corps
1333 Alger Street SE
P.O. Box 7259
Grand Rapids, MI 49510

Editor: G. Richard Broene

DESCRIPTION AND INTERESTS
Cadet Quest strives to help boys grow to become more Christlike through its inspirational fiction and Bible lessons. Circ: 8,000.

Audience: Boys, 9–14 years
Frequency: 7 times each year
Website: www.calvinistcadets.org

FREELANCE POTENTIAL
58% written by nonstaff writers. Publishes 25 freelance submissions yearly; 3% by unpublished writers, 5% by new authors. Receives 360 unsolicited mss yearly.

SUBMISSIONS
Send complete ms. Accepts hard copy, email to submissions@calvinistcadets.org (no attachments), and simultaneous submissions if identified. SASE. Responds in 1 month.

- Articles: 400–1,000 words. Informational and factual articles; profiles; and interviews. Topics include religion, spirituality, stewardship, camping, crafts and hobbies, sports, the environment, and serving God.
- Depts/columns: Word lengths vary. Cadet Corps news items, and Bible lessons.
- Other: Puzzles and cartoons.

SAMPLE ISSUE
24 pages (2% advertising): 2 articles; 1 story; 2 depts/columns; 2 Bible lessons; 1 comic. Sample copy, free with 9x12 SASE ($1.01 postage). Guidelines and theme list available at website.

- "Knowing Amy." Story demonstrates how caring for a special needs child engenders kindness and tolerance in a family.
- "What Is Your Handicap?" Article tells about exceptional individuals who did not let their handicap prevent them from accomplishing great things.
- Sample dept/column: "Project" tells how to make a water rocket.

RIGHTS AND PAYMENT
First and second serial rights. Written material, $.04–$.05 per word. Other material, rates vary. Pays on acceptance. Provides 1 author's copy.

➥EDITOR'S COMMENTS
We have a fiction theme list that explains what we are looking for month by month. We encourage writers to submit based on that list. All selections are made in April.

Calliope

Cobblestone Publishing
30 Grove Street, Suite C
Peterborough, NH 03458

Editors: Rosalie F. Baker

DESCRIPTION AND INTERESTS
Each issue of this world history-themed magazine features lively, well-written stories and articles designed to entertain and inform preteens and teens. Circ: 12,000.

Audience: 9–14 years
Frequency: 9 times each year
Website: www.cobblestonepub.com

FREELANCE POTENTIAL
85% written by nonstaff writers. Publishes 75 freelance submissions yearly; 25% by unpublished writers, 30–40% by authors new to the magazine. Receives 600–900 queries yearly.

SUBMISSIONS
Query with outline, bibliography, and clips or writing samples. All material must relate to upcoming themes. Responds in 4 months.

- Articles: Features, 700–800 words. Sidebars, 300–600 words. Informational articles and profiles. Topics include Western and Eastern world history.
- Fiction: To 800 words. Genres include historical and biographical fiction, adventure, and historical plays.
- Depts/columns: 300–600 words. Current events, archaeology, languages, book reviews.
- Artwork: B/W and color prints or slides. B/W or color line art.
- Other: Puzzles, games, activities, crafts, and recipes, to 700 words.

SAMPLE ISSUE
50 pages (no advertising): 12 articles; 8 depts/columns; 3 activities. Sample copy, $4.95 with 9x12 SASE ($2 postage). Guidelines and theme list available at website.

- "The Snaky Medusa." Article tells the Greek legend of Perseus and the exciting adventures he experienced.
- Sample dept/column: "Fun with Words" explains the funny or strange origins of words related to monster myths.

RIGHTS AND PAYMENT
All rights. Articles and fiction, $.20–$.25 per word. Other material, payment rates vary. Pays on publication. Provides 2 contributor's copies.

➥EDITOR'S COMMENTS
All queries must relate to an upcoming theme. Please see our website for a full list of what we've got coming up.

Camping Magazine

American Camp Association
5000 State Road 67 North
Martinsville, IN 46151-7902

Editor-in-Chief: Harriet Lowe

DESCRIPTION AND INTERESTS
This magazine is a primary source for the most recent trends, the latest research, innovative programming ideas, and up-to-date management tools for the camping industry. It is the official publication of the American Camp Association. Circ: 7,500.

Audience: Camp managers and educators
Frequency: 6 times each year
Website: www.ACAcamps.org

FREELANCE POTENTIAL
98% written by nonstaff writers. Publishes 20 freelance submissions yearly; 50% by unpublished writers, 30% by authors who are new to the magazine. Receives 96 mss yearly.

SUBMISSIONS
Send complete ms. Accepts email submissions to magazine@acacamps.org. Response time varies.

- Articles: 1,500–4,000 words. Informational and how-to articles. Topics include camp management, special education, social issues, careers, health, recreation, crafts, and hobbies.
- Depts/columns: 800–1,000 words. News, opinion pieces, risk management, and building and construction information.
- Artwork: B/W or color prints or slides.

SAMPLE ISSUE
52 pages (20% advertising): 5 articles; 10 depts/columns. Sample copy, $4.50 with 9x12 SASE. Guidelines and editorial calendar available.

- "Ready, Set, Market." Article discusses unique marketing strategies resulting from a survey of special needs camps.
- "One More Song." Personal experience piece from a former counselor reveals favorite memories from songs learned during her camp days.
- "Wish, Wonder, and Surprise." Article explains the many benefits of bringing these three things back into children's lives.

RIGHTS AND PAYMENT
All rights. No payment. Provides 3 copies.

➤EDITOR'S COMMENTS
Many of our articles are written by those in the industry with first-hand knowledge and experience. Submissions should be well written and well researched.

Canadian Children's Literature

Dept. of English, University of Winnipeg
515 Portage Avenue
Winnipeg, Manitoba R3G 1X4
Canada

Editor: Perry Nodelman

DESCRIPTION AND INTERESTS
Canadian Children's Literature is a bilingual academic journal that examines all aspects of children's literature in a range of media. It is read by academics and others who work in the area of children's culture. Circ: 400.

Audience: Educators, scholars, and librarians
Frequency: Twice each year
Website: http://ccl.uwinnipeg.ca

FREELANCE POTENTIAL
99% written by nonstaff writers. Publishes 18–20 freelance submissions yearly; 10% by unpublished writers, 40% by authors who are new to the magazine. Receives 120 unsolicited mss yearly.

SUBMISSIONS
Send complete ms. Prefers email submissions to ccl@uwinnipeg.ca (Microsoft Word or RTF attachments); will accept hard copy. SAE/IRC. Responds in 3 months.

- Articles: 2,000–6,000 words. Informational articles; reviews; profiles; and interviews. Topics include children's literature; film, video, and drama for children; and children's authors.

SAMPLE ISSUE
208 pages (2% advertising): 9 articles. Sample copy, $10. Guidelines and theme list available at website.

- "Many-Coloured Fish in the Sea of Story." Article describes how retelling folktales in all cultural settings can influence children's attitudes and behaviors by teaching important life lessons.
- "Dramatic Responses to *A Short Tree with Tall Ideas*." Article explains why children's plays are sometimes harder to critique than adult performances, and makes suggestions to help in this process.

RIGHTS AND PAYMENT
First serial rights. No payment. Provides 1 contributor's copy.

➤EDITOR'S COMMENTS
This is a peer-reviewed journal. Specialists in English and/or French literature, theater and drama, media studies, education, information science, childhood and cultural studies, and related fields are encouraged to submit their articles.

Canadian Guider

Girl Guides of Canada
50 Merton Street
Toronto, Ontario M4S 1A3
Canada

Submissions: Veveen Gregory

DESCRIPTION AND INTERESTS
This official publication of the Girl Guides of Canada provides information, ideas, and activities to Girl Guide leaders on promoting self-esteem, nurturing healthy lifestyles, and empowering learning in girls in a fun, friendly, and safe environment. Circ: 40,000.

Audience: Girl Guide leaders
Frequency: 3 times each year
Website: www.girlguides.ca

FREELANCE POTENTIAL
75% written by nonstaff writers. Receives 12 queries yearly.

SUBMISSIONS
Query with résumé for articles. Send complete ms for depts/columns. Availability of artwork improves chance of acceptance. SAE/IRC. Responds in 1 month.

- Articles: To 200 words. Informational and how-to articles; profiles; interviews; and personal experience pieces. Topics include leadership and life skills, health and fitness, camping, adventure, nature, the arts, social issues, and contemporary concerns.
- Depts/columns: 50–100 words. "Innovators" features leadership profiles. "Ideas to Go" offers program-related crafts and activities.
- Artwork: B/W and color prints. Digital images at 300 dpi.

SAMPLE ISSUE
48 pages (12% advertising): 15 articles; 9 depts/columns. Sample copy, $3 with 9x12 SAE/IRC. Guidelines and editorial calendar available at website.

- "Protecting the Pale Blue Dot." Article explains how humans are damaging the Earth's environment and discusses how Girl Guides can help protect it.
- Sample dept/column: "Nature Walks and Hikes" offers tips on ways to search for birds, insects, and flora.

RIGHTS AND PAYMENT
All rights. No payment. Provides 2 contributor's copies.

⦿ EDITOR'S COMMENTS
We are looking for articles written by Girl Guides leaders. All material must have national significance with local reference. Content must be written directly to Guiders.

Capper's

1503 SW 42nd Street
Topeka, KS 66609

Editor-in-Chief: Katherine Compton

DESCRIPTION AND INTERESTS
A nationally distributed tabloid publication, *Capper's* emphasizes traditional home and family values for readers who live in the rural midwestern U.S. Circ: 200,000.

Audience: Families
Frequency: Monthly
Website: www.cappers.com

FREELANCE POTENTIAL
90% written by nonstaff writers. Publishes 40–50 freelance submissions yearly; 50% by unpublished writers, 70% by authors who are new to the magazine. Receives 480 mss yearly.

SUBMISSIONS
Send complete ms with photos for articles. Query for fiction. Accepts hard copy. SASE. Responds to queries in 1 month, to mss in 3–4 months.

- Articles: 700 words. General interest, historical, inspirational, and nostalgic articles. Topics include family life, travel, hobbies, and occupations.
- Fiction: To 25,000 words. Serialized novels.
- Depts/columns: 300 words. Personal experience pieces, humor, and essays.
- Artwork: 35mm color slides, transparencies, or prints.
- Other: Jokes, 5–6 per submission.

SAMPLE ISSUE
48 pages (3% advertising): 26 articles; 1 serialized novel; 8 depts/columns. Sample copy, $1.95. Guidelines available.

- "Botanical Garden in Birmingham, AL, is a Beautiful Oasis." Article describes some of the specialty gardens within this popular tourist attraction.
- "Antique Truck Lover Didn't Let Life's Passion Take the Back Seat." Essay recounts a father turning his passion for old cars into a unique limousine business.

RIGHTS AND PAYMENT
Standard rights. Articles, $2.50 per column inch; serialized novels, $75–$300. Pays on publication for nonfiction, on acceptance for fiction. Provides up to 5 contributor's copies.

⦿ EDITOR'S COMMENTS
Our "Heart of the Home" section is the best place for new writers to get published. We also buy some feature stories.

Careers and Colleges

2 LAN Drive, Suite 100
Westford, MA 01886

Editor: Anne Kandra

DESCRIPTION AND INTERESTS
This cheerful publication guides high school students toward a successful future. Articles introduce teens to college life and present them with career choices. Circ: 752,000.

Audience: 15–18 years
Frequency: 3 times each year
Website: www.careersandcolleges.com

FREELANCE POTENTIAL
80% written by nonstaff writers. Publishes 6 freelance submissions yearly; 10% by authors who are new to the magazine. Receives 50 queries yearly.

SUBMISSIONS
Query with clips; or send complete ms. Accepts email to editor@careersandcolleges.com. Responds in 2 months.

- Articles: 800–2,400 words. Informational and how-to articles; profiles; interviews; and personal experience pieces. Topics include post-secondary education, independent living, campus life, career choices, social issues, and personal growth.
- Depts/columns: Staff written.

SAMPLE ISSUE
32 pages (29% advertising): 8 articles; 2 depts/columns. Sample copy available at website. Guidelines available.

- "I Just Play One on TV." Article describes jobs portrayed on top television shows and talks to real people who actually do them.
- "Tech Tools for College." Article recommends computer, audio, and video equipment to make life at school easier.
- "Home Sweet Dorm." Article provides tips for equipping and organizing a dorm room.

RIGHTS AND PAYMENT
First North American serial and electronic rights. Written material, payment rates vary. Pays 2 months after acceptance. Provides 2 contributor's copies.

••◆EDITOR'S COMMENTS
We have a roster of regular contributors, but welcome freelance queries. However, please note that because the magazine is published only three times per year and much of the editorial consists of updates of prior material, freelance opportunities are limited. All articles should help students achieve their goals.

Carolina Parent

5716 Fayetteville Road, Suite 201
Durham, NC 27713

Editor: Crickett Gibbons

DESCRIPTION AND INTERESTS
Each themed issue of this regional tabloid is filled with helpful, informative articles for parents living in the Research Triangle area of North Carolina. Circ: 56,000.

Audience: Parents
Frequency: Monthly
Website: www.carolinaparent.com

FREELANCE POTENTIAL
55% written by nonstaff writers. Publishes 40 freelance submissions yearly; 10% by unpublished writers, 15% by authors who are new to the magazine. Receives 120 queries and unsolicited mss yearly.

SUBMISSIONS
Query with outline and writing samples. New writers, send complete ms. Accepts hard copy and email submissions to editorial@carolinaparent.com (Microsoft Word attachments). SASE. Response time varies.

- Articles: 500–1,200 words. Informational, self-help, and how-to articles; profiles; and personal experience pieces. Topics include college planning, technology, crafts, hobbies, education, health, fitness, humor, music, nature, the environment, parenting and children's issues, recreation, regional news, sports, and travel.
- Depts/columns: Word lengths vary. Family finances, family issues, home and garden, child development, pregnancy, health, news, and events.

SAMPLE ISSUE
80 pages: 4 articles; 18 depts/columns. Guidelines and editorial calendar available.

- "Stress Less." Article provides specific strategies and suggestions for reducing stress and increasing energy.
- Sample dept/column: "Baby and Me" explains that plus-size women should find supportive caregivers for a positive pregnancy experience.

RIGHTS AND PAYMENT
First and electronic rights. Written material, $50–$75. Pays on publication.

••◆EDITOR'S COMMENTS
We need more articles on pregnancy. We are always interested in informative, age-specific submissions for parents of children from birth to age five, and from 11 to 18 years.

Catholic Digest

1 Montauk Avenue, Suite 200
New London, CT 06320

Articles Editor: Kerry Weber

DESCRIPTION AND INTERESTS
Most of the articles and true stories featured in
Catholic Digest contain a faith element or a
connection to the Catholic Church. Poetry and
fiction are not published. Circ: 285,000.

Audience: Adults
Frequency: 11 times each year
Website: www.catholicdigest.com

FREELANCE POTENTIAL
44% written by nonstaff writers. Publishes
100–200 freelance submissions yearly; 12%
by authors who are new to the magazine.
Receives 4,800 unsolicited mss yearly.

SUBMISSIONS
Send complete ms. Accepts hard copy and
email submissions to cdsubmissions@
bayard-inc.com. No simultaneous submissions.
SASE. Responds in 6–8 weeks.

- Articles: 1,000–2,000 words. Informational
  articles; profiles; and personal experience
  pieces. Topics include religion, prayer, spirituality, relationships, family issues, history,
  and nostalgia.
- Depts/columns: 50–500 words. True stories
  about faith, profiles of volunteers.
- Other: Filler, to 1,000 words.

SAMPLE ISSUE
128 pages (13% advertising): 9 articles; 18
depts/columns. Sample copy, free with 6x9
SASE ($1 postage). Guidelines available.

- "Snowflakes, a Sick Calf, and Christmas
  Eve." Article reveals how a lonely young man
  finds new meaning in his life by nurturing a
  sickly newborn calf.
- "Are Protestants Rediscovering Mary?" Article
  explains the recent shift in the Protestant
  view of the Mother of God.
- Sample dept/column: "Love Your Neighbor"
  profiles a young woman serving as a
  Salesian Missionary in Bolivia.

RIGHTS AND PAYMENT
One-time rights. Articles, $100–$400. Depts/
columns, $2 per published line. Pays on publi-
cation. Provides 2 contributor's copies.

➡EDITOR'S COMMENTS
We are looking for topical articles with national
appeal. We appreciate writing that includes
lively dialogue and/or quotes by notable individuals. We do not accept freelance Q&As.

Catholic Forester

P.O. Box 3012
335 Shuman Boulevard
Naperville, IL 60566-7012

Associate Editor: Patricia Baron

DESCRIPTION AND INTERESTS
This full-color magazine is published for members of Catholic Order of Foresters, a national
fraternal insurance society. It offers news
about the society as well as informative articles on a wide variety of topics. Circ: 100,000.

Audience: Catholic Forester Members
Frequency: Quarterly
Website: www.catholicforester.com

FREELANCE POTENTIAL
20% written by nonstaff writers. Publishes 4–8
freelance submissions yearly; 5% by unpublished writers, 20% by authors who are new to
the magazine. Receives 240 unsolicited mss
each year.

SUBMISSIONS
Send complete ms. Accepts hard copy. SASE.
Responds in 3–4 months.

- Articles: 1,000 words. Informational and
  inspirational articles. Topics include money
  management, fitness, health, family life,
  investing, senior issues, careers, parenting,
  and nostalgia.
- Fiction: 1,000 words. Inspirational, humorous, and light fiction.

SAMPLE ISSUE
40 pages (no advertising): 8 articles; 1 story;
2 activities; 6 depts/columns. Sample copy,
free with 9x12 SASE (3 first-class stamps).
Guidelines available.

- "St. Barbara Youth On the Move." Article pro-
  files an active teen youth group located in
  Erlanger, Kentucky.
- "Seven Steps to Repair and Restore Your
  Life." Article describes seven practical and
  spiritual ways to overcome seemingly insur-
  mountable obstacles.
- "Counting Blessings." Story tells of a family
  counting their blessings as they travel on
  Thanksgiving Day.

RIGHTS AND PAYMENT
First North American serial rights. Written material, $.30 per word. Reprints, $50. Pays on
acceptance. Provides 3 contributor's copies.

➡EDITOR'S COMMENTS
Manuscripts should have a good lead, style,
and rhythm that holds the reader's interest.
Look for ways to project personal experiences into informational topics.

Catholic Library World

100 North Street, Suite 224
Pittsfield, MA 01201-5109

Editor: Mary E. Gallagher, SSJ

DESCRIPTION AND INTERESTS
This magazine publishes articles and reviews of interest to library professionals as well as members of the Catholic Library Association. Circ: 1,100.

Audience: Library professionals
Frequency: Quarterly
Website: www.cathla.org

FREELANCE POTENTIAL
90% written by nonstaff writers. Publishes 16–20 freelance submissions yearly. Receives 12 queries and unsolicited mss yearly.

SUBMISSIONS
Query or send complete ms. Accepts hard copy and email submissions to cla@cathla.org (Microsoft Word attachments). SASE. Response time varies.

- Articles: Word lengths vary. Informational articles and reviews. Topics include books, reading, library science, and Catholic Library Association news.
- Depts/columns: 150–300 words. Media and book reviews. Adult book topics include theology, spirituality, pastoral issues, church history, education, literature, library science, philosophy, and reference; children's and young adult book topics include biography, fiction, multicultural issues, picture books, reference, science, social studies, and values.
- Artwork: B/W and color prints or transparencies. Line art.

SAMPLE ISSUE
262 pages (2% advertising): 4 articles; 2 depts/columns; 60 book reviews; 4 media reviews. Sample copy, $15. Reviewers' guidelines available.

- "2008 Regina Medalist: Vera B. Williams." Article profiles the award-winning children's author and illustrator.
- "People of the Book: Establishing a Parish Book Discussion Group." Article offers tips for starting a thriving parish book group.

RIGHTS AND PAYMENT
All rights. No payment. Provides 1 copy.

➔EDITOR'S COMMENTS
We are interested in articles on any and all topics that would interest library professionals, especially those in Catholic libraries. We have specific guidelines for reviews.

Celebrate

2923 Troost Avenue
Kansas City, MO 64109

Submissions: Abigail L. Takala

DESCRIPTION AND INTERESTS
A take-home paper for preschool and kindergarten Sunday school students, *Celebrate* entertains, educates, and enlightens with its Bible stories, activities, songs, and poems. Lessons are designed to equip believers to share the good news of Christ as well as develop a deeper understanding of Christianity. Circ: 40,000.

Audience: 3–6 years
Frequency: Weekly
Website: www.wordaction.com

FREELANCE POTENTIAL
90% written by nonstaff writers. Publishes 100 freelance submissions yearly; 30% by unpublished writers, 35% by authors who are new to the magazine. Receives 200 queries yearly.

SUBMISSIONS
Query. Accepts hard copy and email queries to alt@wordaction.com (Microsoft Word attachments). SASE. Responds in 2–4 weeks.

- Other: Bible stories, word lengths vary. Features stories that show children dealing with the issues related to a Bible story or lesson. Songs, finger plays, action rhymes, crafts, activities. Poetry, 4–8 lines.

SAMPLE ISSUE
4 pages (no advertising): 1 story; 3 activities; 1 poem. Sample copy, free with #10 SASE (1 first-class stamp). Writers' guidelines and theme list available.

- "Deborah Trusts God." Bible story tells of the woman whose trust in God allowed an army of followers to defeat an evil king.
- Sample activity: "Trust Ring" explains how to make a simple flip-book that records times the child has trusted God, as well as when family members have done the same.

RIGHTS AND PAYMENT
Rights vary. Written material, payment rates vary. Payment policy varies.

➔EDITOR'S COMMENTS
We believe in providing curricula that allows people of all ages to discover God's Word. In *Celebrate*, we bring that discovery to the very young. We look for ideas that will inspire young children while sparking their imagination. The material in each week's publication follows a well-planned theme.

Central Penn Parent

101 North Second Street, 2nd Floor
Harrisburg, PA 17101

Editor: Anna Seip

DESCRIPTION AND INTERESTS
Targeting parents in Cumberland, Dauphin, Lancaster, and York counties in Pennsylvania, this tabloid is filled with information on caring for healthy and happy children. Regional activities and resources and a calendar of events are also included in each issue. Circ: 35,000.
Audience: Parents
Frequency: Monthly
Website: www.centralpennparent.com

FREELANCE POTENTIAL
75% written by nonstaff writers. Publishes 60 freelance submissions yearly; 20% by unpublished writers, 10% by authors who are new to the magazine. Receives 1,300 queries yearly.

SUBMISSIONS
All articles are assigned to local writers. Query for reprints only. Accepts email queries to annas@journalpub.com. Availability of artwork improves chance of acceptance. SASE. Responds in 2 weeks.

- Articles: 1,200–1,500 words. Informational articles and reviews. Topics include local family events and activities, health, nutrition, discipline, education, home life, technology, literature, parenting, and travel.
- Depts/columns: 700 words. Family finances, health, infant issues, news, and education.
- Artwork: Color prints and transparencies. Line art.
- Other: Submit seasonal material at least 2 months in advance.

SAMPLE ISSUE
46 pages (50% advertising): 2 articles; 13 depts/columns; 1 calendar. Sample copy, free. Guidelines available.

- "Homeless: One Father's Story." Article recounts a father's struggle to rebuild a life for him and his autistic son.
- Sample dept/column: "Take Note" offers tips for buying a used car.

RIGHTS AND PAYMENT
All rights. Reprints, $35–$50. Pays on publication. Provides author's copies upon request.

•◦EDITOR'S COMMENTS
We prefer to work with local writers who can provide us with informative and reader-friendly pieces that are of interest to parents of newborns to teens.

Characters

P.O. Box 708
Newport, NH 03773

Editor: Cindy Davis

DESCRIPTION AND INTERESTS
This literary journal for middle-grade and young adult readers publishes fiction in a wide range of genres. It gives preference to works written by children and teens. Circ: Unavailable.
Audience: 8–17 years
Frequency: 4 times each school year
Website: www.cdavisnh.net

FREELANCE POTENTIAL
100% written by nonstaff writers. Publishes 100 freelance submissions yearly; 75% by unpublished writers, 50% by authors who are new to the magazine. Receives 600 unsolicited mss yearly.

SUBMISSIONS
Send complete ms with brief biography. Prefers hard copy; will accept email to hotdog@nhvt.net (no attachments). Accepts simultaneous submissions if identified. SASE. Responds in 2–4 weeks.

- Fiction: To 2,000 words. Genres include mystery, contemporary and historical fiction, humor, fantasy, adventure, science fiction, romance, and stories with nature themes.

SAMPLE ISSUE
44 pages (no advertising): 10 stories. Sample copy, $5.75. Guidelines available.

- "Bracism." Story tells of a teenage drama queen who fears she hasn't been invited to the biggest party of the year because she wears braces.
- "Selfless Fashion." Story describes a 16-year-old "skater boy" who cuts his long hair on behalf of his grandmother.
- "Bailey's Witty Scheme." Story tells how a girl manages to shake off her tagalong little brother.

RIGHTS AND PAYMENT
One-time and electronic rights. Written material, $5. Pays on publication. Provides 1 contributor's copy.

•◦EDITOR'S COMMENTS
We receive too many "lost in a new neighborhood" and "sibling rivalry" stories. We love stories where children are faced with some other dilemma—whether physical or cerebral—that they can solve with little intervention from adults. Please note that we publish very few read-to stories, and no poetry.

Charlotte Parent

2125 Southend Drive, Suite 253
Charlotte, NC 28230

Editor: Eve White

DESCRIPTION AND INTERESTS
This magazine has been a valuable, one-stop parenting resource for readers in the metropolitan Charlotte region since 1987. It features local insight, resources, and solutions to the challenges of raising children today. Circ: 55,000.
Audience: Parents
Frequency: Monthly
Website: www.charlotteparent.com

FREELANCE POTENTIAL
50% written by nonstaff writers. Publishes 45 freelance submissions yearly; 15% by unpublished writers, 25% by new authors. Receives 1,000 queries, 800 unsolicited mss yearly.

SUBMISSIONS
Query or send complete ms with résumé and bibliography. Prefers email to editor@ charlotteparent.com; will accept hard copy, Macintosh disk submissions, and simultaneous submissions. SASE. Responds if interested.

- Articles: 500–1,000 words. Informational and how-to articles. Topics include parenting, family life, finances, education, health, fitness, vacations, entertainment, regional activities, and the environment.
- Depts/columns: Word lengths vary. Child development, restaurant and media reviews, and children's health.
- Artwork: High-density Macintosh images.
- Other: Activities. Submit seasonal material 2–3 months in advance.

SAMPLE ISSUE
72 pages (50% advertising): 7 articles; 11 depts/columns; 1 calendar. Sample copy, free with 9x12 SASE (5 first-class stamps). Guidelines and editorial calendar available.

- "Back To School." Article provides a back-to-school checklist for parents that goes above and beyond school supplies.
- "Unlock My Child's Potential." Article explains how to help when a child isn't working up to his or her potential.

RIGHTS AND PAYMENT
First and Internet rights. Written material, payment rates vary. Pays on publication. Provides 1 contributor's copy.

➥EDITOR'S COMMENTS
We look for writers who know how to share the joys and challenges of parenthood.

ChemMatters

American Chemical Society
1155 16th Street NW
Washington, DC 20036

Editor: Pat Pages

DESCRIPTION AND INTERESTS
The mission of ChemMatters is to demystify everyday chemistry and to show that it can be interesting—even exciting. The magazine is read by introductory chemistry students and their teachers in 35 countries. Circ: 40,000.
Audience: YA–Adult
Frequency: Quarterly
Website: www.acs.org/chemmatters

FREELANCE POTENTIAL
90% written by nonstaff writers. Publishes 20 freelance submissions yearly; 70% by unpublished writers. Receives 30–40 queries yearly.

SUBMISSIONS
Query with abstract, outline, source list, possible artwork, estimated word count, writing sample, and résumé. Prefers email queries to chemmatters@acs.org; will accept hard copy. Availability of artwork improves chance of acceptance. SASE. Responds in 2 weeks.

- Articles: 1,400–2,100 words. Informational articles. Topics include the chemical aspects of things we take for granted, such as food, beverages, and biological functions, as well as hot topics like forensic science.
- Depts/columns: 1,400–2,100 words. "ChemHistory," "ChemMystery," and "ChemSumer," which explains the chemistry of everyday products used by teens.
- Artwork: JPEG and GIF images.
- Other: Puzzles and activities.

SAMPLE ISSUE
20 pages (no advertising): 6 articles; 2 depts/columns. Sample copy available at website. Guidelines and theme list available via email request to chemmatters@acs.org.

- "The Quest for a Clean Drink." Article demonstrates how chemistry provides solutions for contaminated drinking water.
- "The Chemistry of Arson Investigation." Article describes which clues to look for to determine the cause of a fire.

RIGHTS AND PAYMENT
All rights. Articles, $500–$1,000 words. Pays on acceptance. Provides 5 contributor's copies.

➥EDITOR'S COMMENTS
Although most of our readers are in high school, ChemMatters is written on an adult level; however, writers should assume nothing.

Chesapeake Family

929 West Street, Suite 307
Annapolis, MD 21401

Editor: Cathy Ashby

DESCRIPTION AND INTERESTS
Chesapeake Family is a free parenting publication serving the Anne Arundel, Calvert, Bowie, Upper Marlboro, and Kent Island areas of Maryland. Its goal is to make readers' lives easier by presenting information they need. Circ: 40,000.

Audience: Parents
Frequency: Monthly
Website: www.chesapeakefamily.com

FREELANCE POTENTIAL
80% written by nonstaff writers. Publishes 20 freelance submissions yearly; 20% by authors who are new to the magazine. Receives 120 unsolicited mss yearly.

SUBMISSIONS
Prefers query with outline; will accept ms. Prefers email to editor@chesapeakefamily.com; will accept hard copy and faxes to 410-280-0255. SASE. Response time varies.

- Articles: 1,000–1,200 words. Informational and how-to articles; profiles; and personal experience pieces. Topics include parenting, the environment, music, regional news, current events, education, entertainment, health, and family travel.
- Fiction: "Just for Kids" features stories and poems by local 4- to 12-year-old children.
- Depts/columns: 700–900 words. Health, education, and child development.
- Other: Submit seasonal material 3–6 months in advance.

SAMPLE ISSUE
68 pages (45% advertising): 5 articles; 9 depts/columns; 1 events calendar. Sample copy, guidelines, and editorial calendar available.

- "How Much Is Too Much?" Article provides guidelines for dispensing cold and flu medication to children.
- "Wishing for a Girl, Expecting a Boy." Essay shares a woman's feelings about having a boy.
- Sample dept/column: "Family Fun" recommends a trip to Mount Vernon.

RIGHTS AND PAYMENT
One-time print and Web rights. Articles, $75–$150. Depts/columns, $50. Reprints, $35. Kill fee, $25. Payment policy varies.

➥EDITOR'S COMMENTS
We generally do not run first-person essays unless they are unusually compelling.

Chess Life

P.O. Box 3967
Crossville, TN 38557-3967

Editor: Daniel Lucas

DESCRIPTION AND INTERESTS
The official publication of the United States Chess Federation, *Chess Life* is read by chess enthusiasts of all ages and skill levels. It covers the history of chess, instruction, strategies, the players, and major and minor chess events throughout the world. Circ: 80,000.

Audience: YA–Adult
Frequency: Monthly
Website: www.uschess.org

FREELANCE POTENTIAL
75% written by nonstaff writers. Publishes 30 freelance submissions yearly; 30% by unpublished writers. Receives 180–420 queries yearly.

SUBMISSIONS
Query with clips or writing samples. Accepts hard copy, IBM disk submissions, and email queries to dlucas@uschess.org. SASE. Responds in 1–3 months.

- Articles: 800–3,000 words. Informational, how-to, and historical articles; profiles; humor; and personal experience and opinion pieces. Topics include chess games and strategies, tournaments and events, and personalities.
- Depts/columns: To 1,000 words. Book and product reviews, short how-to's, and player profiles.
- Artwork: B/W and color prints.
- Other: Chess-oriented cartoons, contests, and games.

SAMPLE ISSUE
72 pages (16% advertising): 5 articles; 11 depts/columns. Sample copy, free with 9x12 SASE. Guidelines available at website.

- "Kaufman: A Perfect Champion." Article reports on IM Larry Kaufman's win in the U.S. senior championship with a perfect five-for-five record.
- Sample dept/column: "A Look at Books" reviews *Hooked on Chess*, the memoir of chess Olympian Bill Hook.

RIGHTS AND PAYMENT
All rights. Written material, $100 per page. Artwork, $15–$100. Kill fee, 30%. Pays on publication. Provides 2 contributor's copies.

➥EDITOR'S COMMENTS
We cannot consider any writer who isn't fully involved in the game of chess.

Chess Life for Kids

P.O. Box 3967
Crossville, TN 38577

Editor: Glenn Petersen

DESCRIPTION AND INTERESTS

Chess Life for Kids is published for scholastic members of the U.S. Chess Federation. It features age-appropriate chess instruction for players ages 12 years and under. It also covers Federation-affiliated activities such as tournaments and other events. Circ: 22,700.

Audience: 12 years and under
Frequency: 6 times each year
Website: www.uschess.org

FREELANCE POTENTIAL

30% written by nonstaff writers. Publishes 12–18 freelance submissions yearly; 10% by unpublished writers. Receives 36 queries, 12 unsolicited mss yearly.

SUBMISSIONS

Query or send complete ms. Accepts email submissions to gpetersen@uschess.org and simultaneous submissions if identified. Responds in 2 weeks.

- Articles: To 1,000 words. Informational and instructional articles; and profiles. Topics include chess instruction, game tips and strategies, tournaments, chess masters, chess camps, and chess lessons.

SAMPLE ISSUE

32 pages: 13 articles.

- "Tales of the Arabian Knights." Article incorporates chess moves and strategies into a story designed to help a king learn how to rule his kingdom.
- "What a Shot!" Article recounts a game played at the English Opening in order to instruct the reader about tactical moves.
- "Attacking the Kingside Castled Position." Article shows that there many different ways to accomplish a common mating plan.

RIGHTS AND PAYMENT

First North American serial rights. Written material, $75 per page. Pays on publication.

⟶EDITOR'S COMMENTS

We are always looking for creative approaches to chess instruction that will appeal to young readers who are still developing as chess players. The best way to determine if your work fits our needs and objectives is to read an issue of our magazine. We also seek information regarding youth chess tournaments, camps, and other activities.

Chicago Parent

141 South Oak Park Avenue
Chicago, IL 60302

Editor: Tamara O'Shaughnessy

DESCRIPTION AND INTERESTS

In this tabloid, Chicago-area parents find timely information on a range of parenting topics, all with a distinctly local focus. Circ: 138,000.

Audience: Parents
Frequency: Monthly
Website: www.chicagoparent.com

FREELANCE POTENTIAL

85% written by nonstaff writers. Publishes 50+ freelance submissions yearly; 30% by unpublished writers, 45% by authors who are new to the magazine. Receives 1,560 queries yearly.

SUBMISSIONS

Query with résumé and clips. Accepts email queries to chiparent@chicagoparent.com. Responds in 6 weeks.

- Articles: 1,500–2,500 words. Informational articles; profiles; personal experience pieces; and humor. Topics include pregnancy, childbirth, parenting, grandparenting, foster care, adoption, day care, child development, health, education, recreation, and family issues.
- Depts/columns: 850 words. Crafts, activities, media reviews, health, travel, family finances, and regional events.
- Other: Cartoons for parents. Submit seasonal material at least 2 months in advance.

SAMPLE ISSUE

128 pages (60% advertising): 10 articles; 19 depts/columns. Sample copy, $3.95. Guidelines and editorial calendar available.

- "Pediatrician Pointers for Parents." Article discusses ways to make the most of the short time with the doctor.
- "Eco-conscious Birthdays." Article offers suggestions for an eco-friendly birthday party.
- Sample dept/column: "Short Stuff" explains how to turn children into birdwatchers.

RIGHTS AND PAYMENT

One-time and northwest Indiana and Illinois exclusive rights. Articles, $125–$350. Depts/columns, $25–$100. Kill fee, 10%. Pays on publication. Provides contributor's copies upon request.

⟶EDITOR'S COMMENTS

The best way for new writers to get started with us is the "Short Stuff" column. Please note that we use Chicago-area writers only.

Childhood Education

Association for Childhood Education
International
17904 Georgia Avenue, Suite 215
Olney, MA 20832

Editor: Anne W. Bauer

DESCRIPTION AND INTERESTS
This is an academic journal serving teachers, teacher educators, parents, child-care workers, administrators, school librarians, and other professionals who are concerned with the education of children from infancy through adolescence. It explores emerging ideas and research on childhood education topics. Circ: 10,000.

Audience: Educators; child-care professionals
Frequency: 6 times each year
Website: www.acei.org

FREELANCE POTENTIAL
98% written by nonstaff writers. Publishes 40 freelance submissions yearly; 75% by authors who are new to the magazine. Receives 120 unsolicited mss yearly.

SUBMISSIONS
Send 4 copies of complete ms. Accepts hard copy, IBM or Macintosh disk submissions, and email submissions to abauer@acei.org. SASE. Responds in 3 months.

- Articles: 1,400–3,500 words. Informational articles. Topics include innovative teaching strategies, the teaching profession, research findings, parenting and family issues, communities, drug education, and safe environments for children.
- Depts/columns: 1,000 words. Research news, education issues, parenting, and book and media reviews.

SAMPLE ISSUE
194 pages: 5 articles; 8 depts/columns. Sample copy, free with 9x12 SASE (3 first-class stamps). Writers' guidelines and editorial calendar available.

- "Forty Years of School Readiness Research: What Have We Learned?" Article reflects on where schools stand in regard to kindergarten readiness preparation.
- "Starting School: A Community Endeavor." Article explains how various segments of the community have the responsibility for preparing young people for school.

RIGHTS AND PAYMENT
All rights. No payment. Provides 5 copies.

➼EDITOR'S COMMENTS
While we welcome manuscripts, please be advised that we use only professionals in the field, and that all articles are peer reviewed.

Children and Families

1651 Prince Street
Alexandria, VA 22310

Editor: Julie Antoniou

DESCRIPTION AND INTERESTS
Published by the National Head Start Association, this award-winning magazine offers creative ideas, practical solutions, and expert advice to help early childhood professionals (primarily Head Start directors) provide high-quality services to children and families. Circ: 12,000.

Audience: Early childhood professionals
Frequency: Quarterly
Website: www.nhsa.org

FREELANCE POTENTIAL
90% written by nonstaff writers. Publishes 25 freelance submissions yearly; 30% by unpublished writers, 70% by authors who are new to the magazine. Receives 24 queries yearly.

SUBMISSIONS
Query with biography, outline, and possible sidebar information. Accepts email queries to julie@nhsa.org and simultaneous submissions if identified. Responds in 1–3 months.

- Articles: 1,800–3,800 words. Informational and how-to articles. Topics include teaching skills, advocacy strategies, problem-solving, administrative issues, school readiness, professional development, special needs and inclusion, parental involvement and partnerships, and child development research.
- Depts/columns: Word lengths vary. News, tips for home visits, teaching tactics, leadership advice, lesson plans, literacy projects, and baby-care issues.

SAMPLE ISSUE
62 pages (20% advertising): 2 articles; 7 depts/columns. Sample copy and guidelines available at website.

- "Bullying: The Basics." Article provides an overview of bullying and tells how early childhood professionals can help prevent it.
- Sample dept/column: "Young Ones" discusses the benefits of breastfeeding.

RIGHTS AND PAYMENT
First rights. No payment. Provides 2+ contributor's copies.

➼EDITOR'S COMMENTS
We do not offer financial compensation, but we do publish author biographical information that includes publications and product or service details.

Children's Digest

Children's Better Health Institute
1100 Waterway Boulevard
Indianapolis, IN 46202

Editor: Danny Lee

DESCRIPTION AND INTERESTS
Published by the Children's Better Health Institute, this magazine features articles meant to encourage children of all cultures to strive for excellence in fitness, academics, medicine, and science. Circ: 60,000.

Audience: 10–12 years
Frequency: 6 times each year
Website: www.childrensdigestmag.org

FREELANCE POTENTIAL
50% written by nonstaff writers. Publishes 25 freelance submissions yearly; 70% by unpublished writers. Receives 1,200 queries yearly.

SUBMISSIONS
Query or send complete ms. Accepts hard copy and email submissions to d.lee@cbhi.org. SASE. Responds in 3 months.

- Articles: To 1,200 words. Informational and how-to articles; profiles; interviews; and personal experience pieces. Topics include health, exercise, safety, hygiene, drug education, and nutrition.
- Fiction: To 1,500 words. Genres include multicultural, ethnic, and science fiction; fantasy; adventure; mystery; humor; and stories about animals and sports.
- Depts/columns: To 1,200 words. Reviews, recipes, and health Q&As.
- Other: Puzzles, games, and activities. Poetry, to 25 lines. Submit seasonal material 8 months in advance.

SAMPLE ISSUE
34 pages (6% advertising): 5 articles; 1 story; 2 depts/columns; 3 poems; 6 activities. Sample copy, $1.25 with 9x12 SASE. Guidelines available at website.

- "Keeping a Road Journal." Article explains how to keep a road journal while traveling with the family.
- "Sam Is Gym-tastic!" Article profiles a young woman who represented the U.S. in gymnastics in the summer Olympics.

RIGHTS AND PAYMENT
All rights. Written material, $.20 per word. Pays prior to publication. Provides 10 copies.

➭EDITOR'S COMMENTS
We are looking for good material written for our age demographic that will inspire kids to be the best they can be.

Children's Ministry

1515 Cascade Avenue
Loveland, CO 80539

Associate Editor: Carmen Kamrath

DESCRIPTION AND INTERESTS
The goal of this magazine is to equip church leaders and volunteers with the skills necessary to teach children about Jesus. Its content can be described as inspirational, motivational, and educational, concentrating on techniques and teaching ideas. Circ: 65,000.

Audience: Children's ministry leaders
Frequency: 6 times each year
Website: www.childrensministry.com

FREELANCE POTENTIAL
60–80% written by nonstaff writers. Publishes 25–35 freelance submissions yearly; 60% by unpublished writers, 60% by authors who are new to the magazine. Receives 2,400 unsolicited mss yearly.

SUBMISSIONS
Send complete ms. Prefers email submissions to ckamrath@cmmag.com; will accept hard copy. SASE. Responds in 2–3 months.

- Articles: 500–1,700 words. Informational and how-to articles; and personal experience pieces. Topics include Christian education, family issues, child development, and faith.
- Depts/columns: 50–300 words. Educational issues, activities, devotionals, family ministry, parenting, crafts, and resources.
- Other: Activities, games, and tips. Submit seasonal material 6–8 months in advance.

SAMPLE ISSUE
138 pages (50% advertising): 10 articles; 15 depts/columns. Sample copy, $2 with 9x12 SASE. Guidelines available.

- "This Is Family Ministry." Article explains family ministry and how it is sweeping through children's ministries today.
- Sample dept/column: "Leading Volunteers" offers ideas for spreading excitement about a program throughout the church community.

RIGHTS AND PAYMENT
All rights. Articles, $40–$400. Depts/columns, $40–$75. Pays on acceptance. Provides 1 contributor's copy.

➭EDITOR'S COMMENTS
We look for articles that can inspire children's ministry leaders to provide the absolute best ministry they can. To do that, articles must portray a sense of excitement while offering instruction.

Children's Playmate

Children's Better Health Institute
1100 Waterway Boulevard
P.O. Box 567
Indianapolis, IN 46206-0567

Editor: Terry Harshman

DESCRIPTION AND INTERESTS
Like its sister publications, *Turtle Magazine* and *Humpty Dumpty's Magazine*, this publication contains easy stories and fun activities geared to its specific age level and intended to encourage reading. Circ: 80,000.

Audience: 6–8 years
Frequency: 6 times each year
Website: www.childrensplaymatemag.org

FREELANCE POTENTIAL
35% written by nonstaff writers. Publishes 40 freelance submissions yearly; 20% by unpublished writers, 50% by authors who are new to the magazine. Receives 900 unsolicited mss each year.

SUBMISSIONS
Send complete ms. Accepts hard copy. SASE. Responds in 2–3 months.

- Articles: To 500 words. Humorous and how-to articles. Topics include health, fitness, nature, the environment, science, hobbies, crafts, multicultural and ethnic subjects, and sports.
- Fiction: To 100 words. Humorous fiction and rebus stories.
- Other: Puzzles, activities, games, and recipes. Poetry, to 20 lines. Submit seasonal material about unusual holidays 8 months in advance.

SAMPLE ISSUE
34 pages (no advertising): 4 articles; 1 story; 9 activities; 3 poems. Sample copy, $1.75 with 9x12 SASE. Guidelines available.

- "Ladybug Love." Article explains the lifestyle and life cycle of the ladybug.
- "Grandpap's Old Fishing Hat." Story tells of a young boy's love of fishing with his grandfather, and his grandfather's ratty fishing hat.
- "Liquid Light." Activity illustrates how light can travel through a tiny stream of water.

RIGHTS AND PAYMENT
All rights; returns book rights upon request. Articles and fiction, $.32 per word. Poetry, $35 per poem. Pays on publication. Provides up to 10 contributor's copies.

➥EDITOR'S COMMENTS
Because of a backlog of material, we are accepting poems, rebuses, and easy recipes only. Please check our website for any change in our needs before submitting.

Children's Voice

Child Welfare League of America
2345 Crystal Drive, Suite 250
Arlington, VA 22202

Managing Editor: Emily Shenk

DESCRIPTION AND INTERESTS
Published by the Child Welfare League of America, this magazine covers national, state, and local news that affects youth and families. It is read by a wide array of child welfare professionals and volunteers. Circ: 25,000.

Audience: Child welfare advocates
Frequency: 6 times each year
Website: www.cwla.org/voice

FREELANCE POTENTIAL
30% written by nonstaff writers. Publishes 5 freelance submissions yearly; 50% by unpublished writers, 50% by authors who are new to the magazine. Receives 10 queries yearly.

SUBMISSIONS
Query. Accepts email to voice@cwla.org (text only). Responds in 1 month.

- Articles: 2,000–2,500 words. Informational and how-to articles; profiles; interviews; and personal experience pieces. Topics include child welfare issues, nonprofit management and leadership, legal issues, and agency programs and practices.
- Depts/columns: 200–500 words. Agency news; public policy alerts; state-level child welfare news; Q&As; reports on special education.

SAMPLE ISSUE
40 pages (20% advertising): 4 articles; 8 depts/columns. Sample copy, $10. Guidelines available at website.

- "Walking the Walk, Not Just Talking the Talk." Article provides eight steps toward implementing evidence-based practice.
- Sample dept/column: "Spotlight On" profiles the nonprofit group Educate Tomorrow, which matches youth with educational mentors.

RIGHTS AND PAYMENT
All rights. No payment. Provides contributor's copies and a 1-year subscription.

➥EDITOR'S COMMENTS
We are especially interested in submissions for features or "Spotlight On" that highlight new, innovative programs with proven, positive results or that spotlight success stories of individual children or families who have been served by the child welfare system. We also encourage suggestions for guest columns, interviews, and new departments.

Children's Writer

Institute of Children's Literature
95 Long Ridge Road
West Redding, CT 06896-1124

Editor: Susan Tierney

DESCRIPTION AND INTERESTS
As the name suggests, this newsletter covers writing and publishing trends for children's book and magazine writers. Circ: 14,000.
Audience: Children's writers
Frequency: Monthly
Website: www.childrenswriter.com

FREELANCE POTENTIAL
100% written by nonstaff writers. Publishes 75 freelance submissions yearly; 10% by unpublished writers, 15% by authors who are new to the magazine. Receives 60+ queries each year.

SUBMISSIONS
Query with outline, synopsis, and résumé. Prefers email submissions through website; will accept hard copy and disk submissions. SASE. Responds in 2 months.

- Articles: 1,500–2,000 words. Informational and how-to articles; publisher profiles; and interviews with editors and writers. Topics include children's book and magazine publishing trends; new markets; genres; writing techniques; research; motivation; and business issues.
- Depts/columns: To 750 words, plus 125-word sidebar. Writing and publishing tips.

SAMPLE ISSUE
12 pages (no advertising): 2 articles; 4 depts/columns. Sample copy, free with #10 SASE (1 first-class stamp). Guidelines available at website or with SASE.

- "Don't Call It Chick Lit (Well, Like, Sometimes)." Article discusses how the term has become overused and dismissive of literature featuring female protagonists.
- Sample dept/column: "Inspiration" features playful ideas for innovative stories.

RIGHTS AND PAYMENT
First North American serial rights. Articles, $135–$350. Pays on publication.

•°EDITOR'S COMMENTS
Our style is largely journalistic, professional, yet encouraging. Contributors should include practical considerations and ideas on which readers may act. Articles must cite solid sources, most particularly editors, using quotes throughout. Preferred sources are publishers currently seeking submissions.

Christian Home & School

3350 East Paris Avenue SE
Grand Rapids, MI 48512-3054

Managing Editor: Paul Brinkerhoff

DESCRIPTION AND INTERESTS
This magazine aims to promote Christian education and address a wide range of parenting topics. It is published by Christian Schools International for parents of children in Christian schools. Circ: 66,000.
Audience: Parents
Frequency: 3 times each year
Website: www.csionline.org

FREELANCE POTENTIAL
95% written by nonstaff writers. Publishes 20–25 freelance submissions yearly; 10% by unpublished writers, 30% by authors who are new to the magazine. Receives 60 queries, 150 unsolicited mss yearly.

SUBMISSIONS
Query or send complete ms. Accepts hard copy, email to pbrinkerhoff@csionline.org, and simultaneous submissions if identified. SASE. Responds in 7–10 days.

- Articles: 1,000–2,000 words. Informational, how-to, and self-help articles; and personal experience pieces. Topics include education, parenting, life skills, decision-making, self-control, discipline, family travel, faith, marriage, and social issues.
- Fiction: Word lengths vary. Publishes stories with Christian themes and stories about Christmas.
- Depts/columns: "Parentstuff," 100–250 words. Reviews and parenting tips, word lengths vary.

SAMPLE ISSUE
34 pages (15% advertising): 5 articles; 10 depts/columns. Sample copy, free with 9x12 SASE ($1.11 postage). Guidelines available.

- "Cultivate a Heart of Gratefulness." Article explains how parents can encourage less getting and more giving in their children.
- Sample dept/column: "Parentstuff" suggests ways to keep Christ in Christmas celebrations and traditions.

RIGHTS AND PAYMENT
First rights. Written material, $50–$250. Pays on publication. Provides 5 author's copies.

•°EDITOR'S COMMENTS
Our readership is more interested in the day-to-day concerns of raising children than theories. Seasonal material is always needed.

The Christian Science Monitor

1 Norway Street
Boston, MA 02115

Home Forum Editor: Judy Lowe

DESCRIPTION AND INTERESTS
This international daily newspaper presents a weekly "Home Forum" section called "Kidspace" that is aimed at children. Regular "Home Forum" articles deal with parenting issues. Circ: 80,000.

Audience: Adults; children, 7–12 years
Frequency: Daily; "Kidspace," weekly
Website: www.csmonitor.com

FREELANCE POTENTIAL
99% written by nonstaff writers. Publishes 1,000 freelance submissions yearly; 10% by unpublished writers, 40% by authors who are new to the magazine. Receives 600 queries, 6,600 unsolicited mss yearly.

SUBMISSIONS
Query with résumé and clips for "Kidspace." Send complete ms for "Home Forum." Accepts email submissions to homeforum@csmonitor.com. Responds in 3 weeks.

- Articles: "Kidspace," 700–900 words, plus sidebars. Informational articles; profiles; and interviews. Topics include occupations and everyday science. "Home Forum" essays, 300–1,000 words. Personal experience pieces and humor. Topics include home, family, and parenting.
- Other: Short bits of information of interest to kids, 150–400 words. Poetry, to 20 lines. Submit seasonal material 1 month in advance.

SAMPLE ISSUE
20 pages (15% advertising): 14 articles; 8 depts/columns. Guidelines available.

- "In the Saddle at the Bison Museum." Article profiles this Arizona museum and includes historical facts about buffalo.
- "A Walnut From Tree to Table." Article describes the growing and manufacturing process of California walnuts.

RIGHTS AND PAYMENT
Exclusive rights. Kidspace, $230+. Essays, $75–$160. Poetry, $20–$40 per poem. Short bits, $70. Pays on publication. Provides 1 contributor's copy.

➦EDITOR'S COMMENTS
We would like to see "chunks" of related text that can be presented individually. More material on science topics is needed.

Christian Work at Home Moms

2602 Hummingbird Circle
Bellevue, NE 68123

Editor: Jill Hart

DESCRIPTION AND INTERESTS
This privately-owned website offers practical as well as spiritual advice to Christian women and men who want to start a successful home-based business. Hits per month: 1.5 million.

Audience: Parents
Frequency: Weekly
Website: www.cwahm.com

FREELANCE POTENTIAL
90% written by nonstaff writers. Publishes 150 freelance submissions yearly; 75% by unpublished writers, 50% by authors who are new to the magazine. Receives 300 unsolicited mss each year.

SUBMISSIONS
Send complete ms. Accepts submissions through website only. Response time varies.

- Articles: 600–800 words. Informational and how-to articles; profiles; interviews; and personal experience pieces. Topics include telecommuting, home businesses, website management and design, search engine optimization, copywriting, money management, blogging, marriage, parenting, spiritual growth, and homeschooling.

SAMPLE ISSUE
Sample copy and writers' guidelines available at website.

- "It's About Time to Revisit the Daily Planner." Article explains the importance of keeping a daily planner for making the best use of the time God has provided, as well as for making time for God.
- "How to Find Your Dream Business." Article reveals that a home business that suits one's interests, abilities, and life circumstances will be successful if it fills a need.
- "Mixing Christianity and Your WAHM Business—Recipe for Disaster or Faithful Formula?" Article suggests ways for Christians who identify their faith with their business to deal with critical individuals.

RIGHTS AND PAYMENT
Electronic rights. No payment.

➦EDITOR'S COMMENTS
We are seeking more submissions of articles on homeschooling, as well as more Christian-themed humorous pieces. We encourage writers to become familiar with our site.

Cicada

70 East Lake Street, Suite 300
Chicago, IL 60601

Executive Editor: Deborah Vetter

DESCRIPTION AND INTERESTS
This literary magazine for young adults features original short stories, poetry, and essays. There is also a section of reader opinions and commentaries on current social and cultural issues of interest to this age group. Circ: 18,500.

Audience: 14–21 years
Frequency: 6 times each year
Website: www.cicadamag.com

FREELANCE POTENTIAL
95% written by nonstaff writers. Publishes 100 freelance submissions yearly; 40% by unpublished writers, 60% by authors who are new to the magazine. Receives 2,000 mss yearly.

SUBMISSIONS
Submission policy is currently in flux. Prospective contributors are advised to check website for submission information.

- Articles: To 5,000 words. Essays and personal experience pieces.
- Fiction: To 5,000 words. Genres include adventure; fantasy; humor; and historical, contemporary, and science fiction. Also features plays and stories presented in sophisticated cartoon format. Novellas, to 15,000 words.
- Depts/columns: "Expressions," 500–1,500 words.
- Other: Cartoons. Poetry, to 25 lines.

SAMPLE ISSUE
48 pages (no advertising): 5 stories; 4 depts/columns; 4 poems. Sample copy, $8.50. Guidelines available at website.

- "Amnesty." Story tells of woman's nine-month cohabitation with a mouse that she fed and cared for before setting free.
- "Stranded." Story revolves around a couple's out-of-body experience on the side of a highway after they die in a car accident.

RIGHTS AND PAYMENT
Rights vary. Written material, payment rates vary. Pays on publication.

☛EDITOR'S COMMENTS
Our magazine has a new look and format, so we encourage prospective writers to study recent issues. Book reviews now appear online only. Our new submissions policy is not in place as of this writing; please visit our website for the latest information.

Circle K

Circle K International
3636 Woodview Trace
Indianapolis, IN 46268-3196

Executive Editor: Kasey Jackson

DESCRIPTION AND INTERESTS
Members of Circle K International, the world's largest collegiate service organization, read this magazine for articles about leadership, career development, and community service, among other topics. Circ: 10,000.

Audience: YA–Adult
Frequency: 6 times each year
Website: www.circlek.org

FREELANCE POTENTIAL
50% written by nonstaff writers. Publishes 12 freelance submissions yearly; 50% by unpublished writers, 50% by authors who are new to the magazine. Receives 48+ queries yearly.

SUBMISSIONS
Query. Accepts hard copy, email queries to ckimagazine@kiwanis.org, and faxes to 317-879-0204. SASE. Responds in 2 weeks.

- Articles: 1,500–2,000 words. Informational and self-help articles. Topics include social issues, collegiate trends, community involvement, leadership, and career development.
- Depts/columns: Word lengths vary. News and information about Circle K activities.
- Artwork: 5x7 or 8x10 glossy prints. TIFF or JPEG images at 300 dpi or higher.

SAMPLE ISSUE
20 pages (no advertising): 6 articles; 5 depts/columns. Sample copy, $.75 with 9x12 SASE ($.75 postage). Guidelines available.

- "Convention Query." Q&A provides guidance for those attending the annual Circle K International Convention for the first time.
- "What Not to Miss: Tips from a Native Oregonian." Article offers things to see and do while visiting Portland, Oregon, site of the 2007 Circle K International Convention.
- Sample dept/column: "Action Notes" contains brief reports of chapter activities.

RIGHTS AND PAYMENT
First North American serial rights. Written material, $150–$400. Artwork, payment rates vary. Pays on acceptance. Provides 3 author's copies.

☛EDITOR'S COMMENTS
In all articles, treatment must be objective and in-depth, and each major point should be substantiated by illustrative examples and quotes from expert sources. Single-source articles and essays are quickly rejected.

City Parent

467 Speers Road
Oakville, Ontario L6K 3S4
Canada

Editor: Jane Muller

DESCRIPTION AND INTERESTS
Distributed free throughout the greater Toronto area, *City Parent* covers all the bases for local families, from day care to night feedings, summer camp to winter sports. Circ: 70,000.

Audience: Parents
Frequency: Monthly
Website: www.cityparent.com

FREELANCE POTENTIAL
60% written by nonstaff writers. Publishes 24–30 freelance submissions yearly; 10% by authors who are new to the magazine. Receives 300+ unsolicited mss yearly.

SUBMISSIONS
Send complete ms. Accepts email submissions to cityparent@haltonsearch.com. Availability of artwork improves chance of acceptance. Responds immediately if interested.

- Articles: 500–1,000 words. Informational articles. Topics include arts and entertainment, health, fitness, multicultural and ethnic issues, recreation, self-help, social issues, travel, and crafts for children.
- Depts/columns: Word lengths vary. Child-development stages, education tips, environmental issues, teen issues, and reviews of parenting and children's books.
- Artwork: Color prints or transparencies.

SAMPLE ISSUE
48 pages (65% advertising): 8 articles; 7 depts/columns. Sample copy available at website. Guidelines and editorial calendar available.

- "Travel Made Easy." Article describes the Kidz Travelmate, which helps parents to tote carseats—and children—through airports.
- "Luminato." Article previews an international arts festival coming to Toronto.
- Sample dept/column: "Snapshots: Toddler" summarizes research on temper tantrums.

RIGHTS AND PAYMENT
First rights. Written material, $50–$100. Pays on publication. Provides 1 contributor's copy.

➥EDITOR'S COMMENTS
We are interested in feature articles, guest columns, and event-related articles that are tied in with a specific event such as a concert or seminar. We also look for material that is based on one of our specified themes, such as summer camp or birthday parties.

The Claremont Review

4980 Wesley Road
Victoria, British Columbia V8Y 1Y9
Canada

The Editors

DESCRIPTION AND INTERESTS
The editors of *The Claremont Review* publish first-class poetry and short stories by writers ages 13 through 19 living anywhere in the English-speaking world. Circ: 500.

Audience: 13–19 years
Frequency: Twice each year
Website: www.theclaremontreview.ca

FREELANCE POTENTIAL
99% written by nonstaff writers. Publishes 150 freelance submissions yearly; 90% by unpublished writers, 90% by authors who are new to the magazine. Receives 540 unsolicited mss yearly.

SUBMISSIONS
Send complete ms with biography. Accepts hard copy. SAE/IRC. Responds in 2–6 weeks.

- Articles: Word lengths vary. Interviews with contemporary authors and editors.
- Fiction: To 5,000 words. Genres include traditional, literary, experimental, and contemporary fiction.
- Artwork: B/W or color prints or transparencies.
- Other: Poetry, no line limit. Plays.

SAMPLE ISSUE
136 pages (2% advertising): 1 article; 1 play; 11 stories; 52 poems. Sample copy, $10 with 9x12 SAE/IRC. Guidelines available at website and inside each issue.

- "Bob Can Hope." Story tells of a strange, anxious man and his many quirks.
- "Lights! Lights! O the Lights and the Lies!" One-act play recounts a philosophical conversation between two college students.
- "Interview with Carla Funk." Q&A with Victoria's first poet laureate discusses her writing.

RIGHTS AND PAYMENT
Rights vary. No payment. Provides 1 copy.

➥EDITOR'S COMMENTS
We publish anything from traditional to postmodern, but with a preference for works that reveal something of the human condition. The editorial staff runs our magazine on a volunteer basis, and all of our editors are writers themselves. We not only publish successful writers, we publish those working toward success. All submissions accompanied by an SAE/IRC receive written comments.

The Clearing House

Heldref Publications
1319 18th Street
Washington, DC 20036

Managing Editor: Melanie Bonsall

DESCRIPTION AND INTERESTS

Informative and practical articles on teaching and administration in middle schools and high schools are found in this magazine. In peer-reviewed articles, educators report their successes in teaching as well as their administrative procedures, school programs, and teacher education practices. Circ: 1,500.

Audience: Educators
Frequency: 6 times each year
Website: www.heldref.org/tch.php

FREELANCE POTENTIAL

100% written by nonstaff writers. Publishes 65 freelance submissions yearly; 60% by authors who are new to the magazine. Receives 121 unsolicited mss yearly.

SUBMISSIONS

Send complete ms. Accepts online submissions at http://mc.manuscriptcentral.com/tch. Responds in 3–4 months.

- Articles: To 2,500 words. Informational and how-to articles. Topics include educational trends and philosophy, pre-service and in-service education, curriculum, learning styles, discipline, guidance and counseling, gifted and special education, teaching techniques, educational testing and measurement, and technology.
- Depts/columns: Word lengths vary. Education news, opinion pieces, and book reviews.

SAMPLE ISSUE

42 pages (no advertising): 9 articles; 1 dept/column. Sample copy, $19. Guidelines available in each issue.

- "Who Is No Child Left Behind Leaving Behind?" Article reviews the effects of this federal legislation.
- "Helping Teachers Become Leaders." Article discusses ways to support teachers in leadership roles.
- Sample dept/column: "Book Review" looks at a title on "why right-brainers will rule the future."

RIGHTS AND PAYMENT

All rights. No payment. Provides free online access for life and discounted author's copies.

➤EDITOR'S COMMENTS

We also publish a few first-person accounts and opinion pieces on controversial issues.

Cleveland Family

35475 Vine Street, Suite 224
Eastlake, OH 44095-3147

Editor: Terri Nighswonger

DESCRIPTION AND INTERESTS

Child-rearing and development, health, and discipline are all covered in this regional magazine for parents in Ohio's Cuyahoga and Lorain counties. It is also a resource for area events and activities. Circ: 70,000.

Audience: Parents
Frequency: Monthly
Website: www.neohiofamily.com

FREELANCE POTENTIAL

50% written by nonstaff writers. Publishes 40–50 freelance submissions yearly; 33% by authors who are new to the magazine. Receives 9,000+ queries yearly.

SUBMISSIONS

Query. Accepts email queries to editor@tntpublications.com. Responds if interested.

- Articles: 500+ words. Informational, self-help, and how-to articles; profiles; and reviews. Topics include the arts, animals, computers, crafts, health, fitness, education, popular culture, sports, the environment, religion, family travel, and regional issues.
- Depts/columns: Word lengths vary. News, advice, education, teen issues, humor, and stepfamilies.
- Artwork: High resolution JPEG and TIFF files.

SAMPLE ISSUE

42 pages (50% advertising): 5 articles; 8 depts/columns. Editorial calendar available.

- "How Dads Can Make a Difference." Article explains research that proves fathers shape the competence and character of their kids.
- "Raising a Child with a Competitive Spirit." Article looks at ways to keep a child's view of competition in perspective.
- Sample dept/column: "Family Health" discusses night terrors and how to prevent and cope with them.

RIGHTS AND PAYMENT

Exclusive rights. Written material, payment rates vary. Pays on publication. Provides 1 contributor's copy.

➤EDITOR'S COMMENTS

We prefer to work with local writers. Our magazine focuses on issues that affect families with children from birth to age 18. We are looking for factual articles with practical information, rather than first-person pieces.

Click

Carus Publishing
70 East Lake Street, Suite 300
Chicago, IL 60601

Editor

DESCRIPTION AND INTERESTS
For more than a decade, *Click* has served to introduce very young children to the world of ideas. Each issue of *Click* explores a single theme in both fiction and nonfiction pieces, accompanied by colorful illustrations and photographs. Circ: 62,000.

Audience: 3–7 years
Frequency: 9 times each year
Website: www.cricketmag.com

FREELANCE POTENTIAL
80% written by nonstaff writers. Of the freelance submissions published yearly, 10% are by authors who are new to the magazine. Receives 48–60 queries yearly.

SUBMISSIONS
Send résumé and clips. All material is commissioned from experienced authors.

- Articles: To 1,000 words. Informational articles; interviews; and photo-essays. Topics include the natural, physical, and social sciences; the arts; technology and science; math; and history.
- Other: Poetry, cartoons, and activities.

SAMPLE ISSUE
36 pages (no advertising): 3 articles; 2 stories; 2 cartoons; 2 activities. Sample copy available for ordering at website.

- "Life On the Ice." Article describes what goes on at Earth's polar ice caps, and why they are called lands of peace and science.
- "Keeping Warm." Article explores the ways that various animals protect themselves from extreme temperatures.
- "Finn's Just Right World." Story follows a family on their trip to a conservatory in the middle of winter, where they discuss global warming and what they can do about it.

RIGHTS AND PAYMENT
Rights vary. Written material, payment rates vary. Payment policy varies.

➥EDITOR'S COMMENTS
All submissions should clearly address a specific theme in depth. Articles should be age-appropriate while factually correct; and stories should contain and explain the nonfiction idea within them in a friendly way. Prospective authors are encouraged to study past issues to understand our expectations.

Cobblestone

Cobblestone Publishing
30 Grove Street, Suite C
Peterborough, NH 03458

Editor: Meg Chorlian

DESCRIPTION AND INTERESTS
Cobblestone magazine helps children discover American history through lively approaches to its subjects. All material in each issue relates to a particular theme. Circ: 27,000.

Audience: 8–14 years
Frequency: 9 times each year
Website: www.cobblestonepub.com

FREELANCE POTENTIAL
85% written by nonstaff writers. Publishes 180 freelance submissions yearly; 20% by unpublished writers, 25% by authors who are new to the magazine. Receives 350 queries yearly.

SUBMISSIONS
Query with outline, bibliography, and clips or writing samples. All queries must relate to upcoming themes. Accepts hard copy. SASE. Responds in 5 months.

- Articles: Features, 700–800 words. Sidebars, 300–600 words. Informational articles; profiles; and interviews. Topics include American history and historical figures.
- Fiction: To 800 words. Genres include historical, multicultural, and biographical fiction; adventure; and retold legends.
- Artwork: Color prints or slides. Line art.
- Other: Puzzles, activities, and games; to 500 words. Poetry, to 100 lines.

SAMPLE ISSUE
48 pages (no advertising): 12 articles; 7 depts/columns; 2 activities. Sample copy, $5.95 with 10x13 SASE ($2 postage). Guidelines and theme list available.

- "Shays Stirs Things Up." Article describes the efforts of Daniel Shays to get the Articles of Confederation revoked.
- Sample dept/column: "The Past Is Present" explains the various places the original Declaration of Independence, the Constitution, and the Bill of Rights have called home.

RIGHTS AND PAYMENT
All rights. Written material, $.20–$.25 per word. Artwork, payment rates vary. Pays on publication. Provides 2 contributor's copies.

➥EDITOR'S COMMENTS
Please obtain a list of our upcoming themes before querying, as everything in an issue must relate to that theme. Please see back issues before contacting us.

College Outlook

20 East Gregory Boulevard
Kansas City, MO 64114-1145

Editor: Kellie Houx

DESCRIPTION AND INTERESTS
Written for high school students preparing for college, this magazine provides information on the college application and selection processes as well as career planning advice and profiles. Circ: 440,000 (spring); 710,000 (fall).

Audience: College-bound students
Frequency: Twice each year
Website: www.collegeoutlook.net

FREELANCE POTENTIAL
40% written by nonstaff writers. Publishes 4–5 freelance submissions yearly; 10% by unpublished writers, 20% by authors who are new to the magazine. Receives 5 queries yearly.

SUBMISSIONS
Query with clips or writing samples. Accepts hard copy. Availability of artwork improves chance of acceptance. SASE. Responds in 1 month.

- Articles: To 1,500 words. Informational and how-to articles; personal experience pieces; and humor. Topics include school selection, financial aid, scholarships, student life, extracurricular activities, money management, and college admissions procedures.
- Artwork: 5x7 B/W and color transparencies.
- Other: Gazette items on campus subjects, including fads, politics, classroom news, current events, leisure activities, and careers.

SAMPLE ISSUE
40 pages (15% advertising): 11 articles. Sample copy, free. Guidelines available.

- "College Admissions Essays: What's the Best Approach?" Article offers tips for writing an essay that will stand out from the rest.
- "Student Learns to Step Out of the Shadows." Personal experience piece tells of a boy's quest to come out from under his older brother's shadow, both on the basketball court and in life.

RIGHTS AND PAYMENT
All rights. All material, payment rates vary. Provides contributor's copies.

➙EDITOR'S COMMENTS
We are looking for entertaining, easy-to-read, and informative pieces that are not biased toward a particular school or program. Writers should have a strong knowledge of their subject matter.

Columbus Parent

7801 North Central Drive
Lewis Center, OH 43035

Editor: Staci Perkins

DESCRIPTION AND INTERESTS
Columbus Parent is available free of charge to parents in the greater Columbus region of Ohio. It contains parenting advice, humor, and articles on topics of interest to anyone caring for children. It also serves as a resource for recreational, educational, and family-based activities. Circ: 125,000.

Audience: Parents
Frequency: Monthly
Website: www.columbusparent.com

FREELANCE POTENTIAL
100% written by nonstaff writers. Publishes 48 freelance submissions yearly; 25% by authors who are new to the magazine. Receives 100 queries yearly.

SUBMISSIONS
Query. Accepts email to columbusparent@ thisweeknews.com. Response time varies.

- Articles: 700 words. Informational, self-help, and how-to articles; profiles; interviews; and reviews. Topics include current events, health, humor, music, recreation, and travel.
- Fiction: 300 words. Humorous stories.
- Depts/columns: 300 words. Local events and people, food, health, book reviews, and travel destinations.
- Artwork: Color prints and transparencies.
- Other: Submit seasonal material 2 months in advance.

SAMPLE ISSUE
56 pages (50% advertising): 5 articles; 16 depts/columns. Sample copy, free. Guidelines and theme list available.

- "20 Things Every Dad Should Know." Article lists 20 important pieces of advice for fathers and fathers-to-be.
- Sample dept/column: "Pediatric Health Source" tells of a boy living with hemophilia.

RIGHTS AND PAYMENT
Rights vary. Written material, $.10–$.20 per word. Pays on publication.

➙EDITOR'S COMMENTS
Because our content is localized to the central Ohio region, we prefer to work with local writers. All information should be backed up by professional resources. If you want to get our attention, offer an idea unlike anything found in other parenting tabloids.

Complete Woman

875 North Michigan Avenue, Suite 3434
Chicago, IL 60611-1901

Executive Editor: Lora Wintz

DESCRIPTION AND INTERESTS
Complete Woman wants to be every woman's friend by offering concise, informative, and entertaining articles written in an honest, approachable tone. Circ: 875,000.

Audience: Women
Frequency: 6 times each year
Website: www.asspub.vflex.com

FREELANCE POTENTIAL
90% written by nonstaff writers. Publishes 75 freelance submissions yearly; 20% by unpublished writers, 30% by authors who are new to the magazine. Receives 720 queries each year.

SUBMISSIONS
Query with clips; or send complete ms. Accepts hard copy and simultaneous submissions if identified. SASE. Responds in 3 months.

- Articles: 800–1,200 words. Self-help articles; confession and personal experience pieces; humor; profiles; and interviews. Topics include health, fitness, beauty, skin care, fashion, dining, relationships, romance, sex, business, self-improvement, and celebrities.
- Depts/columns: Word lengths vary. Careers, new products, beauty tips, and news briefs.

SAMPLE ISSUE
106 pages (15% advertising): 15 articles; 28 depts/columns. Sample copy, $3.99 at newsstands.

- "All About Eva." Article profiles actress Eva Longoria of *Desperate Housewives* fame.
- "How I Started My Own Green-Themed Business." Article features interviews with two women who have blended eco-friendly practices with entrepreneurship.
- Sample dept/column: "Career Wise" includes an interview with recording artist Fergie about her newest line of handbags and accessories.

RIGHTS AND PAYMENT
Rights vary. Written material, payment rates vary. Pays on publication. Provides 1 copy.

➥EDITOR'S COMMENTS
Writers should be able to present useful information in an honest, conversational way that is easy to read during a busy woman's lunch hour, on the train after a hard day at work, or before going to sleep at night. We're not interested in overwhelming her with heavy reading.

Conceive Magazine

622 East Washington Street, Suite 440
Orlando, FL 32801

Editorial Director: Beth Weinhouse

DESCRIPTION AND INTERESTS
This magazine is written for women who are trying to conceive, either naturally or with assisted reproductive technology. It also covers adoption issues. Circ: 200,000.

Audience: Women
Frequency: Quarterly
Website: www.conceiveonline.com

FREELANCE POTENTIAL
75% written by nonstaff writers. Publishes 45 freelance submissions yearly; 5% by unpublished writers, 10% by authors who are new to the magazine. Receives 1,200 queries yearly.

SUBMISSIONS
Query with résumé and clips. Accepts hard copy and email queries to bethweinhouse@ conceivemagazine.com. SASE. Response time varies.

- Articles: Word lengths vary. Informational articles; profiles; and interviews. Topics include family planning, adoption, infertility issues, and baby products.
- Depts/columns: Word lengths vary. Topics include fitness, health, medical updates, beauty, advice from counselors, and personal experiences.

SAMPLE ISSUE
96 pages: 4 articles; 13 depts/columns. Sample copy, $4.99 at newsstands. Guidelines available at website.

- "Cancer Survivors Can." Article explains the options that exist for preserving fertility after cancer treatment.
- "Taking the Long Way to Motherhood." Article profiles two of the women of the music group The Dixie Chicks and their long journey to parenthood.
- Sample dept/column: "A Family Is Born" tells how triplets were born to two sisters.

RIGHTS AND PAYMENT
All or first rights. Written material, $.50–$1 per word. Kill fee varies. Pays on publication.

➥EDITOR'S COMMENTS
We are particularly interested in seeing more articles about how infertility affects couple relationships. Unfortunately, due to staffing and other issues, there may be fewer opportunities for freelancers.

Connect for Kids

Forum for Youth Investment
The Cady-Lee House
7064 Eastern Avenue NW
Washington, DC 20012

Editor: Caitlyn Johnson

DESCRIPTION AND INTERESTS
Connect for Kids believes in putting children first, and is read by like-minded individuals such as parents, youth development workers, and other professionals who work with children. It is an online forum that covers all topics that relate to modern youth, such as health, education, social trends, legislation, and current events. Hits per month: 50,000+.

Audience: Parents; child welfare professionals
Frequency: 26 times each year
Website: www.connectforkids.org

FREELANCE POTENTIAL
40% written by nonstaff writers. Publishes 24 freelance submissions yearly; 25% by authors who are new to the magazine. Receives 150 queries yearly.

SUBMISSIONS
Query. Accepts email queries to caitlin@ connectforkids.org. Response time varies.

- Articles: 900–1,500 words. Informational articles; profiles; reviews; and photo-essays. Topics include adoption, foster care, the arts, child abuse and neglect, health, education, child care and early development, kids and politics, community building, learning disabilities, crime and violence prevention, parent involvement in education, out-of-school time, diversity and awareness, education, family income, volunteering, and mentoring.

SAMPLE ISSUE
Sample issue and writers' guidelines available at website.

- "We Are Speaking Up." Article reports on parents who are becoming activists in the fight for more state-subsidized child care funding.
- "Geography Matters for Child Well-Being." Article reports on a study that found the states with the best health and well-being outcomes from youth are those that invest the most in children's programs.

RIGHTS AND PAYMENT
First rights. No payment.

➵EDITOR'S COMMENTS
Articles must give site visitors the information and tools they need to learn about issues that affect children, families, and communities, and how to take action to improve programs and policies.

Connecticut Parent

420 East Main Street, Suite 18
Branford, CT 06405

Editor & Publisher: Joel MacClaren

DESCRIPTION AND INTERESTS
This regional publication offers parents plenty of useful information on topics such as health and fitness, education, and child-rearing issues, as well as on local attractions and activities for Connecticut families. Circ: 60,000.

Audience: Parents
Frequency: Monthly
Website: www.ctparent.com

FREELANCE POTENTIAL
20% written by nonstaff writers. Publishes 50 freelance submissions yearly; 10% by authors who are new to the magazine. Receives 1,000+ unsolicited mss yearly.

SUBMISSIONS
Send complete ms. Prefers email submissions to ctparent@aol.com; will accept hard copy. SASE. Response time varies.

- Articles: 500–1,000 words. Informational, self-help, and how-to articles; profiles; and interviews. Topics include maternity and childbirth issues, parenting, regional news, family relationships, social issues, education, special education, health, fitness, nutrition, safety, entertainment, and travel.
- Depts/columns: 600 words. Family news, new product information, and media reviews.

SAMPLE ISSUE
68 pages (60% advertising): 7 articles; 8 depts/columns. Sample copy, $5 with 9x12 SASE. Guidelines available.

- "Birthing Issues." Article outlines the pros and cons of patient-choice C-sections.
- "Adult Education." Article explains the family issues that parents need to consider before returning to school.
- Sample dept/column: "Family Fun Calendar" lists the many family-friendly activities, including musical performances and seasonal hikes, in the area.

RIGHTS AND PAYMENT
One-time rights. Written material, payment rates vary. Pays on publication. Provides 1 tearsheet.

➵EDITOR'S COMMENTS
All submissions should be relevant to our target audience—Connecticut parents and families. We would like to see topics covered by writers with a certain level of expertise and/or experience.

Connecticut's County Kids

1175 Post Road East
Westport, CT 06880

Editor: Linda Greco

DESCRIPTION AND INTERESTS
For more than 20 years, this regional tabloid has served parents living in New Haven and Fairfield counties in Connecticut. It features articles on family and parenting issues, and serves as a source of information about area activities, news, and events. Circ: 30,000.
Audience: Parents
Frequency: Monthly
Website: www.countykids.com

FREELANCE POTENTIAL
90% written by nonstaff writers. Publishes 60 freelance submissions yearly; 10% by unpublished writers, 25% by authors who are new to the magazine. Receives 1,200 queries, 960 unsolicited mss yearly.

SUBMISSIONS
Query or send complete ms. Prefers email submissions to countykids@ctcentral.com; will accept hard copy. SASE. Responds only if interested.

- Articles: 600–1,200 words. Informational articles; profiles; and personal experience pieces. Topics include nature, animals, crafts, ethnic subjects, and sports.
- Depts/columns: 500–800 words. Pediatric health, growth, and development; and family and parenting issues.

SAMPLE ISSUE
32 pages (50% advertising): 13 articles; 13 depts/ columns. Sample copy, free with 10x13 SASE. Writers' guidelines and editorial calendar available.

- "Caution Is a Friend to Fido When Temps Take a Dive." Article suggests ways to protect pets during cold weather.
- Sample dept/column: "Reading Corner" reviews six chapter books for elementary school-aged children.

RIGHTS AND PAYMENT
First rights. Written material, payment rates vary. Pays on publication. Provides 2 copies.

➥EDITOR'S COMMENTS
Prospective authors should consult our editorial calendar for themes of upcoming issues. We like good photos and urge writers to include them whenever possible. We will not use articles that have already appeared in other Connecticut publications.

Countdown for Kids

Juvenile Diabetes Research Foundation
120 Wall Street, 19th Floor
New York, NY 10005

Submissions Editor: Jason Dineen

DESCRIPTION AND INTERESTS
Children with Type 1 diabetes and their families turn to this magazine for easy-to-read information and stories on the social and emotional issues of dealing with the illness. It is published by the Juvenile Diabetes Research Foundation in conjunction with *Countdown to a Cure*. Circ: Unavailable.
Audience: 10+ years
Frequency: Quarterly
Website: www.jdrf.org

FREELANCE POTENTIAL
50% written by nonstaff writers. Publishes 6–8 freelance submissions yearly; 10% by authors who are new to the magazine. Receives 120 queries and unsolicited mss yearly.

SUBMISSIONS
Query or send complete ms. Accepts hard copy. SASE. Response time varies.

- Articles: Word lengths vary. Informational, factual, and self-help articles; personal experience pieces; profiles; and interviews. Topics include coping with Type 1 diabetes, health, fitness, careers, college, popular culture, social issues, and diabetes research.
- Depts/columns: Word lengths vary. Diabetes news and information; career profiles; and advice.

SAMPLE ISSUE
12 pages (1% advertising): 3 articles; 2 depts/columns. Sample copy available.

- "Tuned In." Article interviews American Idol finalist Elliott Yamin on his career and his involvement with a diabetes campaign.
- "On Her Toes." Article profiles a teenage girl managing a pre-professional dance career, school, and diabetes.
- "The Power of Now." Article features two brothers who perform circus acts around the country, and includes tips for dealing with diabetes while following dreams.

RIGHTS AND PAYMENT
First North American serial rights. Written material, payment rates vary. Pays on publication. Provides 1 contributor's copy.

➥EDITOR'S COMMENTS
We welcome inspirational articles that help readers deal with issues surrounding diabetes, as well as current research information.

Cousteau Kids

Weekly Reader
1 Reader's Digest Road
Pleasantville, NY 10570

Editor

DESCRIPTION AND INTERESTS
Published by The Cousteau Society and Weekly Reader, this magazine introduces kids to the wonders that exist under water. With stewardship for the planet as a constant theme, readers learn about the oceans, marine life, and Earth's water systems. Circ: 80,000.

Audience: 8–12 years
Frequency: 6 times each year
Website: www.cousteaukids.org

FREELANCE POTENTIAL
10% written by nonstaff writers. Publishes 4 freelance submissions yearly; 50% by authors who are new to the magazine. Receives 48 queries yearly.

SUBMISSIONS
Query. Accepts hard copy and simultaneous submissions if identified. SASE. Responds only if interested.

- Articles: 400–600 words. Shorter pieces, to 250 words. Informational articles. Topics include aquatic organisms, underwater habitats, ocean phenomena, the environment, and the physical properties of water.
- Depts/columns: Staff written.
- Artwork: Color slides.
- Other: Games based on scientific fact, original science experiments, and art projects related to an ocean theme.

SAMPLE ISSUE
24 pages (no advertising): 5 articles; 4 depts/columns; 2 activity pages. Sample copy, $2.50 with 9x12 SASE (3 first-class stamps). Guidelines available.

- "Siren Song." Article profiles the manatee and explains how humans are causing the gentle sea mammals' numbers to decline.
- "The Wonderful, Weird World of Sea Dragons." Article introduces readers to two types of sea dragons.

RIGHTS AND PAYMENT
One-time reprint rights; worldwide translation rights for use in Cousteau Society publications. Articles, $100–$350. Short pieces, $15–$100. Pays on publication. Provides 3 author's copies.

⟡EDITOR'S COMMENTS
Our readers are young, and we want our articles to draw them in and spark a love for the sea and its creatures.

Creative Connections

P.O. Box 98037
135 Davie Street
Vancouver, British Columbia V6Z 2Y0
Canada

Editor: Kalen Marquis

DESCRIPTION AND INTERESTS
Formerly listed as *Mr. Marquis' Museletter*, this literary newsletter publishes articles, essays, fiction, reviews, and poems that tackle psychology, sociology, philosophy, and spirituality in the broadest sense. Circ: 150.

Audience: 2–21 years
Frequency: Quarterly
**Website: http://creative-connections.
    spaces.live.com**

FREELANCE POTENTIAL
80% written by nonstaff writers. Publishes 40 freelance submissions yearly; 75% by unpublished writers, 75% by authors who are new to the magazine. Receives 120 queries, 360 unsolicited mss yearly.

SUBMISSIONS
Query with writing samples; or send complete ms. Accepts hard copy, email submissions to creative-connections@hotmail.com, and simultaneous submissions if identified. SAE/IRC. Responds in 4 months.

- Articles: 300 words. Informational articles; opinion and personal experience pieces; profiles; interviews; and book reviews. Topics include nature, the arts, current events, history, multicultural and ethnic issues, music, and popular culture. Also publishes biographies of painters, writers, and inventors.
- Fiction: 300 words. Genres include adventure; and contemporary, historical, multicultural, and inspirational fiction.
- Artwork: Line art.
- Other: Poetry, 4–16 lines. Accepts seasonal material 6 months in advance.

SAMPLE ISSUE
10 pages (no advertising): 3 articles; 1 story; 8 poems. Sample copy, $2 with #10 SAE/IRC. Guidelines available.

- "On Being Afraid." Essay tells of a girl who overcomes her fear of a dog.
- "For the Love of Learning." Articles offers tips to increase learning and memory skills.

RIGHTS AND PAYMENT
One-time rights. No payment. Provides 1 copy.

⟡EDITOR'S COMMENTS
We will be expanding our mandate to include more wisdom/wonder/wellness topics for all ages. Your thoughts and input are welcome.

Creative Kids

Prufrock Press
P.O. Box 8813
Waco, TX 76714

Editor: Lacy Compton

DESCRIPTION AND INTERESTS

A creative outlet for children, this magazine publishes their original essays, stories, songs, poetry, plays, and artwork. Circ: 3,600.

Audience: 8–16 years
Frequency: Quarterly
Website: www.prufrock.com

FREELANCE POTENTIAL

99% written by nonstaff writers. Publishes 150 freelance submissions yearly; 75% by unpublished writers, 75% by authors who are new to the magazine. Receives 1,800–3,600 unsolicited mss yearly.

SUBMISSIONS

Send complete ms. Accepts hard copy with author's birthday, grade, and school. No adult submissions. SASE. Responds 4–6 weeks.

- Articles: 500–1,200 words. Informational, self-help, and how-to articles; essays; photo-essays; humor; and personal experience pieces. Topics include pets, sports, social issues, travel, and gifted education.
- Fiction: 500–1,200 words. Genres include realistic, inspirational, historical, and multicultural fiction; mystery; suspense; folktales; humor; and problem-solving stories. Also publishes plays.
- Depts/columns: Word lengths vary. Short opinion pieces.
- Artwork: B/W line drawings. Color copies of paintings, colored-pencil sketches, and collages. B/W and color glossy photos.
- Other: Poetry and songs. Puzzles, games, and cartoons. Submit seasonal material 1 year in advance.

SAMPLE ISSUE

34 pages (no advertising): 4 articles; 6 stories; 1 dept/column; 11 poems; 2 cartoons; 7 activities. Sample copy and guidelines available at website.

- "Window of Time." Story tells of a boy who investigates a mysterious noise in his attic.
- Sample dept/column: "Write On" is an editorial about parents limiting video-game time.

RIGHTS AND PAYMENT

Rights vary. No payment. Provides 1 copy.

➥EDITOR'S COMMENTS

We are looking for the very best material by students. All work must be original, and may be submitted by the author, parent, or teacher.

Cricket

Carus Publishing
70 East Lake Street, Suite 300
Chicago, IL 60601

Submissions: Deborah Vetter

DESCRIPTION AND INTERESTS

This literary magazine for young readers publishes original fiction, poems, and articles written by some of the most recognized authors for children, as well as up-and-coming writers. Circ: 55,000.

Audience: 9–14 years
Frequency: 9 times each year
Website: www.cricketmag.com

FREELANCE POTENTIAL

100% written by nonstaff writers. Publishes 100 freelance submissions yearly; 30% by unpublished writers, 50% by new authors. Receives 12,000 unsolicited mss yearly.

SUBMISSIONS

Send complete ms; include bibliography for nonfiction. Accepts hard copy and simultaneous submissions if identified. SASE. Responds in 4–6 months.

- Articles: 200–1,500 words. Informational and how-to articles; biographies; and profiles. Topics include science, art, technology, history, architecture, geography, foreign culture, adventure, and sports.
- Fiction: 200–2,000 words. Genres include humor, mystery, fantasy, science fiction, folktales, fairy tales, mythology, and historical and contemporary fiction.
- Depts/columns: Staff written.
- Other: Poetry, to 25 lines. Puzzles, games, crafts, recipes, and science experiments; word lengths vary.

SAMPLE ISSUE

48 pages (no advertising): 1 article; 6 stories; 2 depts/columns; 3 poems; 1 activity. Sample copy, $5 with 9x12 SASE. Guidelines available at website.

- "The Mothers of Mother's Day." Article explains the history of Mother's Day.
- "Before the Week's Through." Story portrays a few months in the life of a farm family struggling to survive in the Dust Bowl.

RIGHTS AND PAYMENT

Rights vary. Articles and fiction, to $.25 per word. Poetry, to $3 per line. Pays on publication. Provides 6 contributor's copies.

➥EDITOR'S COMMENTS

We don't have a theme list. Appropriate submissions are considered any time of the year.

Crinkles

3520 South 35th Street
Lincoln, NE 68506

Editor: Deborah D. Levitov

DESCRIPTION AND INTERESTS
So named "because learning makes crinkles in your brain," this magazine encourages children to discover new places, people, events, and things through the use of library resources. Each issue focuses on a particular topic, with features that teach readers to apply problem-solving and critical-thinking skills. Circ: 6,000.

Audience: 7–12 years
Frequency: 6 times each year
Website: www.crinkles.com

FREELANCE POTENTIAL
70% written by nonstaff writers. Publishes 2–3 freelance submissions yearly; 10% by unpublished writers, 50% by authors who are new to the magazine. Receives 36 queries each year.

SUBMISSIONS
Query with résumé. Accepts email queries to deborah.levitov@lu.com. Responds in 1 month.

- Articles: Word lengths vary. Informational, factual, and how-to articles. Topics include history, geography, multicultural and ethnic subjects, social issues, science, animals, nature, the arts, and sports.
- Other: Puzzles, mazes, games, and crafts.

SAMPLE ISSUE
52 pages (no advertising): 11 articles; 12 activities. Guidelines available at website.

- "Children's Day in Turkey." Article describes the national holiday that celebrates Turkey's youngest citizens.
- "The Home of St. Nicholas." Article presents a brief biography of the real St. Nicholas, who was born in Turkey.
- "Weave a Story Puzzle." Activity challenges children to piece together a Turkish story.

RIGHTS AND PAYMENT
All rights. Written material, $150. Payment policy varies. Provides contributor's copies upon request.

➡ EDITOR'S COMMENTS
We welcome unsolicited articles and reader contributions to all departments (People, Places, Events, and Things). Please note that unsolicited manuscripts will not be returned. Prompt notification of publication will be made to all contributors, as will appropriate attribution of all contributions.

Crow Toes Quarterly

186-8120 No. 2 Road, Suite 361
Richmond, British Columbia V7C 4C1
Canada

Managing Editor: Christopher Millin

DESCRIPTION AND INTERESTS
Established in 2006, this magazine publishes "playfully dark," intelligent, descriptive literature written for children. It also sponsors a contest in which children can get their own stories published. Circ: 1,000.

Audience: 8–13 years
Frequency: Quarterly
Website: www.crowtoesquarterly.com

FREELANCE POTENTIAL
90% written by nonstaff writers. Publishes 24 freelance submissions yearly; 90% by unpublished writers, 100% by authors who are new to the magazine. Receives 180 unsolicited mss each year.

SUBMISSIONS
Send complete ms. Accepts hard copy. SASE. Responds in 4 months.

- Articles: Word lengths vary. Informational articles. Topics includes the arts and writing.
- Fiction: To 3,000 words. Genres include horror, humor, mystery, suspense, folktales, folklore, fantasy, and science fiction.
- Artwork: B/W and color JPEG images at 300 dpi. Line art.
- Other: Poetry. Seasonal material.

SAMPLE ISSUE
36 pages (no advertising): 5 stories; 3 poems. Sample copy, $7 ($6 Canadian) with 9x12 SAE/IRC. Guidelines available at website.

- "The Baby Sitter's Story." Story features a babysitter telling her young wards a ghost story that spooks them.
- "The Perfect Pitch." Story finds a young boy battling a ghostly monster in the barn with help from his deceased brother.

RIGHTS AND PAYMENT
First Canadian serial rights. No payment. Provides 3 contributor's copies.

➡ EDITOR'S COMMENTS
We are not interested in anything that goes outside *Crow Toes Quarterly*'s style. This can be subjective, but after looking through an issue and visiting our website, it should become apparent what type of writing gets our attention. Please be sure to send only your most carefully edited work. Anything mired in mistakes or unclear writing will be discarded.

Curious Parents

301 North Church Street, Suite 226
Moorestown, NJ 08057

Editor: Matt Stringer

DESCRIPTION AND INTERESTS
The mission of *Curious Parents* is at once simple and necessary: to provide local information for inspired parents in Pennsylvania, New Jersey, and Delaware, and to make them aware of local resources. It features articles that cover family issues, successful parenting, recreation, healthy living, and local entertainment. Circ: 265,000.

Audience: Parents
Frequency: Monthly
Website: www.curiousparents.com

FREELANCE POTENTIAL
60% written by nonstaff writers. Publishes 70–100 freelance submissions yearly; many by unpublished writers and authors who are new to the magazine.

SUBMISSIONS
Send complete ms with brief description and brief author biography. Accepts email submissions to editor@curiousparents.com. Response time varies.

- Articles: Word lengths vary. Informational, how-to, and self-help articles. Topics include crafts, hobbies, current events, recreation, special education, safety, health, family entertainment, networking, parenting, and travel.
- Depts/columns: Word lengths vary. Health issues and automobile safety; book reviews.

SAMPLE ISSUE
40 pages: 12 articles; 3 depts/columns. Sample copy, free with 9x12 SASE. Guidelines available at website.

- "Don't Be a Wet Blanket When it Comes to the Pressures of Potty Training." Article provides tips for determining when a toddler is ready to begin potty training.
- Sample dept/column: "Camp" supports the idea of a language immersion camp to help children learn a second language.

RIGHTS AND PAYMENT
All rights. No payment.

➻EDITOR'S COMMENTS
As a regional magazine, we prefer material that features an angle or source within our tri-state area. We favor writers from the region who can speak authoritatively about the subject at hand.

Current Health 1

Weekly Reader Publishing
1 Reader's Digest Road
Pleasantville, NY 10570-7000

Editor: Erin R. King

DESCRIPTION AND INTERESTS
Written for students in grades four through seven, *Current Health 1* deals with health, safety, and emotional topics in a lively manner to best reach its middle-grade audience. A wide range of contemporary issues is addressed, and each issue includes a teacher's guide. Circ: 163,973.

Audience: Grades 4–7
Frequency: 8 times each year
Website: www.weeklyreader.com/ch1

FREELANCE POTENTIAL
75% written by nonstaff writers. Publishes 40 freelance submissions yearly; 25% by authors who are new to the magazine. Receives 60 queries yearly.

SUBMISSIONS
All articles are assigned. No unsolicited mss. Query with letter of introduction, areas of expertise, publishing credits, and clips. Accepts email queries to currenthealth@weeklyreader.com. Responds in 1–4 months.

- Articles: 850–1,000 words. Informational articles. Topics include nutrition, fitness, disease prevention, drugs, alcohol, emotional well-being, and first aid.
- Depts/columns: Word lengths vary. Physical activities; health news and advice; safety tips; and summaries of medical research.

SAMPLE ISSUE
32 pages (no advertising): 6 articles; 8 depts/columns. Sample copy available. Guidelines provided upon agreement.

- "Salt at Fault." Article discusses the benefits of sodium in moderation and reviews which foods are good and bad sodium sources.
- "Ready or Not?" Article provides suggestions for dealing with peer pressure for those who are not ready to date.
- Sample dept/column: "Pulse" warns against the dangers of a new form of "meth."

RIGHTS AND PAYMENT
All rights. Articles, $150+. Provides 2 contributor's copies.

➻EDITOR'S COMMENTS
Material on the most current ideas, events, and issues is always of interest to us. We are specifically in need of pieces relevant to fourth- and fifth-grade students.

Current Health 2

Weekly Reader Publishing
1 Reader's Digest Road
Pleasantville, NY 10570

Editor: Meredith Matthews

DESCRIPTION AND INTERESTS
This classroom supplement is designed to give middle school and high school students age-appropriate, current information on general health issues in a way that is tailored to their interests. Circ: 195,000.
Audience: Grades 7–12
Frequency: 8 times each year
Website: www.weeklyreader.com/ch2

FREELANCE POTENTIAL
67% written by nonstaff writers. Publishes 36 freelance submissions yearly; 15% by unpublished writers, 30% by authors who are new to the magazine. Receives 24–36 queries each year.

SUBMISSIONS
Query with letter of introduction listing areas of expertise, publishing credits, clips, outline, and list of sources. No unsolicited mss. Accepts email queries to currenthealth@ weeklyreader.com. Responds in 1–4 months.

- Articles: 800–1,000 words. Informational articles on subjects related to middle school and high school curricula. Topics include fitness, exercise, nutrition, disease, psychology, first aid, safety, human sexuality, drug education, risk-taking behavior, relationships, and public health.
- Depts/columns: Word lengths vary. Health news, safety issues, and Q&As.

SAMPLE ISSUE
30 pages (1% advertising): 6 articles; 5 depts/columns. Sample copy available. Guidelines and editorial calendar available.

- "The Menu Maze." Article explains how to decode restaurant menus.
- "Breakout!" Article explains the causes of, and treatments for, acne.
- Sample dept/column: "Pulse" offers insight to a cause of writer's cramp: the brain.

RIGHTS AND PAYMENT
All rights. Articles, $.50 per word. Pays on publication. Provides 2 contributor's copies.

▪◆EDITOR'S COMMENTS
A query that would really get our attention right now would involve relationships and psychology/mental health issues. We are not currently looking for writers for our teen sexuality supplement.

The Dabbling Mum

508 West Main Street
Beresford, SD 57004

Editor & Owner: Alyice Edrich

DESCRIPTION AND INTERESTS
Referring to itself as a "publication for busy parents," this e-zine is dedicated to empowering parents of all types to realize their dreams. It features articles and inspirational essays. It recently went from a monthly to a weekly. Hits per month: 40,000.
Audience: Parents
Frequency: Weekly
Website: www.thedabblingmum.com

FREELANCE POTENTIAL
90% written by nonstaff writers. Publishes 100 freelance submissions yearly; 20% by unpublished writers, 60% by authors who are new to the magazine. Receives 240 queries each year.

SUBMISSIONS
Query with writing samples. Accepts online queries only. Responds in 8–16 weeks.

- Articles: 500–1,500 words. Informational and how-to articles and personal experience pieces. Topics include family life, parenting, women's issues, home businesses, sales and marketing, Christian living, marriage, entertainment, education, child development, teen issues, and contemporary social concerns.

SAMPLE ISSUE
Sample copy and writers' guidelines available at website.

- "Home Portraits." Article shows parents how to take professional-grade portraits of their children.
- "Let's Play Cards." Essay describes one family's passion for playing card games and discusses the social, math, and motor skills that can be learned.
- Sample dept/column: "Building Your Freelance Business" explains why joining a professional organization can help business.

RIGHTS AND PAYMENT
One-month exclusive online rights; indefinite archival rights. Written material, $20–$40; reprints, $5. Pays on acceptance.

▪◆EDITOR'S COMMENTS
We are currently interested in reviewing queries from industry experts, as well as pieces about how to successfully run specific home businesses.

Dallas Child

Lauren Publications
4275 Kellway Circle, Suite 146
Addison, TX 75001

Editorial Director: Shelley Hawes Pate

DESCRIPTION AND INTERESTS

This "ultimate parent guide" has been serving families in Dallas, Collin, and Denton counties in Texas for more than 20 years. It offers an informed, local, and relevant perspective on issues affecting area families. Circ: 80,000.

Audience: Parents
Frequency: Monthly
Website: www.dallaschild.com

FREELANCE POTENTIAL

30% written by nonstaff writers. Publishes 10–20 freelance submissions yearly; 5–10% by authors who are new to the magazine. Receives 396 queries yearly.

SUBMISSIONS

Query with résumé. Accepts hard copy, email to editorial@dallaschild.com, faxed queries to 972-447-0633, and simultaneous submissions if identified. SASE. Responds in 2–3 months.

- Articles: 1,000–2,000 words. Informational, self-help, and how-to articles; profiles; interviews; humor; and personal experience pieces. Topics include parenting, education, child development, family travel, regional news, recreation, entertainment, current events, social issues, multicultural and ethnic subjects, health, fitness, and crafts.
- Depts/columns: 800 words. Local events, travel tips, and health news.

SAMPLE ISSUE

112 pages (14% advertising): 2 articles; 25 depts/columns. Sample copy, free with 9x12 SASE. Guidelines available at website.

- "She Reminds Me of Me." Article categorizes celebrity mothers by personality type and invites readers to identify with one of them.
- Sample dept/column: "Mommy Diaries" profiles a day in the life of a local mother.

RIGHTS AND PAYMENT

First rights. Written material, payment rates vary. Pays on publication. Provides contributor's copies upon request.

➡️EDITOR'S COMMENTS

It is our goal to respond to all queries; however, due to the volume that we receive, this may not be possible. Please, no phone calls. Ideas are given the consideration they deserve. Please note that we prefer to work with writers from the Dallas–Fort Worth Metroplex.

Dance International

667 Davie Street
Vancouver, British Columbia V6B 2G6
Canada

Managing Editor: Maureen Riches

DESCRIPTION AND INTERESTS

Contemporary and classical dance in Canada and beyond is the topic of this magazine, published by the Vancouver Ballet Society. It reports on dancers and dance companies, and features news and reviews. Circ: 4,000.

Audience: YA–Adult
Frequency: Quarterly
Website: www.danceinternational.org

FREELANCE POTENTIAL

85% written by nonstaff writers. Publishes 95 freelance submissions yearly; 9% by authors who are new to the magazine.

SUBMISSIONS

Send complete ms. Accepts email submissions to danceint@direct.ca (attach file) and disk submissions (RTF files). SASE. Responds in 2 months.

- Articles: 1,500 words. Informational articles; profiles; interviews; opinion pieces; and media reviews—all related to dance.
- Depts/columns: 1,000 words. Commentaries; book and performance reviews.

SAMPLE ISSUE

62 pages (10% advertising): 7 articles; 20 depts/columns. Guidelines available.

- "Ballet in Bloom." Article profiles Ballet Kelowna, which at only five seasons old has become one of Vancouver's hottest cultural attractions.
- "Two Dancers, Two Stories." Article profiles two Royal Winnipeg Ballet dancers who have very different backgrounds.
- Sample dept/column: "View from Vancouver" reviews several dance performances held in the city.

RIGHTS AND PAYMENT

First rights. Articles, $100–$150. Depts/columns, $100. Kill fee, 50%. Pays on publication. Provides 2 contributor's copies.

➡️EDITOR'S COMMENTS

Our readers are well-versed in dance and dance performance. They, and we, expect all of our writers to have an extensive knowledge of dance, and to be able to convey that knowledge convincingly. Although we tend to focus on Canadian dance, dancers, and companies, we do cover dance and performances elsewhere in the world.

Dance Magazine

110 William Street, 23rd Floor
New York, NY 10038

Editor-in-Chief: Wendy Perron

DESCRIPTION AND INTERESTS

Read by dance teachers, choreographers, and serious students alike, *Dance Magazine* covers all aspects of dancing with how-to articles, show reviews, and profiles of dancers and dance companies. Circ: 50,000.

Audience: YA–Adult
Frequency: Monthly
Website: www.dancemagazine.com

FREELANCE POTENTIAL

80% written by nonstaff writers. Publishes 200 freelance submissions yearly; 5% by unpublished writers, 25% by authors who are new to the magazine. Receives many queries yearly.

SUBMISSIONS

Query. Accepts hard copy and email queries to wperron@dancemagazine.com. SASE. Response time varies.

- Articles: Word lengths vary. Informational articles; profiles; and interviews. Topics include dance, dance instruction, choreography, the arts, family, and health concerns.
- Fiction: To 4,000 words. Ethnic and multicultural fiction related to dance.
- Depts/columns: Word lengths vary. New product information, reviews, dance news, and instruction.

SAMPLE ISSUE

138 pages (33% advertising): 5 articles; 15 depts/columns. Sample copy, $4.95 with 9x12 SASE. Guidelines available.

- "Enter: Reality." Article follows one student dancer's journey into the professional arena.
- "Brave New Worlds." Article marks the milestone of the American Dance Festival as it celebrates its 75th year.
- Sample dept/column: "Teacher's Wisdom" interviews Eleanor D'Antuono, artistic director of the New York International Ballet Competition, to find out what she looks for in a young dancer.

RIGHTS AND PAYMENT

Rights negotiable. Written material, payment rates vary. Pays on publication.

☛EDITOR'S COMMENTS

Submissions must be from expert dance writers only. We are looking for material covering all types of dance, as well as current events, education, and strategies for success.

Dance Teacher

110 William Street, Floor 23
New York, NY 10038

Managing Editor: Jeni Tu

DESCRIPTION AND INTERESTS

Professional dance teachers and studio owners read this magazine for practical information and business tips. It is the sister publication of *Dance Magazine*. Circ: 25,000.

Audience: Dance teachers and students
Frequency: Monthly
Website: www.dance-teacher.com

FREELANCE POTENTIAL

67% written by nonstaff writers. Publishes 100–120 freelance submissions yearly; 10% by unpublished writers, 10–15% by authors who are new to the magazine. Receives 100 queries yearly.

SUBMISSIONS

Query. Accepts hard copy and email to jtu@dancemedia.com. SASE. Responds in 2 months.

- Articles: 1,000–2,000 words. Informational and how-to articles; and personal experience pieces. Topics include dance education, business, nutrition, health, injuries, performance production, competition, and dance personalities.
- Depts/columns: 700–1,200 words. Fashion, teaching, competition, dance history, media and product reviews, and industry news.

SAMPLE ISSUE

80 pages (50% advertising): 5 articles; 10 depts/columns. Sample copy, free with 9x12 SASE ($1.37 postage). Guidelines and theme list available.

- "Dynamic Duo." Article profiles ballroom dance partners Tony Meredith and Melanie LaPatin, who own a popular New York studio.
- "Finding the Right Balance." Article offers tips on how best to juggle artistic duties and the mundane aspects of owning a business.
- Sample dept/column: "Teaching" suggests popular jazz music selections that can be used for class and performances.

RIGHTS AND PAYMENT

All rights. Articles, $200–$300. Depts/columns, $150–$250. Pays on publication. Provides 1 contributor's copy.

☛EDITOR'S COMMENTS

We approach teaching dance as the business it is. We like articles that deal with business issues, as well as those that tackle the challenge of blending business and art.

Davey and Goliath's Devotions

Augsburg Fortress Publishers
P.O. Box 1209
Minneapolis, MN 55440-1209

Lead Editor: Becky Weaver Carlson

DESCRIPTION AND INTERESTS
Using the Lutheran Church's classic television characters, this quarterly devotional features weekly content designed to help families "on the go" explore and share faith together in fun, active, real-life ways. Circ: 50,000.

Audience: 3–9 years
Frequency: Quarterly
Website: www.augsburgfortress.org/dg/devotions

FREELANCE POTENTIAL
100% written by nonstaff writers. Publishes 8 freelance submissions yearly; 25% by unpublished writers, 50% by authors who are new to the magazine. Receives 20 queries yearly.

SUBMISSIONS
Query with sample content per guidelines at www.augsburgfortress.org/company/submitcongregational.jsp. Accepts email submissions to cllsub@augsburgfortress.org with "Family Devotions" in subject line. All work is assigned. Response time varies.

- Articles: 100–125 words. Bible stories, facts, and prayers.
- Other: Puzzles, mazes, activities, and games.

SAMPLE ISSUE
64 pages (no advertising): 44 depts/columns. Sample copy and guidelines available at website.

- "Sharing 24–7." Article explains how oil was used for anointing, and suggests children donate oil to the homeless.
- "A Brave Woman." Article retells the biblical story of Esther, who convinced her husband, the king, to rescind his order to kill all of the Jews.
- "Here Comes the Spirit." Article tells the story of Pentecost.
- "Bible Activity" encourages children to read Psalm 23, then illustrate the verses.

RIGHTS AND PAYMENT
All rights. Written material, payment rates vary. Pays on acceptance. Provides 2 contributor's copies.

➡EDITOR'S COMMENTS
Each week has four pages of content: The first two-page spread focuses on a Bible story, while the second spread's content is designed to support families as they share faith together and with others in the world.

Delmarva Youth Magazine

1226 North Division Street
Salisbury, MD 21801

Editor: Maria Cook

DESCRIPTION AND INTERESTS
Parents in Maryland's Delmarva Peninsula region read this magazine for information on parenting issues, family activities, and community resources. A calendar of events is included in each issue. Circ: 18,000.

Audience: Parents
Frequency: 6 times each year
Website: www.delmarvayouth.com

FREELANCE POTENTIAL
80% written by nonstaff writers. Publishes 60 freelance submissions yearly; 15% by unpublished writers, 20% by authors who are new to the magazine.

SUBMISSIONS
Query or send complete ms. Accepts email submissions to delmarvayouth@hotmail.com (Microsoft Word attachments). Response time varies.

- Articles: 500–3,000 words. Informational and how-to articles; and interviews. Topics include parenting, family life, family events and activities, travel, education, health, music, sports, and family finance.
- Depts/columns: Word lengths vary. School and camp news; family health and fitness; and the arts.

SAMPLE ISSUE
54 pages (15% advertising): 12 articles; 5 depts/columns; 1 calendar. Sample copy, $2.50. Guidelines available at website.

- "Toxic Plants + Children = Danger." Article explains which household and garden plants are poisonous if ingested.
- Sample dept/column: "Kids on Delmarva" features an area youth racquetball team.

RIGHTS AND PAYMENT
First print and electronic rights. Articles, $25–$150. Pays on publication. Provides 1 contributor's copy.

➡EDITOR'S COMMENTS
Because this is a regional publication, local writers are given preference, but all writers are welcome to submit material. Articles on making a working parent's life easier are of great interest to us as are articles written by fathers. Also welcome are humorous pieces about parenting, "how-to" articles, and practical information on things to see and do.

Devozine

1908 Grand Avenue
P.O. Box 340004
Nashville, TN 37203-0004

Editor: Sandy Miller

DESCRIPTION AND INTERESTS
Written for teens by teens, this devotional magazine features meditations, stories, and prayers based on a variety of themes. Its goal is to help young readers develop a lifelong practice of spending time with God. Circ: 90,000.

Audience: YA
Frequency: 6 times each year
Website: www.devozine.org

FREELANCE POTENTIAL
100% written by nonstaff writers. Publishes 325 freelance submissions yearly; 50% by authors who are new to the magazine.

SUBMISSIONS
Query. Accepts hard copy and email queries to devozine@upperroom.org. SASE. Responds in 4 months.

- Articles: 150–500 words. Informational articles; personal experience pieces; profiles; and reviews. Topics include Christian faith, mentoring, independence, courage, teen parenting, creativity, social issues, and relationships.
- Fiction: 150–250 words. Genres include contemporary and inspirational fiction.
- Depts/columns: 75–100 words. Reviews; new product information.
- Other: Daily meditations, 150–250 words. Prayers and poetry, 10–20 lines. Submit seasonal material 6–8 months in advance.

SAMPLE ISSUE
80 pages (no advertising): 8 articles; 60 devotionals; 6 reviews. Guidelines and theme list available at website.

- "Loving No Matter What." Article shares a young woman's experience of traveling to Africa to spend time with HIV patients.
- "Singer-Songwriter Jake Smith on Learning to Trust." Article describes Smith's relationship with God while surviving and recovering from a hurricane and family illness.

RIGHTS AND PAYMENT
First and second rights. Features, $100. Meditations, $25. Pays on acceptance.

➛EDITOR'S COMMENTS
Devotionals should tell real experiences of people who are struggling to apply faith in their lives. Please note that they must correspond with our themes.

Dig

Cobblestone Publishing
30 Grove Street, Suite C
Peterborough, NH 03458

Editor: Rosalie F. Baker

DESCRIPTION AND INTERESTS
Dig aspires to pique the interest of middle-school students in all things related to archaeology through wondrous photos and lively articles that put them at the scene. Circ: 19,000.

Audience: 9–14 years
Frequency: 9 times each year
Website: www.digonsite.com

FREELANCE POTENTIAL
80% written by nonstaff writers. Publishes 40 freelance submissions yearly; 40% by unpublished writers, 60% by authors who are new to the magazine. Receives 600–900 queries yearly.

SUBMISSIONS
Submissions must relate to an upcoming theme. Query with outline, bibliography, and clips or writing samples. Accepts hard copy. SASE. Responds in 4 months.

- Articles: 300–800 words. Informational articles and photo-essays. Topics include nature, animals, science, and technology.
- Fiction: To 800 words. Genres include historical and biographical fiction, adventure, and retold legends.
- Depts/columns: Word lengths vary. Art, archeology facts, and projects.
- Artwork: B/W and color prints.
- Other: Quizzes. Activities, to 700 words. Poetry, to 100 lines.

SAMPLE ISSUE
32 pages (no advertising): 7 articles; 6 depts/columns. Sample copy, $4.95 ($2 shipping and handling) with 10x13 SASE. Guidelines and theme list available at website.

- "Home Sweet Home." Article explains a massive excavation that unveiled a home and lifestyle in the ancient city of Leptis Magna.
- Sample dept/column: "Stones & Bones" offers bits of news, facts, discoveries, and information about ancient cultures.

RIGHTS AND PAYMENT
All rights. Written material, $.20–$.25 per word. Artwork, $15–$100. Pays on publication. Provides 2 contributor's copies.

➛EDITOR'S COMMENTS
Everything submitted, from article queries to photographs and activities, must relate to one of our planned themes. A full list of upcoming themes is available at our website.

Dimensions

DECA Inc.
1908 Association Drive
Reston, VA 20191-1594

Editor: Lyn Fiscus

DESCRIPTION AND INTERESTS
Though it is an educational journal,
Dimensions takes a direct and conversational
approach to its readership of high school marketing students, rather than an academic tone.
Articles cover various aspects of business,
management, sales, and marketing in a way
that appeals to young adults. Circ: 176,000.

Audience: 14–18 years
Frequency: Quarterly
Website: www.deca.org

FREELANCE POTENTIAL
50% written by nonstaff writers. Publishes
12–16 freelance submissions yearly; 50–75%
by authors who are new to the magazine.

SUBMISSIONS
Query or send complete ms with author bio.
Accepts hard copy, email to deca_dimensions@
deca.org, Macintosh disk submissions (RTF
files), and simultaneous submissions if identified. SASE. Response time varies.

- Articles: 800–1,200 words. Informational
  and how-to articles; profiles; interviews; and
  personal experience pieces. Topics include
  general business, management, marketing,
  sales, leadership development, entrepreneurship, franchising, personal finance, advertising, e-commerce, business technology, and
  career opportunities.
- Depts/columns: 400–600 words. DECA chapter news, opinions, and news briefs.

SAMPLE ISSUE
24 pages (45% advertising): 10 articles; 5 depts/
columns. Sample copy, free with 9x12 SASE.
Guidelines available.

- "Look the Part." Article tells students how to
  be presentable, polished, and professional
  in their style of dress.
- Sample dept/column: "Short Stuff" includes
  the results of a survey on what college students feel their presidential vote is worth.

RIGHTS AND PAYMENT
First serial rights. Written material, payment
rates vary. Pays on publication. Provides 2 contributor's copies.

➥EDITOR'S COMMENTS
We have a small editorial staff, and there can
be delays in responding to queries. Email
queries are generally answered more quickly.

Dimensions of Early Childhood

Southern Early Childhood Association
P.O. Box 55930
Little Rock, AR 72215-5930

Dimensions Manager: Jennifer Bean

DESCRIPTION AND INTERESTS
Published as a refereed journal for early childhood educators, *Dimensions of Early
Childhood* supports high-quality experiences
for young children by addressing both the continued interests and emerging ideas in the
field. Circ: 19,000.

Audience: Early childhood professionals
Frequency: 3 times each year
Website: www.southernearlychildhood.org

FREELANCE POTENTIAL
99% written by nonstaff writers. Publishes 40
freelance submissions yearly; 90% by unpublished writers, 80% by authors new to the magazine. Receives 84 unsolicited mss yearly.

SUBMISSIONS
Send complete ms. Accepts email submissions
to editor@southernearlychildhood.org.
Responds in 3–4 months.

- Articles: Word lengths vary. Informational
  articles. Topics include emergent curriculum
  for children, effective classroom practices,
  theory and research, program administration, family relationships, and resource
  systems.
- Depts/columns: Word lengths vary. Book
  reviews.

SAMPLE ISSUE
40 pages (20% advertising): 4 articles; 3
depts/columns. Sample copy, $5. Writer's
guidelines available.

- "The Earlier, the Better: Early Intervention
  Programs for Infants and Toddlers at Risk."
  Article examines the characteristics of successful programs for developmentally challenged children.
- "Young Learners at Natural History
  Museums." Article discusses the benefits of
  visits to natural history museums as a valuable way to gain first hand experience with
  authentic objects and specimens.

RIGHTS AND PAYMENT
All rights. No payment. Provides 2 copies.

➥EDITOR'S COMMENTS
Our magazine supports constructivist
approaches to child and adult learning,
rather than offering prescriptive techniques.
Collaborative efforts between academics and
practitioners are encouraged.

Discovery Girls

4300 Stevens Creek Boulevard, Suite 190
San Jose, CA 95129

Editor: Sarah Verney

DESCRIPTION AND INTERESTS
The goal of this magazine is to give girls the advice, encouragement, and inspiration needed to navigate the difficult preteen years. Coverage ranges from nutrition and fitness to friendships, entertainment, and social issues. Circ: 120,000.

Audience: Girls, 7–12 years
Frequency: 6 times each year
Website: www.discoverygirls.com

FREELANCE POTENTIAL
25% written by nonstaff writers. Publishes 5–10 freelance submissions yearly; 50% by authors who are new to the magazine.

SUBMISSIONS
Query with sample paragraph; or send complete ms. Accepts hard copy and email submissions to sarah@discoverygirls.com. SASE. Response time varies.

- Articles: Word lengths vary. Informational and how-to articles; and personal experience pieces. Topics include nutrition, fitness, careers, social issues, sports, fashion, popular culture, volunteering, and entertainment.
- Depts/columns: Word lengths vary. Celebrity news, relationship advice, health and beauty, and book reviews.
- Other: Quizzes and contests.

SAMPLE ISSUE
56 pages: 5 articles; 1 story; 9 depts/columns; 5 quizzes/contests.

- "How to Be More Likeable." Article offers tips on becoming more likeable by showing genuine interest in others.
- "Talk to Your BF About Embarrassing Stuff." Article suggests ways to discuss embarrassing or difficult situations.
- Sample dept/column: "Humor" describes the meanings behind some of the more common dreams in a fun and clever way.

RIGHTS AND PAYMENT
All rights. Written material, payment rates vary.

➠EDITOR'S COMMENTS
We do not publish fiction or personal essays unless they are written by girls who are our readers' age. Many times we do not buy the articles submitted by nonstaff writers but if we like their writing, we hire them to develop our ideas.

Dogs for Kids

BowTie, Inc.
P.O. Box 6050
Mission Viejo, CA 92690-6050

Editor: Jackie Franza

DESCRIPTION AND INTERESTS
This magazine is devoted to the education and entertainment of young people who enjoy purebred and mixed-breed dogs, or who just enjoy reading about them. Circ: 40,000.

Audience: 8–13 years
Frequency: 6 times each year
Website: www.dogsforkids.com

FREELANCE POTENTIAL
50% written by nonstaff writers. Publishes 25 freelance submissions yearly; 20% by authors who are new to the magazine. Receives 500 queries yearly.

SUBMISSIONS
Query with writing samples. Accepts email queries to dogsforkids@bowtieinc.com. Responds in 8–10 weeks.

- Articles: 1,000 words. Informational and how-to articles; and profiles of children making a difference with dogs. Topics include pet care and dog-related activities and careers.
- Depts/columns: Staff written.
- Other: Puzzles, activities, games, quizzes.

SAMPLE ISSUE
64 pages (10% advertising): 8 articles; 5 depts/columns; 6 puzzles and games. Sample copy, $2.99 with 9x12 SASE. Writers' guidelines available.

- "How Healthy Is Your Dog?" Article provides many simple ways kids can know if their dog is sick.
- "Start a Bow-Wow Business." Article explains the ins and outs of starting a dog-sitting or dog-walking business.
- "This Girl Rocks." Article profiles a rock star's daughter who is the co-host of a television show about pet care.

RIGHTS AND PAYMENT
First rights. Articles, $300. Pays on publication. Provides 2 contributor's copies.

➠EDITOR'S COMMENTS
We already know kids love dogs; we choose fast-paced, easy-to-read articles that reinforce that existing bond. All pieces should be lively, positive, motivational, and direct—that means using "you" instead of "children" or "dog owners." Please note that we do not accept breed profiles.

Dovetail

*A Journal By and For
Jewish/Christian Families*

775 Simon Greenwell Lane
Boston, KY 40107

Editor: Mary Rosenbaum

DESCRIPTION AND INTERESTS
This online magazine, a publication of the
Dovetail Institute for Interfaith Family
Resources, addresses the issues faced by
families that are blending Christian and Jewish
traditions. Hits per month: Unavailable.

Audience: Interfaith families
Frequency: Quarterly
Website: www.dovetailinstitute.org

FREELANCE POTENTIAL
95% written by nonstaff writers. Publishes 10
freelance submissions yearly; 90% by unpublished writers, 90% by authors who are new to
the magazine. Receives 192 queries and unsolicited mss yearly.

SUBMISSIONS
Query or send complete ms. Accepts hard
copy, Macintosh and text file disk submissions,
email submissions to DI-IFR@bardstown.com,
and simultaneous submissions if identified.
Availability of artwork improves chance of
acceptance. SASE. Responds in 1–2 months.

- Articles: 800–1,000 words. Informational
  articles; profiles; interviews; reviews; and personal experience pieces. Topics include interfaith community, parenting, antisemitism,
  gender roles, religious holidays, family
  issues, social concerns, and education.
- Poetry, line lengths vary.

SAMPLE ISSUE
14 pages: 5 articles. Sample issue, guidelines,
and theme list available at website.

- "When Santa Went Down the Wrong
  Chimney." Article relates a sweet fable about
  a Jewish child's gift to Santa's reindeer.
- "Two Grandmothers to Love." Article provides a review of a book about interfaith
  grandmothers and their role in their grandchildren's lives.

RIGHTS AND PAYMENT
One-time rights. Articles, $25. Reviews, $15.
Pays on publication. Provides download access
to copies.

⇢EDITOR'S COMMENTS
As a nondenominational publication, we are
seeking works that offer an open and tolerant
approach to interfaith marriage and family
life, and that show respect for both the
Jewish and Christian traditions.

Dramatics

Educational Theatre Association
2343 Auburn Avenue
Cincinnati, OH 45219

Editor: Donald Corathers

DESCRIPTION AND INTERESTS
Dramatics is dedicated to the advancement of
secondary school theater. Its objective is to
provide serious theater students (and teachers)
with the knowledge they need to make better
theater and to decide whether to pursue a
career in theater. Circ: 37,000.

Audience: High school students and teachers
Frequency: 9 times each year
Website: www.edta.org

FREELANCE POTENTIAL
80% written by nonstaff writers. Publishes 41
freelance submissions yearly; 25% by unpublished writers, 50% by authors who are new to
the magazine. Receives 480 unsolicited mss
each year.

SUBMISSIONS
Send complete ms. Accepts hard copy and
email submissions to dcorathers@edta.org.
SASE. Responds in 2–4 months.

- Articles: 750–4,000 words. Informational
  articles, book reviews, and interviews.
  Topics include playwriting, musical theater,
  acting, auditions, stage makeup, set design,
  and theater production.
- Fiction: Word lengths vary. Full-length and
  one-act plays for high school audiences.
- Depts/columns: Word lengths vary. Industry
  news, theater-related items, acting techniques.
- Artwork: 5x7 or larger B/W prints; 35mm or
  larger color transparencies. Line art. High-
  resolution JPGs or TIFs.

SAMPLE ISSUE
68 pages (40% advertising): 3 articles; 1 play;
4 depts/columns. Sample copy, $3 with 9x12
SASE. Guidelines available.

- "Upstairs, Downstairs." Article offers complete plans for a stair unit that is strong,
  adaptable, and easy to store.
- Sample dept/column: "Acting in Musical
  Theatre" offers ideas for understanding a
  character's relationship to other characters.

RIGHTS AND PAYMENT
First rights. Written material, $100–$500. Pays
on publication. Provides 5 author's copies.

⇢EDITOR'S COMMENTS
What makes us happy is writers who understand their audience and include student
voices in their pieces, when appropriate.

Earlychildhood News

2 Lower Ragsdale, Suite 200
Monterey, CA 93940

Assistant Editor: Susan Swanson

DESCRIPTION AND INTERESTS
Once a nationally circulated magazine, this resource for early childhood professionals, teachers, and homeschooling parents is now published exclusively online. Hits per month: 50,000.

Audience: Early childhood professionals, teachers, and parents
Frequency: Monthly
Website: www.earlychildhoodnews.com

FREELANCE POTENTIAL
90% written by nonstaff writers. Publishes 10+ freelance submissions yearly; 5% by unpublished writers, 15% by authors who are new to the magazine. Receives 96 queries each year.

SUBMISSIONS
Query with author biography. Accepts email to sswanson@excelligence.com. No simultaneous submissions. Responds in 2 months.

- Articles: 800–1,200 words. Informational, how-to, and research-based articles; and personal experience pieces. Topics include child development, curriculum, family relationships, health and safety, nutrition, behavior management, and professional development.
- Other: Activities and crafts.

SAMPLE ISSUE
Sample copy, guidelines, and editorial calendar available at website.

- "Supporting the Parent-Infant Bond." Article discusses how child-care providers can help children develop self-esteem.
- "Tips for Making a Multi-Use Space Your Own." Article tells how the director of an after-school program transformed a school "cafeterium" into a space she could use.
- "Green Classrooms Are Easier Than You Think!" Article recommends ways to inspire eco-friendly actions in the classroom.

RIGHTS AND PAYMENT
All rights. Written material, $75–$300. Pays on acceptance.

➜EDITOR'S COMMENTS
If your article is research-based, it should contain recent studies (no more than five years old). Always cite original research. If you are building on the work of others, be sure to acknowledge their work.

Early Childhood Today

Scholastic, Inc.
557 Broadway
New York, NY 10012-3999

Editor: Tia Kaul Disick

DESCRIPTION AND INTERESTS
This magazine, now completely online, is geared toward early childhood education professionals. It features informative articles on topics pertinent to the field, as well as resources, advice, and support from its online community of readers. Hits per month: 55,000.

Audience: Early childhood professionals
Frequency: 8 times each year
Website: www.earlychildhoodtoday.com

FREELANCE POTENTIAL
100% written by nonstaff writers. Publishes 10 freelance submissions yearly; 50% by authors who are new to the magazine. Receives 10–15 queries yearly.

SUBMISSIONS
Query. Accepts queries via website and email queries to ect@scholastic.com. Responds in 1 month.

- Articles: Word lengths vary. Informational, educational, and how-to articles. Topics include child advocacy, child development, special needs, communication, physical development, family issues, health, technology, and cultural issues.
- Depts/columns: Word lengths vary. News, teacher tips, health, and ideas for age-appropriate activities.

SAMPLE ISSUE
Sample copy and writers' guidelines available at website.

- "Keep the Cool in School: Tolerance—The Fifth Core Strength." Article provides ideas for promoting tolerance among students.
- "Creating an Emotionally Safe Classroom." Article explains the importance of the first few days of school to a student's emotional well-being.

RIGHTS AND PAYMENT
All rights. Written material, payment rates vary. Pays on acceptance.

➜EDITOR'S COMMENTS
We prefer that our writers have a professional background in early childhood education, and all submissions should reference professional sources. Our new online presence is a bit different from our print version, so please take some time to see what we're all about before querying us with your ideas.

Educational Horizons

Pi Lambda Theta
P.O. Box 6626
4101 East Third Street
Bloomington, IN 47407-6626

Managing Editor

DESCRIPTION AND INTERESTS
Founded in the spirit of academic excellence, this journal addresses educational, social, and cultural issues through research findings and scholarly essays. Circ: 17,000.

Audience: Pi Lambda Theta members
Frequency: Quarterly
Website: www.pilambda.org

FREELANCE POTENTIAL
95% written by nonstaff writers. Publishes 10–15 freelance submissions yearly; 75% by authors who are new to the magazine. Receives 60 queries, 12 unsolicited mss yearly.

SUBMISSIONS
Prefers query with proposed topic and word length; will accept complete ms. Accepts hard copy, disk submissions (.txt files), email to publications@pilambda.org, and simultaneous submissions if identified. SASE. Responds to queries in 1 month, to mss in 3–4 months.

- Articles: Research articles, 2,500–4,000 words. Scholarly essays, 1,000–2,000 words. Topics include educational, social, and cultural issues of significance.
- Depts/columns: 500–750 words. Education topics in the news; multicultural education; legal issues; and book reviews.
- Artwork: Graphs and charts set in Microsoft Excel and saved as separate files.

SAMPLE ISSUE
120 pages (4% advertising): 4 articles; 2 depts/columns. Sample copy, $5 with 9x12 SASE. Guidelines available at website.

- "Student Expression: The Uncertain Future." Article analyzes the *Morse v. Frederick* Supreme Court decision regarding speech.
- Sample dept/column: "Behind Every Silver Lining" examines the current state of public schoolteaching.

RIGHTS AND PAYMENT
First rights. No payment. Provides 5 contributor's copies.

➥EDITOR'S COMMENTS
Ordinarily, guest editors assemble the manuscripts for our publication. Occasionally they request that we run a general call for papers. Acceptance of other submissions depends on unpredictable openings in the schedule.

Educational Leadership

ASCD
1703 North Beauregard Street
Alexandria, VA 22311-1714

Editor-in-Chief: Marge Scherer

DESCRIPTION AND INTERESTS
Read by teachers, principals, and school superintendents from preschool through grade 12, *Educational Leadership* is written primarily by these practitioners themselves. As the flagship publication of the Association for Supervision and Curriculum Development, it offers practical, research-based information and examples from classroom experience. Circ: 180,000.

Audience: Educators
Frequency: 8 times each year
Website: www.ascd.org

FREELANCE POTENTIAL
95% written by nonstaff writers. Publishes 130 freelance submissions yearly; 50% by unpublished writers, 50% by new authors. Receives 900 unsolicited mss yearly.

SUBMISSIONS
Send 2 copies of complete ms. Accepts hard copy. SASE. Responds in 2 months.

- Articles: 1,500–2,500 words. Informational, how-to, and research-based articles; program descriptions; and personal experience and opinion pieces. Topics include reading, assessment, instructional strategies, student achievement, gifted and special education, science, technology, and multicultural issues.
- Depts/columns: Word lengths vary. Opinions, accountability issues, research findings, leadership challenges, principals' perspectives, ASCD news, and policy reviews.
- Artwork: B/W and color prints or slides; digital images at 300 dpi. Line art. Send only upon request.

SAMPLE ISSUE
96 pages (25% advertising): 15 articles; 9 depts/columns. Sample copy, $7. Guidelines and theme list available at website.

- "Differentiating Math Through Expeditions." Article describes a method of pushing students to think for themselves.
- Sample dept/column: "What Research Says About . . ." takes a look at small learning communities.

RIGHTS AND PAYMENT
All or first rights. No payment. Provides 5 copies.

➥EDITOR'S COMMENTS
We look for conversational manuscripts that are insightful and helpful to educators.

Education Forum

60 Mobile Drive
Toronto, Ontario M4A 2P3
Canada

Managing Editor: Marianne Clayton

DESCRIPTION AND INTERESTS
Educators in Ontario read *Education Forum* for important coverage of news, trends, and strategies related to teaching. Circ: 50,000.

Audience: Teachers
Frequency: 3 times each year
Website: www.osstf.on.ca

FREELANCE POTENTIAL
90% written by nonstaff writers. Publishes 35 freelance submissions yearly; 20% by unpublished writers, 80% by authors who are new to the magazine. Receives 48 queries and unsolicited mss yearly.

SUBMISSIONS
Query with clips or writing samples; or send complete ms. Accepts hard copy. No simultaneous submissions. SAE/IRC. Responds in 1–2 months.

- Articles: To 2,500 words. How-to and practical application articles on education trends; discussions of controversial issues; and teaching techniques for use in secondary school classrooms.
- Depts/columns: "Openers," to 300 words; news and opinion pieces. "Forum Picks," word lengths vary; media and software reviews.
- Artwork: B/W and color prints; color transparencies. Line art.
- Other: Classroom activities, puzzles, and games. Submit seasonal material 8 months in advance.

SAMPLE ISSUE
46 pages (18% advertising): 4 articles; 8 depts/columns. Sample copy, free with 9x12 SAE/IRC. Guidelines available.

- "Graphic Novels to the Rescue." Article discusses how illustrated heroes entice reluctant readers to read.
- Sample dept/column: "On a Global Front" looks at the parallel between teacher working conditions and student learning.

RIGHTS AND PAYMENT
First North American serial rights. No payment. Provides 5 contributor's copies.

⇒EDITOR'S COMMENTS
Writers are encouraged to contact us with their well-researched articles on education trends and profiles of successful public school educators.

Education Week

6935 Arlington Road, Suite 100
Bethesda, MD 20814-5233

Executive Editor: Greg Chronister

DESCRIPTION AND INTERESTS
Considered by many to be a must-read for education professionals, this tabloid covers news topics that affect education in preschool through grade 12, as well as trends, legislation, state policies, and technology. It also publishes special themed supplements that revolve around an individual issue. Circ: 50,000.

Audience: Educators
Frequency: 45 times each year
Website: www.edweek.org

FREELANCE POTENTIAL
8% written by nonstaff writers. Publishes 125 freelance submissions yearly; 80% by unpublished writers, 75% by authors who are new to the magazine. Receives 600 unsolicited mss each year.

SUBMISSIONS
Send complete ms. Accepts IBM disk submissions (Word Perfect or Microsoft Word) and Macintosh disk submissions (plain text). SASE. Responds 6–8 weeks.

- Articles: 1,200–1,500 words. Essays about child development and education related to grades K–12 for use in "Commentary" section.
- Depts/columns: Staff written.

SAMPLE ISSUE
40 pages (25% advertising): 10 articles; 12 depts/columns; 3 commentaries. Sample copy, $3 with 9x12 SASE ($1 postage). Writers' guidelines available.

- "Reading First Funds Headed for Extinction." Article breaks the news that federal funding for the Reading First program is doomed.
- "To NAACP, Obama Stresses Parental Theme." Article reports on a speech made by Senator Barack Obama.

RIGHTS AND PAYMENT
First rights. "Commentary," $200. Pays on publication. Provides 2 contributor's copies.

⇒EDITOR'S COMMENTS
We are very much a newspaper in that we have a complete staff that covers any and all news and current events that affect education in this country. We do not use freelancers for our articles. Freelancers with professional experience in education are encouraged to submit essays for our "Commentary" section on topics of interest to educators.

EduGuide

Partnership for Learning
321 North Pine
Lansing, MI 48933

Editor

DESCRIPTION AND INTERESTS
Dedicated to student achievement at any age, *EduGuide* equips parents with the tools needed to help their children succeed. Four editions are published to target the elementary school, middle school, high school, and college years. Circ: 600,000.

Audience: Parents and teens
Frequency: Annually
Website: www.eduguide.org

FREELANCE POTENTIAL
40% written by nonstaff writers. Publishes 25–30 freelance submissions yearly; 10% by unpublished writers, 40% by authors who are new to the magazine.

SUBMISSIONS
Query. No unsolicited mss. Accepts hard copy. SASE. Responds in 4–6 weeks.

- Articles: 500–1,000 words. Informational and how-to articles; profiles; interviews; and personal experience pieces. Topics include the arts, college, careers, computers, gifted education, health, fitness, history, humor, mathematics, music, science, technology, special education, and issues related to elementary and secondary education.
- Depts/columns: Staff written.
- Artwork: Color prints and transparencies. Line art.
- Other: Submit seasonal material 3 months in advance.

SAMPLE ISSUE
32 pages: 6 articles; 1 dept/column. Sample copy, $3 with 9x12 SASE ($1 postage). Guidelines and editorial calendar available.

- "3 Smart Moves for Parents." Article explains how to make the most of the high school years in preparation for college.
- "How I Found My College." Personal experience piece shares tips on choosing the right college and preparing for it.

RIGHTS AND PAYMENT
First or second rights. All material, payment rates vary. Pays on acceptance. Provides 5 contributor's copies.

❧EDITOR'S COMMENTS
We no longer review unsolicited manuscripts. Please check our website for the most up-to-date submission information.

Edutopia

P.O. Box 3494
San Rafael, CA 94912

Executive Editor: Jennifer Sweeney

DESCRIPTION AND INTERESTS
Edutopia is a publication of the George Lucas Educational Foundation, which was formed to address the issue of how slowly our public school system changes. It offers ideas for creating ideal, innovative, and interactive learning environments. Circ: 100,000.

Audience: Teachers, parents, policy makers
Frequency: 8 times each year
Website: www.edutopia.org

FREELANCE POTENTIAL
70% written by nonstaff writers. Publishes 20 freelance submissions yearly; 30% by authors who are new to the magazine. Receives 36–60 queries yearly.

SUBMISSIONS
Query with résumé and clips. Accepts email queries to edit@edutopia.org. Response time varies.

- Articles: 300–2,500 words. Informational and how-to articles; and personal experience pieces. Topics include computers, education, current events, health and fitness, nature, the environment, popular culture, recreation, science, technology, social issues, and travel.
- Depts/columns: 700 words. Health, education, and ethnic and multicultural issues.

SAMPLE ISSUE
56 pages (35% advertising): 3 articles; 15 depts/columns. Sample copy, $4.95. Writers' guidelines available.

- "Building a Better Teacher." Article explains that most teacher training programs leave new teachers unprepared for real-world classrooms, and looks at some innovative programming that addresses the problem.
- "Adopt and Adapt: Sharing Tech for the Classroom." Article looks at the current state of technology in the classroom, and offers a hope for how it can be better, faster.

RIGHTS AND PAYMENT
First North American serial rights. Written material, payment rates vary. Pays on acceptance. Provides 2 contributor's copies.

❧EDITOR'S COMMENTS
We are interested in articles that address how the public education system can move closer to being what we call "public edutopia."

Elementary School Writer

Writer Publications
P.O. Box 718
Grand Rapids, MI 55744-0718

Editor: Emily Benes

DESCRIPTION AND INTERESTS

This newspaper provides elementary and middle school students with their first opportunity to become published writers. Content is written by children and submitted by teachers who use *Elementary School Writer* as a learning tool in the classroom. Circ: Unavailable.

Audience: Elementary and middle school students and teachers
Frequency: 6 times each year
Website: www.writerpublications.com

FREELANCE POTENTIAL

100% written by nonstaff writers. Publishes 300 freelance submissions yearly; 95% by unpublished writers, 75% by authors who are new to the newspaper. Receives 100 unsolicited mss yearly.

SUBMISSIONS

Accepts complete ms via subscribing teachers only. Accepts hard copy, email submissions to writer@mx3.com (ASCII text only), and simultaneous submissions if identified. SASE. Response time varies.

- Articles: To 1,000 words. Informational and how-to articles; profiles; humor; opinions; and personal experience pieces. Topics include current events, multicultural and ethnic issues, nature, the environment, popular culture, sports, and travel.
- Fiction: To 1,000 words. Genres include humor, science fiction, and stories about nature and sports.
- Other: Poetry, no line limit. Seasonal material.

SAMPLE ISSUE

8 pages (no advertising): 14 articles and essays; 9 stories; 19 poems. Guidelines available in each issue.

- "When I Got My Puppy." Article explains how the author handled the challenges of caring for three dogs at once.
- "Weird Town." Story tells of a mysterious vacant lot where three children disappeared.

RIGHTS AND PAYMENT

One-time rights. No payment.

⟶EDITOR'S COMMENTS

Each submission should include an end statement such as: "This is my original work. It does not contain the words of anyone else without proper credit."

Ellery Queen's Mystery Magazine

475 Park Avenue South, 11th Floor
New York, NY 10016

Editor: Janet Hutchings

DESCRIPTION AND INTERESTS

Every kind of mystery story—from the psychological suspense tale to the private eye case—is found within the pages of *Ellery Queen's Mystery Magazine*, which has been entertaining readers since 1941. Circ: 180,780.

Audience: YA–Adult
Frequency: 10 times each year
Website: www.themysteryplace.com/eqmm

FREELANCE POTENTIAL

100% written by nonstaff writers. Publishes 125 freelance submissions yearly; 7% by unpublished writers, 25% by authors who are new to the magazine. Receives 2,600 unsolicited mss yearly.

SUBMISSIONS

Send complete ms. Accepts hard copy and simultaneous submissions if identified. SASE. Responds in 3 months.

- Fiction: Feature stories, 2,000–12,000 words. Minute Mysteries, 250 words. Novellas by established authors, to 20,000 words. Genres include contemporary and historical crime fiction, psychological thrillers, mystery, suspense, and detective/private-eye stories.
- Other: Poetry.

SAMPLE ISSUE

144 pages (6% advertising): 10 stories; 2 book reviews; 2 poems. Sample copy, $5.50. Guidelines available.

- "Mother Dear." Memoir-style mystery follows a man through life as he learns to live with the guilt of killing his mother.
- "The Problem of the Secret Patient." Story tells how a doctor solves the hospital murder of a war criminal.
- "Suchness." Story relates the experience of a prep school girl who is being stalked.

RIGHTS AND PAYMENT

First and anthology rights. Written material, $.05–$.08 per word. Pays on acceptance. Provides 3 contributor's copies.

⟶EDITOR'S COMMENTS

We especially encourage unpublished fiction writers to submit stories. First-story submissions should be addressed to our Department of First Stories. We look for strong writing, an original and exciting plot, and professional craftsmanship.

Exceptional Parent

416 Main Street
Johnstown, PA 15901

Editor-in-Chief: Dr. Rick Rader

DESCRIPTION AND INTERESTS

Exceptional Parent is devoted to issues that are unique to parents with special needs children. It features informational articles about special care, health issues, finances, education, stress management, as well as inspirational pieces by parents for parents. Circ: 70,000.

Audience: Parents, teachers, professionals
Frequency: Monthly
Website: www.eparent.com

FREELANCE POTENTIAL

95% written by nonstaff writers. Publishes 50–60 freelance submissions yearly; 50% by unpublished writers, 50% by authors who are new to the magazine. Receives 96+ queries each year.

SUBMISSIONS

Query. Accepts hard copy. SASE. Responds in 3–4 weeks.

- Articles: To 2,500 words. Informational articles; profiles; interviews; and personal experience pieces. Topics include the social, psychological, legal, political, technological, financial, and educational concerns of individuals with disabilities and their parents.
- Depts/columns: Word lengths vary. Opinion and personal experience pieces, news, new product information, and media reviews.

SAMPLE ISSUE

100 pages (50% advertising): 17 articles; 16 depts/columns. Sample copy, $4.99 with 9x12 SASE ($2 postage). Guidelines and editorial calendar available.

- "Neonatologists and Pediatricians Highlight Progress and Barrier to Preemie Care." Article explains the latest research in caring for premature babies.
- "Glimpses of William." Essay offers a first-person description of the author's life with an autistic son.

RIGHTS AND PAYMENT

First North American serial rights. No payment. Provides 2 contributor's copies.

➡ EDITOR'S COMMENTS

While we certainly publish a number of medical-related articles from health care professionals, we are especially interested in articles from parents about coping—and thriving—with a special needs child.

Exchange

P.O. Box 3249
Redmond, WA 98073-3249

Editor: Bonnie Neugebauer

DESCRIPTION AND INTERESTS

Child-care professionals read this magazine for timely information on early childhood development, parenting topics, and staff and curriculum development. Circ: 30,000.

Audience: Child-care professionals
Frequency: 6 times each year
Website: www.childcareexchange.com

FREELANCE POTENTIAL

65–75% written by nonstaff writers. Publishes 75 freelance submissions yearly; 50% by unpublished writers, 60% by new authors.

SUBMISSIONS

Send complete ms with brief author biography and article references. Accepts email submissions to bonnien@childcareexchange.com (Microsoft Word attachments or text); requires hard copy in addition to electronic copy. Availability of artwork improves chance of acceptance. SASE. Response time varies.

- Articles: 1,800 words. Informational, how-to, and self-help articles. Topics include child development; education; and social, multicultural, and ethnic issues.
- Depts/columns: Word lengths vary. Staff development and training; parent perspective pieces.
- Artwork: Color prints. Line art.

SAMPLE ISSUE

96 pages: 16 articles; 4 depts/columns. Sample copy, $8. Writers' guidelines and theme list available.

- "Challenges Facing Early Childhood Programs Worldwide." Article examines child development and education issues in 16 different countries.
- "We Grow Brains!" Article encourages peers to be proud of their profession.
- Sample dept/column: "From a Parent's Perspective" discusses ways to prevent struggles with eating, sleeping, and toileting.

RIGHTS AND PAYMENT

All rights. Written material, payment rates vary. Pays on publication. Provides 2 author's copies.

➡ EDITOR'S COMMENTS

We have launched a Mentor Writing project, which offers one-on-one help in getting your submissions reviewed by our staff. Please visit our website for complete details.

Faces

Cobblestone Publishing
30 Grove Street, Suite C
Peterborough, NH 03458

Editor: Elizabeth Carpentiere

DESCRIPTION AND INTERESTS
Faces brings the world to kids through articles and profiles of the countries, people, and cultures of the globe. Each issue is dedicated to a singular country or theme. Circ: 15,000.

Audience: 9–14 years
Frequency: 9 times each year
Website: www.cobblestonepub.com

FREELANCE POTENTIAL
70% written by nonstaff writers. Publishes 80–90 freelance submissions yearly; 25% by unpublished writers, 15% by authors who are new to the magazine. Receives 450 queries each year.

SUBMISSIONS
Query with outline, bibliography, and clips or writing samples. Accepts email queries to facesmag@yahoo.com. Responds in 5 months.

- Articles: 800 words. Sidebars, 300–600 words. Informational articles and personal experience pieces. Topics include culture, geography, the environment, cuisine, special events, travel, history, and social issues.
- Fiction: To 800 words. Stories, legends, and folktales from around the world.
- Depts/columns: Staff written.
- Artwork: Color prints or transparencies.
- Other: Games, crafts, puzzles, and activities, to 700 words. Poetry, to 100 lines.

SAMPLE ISSUE
48 pages (no advertising): 16 articles; 7 depts/columns; 1 activity. Sample copy, $5.95 with 9x12 SASE ($2 postage). Guidelines and theme list available at website.

- "The Great Wall." Article profiles the Great Wall of China as one of the "New Seven Wonders of the World."
- "Growing Up in Zambia: Surrounded by Wonder." Article offers a girl's experience of living in Zambia.

RIGHTS AND PAYMENT
All rights. Articles and fiction, $.20–$.25 per word. Activities and puzzles, payment rates vary. Pays on publication. Provides 2 contributor's copies.

➛EDITOR'S COMMENTS
Lively, original approaches to the subject at hand are our primary concerns when choosing material.

Face Up

Redemptorist Communications
75 Orwell Road
Rathgar, Dublin 6
Ireland

Editor: Gerard Moloney

DESCRIPTION AND INTERESTS
Launched in 2001 by a Catholic publishing house, *Face Up* explores social and spiritual issues in a teen-friendly manner. Its colorful, pop-culture look belies the "deeper" message within its pages. Circ: 12,000.

Audience: 14–18 years
Frequency: 10 times each year
Website: www.faceup.ie

FREELANCE POTENTIAL
50% written by nonstaff writers. Publishes 60 freelance submissions yearly; 25% by unpublished writers, 25% by authors who are new to the magazine. Receives 300 unsolicited mss yearly.

SUBMISSIONS
Send complete ms. Accepts email submissions to info@faceup.ie. Responds in 1 month.

- Articles: 900 words. Informational and how-to articles; profiles; interviews; and personal experience pieces. Topics include college, careers, current events, relationships, health, fitness, music, popular culture, celebrities, sports, and multicultural, ethnic, and social issues.
- Depts/columns: 500 words. Opinions, essays, reviews, advice, self-help, profiles, interviews, and "Words of Wisdom."
- Other: Quizzes and crossword puzzles.

SAMPLE ISSUE
48 pages (5% advertising): 4 articles; 18 depts/columns; 1 puzzle. Sample copy available at website. Writers' guidelines and editorial calendar available.

- "The Wrong Kind of Retail Therapy." Article reveals the lifelong consequences of shoplifting and explains how it can quickly become a habit that spirals out of control.
- Sample dept/column: "Entertainment" includes a Q&A with singer Rihanna about how she handles the pressures of fame.

RIGHTS AND PAYMENT
Rights vary. Written material, payment rates vary. Pays on publication. Provides 2 contributor's copies.

➛EDITOR'S COMMENTS
We are an official ministry of the Irish Redemptorists. Our mission is to spread the word of God through the print media.

Faith & Family

432 Washington Avenue
North Haven, CT 06473

Assistant Editor: Robyn Lee

DESCRIPTION AND INTERESTS
Faith & Family is a color-photo filled magazine featuring contemporary, inspirational writing that speaks specifically to Catholic moms. Circ: 35,000.
Audience: Catholic families
Frequency: Quarterly
Website: www.faithandfamilymag.com

FREELANCE POTENTIAL
75% written by nonstaff writers. Publishes 35 freelance submissions yearly; 15% by unpublished writers, 10% by authors who are new to the magazine. Receives 300 queries yearly.

SUBMISSIONS
Query. Accepts email queries to editor@faithandfamilymag.com. Responds in 2–3 months.

- Articles: 600–2,000 words. Informational, how-to, and self-help articles; profiles; interviews; personal experience pieces; and media reviews. Topics include family life, parenting, marriage, religion, and social and political issues.
- Depts/columns: Word lengths vary. Tips for home, garden, and food; entertainment reviews; spirituality.
- Artwork: Color prints or slides.

SAMPLE ISSUE
96 pages (30% advertising): 6 articles; 15 depts/columns. Sample copy, $4.50. Guidelines available.

- "Peaceful Coexistence." Article shares advice from a mom of nine children on the topic of sibling rivalry.
- "Maxed-Out Moms." Article shows how four mothers have balanced their volunteer work and family life.
- Sample dept/column: "The Back Porch" tells how a family dealt with the death of their pet speckled hen.

RIGHTS AND PAYMENT
First North American serial rights. Written material, $.33 per word. Pays on publication.

➡EDITOR'S COMMENTS
We are always looking for well-written articles on current issues and events of interest to Catholic moms, presented from a Catholic perspective. Our focus is on practical spirituality rather than academic theology.

Faith Today

M.I.P. Box 3745
Markham, ON L3R OY4
Canada

Senior Editor: Bill Fledderus

DESCRIPTION AND INTERESTS
Faith Today seeks to provide analysis and interpretation of today's issues for Canada's Evangelical Christians while fostering a sense of community. Each issue features a mix of how-to and informational articles and profiles of individuals and ministries. Circ: 20,000.
Audience: Christians
Frequency: 6 times each year
Website: www.faithtoday.ca

FREELANCE POTENTIAL
60% written by nonstaff writers. Publishes 120 freelance submissions yearly; 1% by unpublished writers, 10% by authors who are new to the magazine. Receives 120 queries yearly.

SUBMISSIONS
Query. Accepts email queries to fteditor@efc-canada.com (Microsoft Word attachments). Responds in 3 weeks.

- Articles: 400–1,500 words. Informational articles; profiles; interviews; and opinion pieces. Topics include Christianity in Canada, issues and ministry initiatives of interest to Christians, prayer, and the Bible.
- Depts/columns: To 900 words. News, arts and culture, theology, and short profiles.

SAMPLE ISSUE
54 pages (20% advertising): 9 articles; 10 depts/columns. Sample copy, free. Guidelines available at website.

- "Woman with a Mission." Article interviews Eileen Stewart-Rhude, a leading woman in several Canadian ministries.
- "Slavery Lives Again." Article discusses the current global slavery problem and how Canadian groups that are helping to free slaves.
- Sample dept/column: "God at Work in Denominations" explains how trust is the key to a feeling of unity within the church.

RIGHTS AND PAYMENT
First or one-time electronic rights. Written material, $65–$1,800. Kill fee, 30–50%. Pays within 2 months of acceptance. Provides 2 contributor's copies.

➡EDITOR'S COMMENTS
We do not accept unsolicited manuscripts but we do use freelance writers for most of our articles and news reports. Only content focusing on Canadians will be considered.

Families On the Go

Life Media
P.O. Box 55445
St. Petersburg, FL 33732

Editor: Barbara Doyle

DESCRIPTION AND INTERESTS
This free family magazine has four editions throughout the Tampa Bay region of Florida. Appearing online as well as in print, its goal is to provide families with information about parenting, education, health, and relationship issues. It also publishes localized recreation and entertainment information. Circ: 120,000.

Audience: Families
Frequency: 6 times each year
Website: www.familiesonthego.org

FREELANCE POTENTIAL
80% written by nonstaff writers. Publishes 50 freelance submissions yearly; 25% by unpublished writers, 20% by authors who are new to the magazine.

SUBMISSIONS
Query or send complete ms. Accepts hard copy and email submissions to editor@familiesonthego.org (Microsoft Word attachments). SASE. Responds only if interested.

- Articles: 350–750 words. Informational articles. Topics include health and wellness, parenting, education, family relationships, home, garden, the arts, travel, and entertainment.
- Depts/columns: Word lengths vary. Community news.

SAMPLE ISSUE
Sample copy, free with 9x12 SASE (4 first-class stamps); also available at website. Guidelines available at website.

- "Outfit Your Child for Back-to-School Success." Article explains learning styles and how parents can prepare their child for a successful school year.
- Sample dept/column: "Health & Wellness" offers tips for preventing kids' eye injuries.

RIGHTS AND PAYMENT
Exclusive regional rights. Written material, payment rates vary. Pays on publication. Provides 2 contributor's copies.

➥ EDITOR'S COMMENTS
We are very interested in receiving queries regarding articles for our publication. Please note that articles may be used in one or several of the *Families On the Go* magazine editions, or in its online version. We are most interested in articles about health and wellness, parenting, education, and the arts.

Family Circle

Meredith Corporation
375 Lexington Avenue, 9th Floor
New York, NY 10017

Articles Editor: Darcy Jacobs

DESCRIPTION AND INTERESTS
Since 1932, this magazine has been guiding women through the ups and downs of family life with its articles on food, home decorating, and health. Circ: 4.2 million.

Audience: Families
Frequency: 15 times each year
Website: www.familycircle.com

FREELANCE POTENTIAL
80% written by nonstaff writers. Publishes many freelance submissions yearly. Receives hundreds of queries yearly.

SUBMISSIONS
Query with clips (including 1 from a national magazine) and author bio. Accepts hard copy. No simultaneous submissions. SASE. Responds in 6–8 weeks if interested.

- Articles: 1,000–2,000 words. Informational and how-to articles; profiles; and personal experience pieces. Topics include parenting, relationships, health, safety, fitness, home decor, travel, fashion, and cooking.
- Depts/columns: 750 words. "My Hometown," "My Family Life," "Good Works," recipes, beauty tips, shopping tips, fitness routines, and advice.

SAMPLE ISSUE
210 pages (48% advertising): 18 articles; 1 story; 20 depts/columns. Sample copy, $1.99 at newsstands. Guidelines available.

- "Shelf Help." Article provides suggestions for solving storage dilemmas by creating elegant, clutter-free displays.
- "Sibling Harmony." Article tells how to help feuding children negotiate a lasting peace.
- Sample dept/column: "Good Works" profiles a woman who founded the Emancipation Network to benefit victims of human-trafficking.

RIGHTS AND PAYMENT
All rights. Written material, payment rates vary. Kill fee, 25%. Pays on acceptance. Provides 1 contributor's copy.

➥ EDITOR'S COMMENTS
Writers must have national magazine experience in order to be considered for publication in *Family Circle*. We require well-written, well-reported stories that offer intelligent and accessible service. We also print the winner of our annual fiction contest each March.

The Family Digest

P.O. Box 40137
Fort Wayne, IN 46804

Manuscript Editor: Corine B. Erlandson

DESCRIPTION AND INTERESTS
The Family Digest is dedicated to the joy and fulfillment of Catholic family life. Its articles encourage a family's relationship to its parish, and affirm the ways in which faith can be expressed in daily life. Circ: 150,000.

Audience: Families
Frequency: 6 times each year

FREELANCE POTENTIAL
95% written by nonstaff writers. Publishes 60 freelance submissions yearly; 30% by authors who are new to the magazine. Receives 500 unsolicited mss yearly.

SUBMISSIONS
Send complete ms. Accepts hard copy. No simultaneous submissions; previously published material will be considered. SASE. Responds in 1–2 months.

- Articles: 700–1,200 words. Informational, self-help, how-to, and inspirational articles; and personal experience pieces. Topics include parish and family life, spirituality, Catholic traditions, the saints, prayer life, and seasonal material.
- Depts/columns: Staff written.
- Other: Humorous anecdotes, 25–100 words. Cartoons. Submit seasonal material 7 months in advance.

SAMPLE ISSUE
48 pages (no advertising): 10 articles; 5 depts/columns. Sample copy, free with 6x9 SASE (2 first-class stamps). Writers' guidelines and theme list available.

- "A Moment for Cowardice or Courage." Article uses an incident on a school bus as a parable for the necessity to do what's right, regardless of whether it is popular.
- "Parenting Our Adult Children." Article offers advice for dealing with adult children.

RIGHTS AND PAYMENT
First North American serial rights. Articles, $40–$60. Anecdotes, $25. Cartoons, $40. Pays 1–2 months after acceptance. Provides 2 contributor's copies.

➥ EDITOR'S COMMENTS
Reading the articles in our magazine will give you the best sense of the types of articles that we accept and publish. If it is of interest to a Catholic family, it is of interest to us.

FamilyFun

47 Pleasant Street
Northampton, MA 01060

Department Editor

DESCRIPTION AND INTERESTS
FamilyFun targets parents of children ages three through twelve with lively features on family cooking, vacations, parties, holidays, and crafts. Circ: 2 million.

Audience: Parents
Frequency: 10 times each year
Website: www.familyfun.com

FREELANCE POTENTIAL
50% written by nonstaff writers. Publishes 100+ freelance submissions yearly; 1% by unpublished writers, 2% by authors who are new to the magazine. Receives thousands of queries yearly.

SUBMISSIONS
Query for features. Send complete ms for depts/columns. Accepts hard copy and email submissions to queries.FamilyFun@disney.com. SASE. Responds in 4–6 weeks.

- Articles: 850–3,000 words. Informational and how-to articles. Topics include food, crafts, parties, holidays, sports, and games.
- Depts/columns: 50–1,200 words. See writers' guidelines for specifics. Crafts, nature activities, recipes, family getaways and traditions, household hints, healthy fun, home decorating and gardening tips, and product reviews.
- Other: Submit seasonal material 6 months in advance.

SAMPLE ISSUE
126 pages (47% advertising): 5 articles; 11 depts/columns. Sample copy, $3.95 at newsstands. Guidelines available at website.

- "Outdoor Science." Article offers experiments that can be carried out in the backyard.
- Sample dept/column: "Family Getaways" provides tours of three colonial Virginia towns.

RIGHTS AND PAYMENT
First serial rights. Written material, payment rates vary. Pays on acceptance.

➥ EDITOR'S COMMENTS
We are currently interested in articles about trips that writers take with their families, as well as detailed craft and activity ideas with photographs. Please check our guidelines for detailed submission requirements. We also pay for good ideas.

Family Works Magazine

4 Joseph Court
San Rafael, CA 94903

Editor: Lew Tremaine

DESCRIPTION AND INTERESTS
This regional magazine for parents living in California's Marin and Sonoma counties publishes insightful articles on raising children, improving family communication, and building compassion. It also offers local information regarding education, health, and recreation, among other issues. Circ: 20,000.
Audience: Parents
Frequency: 6 times each year
Website: www.familyworks.org

FREELANCE POTENTIAL
80% written by nonstaff writers. Publishes 75 freelance submissions yearly; 25% by authors who are new to the magazine. Receives 100+ unsolicited mss yearly.

SUBMISSIONS
Send complete ms. Accepts hard copy, disk submissions, and email submissions to familynews@familyworks.org. Availability of artwork improves chance of acceptance. SASE. Responds in 1 month.

- Articles: 1,000 words. Informational articles; profiles; and interviews. Topics include parenting, family issues, recreation, education, finance, crafts, hobbies, sports, health, fitness, nature, and the environment.
- Depts/columns: Word lengths vary. Community news, reviews, and recipes.
- Artwork: B/W and color prints.

SAMPLE ISSUE
24 pages (46% advertising): 10 articles. Sample copy, free. Guidelines available.

- "When the Unspeakable Happens." Article tells how to help a friend survive the soul-shattering loss of a child.
- "Five Tips to Ease the 'Summer to School' Transition." Article provides tips to keep kids' minds active during the summer.
- "Eight Tips for Improving Communication with Your Kids." Article helps parents gain a more peaceful relationship with their kids.

RIGHTS AND PAYMENT
One-time rights. No payment. Provides 3 contributor's copies.

↝EDITOR'S COMMENTS
We seek articles that will interest readers of all ages and stages of personal and family development.

Faze

4936 Yonge Street, Suite 2400
Toronto, Ontario M2N 6S3
Canada

Editor-in-Chief: Lorraine Zander

DESCRIPTION AND INTERESTS
Canadian teens read this magazine for its lively content on pop culture and celebrities as well as for its insightful and inspiring articles on health, social issues, and contemporary topics. Circ: 150,000.
Audience: 13–18 years
Frequency: 5 times each year
Website: www.fazeteen.com

FREELANCE POTENTIAL
75% written by nonstaff writers. Publishes 10 freelance submissions yearly; 10% by unpublished writers, 10% by authors who are new to the magazine. Receives 100 queries yearly.

SUBMISSIONS
Query. Accepts email queries to editor@ fazeteen.com. Response time varies.

- Articles: Word lengths vary. Informational and factual articles; profiles; interviews; and personal experience pieces. Topics include current affairs, real-life and social issues, famous people, entertainment, science, travel, business, technology, and health.
- Depts/columns: Word lengths vary. Short profiles; career descriptions; new products.

SAMPLE ISSUE
66 pages (30% advertising): 9 articles; 8 depts/columns. Sample copy, $3.50 Canadian. Guidelines available at website.

- "Underneath It All." Article talks with actress Jessica Alba about keeping her values while becoming an A-list Hollywood star.
- "What I Learned in an Islamic School." Article details a teenager's choice to be a practicing Muslim.
- Sample dept/column: "Health & Fitness Body Talk" offers tidbits on carbohydrates, exercise benefits, and new fitness trends.

RIGHTS AND PAYMENT
All rights. Written material, 50–$250. Payment policy varies. Provides 1 contributor's copy.

↝EDITOR'S COMMENTS
Faze is a refreshing alternative to the typical pop culture teen magazine. We usually use the same group of writers, but are interested in hearing from younger writers with good ideas. Our editorial content is determined by our teen panel. We are in the process of updating our online presence.

Fertility Today

P.O. Box 117
Laurel, MD 20725

Editor: Diana Broomfield, M.D.

DESCRIPTION AND INTERESTS
This magazine looks not only at the medical and legal issues of reproductive health, but also the spiritual, emotional, and physical aspects of infertility. Though written primarily by medical specialists, it accepts submissions of personal stories. Circ: 175,000.

Audience: Adults
Frequency: Quarterly
Website: www.fertilitytoday.org

FREELANCE POTENTIAL
75% written by nonstaff writers. Publishes 150 freelance submissions yearly; 15% by authors who are new to the magazine, 90% by experts. Receives 144 queries yearly.

SUBMISSIONS
Query with author biography; physicians should also include address of practice. Accepts email queries to articles@ fertilitytoday.org. Responds in 2 months.

- Articles: 800–1,800 words. Informational articles; opinion pieces; profiles; interviews; and personal experience pieces. Topics include fertility issues and treatments, and male and female reproductive health.
- Depts/columns: 800–1,800 words. "Exercise and Nutrition," "Adoption/Child-Free Living," "Mind, Body & Soul," "My Story," and reviews of books on fertility topics. "Health Forum" written by physicians.

SAMPLE ISSUE
96 pages (25% advertising): 2 articles; 11 depts/columns. Sample copy, $6.95. Guidelines and editorial calendar available at website.

- "Nicole Kidman: Planting Seeds of Change." Article uses the actor's pregnancy at age 40 to point out how lifestyle changes can positively impact fertility.
- Sample dept/column: "My Story" is a personal essay detailing a couple's struggles to conceive a child, and their ultimate success.

RIGHTS AND PAYMENT
All rights. Written material, $.50 per word. Kill fee, $100. Pays on acceptance. Provides 3 contributor's copies.

⇢EDITOR'S COMMENTS
Articles should be reader-friendly, as the target audience is the health consumer, not the health care professional.

FitPregnancy

21100 Erwin Street
Woodland Hills, CA 91367

Executive Editor: Sharon Cohen

DESCRIPTION AND INTERESTS
FitPregnancy is about more than prenatal fitness—in addition to its articles on health, nutrition, and exercise, it also offers expectant women advice on fashion, beauty, childbirth, breastfeeding, baby care, and handling the emotions of motherhood. Circ: 500,000.

Audience: Women
Frequency: 6 times each year
Website: www.fitpregnancy.com

FREELANCE POTENTIAL
40% written by nonstaff writers. Publishes 50 freelance submissions yearly; 30% by authors who are new to the magazine. Receives 360 queries yearly.

SUBMISSIONS
Query with clips. Accepts email to scohen@ fitpregnancy.com. Responds in 1 month.

- Articles: 1,000–1,800 words. Informational articles; profiles; and personal experience pieces. Topics include prenatal fitness and nutrition, postpartum issues, breastfeeding, baby care, psychology, and health.
- Depts/columns: 550–1,000 words. Essays by fathers, family issues, prenatal health, newborn health, psychology, childbirth, prenatal nutrition, and relevant news briefs. "Time Out," 550 words.
- Other: Recipes and meal plans.

SAMPLE ISSUE
128 pages (42% advertising): 5 articles; 12 depts/columns. Sample copy, $4.95 at newsstands. Guidelines available at website.

- "Too Early to Test?" Article offers the nine most common early-pregnancy clues.
- "Tell Me What to Eat." Article provides specifics on prenatal nutrition.
- Sample dept/column: "Labor & Delivery" describes six labor complications and how they are handled.

RIGHTS AND PAYMENT
Rights vary. Written material, payment rates vary. Pays on publication. Provides 2 contributor's copies.

⇢EDITOR'S COMMENTS
Due to the number of queries we receive, we ask that you do not follow up on your idea more than once. Please be patient; we will contact you if we're interested.

FLW Outdoors

30 Gamble Lane
Benton, KY 42054

Managing Editor: Lindsey Waddy

DESCRIPTION AND INTERESTS

This is a magazine about fishing—all kinds of fishing at all skill levels and age levels. In its different editions, *FLW Outdoors* tackles various regions and types of fish. It features articles on technique, boat technology, and tournament information. A pull-out section for kids is included. Circ: 50,000.

Audience: Adults; children, 5–12 years
Frequency: 8 times each year
Website: www.flwoutdoors.com

FREELANCE POTENTIAL

50% written by nonstaff writers. Of the freelance submissions published yearly, 20% are by authors who are new to the magazine. Receives 300 queries yearly.

SUBMISSIONS

Query with writing sample. Accepts email submissions to info@flwoutdoors.com. Responds in 1 week.

- Articles: 200 words. Informational and how-to articles; profiles; and humor. Topics include fish, fishing techniques, fishing gear, nature, and the environment.
- Fiction: To 500 words. Genres include adventure and nature stories.
- Depts/columns: Word lengths vary. Tournaments, boat technology, fishing destinations, product reviews, and environmental issues.
- Other: Puzzles.

SAMPLE ISSUE

88 pages: 5 articles; 15 depts/columns; 1 special section for kids. Sample copy, $3.95 at newsstands. Guidelines available.

- "Proving Grounds." Article explains the importance of conquering the smaller lakes before attempting to go to big water.
- "Roaming with Reds." Article explores some of the best redfishing hot spots in the nation.

RIGHTS AND PAYMENT

First North American serial rights. Written material, $200–$500. Payment policy varies.

➤EDITOR'S COMMENTS

We like short, lively pieces that put the reader in the action. We welcome pieces on all types of fishing, from lakes and rivers to saltwater. We are particularly looking for pieces about fishing with children, or articles that inspire children to take up the hobby.

Focus on the Family Clubhouse

8605 Explorer Drive
Colorado Springs, CO 80920

Assistant Editor: Joanna Lutz

DESCRIPTION AND INTERESTS

This magazine from Focus on the Family targets middle-grade readers with articles, stories, puzzles, and games rooted in Christian values. Circ: 90,000.

Audience: 8–14 years
Frequency: Monthly
Website: www.clubhousemagazine.com

FREELANCE POTENTIAL

75% written by nonstaff writers. Publishes 70 freelance submissions yearly; 5% by unpublished writers, 15% by authors who are new to the magazine. Receives 900 unsolicited mss each year.

SUBMISSIONS

Send complete ms. Accepts hard copy. SASE. Responds in 6–8 weeks.

- Articles: 800–1,000 words. Informational, how-to, and factual articles; interviews; personal experience pieces; and humor. Topics include sports, nature, travel, history, religion, current events, multicultural issues, and noteworthy Christians.
- Fiction: Humor, 500 words. Historical Christian fiction, 900–1,600 words. Choose-your-own adventure stories, 1,600–1,800 words. Mysteries and contemporary, multicultural, fantasy, and science fiction, 1,600 words.
- Depts/columns: 300 words. Short, humorous news articles emphasizing biblical lessons.
- Other: Activities, quizzes, jokes, and recipes.

SAMPLE ISSUE

24 pages (5% advertising): 2 articles; 3 stories; 8 depts/columns. Sample copy, $1.50 with 9x12 SASE (2 first-class stamps). Writers' guidelines available.

- "The Twin Rocks." Choose-your-own adventure story has readers navigate a raging river while encouraging patience and compassion.
- Sample dept/column: "Travel Fun" describes games to play on family vacations.

RIGHTS AND PAYMENT

First rights. Written material, to $150. Pays on acceptance. Provides 5 contributor's copies.

➤EDITOR'S COMMENTS

We'd like to receive more articles about real kids making a difference; good stewardship for tweens; and tips on friendship and sibling relationships. We also seek tween fiction.

Focus on the Family Clubhouse Jr.

8605 Explorer Drive
Colorado Springs, CO 80920

Editorial Assistant: Jamie Dangers

DESCRIPTION AND INTERESTS
This fun magazine from Focus on the Family is designed to inspire, entertain, and teach Christian values to children. Circ: 65,000.

Audience: 4–8 years
Frequency: Monthly
Website: www.clubhousejr.com

FREELANCE POTENTIAL
45% written by nonstaff writers. Publishes 10–15 freelance submissions yearly; 5% by unpublished writers, 10% by authors who are new to the magazine. Receives 720 unsolicited mss yearly.

SUBMISSIONS
Send complete ms. Accepts hard copy. No simultaneous submissions. SASE. Responds in 4–6 weeks.

- Articles: To 600 words. Informational articles. Topics include Christian entertainment, recreation, fitness, health, nature, the environment, hobbies, and multicultural issues.
- Fiction: 250–1,000 words. Genres include Bible stories; humor; folktales; and religious, contemporary, and historical fiction.
- Depts/columns: Word lengths vary. Personal anecdotes; crafts and activities; comic strips; and humor.
- Other: Poetry with biblical themes. Submit seasonal material 6 months in advance.

SAMPLE ISSUE
24 pages (no advertising): 3 articles; 1 story; 1 poem; 3 depts/columns; 1 activity. Sample copy, $1.50 with 9x12 SASE (2 first-class stamps). Guidelines available.

- "Veggie Girl." Article profiles a young girl whose dad is the voice of "Veggie Tales'" Larry the Cucumber, and who herself has started doing voices for the show.
- Sample dept/column: "Kids Like You" presents short items about children using their God-given talents in an artistic way.

RIGHTS AND PAYMENT
First North American serial rights. Written material, to $150. Pays on acceptance. Provides 2 contributor's copies.

❖EDITOR'S COMMENTS
We are always interested in craft projects, recipes, science experiments, games, and factual and fun articles about nature.

Ft. Myers & SW Florida Magazine

15880 Summerlin Road, Suite 189
Fort Myers, FL 33908

Publisher: Andrew Elias

DESCRIPTION AND INTERESTS
This regional tabloid celebrates the arts and living in Southwest Florida with coverage on arts and entertainment as well as health, education, travel, and cuisine. Circ: 20,000.

Audience: Adults
Frequency: 6 times each year
Website: www.ftmyersmagazine.com

FREELANCE POTENTIAL
50% written by nonstaff writers. Publishes 15–25 freelance submissions yearly; 80% by unpublished writers, 75% by authors who are new to the magazine. Receives 2,400–3,000 unsolicited mss yearly.

SUBMISSIONS
Send complete ms. Accepts email submissions to ftmyers@optonline.net (Microsoft Word attachments or pasted in body of email). Responds in 1–6 weeks.

- Articles: 500–2,000 words. Informational articles; profiles; interviews; reviews; and local news. Topics include the arts, media, entertainment, travel, computers, crafts, current events, health and fitness, history, popular culture, recreation, social and environmental issues, and parenting.
- Depts/columns: Word lengths vary. Sports, recreation, and book reviews.
- Artwork: JPEG, TIFF, or PDF images.

SAMPLE ISSUE
28 pages (40% advertising): 1 article; 6 depts/columns; 1 calendar. Sample copy, $3 with 9x12 SASE ($.77 postage). Guidelines and editorial calendar available at website.

- "Classical Lass." Article profiles mezzo soprano Janelle McCoy, who founded the Chamber Music Society of Southwest Florida.
- Sample dept/column: "Food & Drink" reviews different types of olive oils.

RIGHTS AND PAYMENT
One-time rights. Written material, $.10 per word. Artwork, $20–$50. Pays 30 days from publication. Provides 2 contributor's copies.

❖EDITOR'S COMMENTS
Our readers are successful, educated, and active. Submissions should relate to this audience. Writers are asked to provide photographs to accompany their articles whenever possible.

Fort Worth Child

Lauren Publications
4275 Kellway Circle, Suite 146
Addison, TX 75001

Editorial Director: Shelley Hawes Pate

DESCRIPTION AND INTERESTS
Parenting trials and tribulations, child development topics, and education issues—it's all gathered here for parents living in greater Fort Worth. Each issue of the magazine also spotlights local goods, services, and family resources. Circ: 40,000.
Audience: Parents
Frequency: Monthly
Website: www.fortworthchild.com

FREELANCE POTENTIAL
25% written by nonstaff writers. Publishes 12–15 freelance submissions yearly; 20% by authors who are new to the magazine. Receives 240 queries yearly.

SUBMISSIONS
Query with résumé. Accepts hard copy, faxes to 972-447-0633, email queries to editorial@dallaschild.com, and simultaneous submissions if identified. SASE. Response time varies.

- Articles: 1,000–2,500 words. Informational, self-help, and how-to articles; humor; profiles; and personal experience pieces. Topics include parenting, education, health, nutrition, fitness, travel, recreation, crafts, hobbies, and regional news.
- Depts/columns: 800 words. Family activities, health and safety, legislation and advocacy, news, trends, education, child development, humor, and reviews.

SAMPLE ISSUE
62 pages (14% advertising): 2 articles; 11 depts/columns. Sample copy, free with 9x12 SASE. Guidelines available at website.

- "Sensing Trouble." Article explains how to spot and treat Sensory Integration Disorder.
- Sample dept/column: "Daddy Diaries" offers a glimpse into a typical day for the father of a child with special needs.

RIGHTS AND PAYMENT
First rights. Written material, payment rates vary. Pays on publication. Provides contributor's copies upon request.

◆◆EDITOR'S COMMENTS
We welcome ideas for articles that would inform or inspire Forth Worth-area mothers and fathers. We give preference to stories with local perspectives that share new and exciting information.

Fostering Families Today

541 East Garden Drive, Unit N
Windsor, CO 80550

Editor: Richard Fischer

DESCRIPTION AND INTERESTS
Adoptive and foster parents—and those who work with them—read this magazine for guidance through the emotional, legal, and medical aspects of bringing a child into their home. Circ: 26,000.
Audience: Adoptive and foster parents
Frequency: 6 times each year
Website: www.fosteringfamiliestoday.com

FREELANCE POTENTIAL
85% written by nonstaff writers. Publishes 40–45 freelance submissions yearly; 30% by unpublished writers, 30% by authors who are new to the magazine, 30% by experts. Receives 72–120 unsolicited mss yearly.

SUBMISSIONS
Send complete ms. Accepts hard copy and email submissions to louis@adoptinfo.net. SASE. Response time varies.

- Articles: 500–1,200 words. Informational and how-to articles; profiles; and personal experience pieces. Topics include adoption, foster parenting, child development, relevant research, health, education, and legal issues.
- Depts/columns: Word lengths vary. News, opinions, advice, profiles, legislation, book reviews, and child advocacy.
- Other: Poetry.

SAMPLE ISSUE
62 pages (no advertising): 14 articles; 8 depts/columns. Sample copy and guidelines available at website.

- "Laughter, Bonding, and Adopting Older Children." Article posits that shared laughter between parent and child is an effective way to cement a new relationship.
- Sample dept/column: "In the Spotlight" includes an article on building a mutually supportive relationship between foster parents and court-appointed child advocates.

RIGHTS AND PAYMENT
Non-exclusive print and electronic rights. No payment. Provides 3 contributor's copies and a 1-year subscription.

◆◆EDITOR'S COMMENTS
Authors whose stories are accepted must return a signed permission agreement, available at our website. Story credits will contain your name and the brief bio you submit.

Fox Valley Kids

P.O. Box 12264
Green Bay, WI 54307

Publisher: Dr. Brookh Lyonns

DESCRIPTION AND INTERESTS
This newly-revamped tabloid depicts Wisconsin family life through coverage of parenting issues, regional activities, healthy living, games, and crafts. Circ: 40,000.
Audience: Families
Frequency: Monthly
Website: www.newandfoxvalleykids.com

FREELANCE POTENTIAL
100% written by nonstaff writers. Publishes 12+ freelance submissions yearly; 5% by unpublished writers, 50% by authors who are new to the magazine. Receives 60–240 queries and unsolicited mss yearly.

SUBMISSIONS
Query or send complete ms. Accepts email to brookh@bigchoicepublishing.com. Response time varies.

- Articles: To 750 words. Informational, how-to, and humorous articles. Topics include parenting, education, recreation, crafts, hobbies, music, the arts, health, fitness, sports, pets, family travel, and home improvement.
- Depts/columns: To 750 words. Reader stories, news briefs, and crafts.
- Other: Submit seasonal material at least 4 months in advance.

SAMPLE ISSUE
12 pages (50% advertising): 7 articles; 6 depts/columns. Sample copy available at website. Guidelines and editorial calendar available.

- "It's Time to Start Planning for Camp Season!" Article suggests questions to ask when researching summer camps, as well as ways to offset the cost.
- Sample dept/column: "Sharing What We Know" contains readers' accounts of their winter activities.

RIGHTS AND PAYMENT
Rights negotiable. All material, payment rates vary. Pays on publication. Provides 1 contributor's copy upon request.

⬦•EDITOR'S COMMENTS
We are focusing our attention on ways to keep families healthy in mind, body, and spirit. We're interested in more stories that take a "green" view as well as those that support parenting without judgment. Our goal is to support parents as they lead their lives.

The Friend

The Church of Jesus Christ of Latter-day Saints
50 East North Temple, 24th Floor
Salt Lake City, UT 84150

Managing Editor: Vivian Paulsen

DESCRIPTION AND INTERESTS
The Church of Jesus Christ of Latter-day Saints publishes this magazine for children ages 12 and under in order to share the Gospel message with its young readers. Circ: 285,000.
Audience: 3–11 years
Frequency: Monthly
Website: www.lds.org

FREELANCE POTENTIAL
60% written by nonstaff writers.

SUBMISSIONS
Send complete ms. Accepts hard copy. SASE. Responds in 2 months.

- Articles: To 1,200 words. Informational and factual articles; profiles; personal experience pieces; and true stories. Topics include spirituality, the Mormon church, personal faith, and conflict resolution.
- Depts/columns: Word lengths vary. Profiles of children from different countries.
- Other: Poetry, word lengths vary. Puzzles, activities, crafts, and cartoons. Submit seasonal material 8 months in advance.

SAMPLE ISSUE
50 pages (no advertising): 3 articles; 7 true stories; 4 depts/columns; 1 poem; 8 activities; 4 comics. Sample copy, $1.50 with 9x12 SASE (4 first-class stamps). Guidelines available.

- "You Are a Child of God." True story recounts the author's childhood experience of being touched by the Holy Spirit.
- "Mac the Prayer Cat." True story tells of a cat that never misses a chance to join in the family prayer circle.
- "Ties that Bind." True story describes how a young boy realizes that all the families in his ward are like one big family, ready to help each other at any time.

RIGHTS AND PAYMENT
First rights. Written material, $100–$250. Poetry, $25. Activities, $20. Pays on acceptance. Provides 2 contributor's copies.

⬦•EDITOR'S COMMENTS
We are always interested in uplifting stories based on real-life events. Stories should show how the Gospel of Jesus Christ is at work in the lives of Mormon children. It is most important that the reader experience the story from a child's perspective.

Fuel

401 Richmond Street West, Suite 245
Toronto, Ontario M5V 1X3
Canada

Editorial Assistant: Nick Aveling

DESCRIPTION AND INTERESTS
This magazine for Canadian teens—mainly
boys—has recently narrowed its focus to
music, music, and more music. Like its sister
publication VerveGirl, it is distributed to
students free of charge through the school
system. Circ: 100,000.
Audience: 14–18 years
Frequency: Twice each year
Website: www.fuelpowered.com

FREELANCE POTENTIAL
60% written by nonstaff writers. Publishes 12
freelance submissions yearly; 10% by authors
who are new to the magazine, 10% by experts.
Receives 36 queries yearly.

SUBMISSIONS
Query. Accepts hard copy and email queries to
nick@youthculture.com. SAE/IRC. Response
time varies.

- Articles: 1,200–2,000 words. Informational
articles; profiles; interviews; personal experience pieces; and reviews. Topics include
music, college, careers, grooming, and health.
- Depts/columns: Word lengths vary. "Pulse,"
"Hot Stuff," "Between the Lines," "On the
Job," "Advice," and "Must-Haves."

SAMPLE ISSUE
22 pages: 1 article; 8 depts/columns. Sample
copy available at website. Guidelines available.

- "A Pretty Odd Journey." Article reviews the
long-awaited second album from the band
Panic at the Disco.
- "'80s Glam Metal Legends Rock the Boat."
Article describes the 2008 Motley Cruise,
featuring shipboard performances by Vince
Neil, RATT, Skid Row, and Slaughter.
- Sample dept/column: "On the Job" tells how
to make a great workplace impression.

RIGHTS AND PAYMENT
Rights vary. Written material, payment rates
vary. Pays on publication.

☛EDITOR'S COMMENTS
We have reduced our frequency from monthly
to only twice a year, so we now need drasti-
cally less copy than we did previously. Also,
we focus almost exclusively on music now,
so the submissions we are most likely to consider are those centering on popular and
indie music.

Fun For Kidz

P.O. Box 227
Bluffton, OH 48517-0227

Editor: Marilyn Edwards

DESCRIPTION AND INTERESTS
Each themed issue of Fun For Kidz is designed
to develop the interests and encourage the
literacy of its young readers. Articles, poems,
games, and activities fill its pages. Circ: 7,000.
Audience: 5–14 years
Frequency: 6 times each year
Website: www.funforkidz.com

FREELANCE POTENTIAL
90% written by nonstaff writers. Publishes
100–150 freelance submissions yearly; 40%
by unpublished writers, 40% by authors who
are new to the magazine. Receives 3,000
unsolicited mss yearly.

SUBMISSIONS
Send complete ms. Accepts hard copy and
simultaneous submissions if identified.
Availability of artwork improves chance of
acceptance. SASE. Responds in 4–6 weeks.

- Articles: 500 words. Informational and how-
to articles. Topics include nature, animals,
pets, careers, cooking, and sports.
- Fiction: 500 words. Animal stories, humorous fiction, and adventure.
- Depts/columns: Word lengths vary. Puzzles,
science, and collecting.
- Artwork: B/W and color prints or transparencies. Line art.
- Other: Activities, filler, games, jokes, puzzles. Submit seasonal material 6–12 months
in advance.

SAMPLE ISSUE
48 pages (no advertising): 1 article; 3 stories;
4 depts/columns; 3 cartoons; 6 activities;
2 poems; 8 puzzles. Sample copy, $6.
Guidelines and theme list available.

- "A Walk in the Park." Article spotlights the
Beverly Cleary Children's Sculpture Garden
in Portland, Oregon.
- Sample dept/column: "Workshop" explains
how to make a snow sensor.

RIGHTS AND PAYMENT
First or reprint rights. Written material, $.05
per word. Artwork, payment rates vary. Pays on
publication. Provides 1 contributor's copy.

☛EDITOR'S COMMENTS
All material should captivate our readers
with lively writing and images designed to
increase literacy.

Games

6198 Butler Pike, Suite 200
Blue Bell, PA 19422-2600

Editor-in-Chief: R. Wayne Schmittberger

DESCRIPTION AND INTERESTS
Readers of all ages looking to exercise their minds read this magazine for its visual and verbal puzzles and quizzes. Also included are informational articles, game reviews, and contests. Circ: 75,000.
Audience: YA–Adult
Frequency: 10 times each year
Website: www.gamesmagazine-online.com

FREELANCE POTENTIAL
86% written by nonstaff writers. Publishes 200+ freelance submissions yearly; 10% by unpublished writers, 20% by authors who are new to the magazine. Receives 960 queries and unsolicited mss yearly.

SUBMISSIONS
Query with outline; or send complete ms. Accepts hard copy and email submissions to games@kappapublishing.com. SASE. Responds in 6–8 weeks.

- Articles: 1,500–3,000 words. Informational articles; profiles; and humor. Topics include game-related events and people, wordplay, and human ingenuity. Game reviews by assignment only.
- Depts/columns: Staff written, except for "Gamebits."
- Other: Visual and verbal puzzles, quizzes, contests, two-play games, and adventures.

SAMPLE ISSUE
80 pages (8% advertising): 1 article; 3 depts/columns; 20 activities. Sample copy, $4.50 with 9x12 SASE ($1.24 postage). Writers' guidelines available.

- "Tower of Scrabble." Article explains how and why the board game was translated into 28 other languages.
- Sample dept/column: "Gamebits" introduces a new type of playing cards.

RIGHTS AND PAYMENT
All North American serial rights. Articles, $500–$1,200. "Gamebits," $100–$250. Pays on publication. Provides 1 contributor's copy.

➥EDITOR'S COMMENTS
We sometimes buy one-time rights for photos; we'll negotiate in those situations. Features can cover almost any subject, from game- or puzzle-related events or people, to mystery, wordplay, and humor.

Genesee Valley Parent

1 Grove Street, Suite 204
Pittsford, NY 14534

Managing Editor: Barbara Melnyk

DESCRIPTION AND INTERESTS
Parents in the Rochester area of New York pick up this free magazine for information on keeping their families healthy and happy. In addition to covering parenting issues about raising children, it is also a resource for local entertainment and services. Circ: 37,000.
Audience: Parents
Frequency: Monthly
Website: www.gvparent.com

FREELANCE POTENTIAL
75% written by nonstaff writers. Publishes 50 freelance submissions yearly; 5% by authors who are new to the magazine. Receives 240 queries yearly.

SUBMISSIONS
Query with clips or writing samples. Accepts hard copy and simultaneous submissions if identified. SASE. Responds in 1–3 months.

- Articles: 700–1,200 words. Informational and how-to articles; profiles; reviews; humor; and personal experience pieces. Topics include regional family events, local goods and services, special and gifted education, social issues, family problems, health and fitness, and parenting.
- Depts/columns: 500–600 words. Family health, teen issues, toddler issues, and short news items.
- Other: Submit seasonal material 4 months in advance.

SAMPLE ISSUE
68 pages (50% advertising): 4 articles; 9 depts/columns; 1 calendar. Guidelines and editorial calendar available.

- "How to Think and Act Like a Co-Parent." Article explains how parents can, and should, act as a team when it comes to raising their children.
- Sample dept/column: "Tot Time" offers ways to set the stage for raising a healthy baby.

RIGHTS AND PAYMENT
Second rights. Articles, $30–$45. Depts/columns, $25–$30. Pays on publication. Provides 1 tearsheet.

➥EDITOR'S COMMENTS
Any article idea that can be useful to parents is welcome here. We put an emphasis on local information and resources.

GeoParent

16101 North 82nd Street, Suite A-9
Scottsdale, AZ 85260

Editors: Betsy Bailey & Nancy Price

DESCRIPTION AND INTERESTS
Part of the SheKnows® family of websites, *GeoParent* is a highly informative and interactive e-zine featuring articles and tips on parenting issues, baby and child development, childhood stages, nutrition, education, and developing healthy relationships with one's children. Hits per month: Unavailable.

Audience: Parents
Frequency: Weekly
Website: www.geoparent.com

FREELANCE POTENTIAL
90% written by nonstaff writers. Publishes 50 freelance submissions yearly. Receives 50 queries and unsolicited mss yearly.

SUBMISSIONS
Prefers query; will accept complete ms. Accepts hard copy, email submissions to content@coincide.com, and submissions via the website. SASE. Response time varies.

- Articles: 500–2,500 words. Informational articles; and advice. Topics include parenting, child development, family issues, pregnancy and childbirth, infancy, child care, nutrition, health, education, and gifted and special education.
- Depts/columns: Word lengths vary. Parenting tips and advice.

SAMPLE ISSUE
Sample issue and writers' guidelines available at website.

- "Teaching Kids About Diversity." Article provides ways parents can introduce the concept of diversity to their kids, and includes a list of do's and don'ts for when they and their child encounter a disabled person.
- Sample dept/column: "GeoParent Tips & Tricks" offers ideas for getting back-to-school supplies while sticking to a budget.

RIGHTS AND PAYMENT
Rights vary. Written material, $25–$50; $10 for reprints. Pays on publication.

➔EDITOR'S COMMENTS
We're looking for pieces that can draw readers in quickly and offer opportunities for them to interact with the article, as well as each other. Your experience with children is your best source; however, we also expect your piece to be well researched.

Georgia Family Magazine

523 Sioux Drive
Macon, GA 31210

Publisher: Olya Fessard

DESCRIPTION AND INTERESTS
Georgia Family was founded by a former learning disabilities teacher who wanted to provide more information for parents about educational strategies. It has since grown into a general parenting and family magazine with articles on topics of interest to all. It also includes regional calendars of family-friendly events. Circ: 20,000.

Audience: Families
Frequency: Monthly
Website: www.georgiafamily.com

FREELANCE POTENTIAL
60% written by nonstaff writers. Publishes 100–125 freelance submissions yearly.

SUBMISSIONS
Send complete ms. Accepts email submissions to publisher@georgiafamily.com. Response time varies.

- Articles: Word lengths vary. Informational articles. Topics include parenting, family issues, education, gifted and special education, health, and fitness.
- Depts/columns: Word lengths vary. Financial and money matters, science and technology, travel, the arts, etiquette, home and garden, and media and new product reviews.

SAMPLE ISSUE
64 pages: 4 articles; 14 depts/columns. Guidelines available at website.

- "Eight Tips for Yourself for Mother's Day." Article lists eight things a busy mother can do to make herself happier and healthier.
- "Government for the People—Even Kids." Article explains how to discuss the concept of government and politics with your kids.
- "Check It Out." Article explains the importance of eye examinations for young children, and signs of a vision problem.

RIGHTS AND PAYMENT
One-time, reprint, and electronic rights. Articles, $20–$90. Reprints, $10–$30. Pays on publication. Provides 1 tearsheet.

➔EDITOR'S COMMENTS
The best places for freelancers are our health, dining out, travel, and women's health departments. We prefer articles that can be localized to parenting in central Georgia, but we won't ignore general topics, either.

Gifted Education Press Quarterly

10201 Yuma Court
P.O. Box 1586
Manassas, VA 20109

Editor & Publisher: Maurice D. Fisher

DESCRIPTION AND INTERESTS
Available in both print and electronic form, this scholarly newsletter addresses issues surrounding the education of gifted children. Circ: 13,000.
Audience: Educators and parents
Frequency: Quarterly
Website: www.giftededpress.com

FREELANCE POTENTIAL
90% written by nonstaff writers. Publishes 15 freelance submissions yearly; 20% by unpublished writers, 80% by authors who are new to the magazine. Receives 30 queries yearly.

SUBMISSIONS
Query. Accepts hard copy. SASE. Responds in 2 weeks.

- Articles: 3,000–5,000 words. Informational, how-to, and research articles; personal experience pieces; profiles; interviews; and scholarly essays. Topics include gifted education; multicultural, ethnic, and social issues; homeschooling; multiple intelligence; parent advocates; academic subjects; the environment; and popular culture.

SAMPLE ISSUE
18 pages (no advertising): 5 articles. Sample copy, $4 with 9x12 SASE.

- "A Never-Ending Debate: Cultural Bias in Testing." Article explains how test bias can affect the assessment of gifted minority students.
- "Gifted Children and Shakespeare: Using Monologues and Scenes." Article presents both theoretical and practical ideas for teaching Shakespeare to advanced students.
- "John Grisham: Literary Entertainments for Gifted Students." Article discusses the value of this author's novels in teaching gifted students about the legal system.

RIGHTS AND PAYMENT
All rights. No payment. Provides 5 copies.

➥EDITOR'S COMMENTS
We are seeking articles on the impact of "No Child Left Behind" legislation on gifted education programs; the future of gifted education in the United States; mathematics, science, humanities, and literature education for the gifted; the study of aeronautics and rocket design; and innovative summer programs.

Girls' Life

4529 Harford Road
Baltimore, MD 21214

Associate Editor: Mandy Forr

DESCRIPTION AND INTERESTS
All things of interest to pre-teen and teenage girls can be found in *Girls' Life*, including trends, fashion, beauty tips, friendship, emotional issues, TV and movie gossip, and boys. Circ: 400,000.
Audience: 10–14 years
Frequency: 6 times each year
Website: www.girlslife.com

FREELANCE POTENTIAL
35% written by nonstaff writers. Publishes 35–40 freelance submissions yearly; 10% by unpublished writers, 20% by authors new to the magazine. Receives 1,200 queries yearly.

SUBMISSIONS
Query with outline. Accepts hard copy and email queries to mandy@girlslife.com. SASE. Responds in 3 months.

- Articles: 1,200–2,500 words. Informational, service-oriented articles. Topics include self-esteem, health, friendship, relationships, sibling rivalry, school issues, facing challenges, and setting goals.
- Fiction: 2,000–2,500 words. Stories featuring girls.
- Depts/columns: 300–800 words. Celebrity spotlights; profiles of real girls; advice about friendship, beauty, and dating; fashion trends; decorating tips; cooking; crafts; and media reviews.
- Other: Quizzes; fashion spreads.

SAMPLE ISSUE
88 pages (30% advertising): 11 articles; 1 story; 18 depts/columns; 1 quiz. Sample copy, $5. Guidelines and editorial calendar available at website.

- "When You Least Expect It." Article discusses the issues of teen sex and pregnancy, and profiles girls who got pregnant as teens.
- Sample dept/column: "Friends" offers tips for distinguishing bad friends from fun friends.

RIGHTS AND PAYMENT
All or first rights. Written material, payment rates vary. Pays on publication. Provides 1 contributor's copy.

➥EDITOR'S COMMENTS
We are currently interested in material that offers new approaches to topics like friendship, family, and relationships.

Green Teacher

95 Robert Street
Toronto, Ontario M5S 2K5
Canada

Co-Editor: Tim Grant

DESCRIPTION AND INTERESTS
Published on acid-free paper, *Green Teacher* promotes environmental education and global awareness both inside and outside of schools. Circ: 7,500.

Audience: Teachers, grades K–12
Frequency: Quarterly
Website: www.greenteacher.com

FREELANCE POTENTIAL
60% written by nonstaff writers. Publishes 30 freelance submissions yearly; 80% by unpublished writers. Receives 120 queries and unsolicited mss yearly.

SUBMISSIONS
Prefers query with outline; will accept complete ms with 8–10 photos. Prefers email submissions to tim@greenteacher.com (Microsoft Word, RTF, or TXT attachments); will accept IBM disk submissions with hard copy. Availability of artwork improves chance of acceptance. SAE/IRC. Responds in 2 months.

- Articles: 1,500–3,000 words. Informational and how-to articles; opinion pieces; and lesson plans. Topics include environmental and global education.
- Depts/columns: Word lengths vary. Resources, reviews, and announcements.
- Artwork: JPEG and TIFF images at 300 dpi; B/W and color prints or slides. Line art.
- Other: Submit Earth Day material 6 months in advance.

SAMPLE ISSUE
48 pages (12% advertising): 12 articles; 2 depts/columns. Sample copy and guidelines available at website.

- "Using the Arts to Reclaim Community Pride." Article outlines how a Tennessee high school demonstrated the Copper Basin's history of environmental degradation.
- Sample dept/column: "Resources" reviews several books about the environment.

RIGHTS AND PAYMENT
Rights negotiable. No payment. Provides 5 contributor's copies and a free subscription.

➡️ EDITOR'S COMMENTS
Our primary purpose is not simply to inform readers about environmental and global issues, but rather to provide ideas for *teaching* about these issues.

Grit

1503 SW 42nd Street
Topeka, KS 66609-1265

Editor-in-Chief: K. C. Compton

DESCRIPTION AND INTERESTS
Grit celebrates the remarkable qualities of character, pluck, and perseverance, and caters to its readers' love of the rural life. It publishes articles on topics of interest to individuals who live—or who aspire to live—the rural life. Circ: 150,000.

Audience: Families
Frequency: 6 times each year
Website: www.grit.com

FREELANCE POTENTIAL
90% written by nonstaff writers. Publishes 80–90 freelance submissions yearly; 50% by unpublished writers, 50% by authors who are new to the magazine. Receives 2,400 queries each year.

SUBMISSIONS
Query. Prefers email queries to grit@grit.com (include "Query" in subject line); will accept hard copy. SASE. Response time varies.

- Articles: 800–1,500 words. Informational and how-to articles; profiles; and personal experience pieces. Topics include American history, rural family lifestyles, parenting, pets, crafts, community involvement, farming, gardening, and antiques.
- Depts/columns: 500–1,500 words. Farm economics, comfort food, technology, equipment, and medical advice.
- Artwork: 35mm color slides and prints; digital images. B/W prints for nostalgia pieces.

SAMPLE ISSUE
96 pages (50% advertising): 4 articles; 11 depts/columns. Sample copy and editorial calendar, $4 with 9x12 SASE. Guidelines available at website.

- "How to Fish Kid-Style." Article explains how to make an afternoon fishing with the kids a thoroughly enjoyable experience.
- Sample dept/column: "Food" offers a nostalgic look at comfort foods.

RIGHTS AND PAYMENT
Shared rights. All material, payment rates vary. Pays on publication. Provides 3 contributor's copies.

➡️ EDITOR'S COMMENTS
We intend to be an authoritative and sometimes playful voice for those who enjoy the rural or small-town lifestyle.

Group

Group Publishing, Inc.
P.O. Box 481
Loveland, CO 80539-0481

Associate Editor: Scott Firestone

DESCRIPTION AND INTERESTS
Group is an interdenominational magazine for youth ministry leaders. It prides itself on its tireless commitment to ignite incredible things in youth ministry. Each issue is filled with articles and resources to help youth ministry leaders promote a personal relationship with Jesus within their members. Circ: 40,000.
Audience: Youth ministry leaders
Frequency: 6 times each year
Website: www.youthministry.com

FREELANCE POTENTIAL
60% written by nonstaff writers. Publishes 200 freelance submissions yearly; 50% by unpublished writers, 80% by authors who are new to the magazine. Receives 300 queries yearly.

SUBMISSIONS
Query with outline and clips or writing samples; state availability of artwork. Accepts hard copy. SASE. Responds in 2–3 months.

- Articles: 500–1,700 words. Informational and how-to articles. Topics include youth ministry strategies, recruiting and training adult leaders, understanding youth culture, professionalism, time management, leadership skills, and the professional and spiritual growth of youth ministers.
- Depts/columns: "Try This One," to 300 words. "Hands-On Help," to 175 words.
- Artwork: B/W or color illustration samples. No prints.

SAMPLE ISSUE
82 pages (30% advertising): 3 articles; 23 depts/columns. Sample copy, $2 with 9x12 SASE. Guidelines available.

- "Parents Who Are MIA in Their Kids' Lives." Article discusses the issue of parents who do not lead in the discipling of their children, and what complainers should do about it.
- "Protecting Your Family in Ministry." Article discusses the importance of leading a ministry while still making time for one's family.

RIGHTS AND PAYMENT
All rights. Articles, $125–$225. Depts/columns, $50. Pays on acceptance.

➻EDITOR'S COMMENTS
Our sole mission is to help churches develop their youth ministries, and to serve individual youth ministers in their endeavors.

Guide

Review and Herald Publishing Association
55 West Oak Ridge Drive
Hagerstown, MD 21740

Associate Editor: Rachel Whitaker

DESCRIPTION AND INTERESTS
Guide is a Christian story magazine offering tweens and young teens uplifting articles and stories that affirm Christian values. Spanish-language pieces are sometimes featured. Circ: 29,000.
Audience: 10–14 years
Frequency: Weekly
Website: www.guidemagazine.org

FREELANCE POTENTIAL
95% written by nonstaff writers. Publishes 250 freelance submissions yearly; 15% by unpublished writers, 20% by authors who are new to the magazine. Receives 840 unsolicited mss each year.

SUBMISSIONS
Send complete ms. Prefers email submissions to guide@rhpa.org; will accept hard copy and simultaneous submissions if identified. SASE. Responds in 4–6 weeks.

- Articles: To 1,200 words. True stories with inspirational and personal growth themes, true adventure pieces, and humor. Nature articles with a religious emphasis, 750 words.
- Other: Puzzles, activities, and games. Submit seasonal material about Thanksgiving, Christmas, Mother's Day, and Father's Day 8 months in advance.

SAMPLE ISSUE
32 pages: 1 article; 3 stories; 2 Bible lessons; 1 cartoon; 3 activities. Sample copy, free with 9x12 SASE (2 first-class stamps). Writers' guidelines available.

- "Evil Winds." Story is a cautionary tale about a young Christian girl who is pressured to play with a Ouija board.
- "Who Ya Gonna Believe?" Article explains how secular music degrades the values and morals of young listeners.

RIGHTS AND PAYMENT
First and reprint rights. Written material, $.06–$.12 per word. Pays on acceptance. Provides 3 contributor's copies.

➻EDITOR'S COMMENTS
Everything we publish must ring true; we do not publish fiction. We are seeking "everyday kid" stories showing God's involvement in school, family, and friendship situations. Stories must be written from a child's perspective.

Gumbo Teen Magazine

1818 North Dr. Martin Luther King Drive
Milwaukee, WI 53212

Managing Editor: Carrie Trousil

DESCRIPTION AND INTERESTS
This award-winning magazine is not your average teen zine. It is written by teens themselves, for readers who are just like them. It features profiles of teens, and articles on current events, global issues, and such topics as careers, education, money, health, fashion, sports, and entertainment. Circ: 25,000.

Audience: 13–19 years
Frequency: 6 times each year
Website: www.gumboteenmagazine.com

FREELANCE POTENTIAL
15% written by nonstaff writers. Publishes 20 freelance submissions yearly; 90% by unpublished writers, 50% by authors who are new to the magazine. Receives 50 queries yearly.

SUBMISSIONS
Query. Accepts email queries to carrie@mygumbo.com. Responds in 6 weeks.

- Articles: 500–1,000 words. Informational articles; profiles; interviews; and personal experience pieces. Topics include entertainment, popular culture, current events, health, and multicultural and social issues.
- Depts/columns: Word lengths vary. Fashion, travel, African American history, careers, Q&As, living legends, school, health, book and music reviews, news, and technology.
- Other: Poetry, to 500 words. Submit seasonal material 6 months in advance.

SAMPLE ISSUE
58 pages (30% advertising): 3 articles; 20 depts/columns. Sample copy, $3 with 8x10 SASE ($1.23 postage). Guidelines available.

- "She's 'Just' Funny." Article profiles a young actress on Nickelodeon's *Just Jordan*.
- Sample dept/column: "Electronics" offers news of the hottest electronics and games for the holiday season.

RIGHTS AND PAYMENT
One-time rights. Written material, $25. Pays on publication. Provides 10 contributor's copies.

➥EDITOR'S COMMENTS
We welcome article ideas from writers ages 13 through 19, which is also the age range of our readership. We usually have a plethora of entertainment articles. What will get our attention are articles on weightier subjects and challenges that teens face every day.

Gwinnett Parents Magazine

3651 Peachtree Parkway, Suite 325
Suwanee, GA 30024

Editor: Terrie Carter

DESCRIPTION AND INTERESTS
Families living in Gwinnett County, Georgia, read this magazine for parenting articles, regional events and news coverage, and other relevant information. Circ: Unavailable.

Audience: Parents
Frequency: Monthly
Website: www.gwinnettparents.com

FREELANCE POTENTIAL
75% written by nonstaff writers. Publishes several freelance submissions yearly; 5% by unpublished writers, 14% by new authors.

SUBMISSIONS
Query or send complete ms. Accepts email submissions to editor@gwinnettparents.com (include "Editorial Submission" in subject line). Responds in 3–4 weeks.

- Articles: 500–1,000 words. Profiles, 350–450 words. Informational and self-help articles; profiles; and personal experience pieces. Topics include education, recreation, sports, health, working parents, and family finances.
- Depts/columns: Word lengths vary. Parenting advice, health, education, community news, recipes, home improvement, and reviews.

SAMPLE ISSUE
60 pages (50% advertising): 5 articles; 15 depts/columns. Sample copy, $4 with 9x12 SASE ($.77 postage). Guidelines available.

- "What Does My New Baby Really Need?" Article guides new parents as they shop for their new baby's clothing.
- "Top Ten Tips to Get Your Baby Sleeping Better." Article offers techniques for dealing with sleep problems in newborns.
- Sample dept/column: "Gwinnett Public Schools" announces the annual review of the district's Academic Knowledge and Skills curriculum.

RIGHTS AND PAYMENT
First and nonexclusive online archival rights. Written material, $25–$75. Profiles, $25–$50. Pays on publication.

➥EDITOR'S COMMENTS
We use freelancers to write profiles of individuals to be featured in the magazine. We are also seeking freelance writers to assist in editorial content and to work with professionals who need help telling their story.

Happiness

P.O. Box 388
Portland, TN 37148

Editor: Sue Fuller

DESCRIPTION AND INTERESTS
Happiness is a family-oriented television guide that, along with a TV schedule, features self-improvement articles, poems, reader contributions, children's stories, and puzzles. It publishes material with an uplifting, positive tone. Circ: 150,000.

Audience: Families
Frequency: Weekly
Website: www.happiness.com

FREELANCE POTENTIAL
75% written by nonstaff writers. Of the freelance submissions published yearly, 25% are by unpublished writers, 25% by authors who are new to the magazine.

SUBMISSIONS
Send complete ms. Accepts hard copy. Availability of artwork improves chance of acceptance. SASE. Responds in 3 months

- Articles: 500 words. Informational, self-help, and how-to articles; humor; and personal experience pieces. Topics include careers, education, health, fitness, hobbies, animals, pets, nature, the environment, recreation, and travel.
- Depts/columns: 25-75 words. Cooking, health, humor, and tips from readers.
- Artwork: Color prints.
- Other: Puzzles and games. Poetry. Submit seasonal material 4 months in advance.

SAMPLE ISSUE
16 pages (no advertising): 2 articles; 8 depts/columns; 10 activities. Guidelines available.

- "Random Acts of Kindness." Article details one woman's experiences receiving and doing good deeds.
- Sample dept/column: "Happiness Within" includes a lesson learned about a hastily finished job.

RIGHTS AND PAYMENT
First rights. All material, payment rates vary. Pays on publication.

⇢EDITOR'S COMMENTS
Articles on true-to-life, heartwarming experiences are always needed. We look for pieces that emphasize a more fulfilling or happier life in health, personality, and behavior. Everything submitted should inspire our readers to adopt a more positive outlook.

High Adventure

General Council of the Assemblies of God
1445 North Boonville Avenue
Springfield, MO 65802-1894

Editor: John Hicks

DESCRIPTION AND INTERESTS
This periodical written for school-age boys encourages the ongoing development of outdoor skills, promotes physical fitness and health awareness, expands leadership abilities, and fosters a growing personal relationship with Jesus Christ. It is published by the Royal Rangers Ministries of the Assemblies of God. Circ: 87,000.

Audience: Boys, grades K–12
Frequency: Quarterly
Web: www.royalrangers.ag.org/
   highadventure

FREELANCE POTENTIAL
40% written by nonstaff writers. Publishes 6–10 freelance submissions yearly. Receives 150–200 unsolicited mss yearly.

SUBMISSIONS
Send complete ms. Accepts hard copy, IBM disk submissions, email submissions to ranger@ag.org, and simultaneous submissions if identified. SASE. Responds in 1–2 weeks.

- Articles: 1,000 words. Informational, self-help, and how-to articles. Topics include nature, wildlife, Christian youth issues, and salvation. Also publishes adventure pieces, historical biographies, and personal testimonies.
- Depts/columns: Word lengths vary. News.
- Artwork: Color prints.
- Other: Puzzles, games, activities, and jokes. Submit seasonal material 6 months in advance.

SAMPLE ISSUE
16 pages (no advertising): 3 articles; 3 depts/columns; 1 comic. Sample copy, free with 9x12 SASE. Guidelines available.

- "Rangers, Is it More Than Camping?" Article profiles the Royal Rangers program and explains the various age levels and activities.
- "Fundraising Starts with a Good Plan." Article provides a blueprint for a successful fundraising effort.

RIGHTS AND PAYMENT
First or all rights. All material, payment rates vary. Pays on publication. Provides 2 copies.

⇢EDITOR'S COMMENTS
All articles must be in keeping with our beliefs and standards of conduct and morality.

Higher Things

Dare To Be Lutheran

Good Shepherd Lutheran Church
5009 Cassia Street
Boise, ID 83705

Managing Editor: Adriane Dorr

DESCRIPTION AND INTERESTS

As its subtitle suggests, this magazine for teens is aimed at enriching the faith of Lutheran youth. It publishes articles that focus on God's work in Christ for salvation and how the Gospel message can be applied to young people's lives. Circ: Unavailable.

Audience: 13–19 years
Frequency: Quarterly
Website: www.higherthings.org

FREELANCE POTENTIAL

60% written by nonstaff writers. Publishes 36 freelance submissions yearly; 60% by unpublished writers, 40% by authors who are new to the magazine. Receives 36 queries, 72 unsolicited mss yearly.

SUBMISSIONS

Query or send complete ms. Accepts hard copy and email submissions to submissions@ higherthings.org. SASE. Response time varies.

- Articles: 500–800 words. Informational and how-to articles; profiles; interviews; and personal experience pieces. Topics include religion, current events, recreation, social issues, and travel.
- Depts/columns: Staff written.

SAMPLE ISSUE

32 pages (10% advertising): 8 articles; 5 depts/columns. Sample copy, $3. Guidelines and theme list available.

- "Misfits." Article tells readers that they are not alone, even when they feel outcast from the crowd.
- "Dating." Article discusses a Christian approach to romance and dating, supported by Scripture.
- "Islam 101." Article provides an overview of what it deems "a dangerous religion."

RIGHTS AND PAYMENT

Rights vary. No payment. Provides contributor's copies.

➡EDITOR'S COMMENTS

Be as succinct and to the point as possible. Keep your writing exciting and fresh. Just because the subject matter may be difficult doesn't mean your writing style has to suffer. Also, pay attention to your vocabulary—you don't need to use big words to explain complicated concepts.

Highlights for Children

803 Church Street
Honesdale, PA 18431

Manuscript Submissions

DESCRIPTION AND INTERESTS

Each colorful issue of *Highlights for Children* offers "Fun with a Purpose" through articles, stories, and activities. Circ: 2 million.

Audience: Up to 12 years
Frequency: Monthly
Website: www.highlights.com

FREELANCE POTENTIAL

99% written by nonstaff writers. Publishes 200 freelance submissions yearly; 40% by unpublished writers, 60% by authors who are new to the magazine. Receives 6,500 mss yearly.

SUBMISSIONS

Send complete ms for fiction. Query or send complete ms for nonfiction. Accepts hard copy. SASE. Responds to queries in 2–4 weeks, to mss in 4–6 weeks.

- Articles: To 500 words for 3–7 years; to 800 words for 8–12 years. Informational articles; interviews; profiles; and personal experience pieces. Topics include nature, animals, science, crafts, hobbies, world cultures, history, arts, and sports.
- Fiction: To 500 words for 3–7 years; to 800 words for 8–12 years. Genres include adventure, mystery, sports stories, multicultural fiction, and retellings of traditional stories.
- Depts/columns: Word lengths vary. Science experiments and crafts.
- Other: Puzzles and games. Poetry, to 10 lines.

SAMPLE ISSUE

44 pages (no advertising): 4 articles; 3 stories; 3 poems; 1 rebus; 9 activities; 20 depts/columns. Sample copy, free with 9x12 SASE (4 first-class stamps). Guidelines available.

- "Dipped in Chocolate." Article recounts a humorous anecdote about Milton Hershey.
- "The Tennin's Robe." Story is a retelling of a Japanese play in which a magical creature saves a young boy's mother.

RIGHTS AND PAYMENT

All rights. Written material, payment rates vary. Pays on acceptance. Provides 2 copies.

➡EDITOR'S COMMENTS

We will publish only wholesome material. We prefer stories that teach by positive example rather than preach. All submissions must be age-appropriate and sensitive to the diversity of our readership.

High School Writer
Junior High Edition

Writer Publications
P.O. Box 718
Grand Rapids, MI 55744-0718

Editor: Emily Benes

DESCRIPTION AND INTERESTS
Dedicated to showcasing the writing of junior high school students, this tabloid has been providing a real audience for student writers for 23 years. It publishes articles, essays, fiction, and poetry. Circ: 44,000.

Audience: Junior high school students
Frequency: 6 times each year
Website: www.writerpublications.com

FREELANCE POTENTIAL
100% written by nonstaff writers. Publishes 1,000 freelance submissions yearly; 1% by unpublished writers, 99% by authors who are new to the magazine. Receives 100 unsolicited mss yearly.

SUBMISSIONS
Accepts submissions from students of subscribing teachers in junior high school only. Send complete ms. Accepts hard copy and simultaneous submissions if identified. SASE. Response time varies.

- Articles: To 2,000 words. Informational and how-to articles; personal experience pieces; and profiles. Topics include family, religion, health, social issues, careers, college, multicultural issues, travel, nature, the environment, science, and computers.
- Fiction: To 2,000 words. Genres include historical and contemporary fiction, science fiction, drama, adventure, mystery, humor, fantasy, and sports and nature stories.
- Other: Poetry, no line limit.

SAMPLE ISSUE
8 pages (no advertising): 18 stories; 5 articles; 16 poems. Sample copy, free. Guidelines available in each issue.

- "Boy Was I Wrong." Personal experience piece shares a girl's anticipation of getting a new puppy.
- "Frog Intelligence." Article discusses the reasons why frogs are smart.
- "How to Save a Life." Article explains why donating blood is so important.

RIGHTS AND PAYMENT
One-time rights. No payment. Provides 2 contributor's copies.

⚬EDITOR'S COMMENTS
We welcome original submissions on almost every topic as long as it is in good taste.

Home Education Magazine

P.O. Box 1083
Tonasket, WA 98855

Articles Editor: Carol Narigon

DESCRIPTION AND INTERESTS
Founded in 1983, this magazine is dedicated to serving the needs of parents who homeschool, or anyone who simply enjoys learning and living. It strives to empower families to be able to successfully homeschool if they choose. Circ: 110,000.

Audience: Parents
Frequency: 6 times each year
Website: www.homeedmag.com

FREELANCE POTENTIAL
40% written by nonstaff writers. Publishes 40 freelance submissions yearly; 25% by unpublished writers, 40% by authors who are new to the magazine. Receives 720 queries yearly.

SUBMISSIONS
Query or send complete ms. Prefers email to articles@homeedmag.com (Microsoft Word attachments); will accept hard copy. SASE. Responds in 1–2 months.

- Articles: 900–1,700 words. Informational, how-to, and personal experience pieces; interviews; and profiles. Topics include homeschooling, activism, lessons, and parenting issues.
- Depts/columns: Staff written.
- Artwork: B/W and color prints. Digital images at 200 dpi (300 dpi for cover images).

SAMPLE ISSUE
58 pages (13% advertising): 8 articles; 9 depts/columns. Sample copy, $6.50. Guidelines available at website.

- "To Push or Not to Push." Article discusses the dilemma homeschooling parents often face about letting their children set the pace.
- Sample dept/column: "Taking Charge" outlines the impact virtual, or Internet-based, schools might have on homeschooling.

RIGHTS AND PAYMENT
First North American serial and electronic rights. Articles, $50–$100. Artwork, $12.50; $100 for cover art. Pays on acceptance. Provides 1+ contributor's copies.

⚬EDITOR'S COMMENTS
While we have discontinued our poetry section, we have added a cooking section and a photo section. We are interested in receiving the personal accounts of homeschooling experiences by parents or older students.

Home Educator's Family Times

P.O. Box 6442
Brunswick, ME 04011

Editor: Jane R. Boswell

DESCRIPTION AND INTERESTS
As "America's leading homeschooling and family newspaper," this publication provides homeschooling parents with curriculum ideas and teaching tips. It is published both traditionally and electronically. Circ: 25,000. Hits per month: 400,000.
Audience: Parents
Frequency: 6 times each year
Website: www.homeeducator.com/familytimes

FREELANCE POTENTIAL
90% written by nonstaff writers. Publishes 50 freelance submissions yearly; 25% by authors who are new to the magazine.

SUBMISSIONS
Send complete ms with author biography and permission statement from website. Accepts CD submissions and email to famtimes@blazenetme.net (Microsoft Word or text attachments). SASE. Response time varies.

- Articles: 1,000–1,500 words. Informational and how-to articles; opinions; and personal experience pieces. Topics include homeschooling methods and lessons, family life, parenting, pets, reading, art, science, and creative writing.
- Depts/columns: Staff written.

SAMPLE ISSUE
24 pages (41% advertising): 11 articles; 2 depts/columns. Sample copy and guidelines available at website.

- "Parts Is Parts—Intelligence Revisited." Opinion piece discusses learning differences and natural aptitude among children.
- "Careers: A Project for Your Homeschooled Teen." Article provides a formula for weighing the cost of college against the potential income derived from a chosen career.

RIGHTS AND PAYMENT
One-time and electronic reprint rights. No payment.

➡EDITOR'S COMMENTS
When submitting an article, please include a brief biography to appear at the end of the article, along with your email address or website. The byline replaces traditional monetary compensation, and is intended to promote the author's product or service.

Homeschooling Today

P.O. Box 244
Abingdon, VA 24212

Editor-in-Chief: Jim Bob Howard

DESCRIPTION AND INTERESTS
Homeschooling Today is a comprehensive home education magazine that puts an emphasis on literature, fine arts, and Christian living. It exists to support and inform families in the homeschooling process. Circ: 11,500.
Audience: Parents
Frequency: 6 times each year
Website: www.homeschooltoday.com

FREELANCE POTENTIAL
85% written by nonstaff writers. Publishes 60–70 freelance submissions yearly; 6% by unpublished writers, 14% by authors who are new to the magazine. Receives 60–120 queries and unsolicited mss yearly.

SUBMISSIONS
Query for feature articles. Send complete ms for departments. Accepts email submissions to management@homeschooltoday.com (include "Article Submission" in subject line). Responds in 3–6 months.

- Articles: 2,000–2,400 words. Informational, self-help, and how-to articles; profiles; and personal experience pieces. Topics include education, religion, music, technology, special education, the arts, history, mathematics, and science.
- Columns: Staff written.
- Departments: 800–1,200 words. Time management, history, music, religion, the arts, and homeschooling tips. Reviews, 250–500 words.

SAMPLE ISSUE
68 pages: 4 articles; 6 columns; 12 departments. Sample copy, $5.95. Guidelines and theme list available at website.

- "Gifts for the Journey." Article explains ways parents can help prepare children to live a rewarding Christian life.
- Sample dept/column: "Making Melody" offers ideas for starting a child in music lessons.

RIGHTS AND PAYMENT
All rights. Written material, $.08 per word. Pays on publication. Provides 1 contributor's copy.

➡EDITOR'S COMMENTS
Our readers are Christian, so all content must reflect biblical truths. Please refer to our guidelines for a list of departments that are most open to new writers.

Hopscotch

P.O. Box 164
Bluffton, OH 45817-0164

Editor: Marilyn Edwards

DESCRIPTION AND INTERESTS

Designed to challenge young girls to enjoy and make the most of their childhood years, *Hopscotch* features wholesome activities and creative, high-quality fiction. Circ: 16,000.

Audience: Girls, 5–14 years
Frequency: 6 times each year
Website: www.hopscotchmagazine.com

FREELANCE POTENTIAL

80% written by nonstaff writers. Publishes 100–250 freelance submissions yearly; 30% by unpublished writers, 40% by authors who are new to the magazine. Receives 240 queries, 3,600 mss yearly.

SUBMISSIONS

Query or send complete ms. Include artwork with nonfiction. Accepts hard copy and simultaneous submissions if identified. SASE. Responds to queries in 1–2 weeks, to mss in 2–3 months.

- Articles: To 500 words. Informational articles; profiles; and personal experience pieces. Topics include crafts, hobbies, games, sports, pets, nature, careers, and cooking.
- Fiction: To 1,000 words. Genres include mystery, adventure, historical fiction, and multicultural fiction.
- Depts/columns: 500 words. Crafts; cooking.
- Artwork: Prefers B/W prints; will accept color prints.
- Other: Puzzles, activities, and poetry.

SAMPLE ISSUE

48 pages (no advertising): 8 articles; 1 story; 4 depts/columns; 2 poems; 3 comics; 6 puzzles. Sample copy, $6 with 9x12 SASE. Guidelines and theme list available.

- "Pizza Fit for a Queen." Article details the origins of pizza.
- Sample dept/column: "Crafts" describes how to make a stethoscope.

RIGHTS AND PAYMENT

First and second rights. Written material, $.05 per word. Artwork, payment rates vary. Pays on publication. Provides 1 contributor's copy.

➻EDITOR'S COMMENTS

We would like to see more illustrated craft how-to's, creative cooking ideas, and nonfiction with good photo support.

The Horn Book Magazine

56 Roland Street, Suite 200
Boston, MA 02129

Editor-in-Chief: Roger Sutton

DESCRIPTION AND INTERESTS

Focusing on children's and young adult literature, this magazine is read by librarians, teachers, and parents for its in-depth book reviews as well as articles and essays that critique and analyze. Circ: 16,000.

Audience: Parents, teachers, librarians
Frequency: 6 times each year
Website: www.hbook.com

FREELANCE POTENTIAL

70% written by nonstaff writers. Publishes 12–15 freelance submissions yearly; 10% by unpublished writers, 30% by authors who are new to the magazine. Receives 240 queries, 120 unsolicited mss yearly.

SUBMISSIONS

Query or send complete ms. Prefers email to info@hbook.com; will accept hard copy. SASE. Responds in 4 months.

- Articles: To 2,800 words. Interviews with children's authors, illustrators, and editors; critical articles about children's and young adult literature; and book reviews.
- Depts/columns: Word lengths vary. Perspectives from illustrators; children's publishing updates; and special columns.

SAMPLE ISSUE

128 pages (20% advertising): 3 articles; 6 depts/columns; 83 reviews. Sample copy, free with 9x12 SASE. Guidelines and editorial calendar available.

- "Daily Deep Thoughts." Article reveals various humorous thoughts of a children's book writer about the industry.
- "Horn Book Fanfare." Article reveals editors' picks for best books of 2007.
- Sample dept/column: "What Makes a Good Mother Goose" discusses the many different Mother Goose collections available.

RIGHTS AND PAYMENT

All rights. Written material, payment rates vary. Pays on publication. Provides 1 author's copy.

➻EDITOR'S COMMENTS

A distinguished cast of columnists write for our magazine about young adult literature, international publishing, popular reading, and more. We are looking for material that celebrates the best in children's and young adult literature and exposes the worst.

Horsemen's Yankee Pedlar

85 Leicester Street
North Oxford, MA 01537

Editor: Molly Johns

DESCRIPTION AND INTERESTS
This magazine is all about horses and the people who ride and show them. Concerned with equestrian sports in the northeastern U.S., it covers all breeds of horses and disciplines of the sport. Its articles are of interest to horse owners, riders, trainers, and traders. Circ: 50,000.

Audience: YA–Adult
Frequency: Monthly
Website: www.pedlar.com

FREELANCE POTENTIAL
50% written by nonstaff writers. Publishes 40 freelance submissions yearly; 5% by authors who are new to the magazine. Receives 360 queries, 240 unsolicited mss yearly.

SUBMISSIONS
Query or send complete ms. Accepts hard copy and simultaneous submissions if identified. SASE. Responds to queries in 1–2 weeks, to mss in 2–3 months.

- Articles: 500–800 words. Informational and how-to articles; interviews; reviews; and personal experience pieces. Topics include horse breeds, disciplines, training, health care, and equestrian management.
- Depts/columns: Word lengths vary. News, book reviews, business issues, nutrition, and legal issues.
- Artwork: B/W and color prints.

SAMPLE ISSUE
226 pages (75% advertising): 39 articles; 15 depts/columns. Sample copy, $3.99 with 9x12 SASE (7 first-class stamps). Guidelines available.

- "The Winning Equation for Pony Hunters." Article talks to trainers and riders about what it takes to have a winning pony.
- "You Can Answer the Toughest Hunter/Jumper Questions." Article tackles some important questions regarding success in the ring.

RIGHTS AND PAYMENT
First North American serial rights. Written material, $2 per published column inch. Show coverage, $75 per day. Pays 30 days after publication. Provides 1 tearsheet.

➤EDITOR'S COMMENTS
Writers must have a background in horse racing, showing, or training.

Horsepower

P.O. Box 670
Aurora, Ontario L4G 4J9
Canada

Managing Editor: Susan Stafford

DESCRIPTION AND INTERESTS
Horsepower was the first magazine written specifically for horse-crazy kids in North America. Published as a special pullout of *Horse-Canada* magazine, it features training, horse care, and showing information, as well as breed profiles. Circ: 10,000.

Audience: 9–15 years
Frequency: 6 times each year
Website: www.horse-canada.com

FREELANCE POTENTIAL
20% written by nonstaff writers. Publishes 10 freelance submissions yearly; 10% by unpublished writers, 10% by authors who are new to the magazine. Receives 30 queries and unsolicited mss yearly.

SUBMISSIONS
Query with outline/synopsis; or send ms with résumé. Accepts hard copy, disk submissions (ASCII or WordPerfect), and email to info@horse-canada.com. SAE/IRC. Responds to queries in 1–2 weeks, to mss in 2–3 months.

- Articles: 500–1,000 words. Informational and how-to articles; profiles; and humor. Topics include breeds, training, stable skills, equine health, and tack.
- Fiction: 500 words. Adventure and humorous stories; sports stories related to horses.
- Depts/columns: Staff written.
- Artwork: B/W and color prints.
- Other: Activities, games, and puzzles.

SAMPLE ISSUE
14 pages (20% advertising): 3 articles; 5 depts/columns; 2 puzzles, 2 activities. Sample copy, $3.95. Guidelines and theme list available.

- "Breed Profile." Article profiles the Chincoteague "ponies."
- "Stay Safe On the Trails." Article provides safety tips for trail riding.
- "My Horse Is Lame. What Do I Do?" Article explains how to determine the site and cause of lameness.

RIGHTS AND PAYMENT
First North American serial rights. Written material, $50–$90. Artwork, $10–$75. Pays on publication. Provides 1 contributor's copy.

➤EDITOR'S COMMENTS
All horse-related subjects are welcome, but they must be child-friendly.

Hudson Valley Parent

174 South Street
Newburgh, NY 12550

Editor: Leah Black

DESCRIPTION AND INTERESTS
Hudson Valley Parent provides reliable local information to young families in Orange, Dutchess, Rockland, Ulster, Sullivan, and northern Putnam counties of New York. It covers all areas of concern to parents, from health and education to family fun. Circ: 50,000.

Audience: Parents
Frequency: Monthly
Website: www.hvparent.com

FREELANCE POTENTIAL
98% written by nonstaff writers. Publishes 130 freelance submissions yearly; 5% by unpublished writers, 20% by authors who are new to the magazine. Receives 240 queries, 120 unsolicited mss yearly.

SUBMISSIONS
Query with writing samples; or send complete ms with sidebar and author bio. Accepts email submissions to editor@excitingread.com. Responds in 3–6 weeks.

- Articles: 700–1,200 words. Informational and how-to articles. Topics include child care and development, discipline, education, learning disabilities, family health, recreation, travel, and entertainment.
- Depts/columns: 700 words. Health, education, behavior, and kid-friendly recipes.
- Other: Submit seasonal material 6 months in advance.

SAMPLE ISSUE
58 pages (50% advertising): 5 articles; 1 special section; 7 depts/columns. Sample copy, free with 9x12 SASE. Guidelines and editorial calendar available.

- "Keeping Kids Clean." Article stresses the importance of proper handwashing and provides ways to create a handwashing routine.
- Sample dept/column: "Child Behavior" gives advice on helping young children adjust to a new sibling.

RIGHTS AND PAYMENT
One-time rights. Articles, $50–$120. Reprints, $25–$35. Pays on publication. Provides 2 contributor's copies.

☛EDITOR'S COMMENTS
Any unsolicited, non-locally slanted feature will be treated as a reprint. We prefer articles with local information using local sources.

Humpty Dumpty's Magazine

Children's Better Health Institute
1100 Waterway Boulevard
P.O. Box 567
Indianapolis, IN 46206-0567

Editor: Julia Goodman

DESCRIPTION AND INTERESTS
Whether it is read by new readers or read to non-reading children, this magazine is meant to entertain and engage young children while providing them with the information they need to stay healthy. Circ: 236,000.

Audience: 4–6 years
Frequency: 6 times each year
Website: www.humptydumptymag.org

FREELANCE POTENTIAL
10% written by nonstaff writers. Publishes 10–12 freelance submissions yearly; 5% by unpublished writers, 12% by authors who are new to the magazine.

SUBMISSIONS
Send complete ms. Accepts hard copy. SASE. Responds in 2–3 months.

- Articles: To 350 words. Informational and how-to articles. Topics include health, fitness, sports, science, nature, animals, crafts, and hobbies.
- Fiction: To 350 words. Genres include early reader contemporary and multicultural fiction; stories about sports; fantasy; folktales; mystery; drama; and humor.
- Depts/columns: Word lengths vary. Recipes, health advice, and book excerpts.
- Other: Puzzles, activities, and games. Poetry, line lengths vary. Submit seasonal material 8 months in advance.

SAMPLE ISSUE
36 pages (2% advertising): 4 articles; 3 stories; 1 dept/column; 8 activities; 1 poem; 1 recipe. Sample copy, $1.25. Guidelines available.

- "A Special Picnic." Rebus story tells of a Fourth of July surprise for a family on a holiday outing.
- "Counting on Frogs." Articles recounts a frog hunt that begins with finding tadpoles and ends with fully grown frogs.

RIGHTS AND PAYMENT
All rights. Written material, $.22 per word. Pays on publication. Provides 10 author's copies.

☛EDITOR'S COMMENTS
Material sent to us should be short in length, but long in fun. We are designed to be a tool used by a parent and child together, and all material must be engaging for young children, and must support healthy living.

Ignite Your Faith

Christianity Today
465 Gunderson Drive
Carol Stream, IL 60188

Editor: Chris Lutes

DESCRIPTION AND INTERESTS
Written for teenagers and young adults in college, this magazine empowers readers to become fully devoted followers of Christ. It has articles on faith, religion, and contemporary social issues. Circ: 100,000.

Audience: 13–19 years
Frequency: 9 times each year
Website: www.igniteyourfaith.com

FREELANCE POTENTIAL
60% written by nonstaff writers. Publishes 15–20 freelance submissions yearly; 10% by unpublished writers, 20% by authors who are new to the magazine. Receives 300 queries each year.

SUBMISSIONS
Query with 1-page synopsis. No unsolicited mss. Accepts hard copy and simultaneous submissions if identified. SASE. Responds in 3–6 weeks.

- Articles: 700–1,500 words. Personal experience pieces and humor. Topics include Christian values, beliefs, and education.
- Fiction: 1,000–1,500 words. Genres include contemporary fiction with religious themes.
- Depts/columns: Staff written.

SAMPLE ISSUE
56 pages (30% advertising): 7 articles; 8 depts/columns. Sample copy, $3 with 9x12 SASE (3 first-class stamps). Guidelines available.

- "If Anybody Knew the Real Me." First-person account tells of a wild college boy who changed his life when he found Jesus.
- "I Want to Bring Hope." Article profiles Christian musician Britt Nicole.
- "Why Didn't He Hate Me?" Article presents a first-person account of a high school girl who killed a woman in a car accident, and was treated with mercy and sympathy by the woman's husband.

RIGHTS AND PAYMENT
First rights. Written material, $.20–$.25 per word. Pays on acceptance. Provides 2 contributor's copies.

➦EDITOR'S COMMENTS
We want writers who understand and empathize with a Christian worldview, but who also have an ability to write in a way that is respectful, not preachy, to our readers.

Illuminata

5486 Fairway Drive
Zachary, LA 70791

Editor-in-Chief: Bret Funk

DESCRIPTION AND INTERESTS
This online magazine, produced by book publisher Tyrannosaurus Press, targets authors and fans of the fantasy and science fiction genres. In addition to writing tips, articles, and reviews, it publishes original works of fiction. Hits per month: 600.

Audience: YA–Adult
Frequency: Quarterly
Website: www.tyrannosauruspress.com

FREELANCE POTENTIAL
25% written by nonstaff writers. Publishes 5–10 freelance submissions yearly; 95% by unpublished writers, 50% by authors new to the magazine. Receives 10 queries yearly.

SUBMISSIONS
Query. Accepts email queries to info@tyrannosauruspress.com (no attachments). Responds in 1–3 months.

- Articles: 1–2 pages in length. Informational articles. Topics include writing science fiction, fantasy, and horror.
- Fiction: Word lengths vary. Genres include science fiction, fantasy, and horror.
- Depts/columns: "Reviews," 500–1,000 words. Reviews of science fiction and fantasy books and stories.

SAMPLE ISSUE
26 pages (no advertising): 5 articles; 1 story; 5 depts/columns. Sample copy and guidelines available at website.

- "Names in Fiction." Article by an author of fantasy fiction discusses how to create character names.
- "Suspension of Disbelief: Finding Your Audience." Article explains how to break into fantasy writing by finding the right audience.
- Sample dept/column: "Reviews" takes a look at science fiction and fantasy books.

RIGHTS AND PAYMENT
Rights vary. No payment.

➦EDITOR'S COMMENTS
Please do not send print submissions. All contact must be via the Internet or email. Queries should be no longer than three single-spaced lines. Email queries must contain full contact information: address, phone number, and email address. Shorter works have a better chance at publication.

Imagination-Café

P.O. Box 1536
Valparaiso, IN 46384

Editor: Rosanne Tolin

DESCRIPTION AND INTERESTS
The menu at Imagination-Café features articles on everything from sports to science; homework help; interviews; and writing contests; with a side of quizzes and activities. Hits per month: Unavailable.

Audience: 7–12 years
Frequency: Updated daily
Website: www.imagination-café.com

FREELANCE POTENTIAL
75% written by nonstaff writers. Publishes 40–50 freelance submissions yearly.

SUBMISSIONS
Prefers complete ms; will accept query. Accepts email to editor@imagination-café.com (no attachments). Response time varies.

- Articles: Word lengths vary. Informational and how-to articles; profiles; interviews; and reviews. Topics include animals, careers, crafts, hobbies, history, science, technology, sports, and celebrities.
- Depts/columns: Word lengths vary. "Cool Careers," "Before They Were Famous," "Celebrity Screw-Ups," and "School Strategies."
- Other: Puzzles, mazes, word games, quizzes, and recipes.

SAMPLE ISSUE
Sample articles and writers' guidelines available at website.

- "The Need for Speed." Article describes the sport of lacrosse.
- "Daisy Magic." Article explains how to perform a science experiment using fresh-cut flowers and food coloring.
- Sample dept/column: "Celebrity Screw-Ups" contains an anecdote about Michael Jackson making a *faux pas* at the 2002 MTV Video Music Awards.

RIGHTS AND PAYMENT
Non-exclusive print and electronic rights. Written material, $20–$100. Pays on acceptance.

➻EDITOR'S COMMENTS
Subjects should be kid-relevant—snappy writing, subheads, and sidebars are pluses. When writing about most embarrassing moments, please keep it clean! Also note that we prefer material that has not been previously published elsewhere, and that we do not maintain an editorial calendar.

Indy's Child

1901 Broad Ripple Avenue
Indianapolis, IN 46220

Managing Editor: Lynette Rowland

DESCRIPTION AND INTERESTS
This regional tabloid offers parents in central Indiana topical articles on family issues, as well as travel ideas and coverage of family events in the midwestern U.S. Circ: 50,000.

Audience: Parents
Frequency: Monthly
Website: www.indyschild.com

FREELANCE POTENTIAL
95% written by nonstaff writers. Publishes 240+ freelance submissions yearly; 35% by unpublished writers, 60% by authors who are new to the magazine. Receives 600 unsolicited mss yearly.

SUBMISSIONS
Send complete ms with artwork, author bio, and photo. Accepts email submissions to editor@indyschild.com; include "Manuscript: Topic of Article" in subject line. Responds only if interested.

- Articles: 800–1,500 words. Informational and how-to articles; and profiles. Topics include parenting and child development, hobbies and talents, safety, sports and recreation, home improvement, entertainment, travel, and social issues.
- Depts/columns: To 1,000 words. Media and product reviews, women's health, museum notes, college information, local profiles, and "Tweens and Teens."
- Artwork: Color digital images.
- Other: Submit seasonal material 2 months in advance.

SAMPLE ISSUE
54 pages (50% advertising): 7 articles; 13 depts/columns. Sample copy, guidelines, and editorial calendar available at website.

- "This Little Piggy Went to College." Article provides a review of financial considerations for families with college-bound teens.
- Sample dept/column: "Inspired" profiles the "oldest worker in America," who is 101!

RIGHTS AND PAYMENT
First or second rights. Written material, $100–$250. Reprints, $25–$50. Pays on publication.

➻EDITOR'S COMMENTS
We need more coverage of parenting and school issues, sports, and fashion. We especially need reprints.

Inland Empire Family

1451 Quail Street, Suite 201
Newport Beach, CA 92660

Editor: Lynn Armitage

DESCRIPTION AND INTERESTS
The "Inland Empire" of the title refers to a region of southern California encompassing Riverside and San Bernardino counties. Like its sister publication *OC Family*, this magazine tackles parenting issues from a regional perspective, and includes an extensive list of local resources and events. Circ: 55,000.
Audience: Parents
Frequency: Monthly
Website: www.inlandempirefamily.com

FREELANCE POTENTIAL
95% written by nonstaff writers. Publishes several freelance submissions yearly.

SUBMISSIONS
Query. Accepts hard copy. SASE. Response time varies.

- Articles: Word lengths vary. Informational and how-to articles. Topics include parenting, child care and development, education, entertainment, sports, recreation, health, nutrition, travel, summer camp, and pets.
- Depts/columns: Word lengths vary. Parenting advice by age group; self-help for mothers and couples; children's health, fashion, and essays; food and dining; media reviews; and education issues.

SAMPLE ISSUE
106 pages: 1 article; 15 depts/columns. Sample copy and guidelines available.

- "Happy, Happy Birthdays." Article suggests age-appropriate themes, activities, refreshments, and favors for children's parties.
- Sample dept/column: "Passages: Middle Years" discusses the Webkinz fad-turned-phenomenon.
- Sample dept/column: "Mom's Day Off" describes new beauty products.

RIGHTS AND PAYMENT
All rights. Articles, $100–$500. Depts/columns, payment rates vary. Kill fee, $50. Pays within 45 days of publication.

➥EDITOR'S COMMENTS
Accuracy and fairness are of paramount importance in any article we publish. If there is a conflict, it should be discussed immediately with the assigning editor. Comportment is vital: Be polite and professional at all times. Be careful about sourcing your copy.

InsideOut

United Pentecostal Church International
8855 Dunn Road
Hazelwood, MO 63042-2299

Submissions: Tamra Schultz

DESCRIPTION AND INTERESTS
The primary objective of this magazine is to inspire Christian young adults to live their faith. Circ: 5,200.
Audience: 12–18 years
Frequency: 6 times each year
Website: http://pentecostalyouth.org

FREELANCE POTENTIAL
20% written by nonstaff writers. Publishes 10-30 freelance submissions yearly; 20% by unpublished writers, 80% by authors who are new to the magazine. Receives 240 unsolicited mss yearly.

SUBMISSIONS
Send complete ms. Accepts email submissions to tschultz@upci.org and simultaneous submissions if identified. Responds in 3 months.

- Articles: Features, 1,200–1,800 words. Shorter articles, 600–800 words. Profiles and personal experience pieces. Topics include religion, missionary-related subjects, spiritual growth, social issues, and current events.
- Fiction: 600–900 words. Real-life fiction with Christian themes. Genres include humor and romance.
- Depts/columns: Word length varies. Book and music reviews; opinion and first-person pieces by teens; and church events.

SAMPLE ISSUE
18 pages (11% advertising): 5 articles; 6 depts/columns. Sample copy, free with 9x12 SASE (2 first-class stamps). Guidelines available.

- "The Impact of Media on Youth Culture." Article explains that moderation as regards media exposure is Biblically sound.
- "Prevent Identity Theft." Article offers advice for guarding one's Christian identity through prayer, Bible reading, worship services, and fellowship with other Christians.
- Sample dept/column: "Infuze" features eight weekly devotions.

RIGHTS AND PAYMENT
All rights. Articles, $.07 per word. Filler, $5–$10. Pays on publication. Provides 1 copy.

➥EDITOR'S COMMENTS
We are looking for fresh, teen-friendly writing that reflects the Gospel, using the Cambridge King James version of the Bible. Read an issue to see if your work fits our style.

Insight

55 West Oak Ridge Drive
Hagerstown, MD 21740-7390

Editor: Dwain Neilson Esmond

DESCRIPTION AND INTERESTS
Insight's readership is primarily American and Canadian teenagers of the Seventh-day Adventist Church, from a wide variety of ethnic and schooling backgrounds. Circ: 12,000.

Audience: 13–19 years
Frequency: Weekly
Website: www.insightmagazine.org

FREELANCE POTENTIAL
99% written by nonstaff writers. Publishes 150+ freelance submissions yearly; 50% by unpublished writers, 70% by authors who are new to the magazine. Receives 996 unsolicited mss yearly.

SUBMISSIONS
Send complete ms. Accepts hard copy and email submissions to insight@rhpa.org (Microsoft Word attachments). SASE. Responds in 1–3 months.

- Articles: 500–1,500 words. Informational articles; profiles; biographies; reports on volunteer and mission trips; and humor. Topics include social issues, religion, and careers.
- Depts/columns: Word lengths vary. True-to-life stories and personal experience pieces.
- Other: Submit seasonal material 6 months in advance.

SAMPLE ISSUE
16 pages (2% advertising): 3 articles; 4 depts/columns. Sample copy, $2 with 9x12 SASE (2 first-class stamps). Guidelines available.

- "Guy+Guitar=Girls." Article is one man's humorous reflection on what really appeals to the opposite sex.
- "Practical Purity." Article encourages unmarried couples to abstain from sex by focusing on the positive aspects of abstaining.
- Sample dept/column: "On the Case" explains the Adventist health message and provides practical ways to live it.

RIGHTS AND PAYMENT
First rights. Written material, $50–$125. Pays on acceptance. Provides 3 author's copies.

⇢EDITOR'S COMMENTS
We are always looking for profiles of outstanding Christians, whether they are celebrities or everyday members of the Seventh-day Adventist Church. All submissions should have a strong spiritual message.

InSite

P.O. Box 62189
Colorado Springs, CO 80962-2189

Editor: Alison Philips

DESCRIPTION AND INTERESTS
InSite is read by those working in Christian camps and conferences. Its goal is to inform and inspire them to be the most effective in their outreach efforts. It covers camper needs, operations, fundraising, programming, health and safety, and other issues. Circ: 9,000.

Audience: Adults
Frequency: 6 times each year
Website: www.ccca.org

FREELANCE POTENTIAL
90% written by nonstaff writers. Publishes 40 freelance submissions yearly; 15% by unpublished writers, 22% by authors who are new to the magazine. Receives 20 queries yearly.

SUBMISSIONS
Query with résumé and writing samples. Accepts email queries to editor@ccca.org. Availability of artwork improves chance of acceptance. Responds in 1 month.

- Articles: 800–1,500 words. Informational and how-to articles; profiles; and interviews. Topics include Christian camp and conference operations, camp programs, leadership, recreation, religion, social issues, crafts, hobbies, health and fitness, multicultural and ethnic issues, nature, popular culture, and sports.
- Depts/columns: Staff written.
- Artwork: Color prints and digital images.
- Other: Submit seasonal material 6 months in advance.

SAMPLE ISSUE
42 pages (25% advertising): 6 articles; 8 depts/columns. Sample copy, $4.95 with 9x12 SASE ($1.40 postage). Guidelines and editorial calendar available at website.

- "Think Mission." Article explores ways that camps can partner with rescue missions.
- "Hope for Hurting Kids." Article tells how to help abused kids when they come to camp.

RIGHTS AND PAYMENT
First rights. Written material, $.16 per word. Artwork, $25–$250. Pays on publication. Provides 1 contributor's copy.

⇢EDITOR'S COMMENTS
In addition to our informational articles, we also welcome profiles of influential camp and conference professionals.

Instructor

Scholastic Inc.
557 Broadway
New York, NY 10012

Editorial Director: Dana Truby

DESCRIPTION AND INTERESTS
An essential resource for new and experienced teachers of kindergarten through eighth grade, this magazine focuses on helping students succeed, building connections between school and home, and the latest classroom strategies and materials. Circ: 200,000+.
Audience: Teachers, grades K–8
Frequency: 6 times each year
Website: www.teacher.scholastic.com

FREELANCE POTENTIAL
80% written by nonstaff writers. Publishes 25 freelance submissions yearly; 10% by unpublished writers. Receives 100 queries yearly.

SUBMISSIONS
Query. Accepts email queries to dtruby@scholastic.com. Availability of artwork improves chance of acceptance. SASE. Responds in 3–4 months.

- Articles: 1,200 words. Informational and how-to articles; and personal experience pieces. Topics include workplace issues; technology; learning issues; literacy; and teaching of such subjects as reading, writing, mathematics, science, and social studies.
- Depts/columns: News items, Q&As, technology, and "Teachers' Picks," word lengths vary. Classroom activities, to 250 words. Humorous or poignant personal essays, to 400 words. Book reviews.

SAMPLE ISSUE
72 pages (40% advertising): 4 articles; 4 activities; 7 depts/columns. Sample copy, $3 with 9x12 SASE ($.77 postage). Guidelines and theme list available at website.

- "Bring an Author Home." Article explains how to plan a memorable and successful visit from a children's author.
- Sample dept/column: "Your Career" discusses how to thrive on a teacher's salary.

RIGHTS AND PAYMENT
All rights. Articles, $600. Depts/columns, $250–$300. Pays on publication. Provides 2 contributor's copies.

➠EDITOR'S COMMENTS
If you are a teacher with a timely and relevant topic or an idea that has been effective with your students, we would love to hear from you.

InTeen

1551 Regency Court
Calumet City, IL 60409

Editor: LaTonya Taylor

DESCRIPTION AND INTERESTS
InTeen is an award-winning student magazine published by the African American Christian Publishing and Communications Company. It is geared toward urban teens, and it offers inspirational and enlightening Christian articles and Bible lessons specifically geared toward youth. Circ: 75,000.
Audience: 15–17 years
Frequency: Quarterly
Website: www.urbanministries.com

FREELANCE POTENTIAL
90% written by nonstaff writers. Publishes 52 freelance submissions yearly.

SUBMISSIONS
All material is written on assignment. Send résumé with writing samples. SASE. Responds in 3–6 months.

- Articles: Word lengths vary. Bible study guides and lessons; how-to articles; profiles; interviews; and reviews. Topics include religion, college and careers, black history, music, social issues, and multicultural and ethnic issues. Also publishes biographies.
- Fiction: Word lengths vary. Stories may be included in Bible lessons. Genres include inspirational, multicultural, and ethnic fiction; and real-life and problem-solving stories.
- Other: Puzzles, activities, and poetry. Submit seasonal material 1 year in advance.

SAMPLE ISSUE
48 pages (no advertising): 7 articles; 2 poems; 14 Bible studies. Guidelines available.

- "How We See God." Article explains that looking at Jesus through the Gospels helps us to see what God looks like.
- "Get the Truth." Bible study shows, through a brief vignette, that Jesus, as God's Son, has the only truth.

RIGHTS AND PAYMENT
All rights. All material, payment rates vary. Pays 2 months after acceptance. Provides 2 contributor's copies.

➠EDITOR'S COMMENTS
We are especially open to working with new writers at this time. We encourage prospective writers to become familiar with our editorial mission by reading several issues of the magazine. All work is assigned.

InTeen Teacher

1551 Regency Court
Calumet City, IL 60409

Editor: LaTonya Taylor

DESCRIPTION AND INTERESTS
InTeen Teacher is published in conjunction with *InTeen,* the student magazine for urban Christian youth. It contains the Bible study teaching plans and guides to be used by the teachers to support the lessons appearing in *InTeen*. It also offers thought-provoking articles to help teachers better serve their students. Circ: 75,000.
Audience: Religious educators
Frequency: Quarterly
Website: www.urbanministries.com

FREELANCE POTENTIAL
90% written by nonstaff writers. Publishes 52 freelance submissions yearly.

SUBMISSIONS
All material is written on assignment. Send résumé with writing samples. SASE. Responds in 3–6 months.

- Articles: Word lengths vary. Bible study plans and guides for teaching Christian values to African American teens; and how-to articles.
- Fiction: Word lengths vary. Stories may be included as part of study plans. Genres include inspirational, multicultural, and ethnic fiction; and real-life and problem-solving stories.
- Other: Puzzles, activities, and poetry. Submit seasonal material 1 year in advance.

SAMPLE ISSUE
96 pages (no advertising): 2 articles; 14 teaching plans; 14 Bible study guides. Writers' guidelines available.

- "The Two Sides of Justice." Article explains that urban youth need to invest in their futures by having discipline, a strong work ethic, and hope.
- "Christ as Intercessor." Bible study guide uses Hebrews 7 as the basis for recognizing the role of intercessor in Christ and in self, and the need for prayer.

RIGHTS AND PAYMENT
All rights. Written material, payment rates vary. Pays 2 months after acceptance. Provides 2 contributor's copies.

➥EDITOR'S COMMENTS
We need more articles that help Christian educators who minister to African American teens, and we are especially open to working with new writers.

International Gymnast

P.O. Box 721020
Norman, OK 73070

Editor: Dwight Normile

DESCRIPTION AND INTERESTS
All things related to the sport of gymnastics can be found in this magazine. It features profiles of successful and inspirational gymnasts, covers meets, and presents techniques and training tips. It is geared toward the gymnast with the goal of high-level competition. Circ: 14,000.
Audience: Gymnasts, 10–16 years
Frequency: 10 times each year
Website: www.intlgymnast.com

FREELANCE POTENTIAL
10% written by nonstaff writers. Publishes 5 freelance submissions yearly; 50% by authors who are new to the magazine. Receives 12 unsolicited mss yearly.

SUBMISSIONS
Send complete ms. Accepts hard copy and simultaneous submissions if identified. SASE. Responds in 1 month.

- Articles: 1,000–2,250 words. Informational articles; profiles; and interviews. Topics include gymnastics competitions, coaching, and personalities involved in the sport around the world.
- Fiction: To 1,500 words. Gymnastics stories.
- Depts/columns: 700–1,000 words. News, training tips, and opinion pieces.
- Artwork: B/W prints. 35mm color slides for cover art.

SAMPLE ISSUE
46 pages (14% advertising): 4 articles; 12 depts/columns. Sample copy, $5 with 9x12 SASE. Guidelines available.

- "U.S. Olympic Trials." Article reports on the Olympic Trials, where two gymnasts came out on top and others had their Olympic dreams squashed.
- Sample dept/column: "Chalk Talk" presents an interview with former gymnastics champion Dianne Durham.

RIGHTS AND PAYMENT
All rights. Written material, $15–$25. Artwork, $5–$50. Pays on publication. Provides 1 contributor's copy.

➥EDITOR'S COMMENTS
Writers must be very familiar with our sport. An article that takes us behind the scenes of a particular competition or provides some other insight will usually get our attention.

Jack And Jill

Children's Better Health Institute
1100 Waterway Boulevard
P.O. Box 567
Indianapolis, IN 46206-0567

Editor: Daniel Lee

DESCRIPTION AND INTERESTS
Stories, poetry, games, and puzzles are found in the pages of this entertaining health and nutrition magazine for second- and third-grade students. It has been published since 1938. Circ: 200,000.

Audience: 7–10 years
Frequency: 6 times each year
Website: www.jackandjillmag.org

FREELANCE POTENTIAL
10% written by nonstaff writers. Publishes 10 freelance submissions yearly; 70% by authors who are new to the magazine. Receives 1,200 unsolicited mss yearly.

SUBMISSIONS
Send complete ms. Accepts hard copy. SASE. Responds in 3 months.

- Articles: 500–600 words. Informational and how-to articles; humor; profiles; and biographies. Topics include sports, health, exercise, safety, nutrition, and hygiene.
- Fiction: 500–900 words. Genres include mystery, fantasy, folktales, humor, science fiction, and stories about sports and animals.
- Artwork: Submit sketches to Jennifer Saulovic, art director; submit photos to Daniel Lee, editor.
- Other: Poetry, games, puzzles, activities, and cartoons. Submit seasonal material 8 months in advance.

SAMPLE ISSUE
36 pages (4% advertising): 4 articles; 1 story; 7 activities; 2 cartoons; 3 poems. Sample copy, $6.50 ($2 postage). Guidelines available.

- "Animal Magnetism." Article spotlights a biologist from Disney's Animal Kingdom and her work with the park's giraffes and elephants.
- "Bad Bananas." Humorous story features a girl who learns that very ripe bananas can be turned into banana bread.

RIGHTS AND PAYMENT
All rights. Articles and fiction, $.17 per word. Other material, payment rates vary. Pays on publication. Provides 10 contributor's copies.

☛EDITOR'S COMMENTS
We are always looking for fun and interesting material that will teach kids how to live a healthy lifestyle. Humorous stories are especially welcome.

Jakes Magazine

P.O. Box 530
Edgefield, SC 29824

Editor: Matt Lindler

DESCRIPTION AND INTERESTS
As the National Wild Turkey Federation's youth member publication, *Jakes* is dedicated to informing, educating, and involving its readers in wildlife conservation while passing on the hunting tradition. Profiles, news, and stories tailored to the young conservationist fill its pages. *Xtreme JAKES*, a special section for teens, is included in each issue. Circ: 200,000.

Audience: 10–17 years
Frequency: Quarterly
Website: www.nwtf.org/jakes

FREELANCE POTENTIAL
50% written by nonstaff writers. Publishes 30 freelance submissions yearly; 10% by unpublished writers, 30% by authors who are new to the magazine. Receives 150–200 queries and unsolicited mss yearly.

SUBMISSIONS
Query or send complete ms. Accepts material between May and December only. Accepts hard copy, email to MLindler@nwtf.org, and simultaneous submissions if identified. SASE. Response time varies.

- Articles: 600–1,200 words. Informational articles; profiles; and personal experience pieces. Topics include nature, the environment, animals, pets, hunting, fishing, and other outdoor and extreme sports.
- Fiction: 800–1,200 words. Historical fiction.

SAMPLE ISSUE
30 pages: 7 articles; 8 depts/columns. Guidelines available at website.

- "The Great Seed Spitter." Article discusses the Brazilian Tapir and its contribution to the conservation of the rainforest.
- Sample dept/column: "Xtreme Spotlight" profiles the five-time champion of the NWTF's Calling Championships.

RIGHTS AND PAYMENT
Rights vary. Written material, $100–$300. Pays on publication. Provides 2 contributor's copies.

☛EDITOR'S COMMENTS
Writers should keep in mind that stories should be fun to read and interesting to read. Article ideas for *Xtreme JAKES* must be outdoor-related with an emphasis on education and entertainment.

James Hubbard's My Family Doctor

!

The Magazine That Makes House Calls

P.O. Box 38790
Colorado Springs, CO 80906

Managing Editor: Leigh Ann Hubbard

DESCRIPTION AND INTERESTS
Written and published by licensed health-care professionals, this magazine provides its readers with evidence-based, practical, and reliable health information. It covers all types of traditional medicine, including psychiatry and nutrition, and also reports non-biased views of complementary and alternative medicine. Circ: 60,000.

Audience: Adults
Frequency: Quarterly
Website: www.myfamilydoctormag.com

FREELANCE POTENTIAL
90% written by nonstaff writers. Publishes 114 freelance submissions yearly; 8% by unpublished writers, 11% by authors who are new to the magazine.

SUBMISSIONS
Query with writing samples. Prefers email queries to managingeditor@familydoctormag.com. Response time varies.

- Articles: 400–1,200 words. Informational articles. Topics include health, fitness, nutrition, and preventative medicine.
- Depts/columns: 200–400 words. Medical studies and breakthroughs.
- Other: Filler, 200–400 words.

SAMPLE ISSUE
68 pages (2% advertising): 10 articles; 16 depts/columns. Sample copy, $4.95 at newsstands. Guidelines available.

- "How to Help a Loved One." Article explains how to deal with a loved one who suffers from depression, alcoholism or addiction.
- "Doctors Debate Universal Health Care." Article looks at the pros and cons, myths and truths, of universal health care.
- Sample dept/column: "Recipes" includes a recipe for healthier oven-fried chicken.

RIGHTS AND PAYMENT
First North American serial, exclusive syndication for 6 months, and electronic rights. Written material, to $.30 per word. Pays on publication.

➤EDITOR'S COMMENTS
We are continually on the lookout for M.D.s, D.O.s, R.D.s, and Pharm. D.s to write for us. Even if you don't have an idea for a particular story, you can contact us. Include information on your medical background.

Journal of Adolescent & Adult Literacy

International Reading Association
800 Barksdale Road
P.O. Box 8139
Newark, DE 19714-8139

Editorial Assistant: James Henderson

DESCRIPTION AND INTERESTS
This professional journal highlights research-based practice aimed at improving achievement among literacy learners ages 12 and older. Its peer-reviewed articles are read by a host of professionals involved in literacy education. Circ: 16,000.

Audience: Reading education professionals
Frequency: 8 times each year
Website: www.reading.org

FREELANCE POTENTIAL
95% written by nonstaff writers. Publishes 50 freelance submissions yearly; 30% by unpublished writers, 50% by authors who are new to the magazine. Receives 300 unsolicited mss each year.

SUBMISSIONS
Send complete ms. Accepts electronic submissions via http://mc.manuscriptcentral.com/jaal. Responds in 2–3 months.

- Articles: 5,000–6,000 words. Informational and how-to articles; and personal experience pieces. Topics include reading theory, research, and practice; and trends in teaching literacy.
- Depts/columns: Word lengths vary. Opinion pieces, reviews, and technology information.

SAMPLE ISSUE
84 pages (7% advertising): 5 articles; 4 depts/columns. Sample copy, $10. Guidelines available at website.

- "Relay Writing in an Adolescent Online Community." Article reports on relay writing using an online community, and how it can boost literacy skills.
- Sample dept/column: "Reviews" takes a look at several books for adolescents.

RIGHTS AND PAYMENT
All rights. No payment. Provides 5 contributor's copies for articles, 2 for depts/columns.

➤EDITOR'S COMMENTS
Your manuscript will be judged on its originality, significance, scholarship, audience appropriateness, and writing style. All manuscripts must be thoroughly researched, as they will be reviewed by two editorial board members and a guest reviewer. Our guidelines (found under "Publications," then "For Authors") explains the online submission process.

Journal of Adventist Education

12501 Old Columbia Pike
Silver Spring, MD 20904-6600

Editor: Beverly J. Rumble

DESCRIPTION AND INTERESTS
Written for Seventh-day Adventist teachers and educational administrators throughout the world, this magazine emphasizes practical methods of enhancing teaching skills and learning in the classroom. Circ: 15,000.

Audience: Teachers and administrators
Frequency: 5 times each year
Web: http://education.gc.adventist.org/JAE

FREELANCE POTENTIAL
90% written by nonstaff writers. Publishes 30 freelance submissions yearly. Receives 24–48 queries yearly.

SUBMISSIONS
Query. Accepts email queries to rumbleb@gc.adventist.org (Microsoft Word attachments). Availability of artwork improves chance of acceptance. Responds in 3–6 weeks.

- Articles: To 2,000 words. Informational and how-to articles. Topics include parochial, gifted, and special education; new teaching methods and educational approaches; school administration and supervision; classroom management; religion; mathematics; science; and technology.
- Depts/columns: Staff written.
- Artwork: Color prints and slides. JPEG or TIFF images at 300 dpi. Charts and graphs.

SAMPLE ISSUE
48 pages (5% advertising): 9 articles; 1 dept/column. Sample copy, $3.50 with 9x12 SASE ($.68 postage). Guidelines available.

- "Conscientiously Opposed to Bearing Arms." Article provides a historical review of the Seventh-day Adventist position on war.
- "Creating a Culture of Peace in the Elementary Classroom." Article reports on one teacher's efforts to foster compassion and social justice.

RIGHTS AND PAYMENT
First North American serial rights. Articles, to $100. Artwork, payment rates vary. Pays on publication. Provides 2 contributor's copies.

➡️EDITOR'S COMMENTS
Please note that all articles will be posted on our website following print publication. Also, keep in mind that our constituency is international as you explore the implications or applications of a given topic.

Journal of School Health

American School Health Association
7263 State Route 43
P.O. Box 708
Kent, OH 44240-0708

Editor: James H. Price

DESCRIPTION AND INTERESTS
Written for school administration and school health professionals, this journal focuses on healthy students in preschool through grade 12. It covers all topics related to maintaining a healthy school environment. Circ: 5,000.

Audience: School health professionals
Frequency: 10 times each year
Website: www.ashaweb.org

FREELANCE POTENTIAL
95% written by nonstaff writers. Publishes 60 freelance submissions yearly; 90% by authors who are new to the magazine. Receives 120 queries yearly.

SUBMISSIONS
Query or send complete ms. Accepts email submissions via www.manuscriptcentral.com/josh. Responds to queries in 2 weeks, to mss in 3–4 months.

- Articles: 2,500 words. Informational articles; research papers; commentaries; and practical application pieces. Topics include teaching techniques, health services in the school system, nursing, medicine, substance abuse, nutrition, counseling, and ADD/AHD.

SAMPLE ISSUE
58 pages (no advertising): 3 articles; 4 research papers; 1 commentary. Sample copy, $8.50 with 9x12 SASE. Writers' guidelines available at website.

- "A CDC Review of School Laws and Policies Concerning Child and Adolescent Health." Article provides a review of the laws and policies currently in place regarding schools and children's health issues.
- "Promoting Healthy Vision to Students." Article analyzes progress and changes in student vision programs, research, and policy.

RIGHTS AND PAYMENT
All rights. No payment. Provides 2 contributor's copies.

➡️EDITOR'S COMMENTS
Our audience is comprised of school health professionals, so we demand thorough research and bibliographies as well as abstracts for research papers. Writers also must be professionals within the field of school health. All work must be submitted through the submissions website.

JUCO Review

1755 Telstar Drive, Suite 103
Colorado Springs, CO 80920

Executive Editor: Wayne Baker

DESCRIPTION AND INTERESTS
This publication of the National Junior College Athletic Association provides information about the association's athletic programs, successes, personnel, and tournaments. Circ: 2,850.

Audience: YA–Adult
Frequency: 10 times each year
Website: www.njcaa.org

FREELANCE POTENTIAL
30–40% written by nonstaff writers. Publishes 5–7 freelance submissions yearly. Receives 12 unsolicited mss yearly.

SUBMISSIONS
Send complete ms. Accepts hard copy. Availability of artwork improves chance of acceptance. SASE. Responds in 2 months.

- Articles: 1,500–2,000 words. Informational articles. Topics include sports, college, careers, health, fitness, and NJCAA news.
- Artwork: B/W prints and transparencies.

SAMPLE ISSUE
22 pages (25% advertising): 14 articles. Sample copy, $4 for current issue; $3 for back issue with 9x12 SASE. Editorial calendar available.

- "Men's Basketball Coaches Association of the Year Awards." Article profiles the winning coaches in Division I, II, and III.
- "College of the Month." Article profiles Scottsdale Community College, named the NJCAA College of the Month.
- "Academy Offers One-of-a-Kind Associate's to Bachelor's Program." Article explains a program that allows community college athletes to enroll in the U.S. Sports Academy, then transfer their credits to a bachelor degree program.

RIGHTS AND PAYMENT
One-time rights. No payment. Provides 3 contributor's copies.

⟶EDITOR'S COMMENTS
Community college athletics is different from NCAA athletics, and anyone who writes for us should be keenly aware of those differences. That said, we cover NJCAA athletics with the same excitement and fever. We expect writers not only to be good sports reporters, but to have a connection with and understanding of the junior college and community college system, programs, and athletes.

Junior Baseball

14 Woodway Lane
Wilton, CT 06897

Editor/Publisher: Jim Beecher

DESCRIPTION AND INTERESTS
Young baseball players and their parents and coaches are the audience of this magazine. Technical tips, interviews, and articles written for specific age groups can be found in its pages. Circ: 50,000.

Audience: 7–17 years; parents and coaches
Frequency: 6 times each year
Website: www.juniorbaseball.com

FREELANCE POTENTIAL
50% written by nonstaff writers. Publishes 20 freelance submissions yearly; 10% by unpublished writers, 20% by authors who are new to the magazine. Receives 50 queries and unsolicited mss yearly.

SUBMISSIONS
Query with writing samples; or send complete ms with color photos. Accepts email submissions to jim@juniorbaseball.com (Microsoft Word attachments or text). Availability of artwork improves chance of acceptance. SASE. Responds in 1–2 weeks.

- Articles: 750–1,500 words. Informational and how-to articles; profiles; and interviews. Topics include playing tips, teams and leagues, and player safety.
- Depts/columns: "Player's Story," 500 words. "In the Spotlight" news and reviews, 50–100 words. "Hot Prospects," 500–1,000 words. "Coaches' Clinic," 100–1,000 words.
- Artwork: 4x5, 5x7, and 8x10 color prints. Color digital images at 300 dpi.

SAMPLE ISSUE
34 pages (30% advertising): 3 articles; 12 depts/columns. Sample copy, $3.95 with 9x12 SASE ($1.35 postage). Guidelines available.

- "Get a Good Pitch to Hit!" Article offers a simple way to teach youngsters the strike zone.
- Sample dept/column: "Player's Story" is an essay by a young boy about coming back after an injury.

RIGHTS AND PAYMENT
All rights. Articles, $.20 per word. Depts/columns, $25–$100. Artwork, $50–$150. Pays on publication. Provides 1 contributor's copy.

⟶EDITOR'S COMMENTS
Writers should be knowledgeable about baseball. Material should include facts, figures, and anecdotes.

Junior Shooters

7154 West State Street, #377
Boise, ID 83714

Editor: Andrew Fink

DESCRIPTION AND INTERESTS
Junior Shooters is the first magazine of its kind to promote junior involvement in hunting and shooting and its many disciplines, all in one publication. It covers shooting events, venues, gear, and shooting organizations. Circ: 40,000.

Audience: 8–21 years
Frequency: Quarterly
Website: www.juniorshooters.net

FREELANCE POTENTIAL
60% written by nonstaff writers. Publishes several freelance submissions yearly; 60% by unpublished writers, 40% by authors who are new to the magazine.

SUBMISSIONS
Send complete ms. Accepts email submissions to articles@juniorshooters.net and CD submissions (PC format) in Microsoft Word; all submissions should be accompanied by form found at website. Materials are not returned. Response time varies.

- Articles: Word lengths vary. Informational articles; profiles; and personal experience pieces. Topics include all disciplines of shooting sports, techniques, training, coaching, products, and gear.
- Depts/columns: Word lengths vary. Shooting tips, gun safety, new products and gear.
- Artwork: High-resolution digital images.

SAMPLE ISSUE
Sample copy available. Guidelines and editorial calendar available at website.

- "Step Up to an Air Rifle." Article explains the under-appreciated sport of air rifle shooting.
- Sample dept/column: "Ty's Tidbits" offers the author's experiences as a junior shooter.

RIGHTS AND PAYMENT
Non-exclusive rights. No payment.

⚬EDITOR'S COMMENTS
We want articles about juniors, by juniors. Our goal is to have 60 percent of our articles written by young people. We are also interested in articles written by coaches, members of shooting organizations, parents, manufacturers, and anyone else who would like to provide information about supporting juniors in the shooting sports. All work must be accompanied by the submission form found at our website.

JuniorWay

P.O. Box 436987
Chicago, IL 60643

Editor: Katherine Steward

DESCRIPTION AND INTERESTS
This publication is used in Christian religious education classes in urban settings. It features Bible lessons that can be discussed in the classroom and at home. Circ: 75,000.

Audience: 9–11 years
Frequency: Quarterly
Website: www.urbanministries.com

FREELANCE POTENTIAL
95% written by nonstaff writers. Publishes 52 freelance submissions yearly. Receives 240 queries yearly.

SUBMISSIONS
Query with résumé and writing samples. All material is written on assignment. Response time varies.

- Articles: Word lengths vary. Bible lessons, personal experience pieces, and humor. Topics include religion, relationships, social issues, hobbies, crafts, sports, recreation, and multicultural subjects.
- Fiction: Word lengths vary. Inspirational stories with multicultural or ethnic themes, adventure stories, humor, and folktales.
- Artwork: B/W and color prints and transparencies.
- Other: Puzzles, activities, games, and jokes. Poetry. Seasonal material about Vacation Bible School.

SAMPLE ISSUE
34 pages (no advertising): 13 Bible lessons; 1 activities insert; 1 comic. Sample copy, free. Guidelines and theme list available.

- "A Reminder That God Is with Us." Story and lesson tell about the courage and faith of Corrie Ten Boom.
- "I'm Counting on You." Story and lesson explain the importance of keeping promises.
- "Why Me?" Story and lesson show that God's plan for us may not be what we expected.

RIGHTS AND PAYMENT
All rights. All material, payment rates vary. Pays on publication.

⚬EDITOR'S COMMENTS
Prospective authors must understand our mission to young African American children. If you have a background in religious education, send with your ideas for teaching Bible lessons along with samples of your writing.

Justine

6263 Poplar Avenue, Suite 1154
Memphis, TN 38119

Editorial Director/Publisher: Jana Petty

DESCRIPTION AND INTERESTS
Wholesome and entertaining articles that focus
on healthy relationships and building self confidence fill the pages of this teen magazine for
girls. Fashion and beauty tips, nutrition, craft
and decorating projects, celebrity news, and
media reviews are among the topics covered.
Circ: 250,000.

Audience: 13–18 years
Frequency: 6 times each year
Website: www.justinemagazine.com

FREELANCE POTENTIAL
20% written by nonstaff writers. Publishes 25
freelance submissions yearly; 25% by unpublished writers, 90% by authors who are new to
the magazine. Receives 100 queries yearly.

SUBMISSIONS
Query with résumé and clips. Accepts hard
copy. SASE. Response time varies.

- Articles: Word lengths vary. Informational
  articles; profiles; and personal experience
  pieces. Topics include room decorating,
  beauty, health, nutrition, family issues, recreation, travel, and fashion.
- Depts/columns: Word lengths vary. Media
  and book reviews, entertainment, exercise,
  and community service.

SAMPLE ISSUE
96 pages: 5 articles; 32 depts/columns; 1 quiz.
Sample copy, $2.99 at newsstands.

- "The Belles of Broadway." Article interviews
  three up-and-coming teen Broadway stars on
  how they got their start.
- Sample dept/column: "Just' Life" features
  the SADD (Students Against Destructive
  Decisions) Student of the Year.
- Sample dept/column: "Brown-Bagging It"
  offers healthy and tasty suggestions for
  school lunches.

RIGHTS AND PAYMENT
Rights vary. Written material, payment rates
vary. Pays 30 days after publication.

➔EDITOR'S COMMENTS
We set ourselves apart from other teen magazines by providing tasteful yet up-to-the-
minute content. If you have something new,
stylish, and entertaining, and your writing
style matches our audience, we would love
to hear from you.

JVibe

90 Oak Street
P.O. Box 9129
Newton, MA 02464

Editor-in-Chief: Lindsey Silken

DESCRIPTION AND INTERESTS
This magazine for Jewish teens is filled with
celebrity interviews; the latest in music, movies,
and sports; conversations with teens; and information on what's hot in Israel. Circ: 15,000.

Audience: 9–18 years
Frequency: 6 times each year
Website: www.jvibe.com

FREELANCE POTENTIAL
90–95% written by nonstaff writers. Publishes
85 freelance submissions yearly; 10% by
unpublished writers, 20% by authors who are
new to the magazine. Receives 120 queries
each year.

SUBMISSIONS
Query. Accepts email queries to editor@
jvibe.com. Responds in 1 week.

- Articles: 1,400 words. Informational articles;
  profiles; interviews; reviews; and personal
  experience pieces. Topics include religion,
  Israel, popular culture, music, sports, current
  events, humor, college, careers, and the arts.
- Depts/columns: Staff written.
- Other: Submit seasonal material 3+ months
  in advance.

SAMPLE ISSUE
40 pages (10% advertising). Sample copy,
guidelines, and editorial calendar available.

- "Sixty Six Is Hot in '08." Article reviews the
  film Sixty Six, which features a 12-year-old
  striving to plan the greatest bar mitzvah celebration ever.
- "Sex: What Are Your Parents Saying?" Article
  interviews teens and their parents about
  their thoughts and values regarding sex.
- "Super Summer Reading." Article reviews 17
  books that were voted by the advisory board
  as bona fide great summer reads.

RIGHTS AND PAYMENT
All rights. Written material, payment rates vary.
Pays on publication. Provides 2 contributor's
copies.

➔EDITOR'S COMMENTS
We would really like to get more material on
politics, the environment, and current events.
Because we write for teens, we can never get
enough articles based on pop culture.
Feature articles present the best opportunity
for new writers.

Kaboose.com

505 University Avenue
Toronto, Ontario M5G 1X3
Canada

Editor: Leigh Felesky

DESCRIPTION AND INTERESTS
Founded in 1999, this online magazine provides parents—mostly mothers—with information on topics ranging from pregnancy and parenting to education, birthday parties, and scrapbooking. Hits per month: Unavailable.

Audience: Mothers
Frequency: Unavailable
Website: www.kaboose.com

FREELANCE POTENTIAL
95% written by nonstaff writers. Publishes 30 freelance submissions yearly; 10% by unpublished writers, 30% by authors who are new to the magazine. Receives 150 queries yearly.

SUBMISSIONS
Query with outline. Accepts email queries to leigh.felesky@kaboose.com. Responds in 6 weeks.

- Articles: 2,000 words. Informational and how-to articles. Topics include parenting, pregnancy, child care, health, fitness, food, kids' activities, celebrity updates, style, home and garden, crafts, hobbies, and pets.
- Depts/columns: 500 words. Advice, travel, recipes, and entertaining.
- Other: Submit seasonal material for Christmas, Easter, Thanksgiving, and other events that bring families together 6 months in advance.

SAMPLE ISSUE
Sample copy, guidelines, and theme list available at website.

- "The Family Medicine Cabinet." Article explains the medications that should be kept on hand, and the safest way to store them.
- "Egg Carton Carrot Patch." Article tells how to help kids make a cute table decoration using egg cartons.

RIGHTS AND PAYMENT
Rights vary. Written material, $.85 per word. Pays on acceptance. Contributor's copies available at website.

❖EDITOR'S COMMENTS
We are looking for a wide range of material that will appeal to mothers of children ages four to fourteen. The subjects may vary, but a constant requirement is that the material be informative, fun, and provide busy moms with something they didn't know before.

Kaleidoscope
Exploring the Experience of Disability through Literature & Fine Arts

701 South Main Street
Akron, OH 44311-1019

Editor-in-Chief: Gail Willmott

DESCRIPTION AND INTERESTS
Writers for *Kaleidoscope* are given a platform in which to express the experience of disability. Material is written from the perspective of individuals, families, and healthcare professionals. Circ: 1,000.

Audience: YA–Adult
Frequency: Twice each year
Website: www.udsakron.org

FREELANCE POTENTIAL
90% written by nonstaff writers. Publishes 20–35 freelance submissions yearly; 5% by unpublished writers, 75% by authors who are new to the magazine. Receives 120–180 queries, 360 unsolicited mss yearly.

SUBMISSIONS
Query or send complete ms with author bio. Accepts hard copy and email submissions to mshiplett@udsakron.org (Microsoft Word attachments). SASE. Responds to queries in 2 weeks, to mss in 6 months.

- Articles: 5,000 words. Informational articles; personal experience pieces; profiles; interviews; reviews; and humor. Topics include art, literature, biography, multicultural and social issues, and disabilities.
- Fiction: 5,000 words. Genres include folktales, humor, and multicultural and problem-solving fiction.
- Other: Poetry.

SAMPLE ISSUE
64 pages (no advertising): 5 articles; 8 stories; 11 poems. Sample copy, $6 with 9x12 SASE. Guidelines and editorial calendar available.

- "A Portrait of Creativity." Article profiles a watercolor artist struggling with a chronic hand condition.
- "The Freedom of Cages." Essay tells how a woman deals with her middle-age diagnosis of multiple sclerosis.

RIGHTS AND PAYMENT
First rights. Written material, $25–$100. Poetry, $10. Pays on publication. Provides 2 contributor's copies.

❖EDITOR'S COMMENTS
Writers with and without disabilities are welcome to submit their work. We are now accepting material with war themes and are always looking for humor pieces.

Kansas School Naturalist

Emporia State University
Department of Biological Services
Box 4050
Emporia, KS 66801-5087

Editor: John Richard Schrock

DESCRIPTION AND INTERESTS

Teachers, librarians, and others with a keen interest in natural history and nature education are the audience for *Kansas School Naturalist*. Each issue features one lengthy article that relates to the state's natural ecology. A wide range of topics is covered, including astronomy, stream ecology, wildlife, birds, and natural resources. It is published by Emporia State University. Circ: 10,000.

Audience: Teachers; librarians; conservationists
Frequency: Irregular
Website: www.emporia.edu/ksn

FREELANCE POTENTIAL

50% written by nonstaff writers. Of the freelance submissions yearly, 20% are by unpublished writers and 75% are by authors who are new to the magazine.

SUBMISSIONS

Query or send complete ms. Accepts hard copy and email submissions to ksnaturl@emporia.edu. SASE. Response time varies.

- Articles: Word lengths vary. Informational and how-to articles. Topics include natural history, nature, science, technology, astronomy, animals, health, and education—all with a focus on Kansas.
- Artwork: B/W or color prints or transparencies.
- Other: Seasonal material.

SAMPLE ISSUE

14 pages (no advertising): 1 article. Sample copy, free.

- "Checklist of Kansas Ground Spiders." Article comprehensively reviews the 63 species of ground spiders found in Kansas and features background information on the insect's habits, habitats, and benefits to the state's ecology. Color photos of the different Kansas spiders, and an extensive list of references, are included.

RIGHTS AND PAYMENT

All rights. No payment. Provides copies.

➡️EDITOR'S COMMENTS

If you are submitting material for our journal, you must have scientific qualifications on the subject and it must have a relevance to the state of Kansas. Our writing style is conversational but all articles are backed up by exacting, scientific research.

Keeping Family First Online Magazine

P.O. Box 36594
Detroit, MI 48236

Executive Editor: Anita S. Lane

DESCRIPTION AND INTERESTS

Dedicated to mothers and fathers who are committed to building strong families, *Keeping Family First* provides a community of experts and inspirational authors that addresses the specific needs of today's parents. Its content includes parenting advice, solutions-oriented articles, and inspirational interviews. Hits per month: 40,000.

Audience: Parents
Frequency: 6 times each year
Website: www.KeepingFamilyFirst.org

FREELANCE POTENTIAL

100% written by nonstaff writers. Publishes 70 freelance submissions yearly; 56% by unpublished writers, 10% by new authors.

SUBMISSIONS

Query. Accepts email queries online only via www.keepingfamilyfirst.org/submissions.html. Response time varies.

- Articles: Word lengths vary. Informational articles; profiles; and personal experience pieces. Topics include parenting, health, spirituality, education, recreation, home, leisure, and entertainment.
- Depts/columns: Word lengths vary. Parenting tips, family issues, travel, and recreation.

SAMPLE ISSUE

Sample copy and writers' guidelines available at website.

- "The 'Mean Mom's' Summer Survival Guide." Article explains how to become a "mean" mom, focusing on safety and structure, to get your family through the summer.
- "The Transformative Power of Self-Care." Article explains how mothers tend to put themselves last, and how important it is that they give themselves physical, spiritual, and mental care.
- Sample dept/column: "Fashion" offers ideas for updating one's wardrobe on a budget.

RIGHTS AND PAYMENT

Rights vary. No payment.

➡️EDITOR'S COMMENTS

We look for articles that can provide solutions and advice, deliver resources to readers, and offer inspiration for the challenges parents face. We also welcome articles geared specifically toward fathers.

Key Club

Key Club International
3636 Woodview Trace
Indianapolis, IN 46268-3196

Executive Editor: Jack Brockley

DESCRIPTION AND INTERESTS
Formerly published as *Keynoter*, this magazine targets high school students who are members of their local Key Clubs. Academic, self-help, and service- and leadership-related articles are found in each issue. Circ: 240,000.

Audience: 14–18 years
Frequency: Quarterly
Website: www.keyclub.org/magazine

FREELANCE POTENTIAL
20% written by nonstaff writers. Publishes 8 freelance submissions yearly; 5% by unpublished writers, 15% by authors who are new to the magazine. Receives 100 queries yearly.

SUBMISSIONS
Query with outline/synopsis and clips or writing samples. Accepts hard copy, email queries to magazine@kiwanis.org, and simultaneous submissions if identified. SASE. Responds in 1 month.

- Articles: 1,000–1,500 words. Informational, self-help, and service-related articles. Topics include education, teen concerns, community service, leadership, school activities, social issues, and careers.
- Depts/columns: Staff written.
- Artwork: Color prints and illustrations.
- Other: Submit seasonal material about back to school, college, and summer activities 3–7 months in advance.

SAMPLE ISSUE
24 pages (5% advertising): 4 articles; 3 depts/columns. Sample copy, free with 9x12 SASE ($.83 postage). Guidelines available.

- "Building Uganda's Tomorrow." Article chronicles a recent Key Club trip to Uganda to help build a school.
- "Swaziland: Those Kids Need Us." Article details a fundraiser with UNICEF to help fight AIDS in Africa.

RIGHTS AND PAYMENT
First North American serial rights. All material, $150–$350. Pays on acceptance. Provides 3 contributor's copies.

➥EDITOR'S COMMENTS
Material on social and cultural issues of interest to teens is always welcome. We do not publish first-person essays, personal profiles, poetry, or fiction.

Keys for Kids

P.O. Box 1001
Grand Rapids, MI 49510

Editor: Hazel Marett

DESCRIPTION AND INTERESTS
Daily devotionals for children in elementary and middle school make up the majority of this digest-sized magazine. The fun stories are intended to help readers discover who God wants them to be. Circ: 70,000.

Audience: 6–12 years
Frequency: 6 times each year
Website: www.keysforkids.org

FREELANCE POTENTIAL
100% written by nonstaff writers. Publishes 20–30 freelance submissions yearly; 50% by unpublished writers, 90% by authors who are new to the magazine. Receives 60–75 unsolicited mss yearly.

SUBMISSIONS
Send complete ms. Accepts hard copy and email submissions to hazel@cbhministries.org. SASE. Responds in 2 months.

- Articles: 400 words. Devotionals with related Scripture passages and a key thought. Topics include contemporary social issues, family life, trust, friendship, salvation, witnessing, prayer, marriage, and faith.

SAMPLE ISSUE
80 pages: 61 devotionals. Sample copy, free with 9x12 SASE. Guidelines available.

- "Who's To Blame?" Devotional features a boy who finds a way to explain to his friend that God isn't to blame for bad things.
- "The Beautiful Apple." Devotional uses a pretty apple with a worm in it to teach that what's inside us is what counts most.
- "Not Sticky Enough." Devotional features two siblings who need to show more love after their mother hears them speaking badly about their classmates.

RIGHTS AND PAYMENT
First, second, and reprint rights. Written material, $25. Pays on acceptance. Provides 1 contributor's copy.

➥EDITOR'S COMMENTS
Each devotional should suggest an appropriate Scripture passage to read, tell a story with a spiritual application, and present a practical application to the reader's life. We call our magazine *Keys* because each devotional provides a key Bible verse and a key thought that readers can take with them.

Kid Magazine Writers

9 Arrowhead Drive
Ledyard, CT 06339

Editor: Jan Fields

DESCRIPTION AND INTERESTS
This e-zine, written for and by children's magazine writers, is designed to meet the needs of new writers entering the market. Each issue offers specific information on how to get started and includes tips, reviews, and resources from others in the community. Articles are both instructional and motivational. Hits per month: 2,000.

Audience: Children's magazine writers
Frequency: Monthly
Website: www.kidmagwriters.com

FREELANCE POTENTIAL
50% written by nonstaff writers. Publishes 24–28 freelance submissions yearly; 30% by unpublished writers, 50% by authors who are new to the magazine.

SUBMISSIONS
Query or send complete ms. Accepts email submissions to editor@kidmagwriters.com (no attachments). Response time varies.

- Articles: 500+ words. Informational and how-to articles; interviews; essays; and personal experience pieces. Topics include print and online magazines, magazine contests, editors and illustrators, nonfiction and fiction writing, poetry writing, and book and media reviews.

SAMPLE ISSUE
Guidelines available at website.

- "The Absolute Best." Essay recounts a new writer's experience of having her first article published.
- "Stories in Verse." Article explains how to write a story in verse and includes possible magazines to query.
- "Strategy on a Shoestring: Writing on a Short Deadline." Article offers tips on how to write your best when deadlines are tight.

RIGHTS AND PAYMENT
One-time rights. No payment.

➦EDITOR'S COMMENTS
We depend upon the generosity of the children's magazine writing community for our content. We are open to instructive and motivational articles for all of our departments. Please refer to our website for specific information concerning guidelines and the contents of our departments.

The Kids' Ark

P.O. Box 3160
Victoria, TX 77903

Editor: Joy Mygrants

DESCRIPTION AND INTERESTS
The mission of this Christian magazine is to provide children with a biblical foundation on which to base their life choices. Each issue is theme-based and features three fun and adventurous stories along with games, puzzles, and comics. There is also an Internet version of the magazine. Circ: 8,000+.

Audience: 6–12 years
Frequency: Quarterly
Website: www.thekidsark.com

FREELANCE POTENTIAL
90% written by nonstaff writers. Publishes 12 freelance submissions yearly; 80–100% by authors who are new to the magazine.

SUBMISSIONS
Send complete ms. Accepts email submissions to thekidsarksubmissions@yahoo.com (Microsoft Word attachments). Responds in 2 months.

- Fiction: 600 words. Genres include contemporary, historical, and science fiction, all with a Christian base.
- Other: Puzzles, games, and comics.

SAMPLE ISSUE
24 pages: 3 stories; 5 activities; 1 comic. Guidelines and theme list available at website.

- "Randy's Victory." Story tells of a young boy who learns to deal with his anger.
- "Responding to Anger." Story is about two friends who reconcile their differences after an argument.
- "Remembering Yesterday." Story follows a mother and her son as they learn a valuable lesson about unkind people.

RIGHTS AND PAYMENT
First North American, worldwide, and electronic rights, with reprint rights. Fiction, $100. Reprints, $25. Pays on publication.

➦EDITOR'S COMMENTS
All material must match one of our quarterly themes. The first page of your submission should include the theme for which you are submitting as well as your name, phone number, and email address. We do not accept unsolicited poems, games, or comics. Please be sure stories are age-appropriate and include children of all races and ethnic backgrounds. Please refer to our website for a list of upcoming themes.

Kids Life

1426 22nd Avenue
Tuscaloosa, AL 35401

Publisher: Mary Jane Turner

DESCRIPTION AND INTERESTS
Subtitled "Tuscaloosa's Family Magazine," *Kids Life* is available free of charge to residents of western Alabama. It provides information on regional child-related events, activities, and resources, as well as more general parenting articles. Circ: 30,000.
Audience: Parents
Frequency: 6 times each year
Website: www.kidslifemagazine.com

FREELANCE POTENTIAL
75% written by nonstaff writers. Publishes 20 freelance submissions yearly; 50% by unpublished writers, 10% by authors who are new to the magazine. Receives 100 queries and unsolicited mss yearly.

SUBMISSIONS
Query or send complete ms. Accepts hard copy and email submissions to kidslife@comcast.net. SASE. Responds in 2 weeks.

- Articles: 1,000 words. Informational articles and personal experience pieces. Topics include parenting, education, sports, child care, religion, cooking, crafts, health, travel, and current events.
- Depts/columns: Staff written.
- Artwork: Color prints or transparencies. Line art.
- Other: Puzzles, activities, games, and filler.

SAMPLE ISSUE
52 pages (60% advertising): 7 articles; 5 depts/columns. Sample copy, free.

- "Competitive Swimming in Tuscaloosa? Who Knew?" Article describes the Crimson Tide Aquatics swim team program for children.
- "Students Shine at Sprayberry." Article provides an overview of the special education programs offered at a local facility.
- "Building a Better World, One Child at a Time." Article profiles Team Lee Athletics.

RIGHTS AND PAYMENT
Rights vary. Written material, to $30. Pays on publication. Provides 1 contributor's copy.

➦EDITOR'S COMMENTS
We strive to make *Kids Life* a one-stop place for families by covering everything the Tuscaloosa area offers our children, from preschool programs to after-school activities, from family films to community theater.

Kids' Ministry Ideas

55 West Oak Ridge Drive
Hagerstown, MD 21740

Editor: Candy DeVore

DESCRIPTION AND INTERESTS
This Christian publication supports youth ministers by providing affirmation, pertinent articles, program ideas, resource suggestions, and answers to questions, all from a Seventh-day Adventist perspective. Circ: 4,500.
Audience: Sabbath school teachers and Adventist Junior Youth leaders
Frequency: Quarterly
Website: www.kidsministryideas.com

FREELANCE POTENTIAL
100% written by nonstaff writers. Publishes 60 freelance submissions yearly.

SUBMISSIONS
Query or send complete ms. Accepts email to cdevore@rhpa.org and hard copy. SASE. Response time varies.

- Articles: 300–800 words. Informational and how-to articles; and essays. Topics include religious education, youth ministry, family issues, working with volunteers, lesson plans and props, faith, and prayer.
- Depts/columns: Word lengths vary. Leadership training, teaching tips, and crafts.
- Other: Submit seasonal material 6–12 months in advance.

SAMPLE ISSUE
32 pages: 3 articles; 7 depts/columns. Sample copy and guidelines available at website.

- "Kailua's Kids in Action." Article describes the robust children's program at a Seventh-day Adventist church in Hawaii.
- "That Unmentionable Two-Letter Word." Personal essay tells of the author's inability to say "no" to volunteer opportunities.
- Sample dept/column: "Kid Crafts" suggests having children make bookmarks based on Ecclesiastes 3:1.

RIGHTS AND PAYMENT
First North American serial rights. Written material, $20–$100. Pays 5–6 weeks after acceptance. Provides 1 contributor's copy.

➦EDITOR'S COMMENTS
We need your help! We're looking for practical "how-to" material written for or by those who work with children up to the age of 12. Sidebars and lists are encouraged to make information stand out. We welcome submissions from new writers.

Kids VT

P.O. Box 1089
Shelburne, VT 05482

Editor: Susan Holson

DESCRIPTION AND INTERESTS
The "VT" stands for Vermont, the Green Mountain State, where this free newspaper covers all sorts of "kidding around." Its readership consists mainly of young mothers with children under age 12. Circ: 25,000.

Audience: Parents
Frequency: 10 times each year
Website: www.kidsvt.com

FREELANCE POTENTIAL
75% written by nonstaff writers. Publishes 50–60 freelance submissions yearly; 20% by authors who are new to the magazine. Receives 480–960 unsolicited mss yearly.

SUBMISSIONS
Send complete ms. Accepts email submissions to editorial@kidsvt.com (no attachments) and simultaneous submissions if identified. Responds only if interested.

- Articles: 500–1,500 words. Informational articles; profiles; interviews; and humor. Topics include the arts, education, recreation, nature, the environment, music, camps, pregnancy, infancy, and parenting.
- Depts/columns: Word lengths vary. News and media reviews.
- Other: Activities and games. Submit seasonal material 2 months in advance.

SAMPLE ISSUE
32 pages (50% advertising): 9 articles; 12 depts/columns. Guidelines and editorial calendar available at website.

- "Shopping the MyPyramid Way." Article suggests using the USDA's online nutrition guide to compile a weekly grocery list.
- "Keep the Kaboom Safe!" Article provides tips for keeping children safe during July 4th celebrations.
- Sample dept/column: "Let's Party" offers fun activities for babysitters to do with kids.

RIGHTS AND PAYMENT
One-time and reprint rights. Written material, payment rates vary. Pays 30 days after publication. Provides 1–2 contributor's copies.

➠EDITOR'S COMMENTS
Our freelance rates are negotiated prior to publication based on the article's word count and relevance to our readers. Priority is given to local authors.

Kid Zone

801 West Norton Avenue, Suite 200
Muskegon, MI 49441

Editor: Anne Huizenga

DESCRIPTION AND INTERESTS
Each issue of this colorful magazine is filled with kid-friendly crafts, games, puzzles, recipes, jokes, stories, and educational articles—all organized into departments called "zones." Circ: 65,000.

Audience: 4–12 years
Frequency: 6 times each year
Website: www.scottpublications.com

FREELANCE POTENTIAL
90% written by nonstaff writers. Publishes 20–30 freelance submissions yearly; 25% by authors who are new to the magazine. Receives 60–100 queries and unsolicited mss yearly.

SUBMISSIONS
Query or send complete ms. Accepts hard copy and email to ahuizenga@scottpublications.com. Availability of artwork improves chance of acceptance. SASE. Responds to queries in 1 month, to mss in 1 year.

- Articles: 500 words. Informational articles. Topics include animals, food, culture, seasonal subjects, nature, science, and safety.
- Depts/columns: Staff written.
- Artwork: Color prints or transparencies.
- Other: Submit seasonal material 4–6 months in advance.

SAMPLE ISSUE
44 pages (no advertising): 5 articles; 1 story; 9 depts/columns; 1 poem; 10 activities. Sample copy, $4.99. Writers' guidelines and editorial calendar available.

- "Get Your Yard Humming." Article tells all about hummingbirds and how to attract them.
- "A Colossal Gift: The Statue of Liberty." Article presents the history of the statue, along with some fun facts.

RIGHTS AND PAYMENT
World rights. Articles, $20–$50. Artwork, payment rates vary. Pays on publication. Provides 2 contributor's copies.

➠EDITOR'S COMMENTS
We don't take submissions for fictional material, or for poetry, trivia, games, or recipes, unless they are part of a written feature. Our Critter Zone, Chomp Zone, Culture Zone, Fun Zone, and Discovery Zone are the areas most open to freelancers.

Know
The Science Magazine for Curious Kids

501-3960 Quadra Street
Victoria, British Columbia V8X 4A3
Canada

Managing Editor: Adrienne Mason

DESCRIPTION AND INTERESTS
This magazine strives to be fun as well as informative. It features short news items, brief fiction, poetry, interviews, hands-on projects, and articles dedicated to a different theme each issue. Circ: 11,000.

Audience: 6–9 years
Frequency: 6 times each year
Website: www.knowmag.ca

FREELANCE POTENTIAL
50% written by nonstaff writers. Publishes 150 freelance submissions yearly; 5% by unpublished writers, 30% by authors who are new to the magazine. Receives 300 queries, 180 unsolicited mss yearly.

SUBMISSIONS
Query with résumé and clips for nonfiction. Send complete ms for fiction and poetry. Accepts hard copy and email submissions to adrienne@knowmag.ca. SASE. Responds to queries in 1 month, to mss in 3 months.

- Articles: 250 words. Informational and how-to articles; science experiments; and interviews. Topics include chemistry, physics, biology, ecology, zoology, geology, technology, and mathematics.
- Fiction: To 500 words. Theme-related stories.
- Depts/columns: 200–250 words. Science news and discoveries, scientist profiles, astronomy, paleontology, and random facts.
- Other: Poetry. Puzzles, games, and activities.

SAMPLE ISSUE
32 pages (6% advertising): 8 articles; 16 depts/columns. Guidelines, theme list, and sample issue available at website.

- "Tick-Tock Time." Article introduces the basic measurements of time.
- Sample dept/column: "Know How" explains how to make a sundial.

RIGHTS AND PAYMENT
First North American serial rights. Written material, $.40–$.50 Canadian per word. Pays on publication. Provides 2 contributor's copies.

⇝EDITOR'S COMMENTS
We are seeking articles on "hard" sciences such as chemistry, physics, and technology. We also need more articles that cover current science news. Please see our theme list before submitting manuscripts.

Lacrosse

113 West University Parkway
Baltimore, MD 21210

Editor: Paul Krome

DESCRIPTION AND INTERESTS
This magazine, read by players, coaches, officials, fans, and association members, seeks to connect, educate, entertain, and inform the lacrosse community. It is published by US Lacrosse, the sport's national governing body. Circ: 225,000.

Audience: Youth and college lacrosse players
Frequency: Monthly
Website: www.laxmagazine.com

FREELANCE POTENTIAL
40% written by nonstaff writers. Publishes 60 freelance submissions yearly; 5% by unpublished writers, 10% by authors who are new to the magazine. Receives 60 queries yearly.

SUBMISSIONS
Query with clips or résumé. Accepts Macintosh disk submissions (Microsoft Word) and email to pkrome@usalacrosse.org. SASE. Responds to queries in 1 month.

- Articles: 1,200–1,500 words. Informational and how-to articles; profiles; and interviews. Topics include lacrosse gear, training, rules, strategies, leagues, coaching methods, competitions, and events; and sports medicine.
- Depts/columns: 650 words. Organization news and events; tips for players, coaches, and officials; and profiles of personalities.

SAMPLE ISSUE
96 pages (35% advertising): 7 articles; 7 depts/columns. Sample copy, $5. Writer's guidelines available.

- "Tech Savvy." Article explains the importance of using technology for scouting or recruiting.
- "One Decade, One Sport." Article celebrates the 10th anniversary of US Lacrosse.
- Sample dept/column: "Parting Shot" tells of a beloved coach's battle with cancer.

RIGHTS AND PAYMENT
Exclusive rights. Articles, $100–$300. Depts/columns, $100–$150. Pays on publication. Provides 1+ contributor's copies.

⇝EDITOR'S COMMENTS
We would like to hear about what youth leagues around the country are doing, including the coaches and players behind them. Writers should have a strong knowledge of and passion for lacrosse.

Ladies' Home Journal

Meredith Corporation
375 Lexington Avenue, 9th Floor
New York, NY 10017

Deputy Editor: Margot Gilman

DESCRIPTION AND INTERESTS
After 125 years in publication, this women's magazine has become such an institution that some of its departments are trademarked. Content covers marriage, family life, home, beauty, style, and—not to be overlooked—personal fulfillment. Circ: 4.1 million.

Audience: Women
Frequency: Monthly
Website: www.lhj.com

FREELANCE POTENTIAL
85% written by nonstaff writers. Publishes 25 freelance submissions yearly; 1% by unpublished writers, 5% by authors who are new to the magazine. Receives 2,400 queries yearly.

SUBMISSIONS
Query with résumé, outline, and clips or writing samples for nonfiction. Accepts fiction through literary agents only. Accepts hard copy. SASE. Responds in 1–3 months.

- Articles: 1,500–2,000 words. Informational and how-to articles; profiles; interviews; and personal experience pieces. Topics include family issues, parenting, social concerns, fashion, beauty, and women's health.
- Fiction: Word lengths vary. Genres vary.
- Depts/columns: Word lengths vary. Motherhood, marriage, self-help, beauty, home, health, food, news, and lifestyle features.

SAMPLE ISSUE
166 pages (15% advertising): 7 articles; 14 depts/columns. Sample copy, $2.49 at newsstands.

- "Soul Sisters." Interview with *Good Morning America* co-anchors Robin Roberts and Diane Sawyer tells how their friendship helped Roberts beat cancer.
- Sample dept/column: "My Life as a Mom" ponders whether the author is overzealous in her motherly support.

RIGHTS AND PAYMENT
All rights. All material, payment rates vary. Pays on publication. Provides 2 contributor's copies.

➤EDITOR'S COMMENTS
"Never underestimate the power of a woman" has been our slogan since the 1940s. This sentiment should be at the heart of any submission to the magazine—we favor stories of women overcoming challenges.

Ladybug
The Magazine for Young Children

70 East Lake Street, Suite 300
Chicago, IL 60601

Submissions Editor: Jenny Gillespie

DESCRIPTION AND INTERESTS
Ladybug is a reading and listening magazine for very young children. It features original stories, poems, and activities, all illustrated in full color. Circ: 125,000.

Audience: 2–6 years
Frequency: 9 times each year
Website: www.cricketmag.com

FREELANCE POTENTIAL
100% written by nonstaff writers. Publishes 100 freelance submissions yearly. Receives 2,400 unsolicited mss yearly.

SUBMISSIONS
Send complete ms with exact word count. Accepts hard copy and simultaneous submissions. SASE. Responds in 6 months.

- Articles: To 400 words. Informational, humorous, and how-to articles. Topics include nature, family, animals, the environment, and other age-appropriate topics.
- Fiction: To 800 words. Read-aloud, early reader, picture, and rebus stories. Genres include adventure, humor, fantasy, folktales, and contemporary and multicultural fiction.
- Other: Puzzles, learning activities, games, crafts, finger plays, action rhymes, and songs. Poetry, to 20 lines.

SAMPLE ISSUE
34 pages (no advertising): 4 stories; 4 poems; 1 song; 1 finger play; 1 four-page activity section; 1 panel story. Sample copy, $5. Guidelines available at website.

- "Little Penguin's Dream." Story follows a young penguin as he practices various circus acts, and finally becomes the ringmaster.
- "Hiding, Seeking." Story tells a Native American tale of a mother and her son as they travel by way of different animals.

RIGHTS AND PAYMENT
Rights vary. Stories and articles, $.25 per word; $25 minimum. Poems, $3 per line; $25 minimum. Other material, payment rates vary. Pays on publication. Provides 6 copies.

➤EDITOR'S COMMENTS
We are looking for writing that conveys a sense of joy, sparks the imagination, and has a genuine childlike point of view. We urge writers to read several issues of our magazine to familiarize themselves with our style.

The Lamp-Post

1106 West 16th Street
Santa Ana, CA 92706

Senior Editor: David G. Clark

DESCRIPTION AND INTERESTS

The Southern California C.S. Lewis Society publishes this collection of scholarly essays, articles, book reviews, and poems that enhance the reader's understanding of author C.S. Lewis, his works, and his influence. Circ: 100.

Audience: Adults
Frequency: Quarterly

FREELANCE POTENTIAL

95% written by nonstaff writers. Publishes 20 freelance submissions yearly; 10% by unpublished writers, 60% by authors who are new to the magazine. Receives 24–48 unsolicited mss yearly.

SUBMISSIONS

Send complete ms. Accepts email submissions to dgclark@roadrunner.com (Microsoft Word attachments). No simultaneous submissions. Responds in 1 week.

- Articles: To 4,500 words. Informational articles and essays. Topics include C. S. Lewis, his works, and his influence.
- Fiction: To 4,500 words. Stories in the style of C. S. Lewis and the mythopoeic tradition.
- Depts/columns: Word lengths vary. Book reviews.
- Other: Poetry.

SAMPLE ISSUE

36 pages (no advertising): 2 articles; 1 story. Sample copy, $4. Guidelines available in each issue.

- "Two Sonnets by Roger Lancelyn Green." Article analyzes two sonnets written about C. S. Lewis by his former student and friend.
- "Matching Earrings." Story tells of a man whose jewelry purchase reunites him with his estranged daughter.

RIGHTS AND PAYMENT

Rights vary. No payment. Provides 2 contributor's copies.

☛EDITOR'S COMMENTS

While we do accept a few fictional stories and some poetry each year, we are especially interested in thoughtful articles on the philosophy and works of C. S. Lewis. Referenced articles should conform to the *MLA Style Manual*. Please do not send hard copy—we require electronic submissions only.

Language Arts Journal

Ohio State University
333 Arps Hall
1945 North High Street
Columbus, OH 43210

Language Arts Editorial Team

DESCRIPTION AND INTERESTS

Used as a resource by literacy and language arts teachers, this publication provides information and research on all facets of learning and teaching. Published by the National Council of Teachers of English, its primary focus is on issues related to preschool through eighth grade. Circ: 12,000.

Audience: Teachers and teacher-educators
Frequency: 6 times each year
Website: www.ncte.org/pubs/journals/la

FREELANCE POTENTIAL

90% written by nonstaff writers. Publishes 60 freelance submissions yearly; 15% by unpublished writers, 30% by authors who are new to the magazine. Receives 200 unsolicited mss each year.

SUBMISSIONS

Send 6 copies of complete ms; include Microsoft Word file on disk or CD. Accepts hard copy and IBM or Macintosh disk submissions. SASE. Responds in 3–12 months.

- Articles: 2,500–6,500 words. Research articles; position papers; personal experience pieces; and opinion pieces. Topics include language arts, linguistics, and literacy.
- Depts/columns: Word lengths vary. Profiles of children's authors and illustrators; reviews of children's trade books and professional resources; and theme-related research papers.

SAMPLE ISSUE

87 pages (8% advertising): 4 articles; 6 depts/columns. Sample copy, $12.50 sent to NCTE, 1111 W. Kenyon Road, Urbana, IL 61801-1096. Guidelines and theme list available at langarts@osu.edu.

- "Found in Translation." Article reports on a study that connects students' translation skills to academic writing.
- "Learning to Read the Numbers." Article explains how teachers can help students develop strong critical reading skills.

RIGHTS AND PAYMENT

All rights. No payment. Provides 2 copies.

☛EDITOR'S COMMENTS

We are always looking for content that is new to our readers. Of particular interest is poetry that relates to literacy issues.

L.A. Parent

443 East Irving Drive, Suite A
Burbank, CA 91504

Editor: Carolyn Graham

DESCRIPTION AND INTERESTS
L.A. Parent blends informative and entertaining articles with comprehensive listings for local activities, events, and resources. Child rearing, family issues, education, and health and fitness, all with a strong local emphasis, are covered regularly. It is read by parents with children up to the age of 14 years. Circ: 120,000.

Audience: Parents
Frequency: Monthly
Website: www.laparent.com

FREELANCE POTENTIAL
70% written by nonstaff writers. Publishes 20 freelance submissions yearly; 5% by unpublished writers, 10% by authors who are new to the magazine. Receives 120 queries yearly.

SUBMISSIONS
Query with clips. Accepts hard copy. SASE. Responds in 6 months.

- Articles: 400–1,500 words. Practical application and how-to articles; profiles; and interviews. Topics include parenting and family issues, health, fitness, social issues, travel, and gifted and special education.
- Depts/columns: 1,000 words. Family life, technology, travel destinations, and crafts.
- Artwork: B/W or color prints or transparencies.

SAMPLE ISSUE
66 pages (60% advertising): 2 articles; 9 depts/columns; 1 calendar. Sample copy, $3. Guidelines and theme list available.

- "Are We Driving Our Kids Crazy?" Article highlights certain parental behaviors that can cause stress for their children.
- "The Elements of Summer." Article highlights ideas for family activities in the area.
- Sample dept/column: "Woman Wise" discusses the growing issue of woman caring for parents while raising their children.

RIGHTS AND PAYMENT
First serial rights. Written material, payment rates vary. Pays on publication. Provides contributor's copies.

➡️EDITOR'S COMMENTS
Practical parenting information with a local angle is always of interest to us. Articles should be written in a friendly and informal tone. We currently are looking for submissions on education and local outings.

Launch Pad

Teen Missions International
885 East Hall Road
Merritt Island, FL 32953

Editor: Linda Maher

DESCRIPTION AND INTERESTS
Published by Teen Missions International, *Launch Pad* is a full-color tabloid distributed to current and former members and supporters of the organization. It reports on the activities of mission workers, spotlights opportunities for involvement, and offers news of alumni. It also sponsors fundraising activities for member teams. Circ: Unavailable.

Audience: YA–Adult
Frequency: Annually
Website: www.teenmissions.org

FREELANCE POTENTIAL
5% written by nonstaff writers. Publishes 10 freelance submissions yearly; 15% by unpublished writers, 10% by authors who are new to the magazine. Receives 12–24 queries yearly.

SUBMISSIONS
Query. Accepts hard copy. SASE. Response time varies.

- Articles: Word lengths vary. Informational and factual articles; profiles; interviews; personal experience pieces; and photo-essays. Topics include mission work and teen evangelism in different countries.
- Fiction: Word lengths vary. Inspirational, ethnic, and multicultural fiction.
- Depts/columns: Word lengths vary. Alumni news and teen mission opportunities.

SAMPLE ISSUE
8 pages (5% advertising): 4 articles; 6 depts/columns. Sample copy available.

- "Summer Highlights." Article is a collection of personal experience pieces from teen and pre-teen missionaries.
- "Nate Survived 60-Foot Fall!" Article details the accident and recovery of a young boy on a mission in Mozambique.
- Sample dept/column: "Launch Team News" reports on the births, deaths, marriages, and achievements of Launch Team members.

RIGHTS AND PAYMENT
Rights vary. No payment.

➡️EDITOR'S COMMENTS
We are always looking for heartfelt, inspiring testimonials from pre-teen and teen mission workers who are serving abroad. We also are interested in hearing about the good works young people do in their own home towns.

Leadership for Student Activities

1904 Association Drive
Reston, VA 20191-1537

Editor: Mary E. Johnson

DESCRIPTION AND INTERESTS
This publication of the National Association of Student Councils provides information to prepare and empower student leaders to serve their schools and communities. Circ: 30,000.

Audience: YA
Frequency: 9 times each year
Website: www.nasc.us

FREELANCE POTENTIAL
67% written by nonstaff writers. Publishes 18–25 freelance submissions yearly; 75% by unpublished writers, 50% by authors who are new to the magazine. Receives 12–24 queries, 48 unsolicited mss yearly.

SUBMISSIONS
Query with clips; or send complete ms. Accepts hard copy and email submissions to L4SA@att.net. SASE. Responds to queries in 2 weeks, to mss in 1 month.

- Articles: 1,200–1,700 words. Informational and how-to articles; profiles; and interviews. Topics include student activities, leadership development, and careers.
- Depts/columns: Reports on special events, 100–350 words. Advice for and by activity advisors, 1,000–1,500 words. National and regional news, leadership plans, and opinion pieces, word lengths vary.
- Artwork: B/W and color prints or slides
- Other: Submit seasonal material 4 months in advance.

SAMPLE ISSUE
44 pages (21% advertising): 5 articles; 11 depts/columns. Sample copy, free with 9x12 SASE ($1.24 postage). Writers' guidelines and theme list available.

- "Serving Those Who Served." Article describes a variety of ways for schools to honor men and women who fought, and died, for our country.
- Sample dept/column: "Middle Level Activities" offers ways to recognize people's efforts.

RIGHTS AND PAYMENT
All rights. Written material, payment rates vary. Payment policy varies. Provides 5 copies.

❧EDITOR'S COMMENTS
Our magazine serves students, activity advisors, and administrators in secondary schools nationwide.

Leading Edge

4087 JKB
Provo, UT 84602

Fiction or Poetry Director

DESCRIPTION AND INTERESTS
Published by students and alumni of Brigham Young University, this literary journal features science fiction and fantasy stories, as well as related articles and book reviews. It also publishes poetry. Circ: 200.

Audience: YA–Adult
Frequency: Twice each year
Website: www.leadingedgemagazine.com

FREELANCE POTENTIAL
95% written by nonstaff writers. Publishes 18 freelance submissions yearly; most by unpublished writers. Receives 300 unsolicited mss yearly.

SUBMISSIONS
Send complete ms. Accepts hard copy. No simultaneous submissions. SASE. Responds in 3–4 months.

- Articles: Staff written.
- Fiction: To 15,000 words. Genres include science fiction and fantasy.
- Depts/columns: Staff written.
- Other: Poetry dealing with science fiction and fantasy themes; no line limit.

SAMPLE ISSUE
120 pages (no advertising): 6 stories; 3 poems; 3 articles; 4 depts/columns. Sample copy, $5.95. Guidelines available in each issue.

- "Alone." Futuristic story tells of a girl who must endure the execution of her clone upon her 18th birthday.
- "Temp from Hell." Story depicts a world-domination scheme thwarted by an overzealous office temp.

RIGHTS AND PAYMENT
First North American serial rights. Fiction, $.01 per word, minimum $10. Poetry, $10 for first 4 pages, $1.50 per each additional page. Pays on publication. Provides 2 author's copies.

❧EDITOR'S COMMENTS
Although we will consider stories up to 15,000 words, we prefer those under 12,000 words. We will not consider stories with sex, profanity, or excessive violence, nor those that belittle religion or family values. We are interested in helping new authors to improve; therefore, each story is critiqued by at least two members of our staff, and a comment sheet is returned to the author.

Learning and Leading with Technology

175 West Broadway, Suite 300
Eugene, OR 97401-3003

Managing Editor: Davis N. Smith

DESCRIPTION AND INTERESTS

This magazine, published by the International Society for Technology in Education, is filled with practical and innovative ideas for using today's technology to improve teaching and learning in the classroom. Circ: 25,000.

Audience: Educators, grades K–12
Frequency: 8 times each year
Website: www.iste.org/LL

FREELANCE POTENTIAL

90% written by nonstaff writers. Publishes 75 freelance submissions yearly; 50% by unpublished writers, 75% by authors who are new to the magazine. Receives 75 queries yearly.

SUBMISSIONS

Query. Accepts email queries to submissions@iste.org and simultaneous submissions if identified. Response time varies.

- Articles: 600–2,000 words. Informational and how-to articles; and personal experience pieces. Topics include computers and computer science, software, technology, media applications, teaching methods, and telecommunications.
- Depts/columns: Word lengths vary. Research, software reviews, and curriculum ideas.
- Artwork: B/W prints. Line art.

SAMPLE ISSUE

48 pages (20% advertising): 3 articles; 12 depts/columns. Sample copy, free with 9x12 SASE (3 first-class stamps). Guidelines and editorial calendar available at website.

- "Does This Really Work?" Article discusses the keys to implementing new technology while providing evidence of its success.
- "One Size Doesn't Fit All: Customizing Educational Technology Professional Development." Article looks at the evaluation process of ETPD designs.
- Sample dept/column: "Learning Connections" explains how fitness-assessment software can be used in physical education.

RIGHTS AND PAYMENT

All rights; returns limited rights to author upon request. No payment. Provides 3 copies.

EDITOR'S COMMENTS

The articles we publish are generally written by practitioners in the industry. All submissions are editorially reviewed.

Lexington Family Magazine

138 East Reynolds Road, Suite 201
Lexington, KY 40517

Publisher: Dana Tackett

DESCRIPTION AND INTERESTS

This regional tabloid keeps parents in central Kentucky apprised of the latest developments in education, recreation, health, and family entertainment. Circ: 30,000.

Audience: Parents
Frequency: Monthly
Website: www.lexingtonfamily.com

FREELANCE POTENTIAL

50% written by nonstaff writers. Publishes 36 freelance submissions yearly; 40% by authors who are new to the magazine. Receives 250 unsolicited mss yearly.

SUBMISSIONS

Query or send complete ms. Accepts hard copy and email to info@lexingtonfamily.com. SASE. Response time varies.

- Articles: 500–1,500 words. Informational and how-to articles. Topics include the arts, hobbies, current events, education, health, fitness, recreation, regional history, multicultural issues, popular culture, science, technology, family travel, and women's issues.
- Depts/columns: 800 words. News briefs and family health tips.
- Artwork: B/W and color prints. Line art.
- Other: Puzzles, activities, and poetry.

SAMPLE ISSUE

32 pages (50% advertising): 5 articles; 4 depts/columns. Sample copy, free with 9x12 SASE ($1.50 postage). Guidelines and theme list available.

- "The Montessori Difference." Article describes a typical day in a local Montessori preschool and kindergarten classroom.
- Sample dept/column: "Passages" includes an essay by a local mother who instituted a fair-trade fundraiser at her children's school.

RIGHTS AND PAYMENT

All rights. Written material, payment rates vary. Pays on publication. Provides 2 author's copies.

EDITOR'S COMMENTS

Our mission is to serve as a forum for ideas that center on parenting and family issues. Anyone with a parenting story that carries a powerful, humorous, insightful, or simply delightful message is welcome to share that story with us. Many of our most compelling stories have come from our readers.

Library Media Connection

3650 Olentangy River Road, Suite 250
Columbus, OH 43214

Editor: Marlene Woo-Lun

DESCRIPTION AND INTERESTS
A valuable resource for library media specialists of kindergarten through grade 12, *Library Media Connection* features up-to-date reports on research, practical programming ideas, and book and media reviews. Circ: 14,000.
Audience: School librarians/media specialists
Frequency: 7 times each year
Website: www.linworth.com

FREELANCE POTENTIAL
90% written by nonstaff writers. Publishes 215 freelance submissions yearly; 50% by unpublished writers, 50% by authors who are new to the magazine. Receives 144 queries, 144 unsolicited mss yearly.

SUBMISSIONS
Query or send complete ms with résumé. Accepts hard copy, disk submissions (Microsoft Word or ASCII), and email submissions to linworth@linworthpublishing.com. SASE. Responds in 2 weeks.

- Articles: Word lengths vary. Informational and how-to articles; and personal experience and opinion pieces. Topics include library science, research, technology, education, computers, and media services.
- Depts/columns: Word lengths vary. New product information, media reviews, and teaching tips.
- Other: Submit seasonal material 6 months in advance.

SAMPLE ISSUE
82 pages (15% advertising): 7 articles; 9 depts/columns. Sample copy, $11 with 9x12 SASE. Writers' guidelines and theme list available at website.

- "Is Your Library Kid-Friendly?" Article explains the importance of having a strong electronic component in libraries that will appeal to today's students.
- Sample dept/column: "Getting Graphic" reviews the latest releases in graphic novels.

RIGHTS AND PAYMENT
All rights. Written material, payment rates vary. Pays on publication. Provides 4 author's copies.

✒️EDITOR'S COMMENTS
We are looking for practical information and articles that correspond with our upcoming themes, which are found at our website.

LibrarySparks

W5527 State Road 106
P.O. Box 800
Fort Atkinson, WI 53538

Editor: Virginia Harrison

DESCRIPTION AND INTERESTS
Read by elementary school teachers and librarians, this magazine contains engaging activities designed to "spark" a love of literature in children. Circ: Unavailable.
Audience: Librarians and teachers, grades K–6
Frequency: 9 times each year.
Website: www.librarysparks.com

FREELANCE POTENTIAL
100% written by nonstaff writers. Publishes 18 freelance submissions yearly; 25% by authors who are new to the magazine. Receives 60 queries, 24 unsolicited mss each year.

SUBMISSIONS
Query or send complete ms. Accepts hard copy and email submissions to librarysparks@highsmith.com. SASE. Response time varies.

- Articles: Word lengths vary. Informational articles and profiles. Topics include connecting literature to curricula, lesson plans for librarians, teaching library skills, children's authors and illustrators, and ideas for motivating children to read.
- Depts/columns: Word lengths vary. New resources, author profiles, storytelling activities, lesson plans, and helpful hints.
- Other: Reproducible activities and crafts.

SAMPLE ISSUE
58 pages (no advertising): 1 article; 14 depts/columns. Sample copy available at website. Guidelines and editorial calendar available.

- "Dare to Be Scared!" Article provides resources and activities to help young children explore the magic of scary stories.
- Sample dept/column: "Meet the Author" profiles folktale writer Judy Sierra.
- Sample dept/column: "Library Lessons" contains a complete lesson plan for reinforcing research and logical thinking skills with "The Mystery of the Bloody Beagle."

RIGHTS AND PAYMENT
Rights vary. Written material, payment rates vary. Pays on publication. Provides 1 contributor's copy.

✒️EDITOR'S COMMENTS
We publish practical, ready-to-use articles for the elementary and children's librarian. Please see our theme list before submitting.

Lifted Magazine

14781 Memorial Drive, Suite 1747
Houston, TX 77079

Editor: Tiffany Simpson

DESCRIPTION AND INTERESTS
This online magazine for young Christians covers a broad range of lifestyle topics, including faith, travel, college, careers, relationships, entertainment, cooking, politics, and global awareness. It also publishes inspirational fiction. All content is written from a positive, Christian point of view. Hits per month: 2,000.

Audience: YA–Adult
Frequency: 6 times each year
Website: www.liftedmagazine.com

FREELANCE POTENTIAL
90% written by nonstaff writers. Publishes 100 freelance submissions yearly; 50% by unpublished writers, 50% by authors who are new to the magazine.

SUBMISSIONS
Send complete ms. Accepts email submissions to articles@liftedmagazine.com (Microsoft Word or text attachments). Response time varies.

- Articles: 1,000+ words. Informational and how-to articles; Scripture insights; and personal testimonies. Topics include faith, dating, relationships, marriage, college, careers, cooking, politics, the environment, and global awareness. Travel features, 2,000+ words.
- Fiction: 1,000+ words. Inspirational and contemporary fiction.
- Depts/columns: Devotions, 250–500 words. Product, book, and CD reviews, 250 words. Movie and concert reviews, 500–1,000 words.
- Other: Poetry.

SAMPLE ISSUE
Sample copy, guidelines, and editorial calendar available at website.

- "Strategies for More Efficient Textbook Reading." Article provides tips to improve reading comprehension.
- Sample dept/column: "Tough Questions" suggests responses to biblical skeptics.

RIGHTS AND PAYMENT
Rights vary. No payment.

✎EDITOR'S COMMENTS
Should you be submitting something about an event or concert and have photos to accompany the article, they are appreciated and used when possible. We accept manuscript submissions year-round and will add them to the appropriate issue if timely.

The Lion

Lions Clubs International
300 West 22nd Street
Oak Brook, IL 60523-8842

Senior Editor: Jay Copp

DESCRIPTION AND INTERESTS
As the official magazine of Lions Clubs International, *The Lion* publishes the news and activities of the thousands of Lions Clubs International chapters. It also profiles chapter charitable programs, many of which benefit children around the world, and notable Lions Club members. Circ: 600,000.

Audience: Lions Club members
Frequency: 10 times each year
Website: www.lionsclubs.org

FREELANCE POTENTIAL
20% written by nonstaff writers. Publishes 20 freelance submissions yearly; 30% by authors who are new to the magazine. Receives 100 queries and unsolicited mss yearly.

SUBMISSIONS
Prefers query; will accept complete ms. Accepts hard copy and email submissions to jay.copp@lionsclubs.org. SASE. Responds to queries in 10 days, to mss in 2 months.

- Articles: To 1,500 words. Informational articles; profiles; humor; and photo-essays. Topics include Lions Club service projects, disabilities, social issues, and special education.
- Depts/columns: Staff written.
- Artwork: 5x7 or larger color prints; digital JPEG files.

SAMPLE ISSUE
56 pages (6% advertising): 11 articles; 16 depts/columns. Sample copy, free. Guidelines available.

- "The Road Less Traveled." Article reviews several unusual fundraising programs that were very successful.
- "Life After Death." Article explains how Lions Clubs have helped communities of Sri Lanka that were devastated by a tsunami.

RIGHTS AND PAYMENT
All rights. Written material, $100–$700. Pays on acceptance. Provides 4–10 author's copies.

✎EDITOR'S COMMENTS
We welcome freelance articles that depict the service goals and projects of Lions Clubs, as well as other articles on humanitarian subjects that reflect Lions Club ideals. Travel stories, or articles on technology, finance, or consumer issues are also used periodically. No gags, fillers, or puzzles, please.

Listen Magazine

55 West Oak Ridge Drive
Hagerstown, MD 21740

Editor: Céleste Perrino-Walker

DESCRIPTION AND INTERESTS
Positive peer role models, age-appropriate advice, and useful information fill this magazine, which is read by high school students nationwide. Circ: 40,000.

Audience: 12–18 years
Frequency: 9 times each year
Website: www.listenmagazine.org

FREELANCE POTENTIAL
80% written by nonstaff writers. Publishes 90+ freelance submissions yearly; 15% by unpublished writers, 20% by authors who are new to the magazine. Receives 500 unsolicited mss each year.

SUBMISSIONS
Query or send complete ms. Accepts hard copy, email submissions to editor@ listenmagazine.org, and simultaneous submissions if identified. SASE. Response time varies.

- Articles: 800 words. Informational articles; profiles; and self-help pieces. Topics include peer pressure, decision making, self-esteem, self-discipline, family conflict, sports and hobbies, friendship, and healthy choices.
- Depts/columns: 800 words. Opinion pieces; short items on social issues and trends.

SAMPLE ISSUE
32 pages (no advertising): 11 articles; 6 depts/ columns. Sample copy, $2 with 9x12 SASE (2 first-class stamps). Guidelines and editorial calendar available.

- "Cheese for Lunch." Article warns of the dangers of a new illicit drug made from heroin and a common OTC medicine.
- Sample dept/column: "Spotlight" interviews a young political intern.

RIGHTS AND PAYMENT
All rights. Written material, $.05–$.10 per word. Pays on acceptance. Provides 3 contributor's copies.

➠EDITOR'S COMMENTS
We are looking for more true stories, particularly those that show teens making positive choices, as well as more personality features. Writers who are in the know about the latest trends, popular sports and hobbies, and dangerous illicit drugs are encouraged to submit their ideas or well-written articles.

Live

General Council of the Assemblies of God
1445 North Boonville Avenue
Springfield, MO 65802-1894

Editor: Richard Bennett

DESCRIPTION AND INTERESTS
This publication serves as a weekly journal of practical Christian living for young adult and adult Sunday school students. Circ: 50,000.

Audience: 18+ years
Frequency: Quarterly, in weekly sections

FREELANCE POTENTIAL
100% written by nonstaff writers. Publishes 110 freelance submissions yearly; 20% by unpublished writers, 20% by authors who are new to the magazine. Receives 120 queries, 1,440 unsolicited mss yearly.

SUBMISSIONS
Query or send complete ms. Accepts hard copy, email submissions to rl-live@gph.org, and simultaneous submissions if identified. SASE. Responds in 6 weeks.

- Articles: 800–1,100 words. Informational articles; humor; and personal experience pieces. Topics include family issues, parenting, and religious history.
- Fiction: 800–1,100 words. Genres include inspirational fiction, adventure, and stories about family celebrations and traditions.
- Other: Poetry, to 12–25 lines. Filler, 300–600 words. Submit seasonal material 18 months in advance.

SAMPLE ISSUE
8 pages (no advertising): 2 articles. Sample copy, free with #10 SASE ($.41 postage). Guidelines available.

- "The Lady at the Park." Article recounts a woman's brief encounter with a stranger with whom she shared her testimony.
- "Reviving an Old Tradition." Article offers a reminiscence of a family's tradition of waiting until Christmas Eve to place the figure of the baby Jesus in the manger.

RIGHTS AND PAYMENT
First and second rights. Written material, $.10 per word for first rights; $.07 per word for second rights. Pays on acceptance. Provides 2 contributor's copies.

➠EDITOR'S COMMENTS
We seek more true stories that allow readers to apply the lessons conveyed to their own lives. We would also like more articles related to culture from a biblical world view. We continue to be interested in seasonal stories.

Living for the Whole Family

1251 Virginia Avenue
Harrisonburg, VA 22802

Editor: Melodie Davis

DESCRIPTION AND INTERESTS
Written "for the whole family," this free tabloid looks for articles that offer hope and encourage healthy, positive relationships in the home, the workplace, and the community. Circ: 150,000.

Audience: Families
Frequency: Quarterly
Website: www.livingforthewholefamily.com
　　　　www.churchoutreach.com

FREELANCE POTENTIAL
90% written by nonstaff writers. Publishes 50 freelance submissions yearly; 20% by unpublished writers, 50% by authors who are new to the magazine. Receives 480 unsolicited mss yearly.

SUBMISSIONS
Send complete ms. Accepts hard copy and email to melodiemd@msn.com (include title in subject line and email address in body of message). SASE. Responds in 3–4 months.

- Articles: 500 words. Informational, how-to, and inspirational articles; profiles; and personal experience pieces. Topics include family, relationships, marriage, parenting, social issues, spirituality, and community service.
- Depts/columns: Staff written.
- Artwork: B/W and color prints.

SAMPLE ISSUE
28 pages (30% advertising): 12 articles; 2 depts/columns. Sample copy, free with 9x12 SASE (4 first-class stamps). Guidelines available at website.

- "Building Peace Bridges in Harrisonburg." Article profiles an "intentional community" of young adults who encourage an open dialogue on faith.
- "Life with the Gas Queen." First-person essay written for children tells how a girl adjusts to caring for her grandmother after a stroke.

RIGHTS AND PAYMENT
One-time and second rights. Articles, $30–$60. Artwork, $10–$15. Pays 3 months after publication. Provides 2 contributor's copies.

•➔EDITOR'S COMMENTS
We continue to be in need of articles written from a male perspective. Since we go to every home, we do not assume a Christian audience; please take this into account.

Living Safety

Canada Safety Council
1020 Thomas Spratt Place
Ottawa, Ontario K1G 5L5
Canada

President Jack Smith

DESCRIPTION AND INTERESTS
This is a publication that believes an ounce of prevention is worth a pound . . . well, you know the saying. Written primarily for a Canadian audience, it contains articles on safety issues in the home, workplace, recreational venues, and during travel. Circ: 80,000.

Audience: All ages
Frequency: Quarterly
Website: www.safety-council.org

FREELANCE POTENTIAL
75% written by nonstaff writers. Publishes 25 freelance submissions yearly; 65% by unpublished writers, 10% by authors who are new to the magazine. Receives 25 queries yearly.

SUBMISSIONS
Query with résumé and clips or writing samples. Accepts hard copy. SAE/IRC. Responds in 2 weeks.

- Articles: 1,500–2,500 words. Informational articles. Topics include recreational, home, traffic, and school safety; and health issues.
- Depts/columns: Word lengths vary. Safety news, research findings, opinions, and product recalls.
- Other: Children's activities.

SAMPLE ISSUE
30 pages (no advertising): 4 articles; 4 depts/columns; 1 kids' page. Sample copy, free with 9x12 SAE/IRC. Guidelines available.

- "Risk: How Perception Affects Reality." Article discusses how we think nothing of certain actions that actually are more risky than the actions we routinely *do* fear.
- Sample dept/column: "Safety First" offers safety tips for the next family camping trip.

RIGHTS AND PAYMENT
All rights. Articles, to $500. Depts/columns, payment rates vary. Pays on acceptance. Provides 1–5 contributor's copies.

•➔EDITOR'S COMMENTS
We are first and foremost a publication that believes in spreading safety information to our readers, whether it be about safety in the home, during recreation, at work, or on the roadways. That said, we do not publish articles that are meant to spread fear and anxiety. An article must be well-researched, and make its point without being preachy.

Living with Teenagers

One LifeWay Plaza
Nashville, TN 37234-0174

Editor: Bob Bunn

DESCRIPTION AND INTERESTS
Christian parents are the target audience for this magazine that addresses the issues and challenges of raising a teenager. Published by Lifeway Christian Resources, it offers biblical-based discussions and solutions that strengthen and support the family bond. Circ: 35,000.

Audience: Parents
Frequency: Monthly
Website: www.lifeway.com/people

FREELANCE POTENTIAL
90% written by nonstaff writers.

SUBMISSIONS
All material is written on assignment. No queries or unsolicited mss. Submit writing samples if you wish to be considered for an assignment. Accepts hard copy. SASE.

- Articles: 600–2,000 words. Informational, self-help, and how-to articles; profiles; interviews; and reviews. Topics include parenting; family life; education; planning for college; health and fitness; recreation; religion; and social, spiritual, multicultural, and ethnic issues.
- Depts/columns: Staff written.

SAMPLE ISSUE
34 pages (no advertising): 4 articles; 10 depts/columns. Sample copy, free.

- "Out of the Shell." Article examines the difference between shyness and introversion and suggests ways to help teenagers overcome this challenge.
- "Caught in the Sandwich Generation." Article addresses the growing issue of couples caught between raising their children and caring for their aging parents.
- "Technology Overload." Article suggests ways to deal with the distractions caused by students' use of pocket-sized devices.

RIGHTS AND PAYMENT
All rights with nonexclusive license to the writer. Articles, $100–$300. Pays on acceptance. Provides 3 contributor's copies.

➡️EDITOR'S COMMENTS
Ideas for our articles are developed in-house. If you are a trained and highly-qualified Christian writer, please visit our website for more information on our requirements for working with us.

Long Island Woman

P.O. Box 176
Malverne, NY 11565

Publisher: Arie Nadboy

DESCRIPTION AND INTERESTS
Readers of this regional tabloid find beauty, health, lifestyle, and family information within its pages. A calendar of events is also included in each issue. Circ: 40,000.

Audience: Women, 35–65 years
Frequency: Monthly
Website: www.liwomanonline.com

FREELANCE POTENTIAL
100% written by nonstaff writers. Publishes 25 freelance submissions yearly. Receives 1,000 unsolicited mss yearly.

SUBMISSIONS
Send complete ms. Accepts email submissions to editor@liwomanonline.com. Availability of artwork improves chance of acceptance. Responds in 8–10 weeks.

- Articles: 350–2,000 words. Informational and how-to articles; profiles; and interviews. Topics include regional news, lifestyles, family, health, sports, fitness, nutrition, dining, fashion, beauty, business, finance, decorating, gardening, entertainment, media, travel, and celebrities.
- Depts/columns: 500–1,000 words. Book reviews, health advice, personal essays, and profiles.
- Artwork: Electronic B/W and color prints. Line art.
- Other: Submit seasonal material 90 days in advance.

SAMPLE ISSUE
46 pages (60% advertising): 4 articles; 3 depts/columns. Sample copy, $5. Writer's guidelines available.

- "Married and Sleeping Alone." Article recounts one couple's decision to sleep separately for the sake of better sleep.
- Sample dept/column: "FYI" discusses the popularity of girlfriend getaways and explains how they can be helpful in reducing stress.

RIGHTS AND PAYMENT
One-time and electronic rights. Written material, $70–$200. Kill fee, 20%. Pays on publication. Provides 1 tearsheet.

➡️EDITOR'S COMMENTS
If you have a current topic of special interest to women in our region, we would like to hear from you.

The Magazine

643 Queen Street East
Toronto, Ontario M4M 1G4
Canada

Editor: Karen Wong

DESCRIPTION AND INTERESTS

This magazine for pre-teens and teens contains an increasing number of articles written by its readers. It also offers writing contests for its young readers. Its main focus is popular culture, including celebrities, movies, and music. Circ: 89,000.

Audience: 10–18 years
Frequency: Monthly
Website: www.themagazine.ca

FREELANCE POTENTIAL

40% written by nonstaff writers. Publishes 20 freelance submissions yearly; 60% by unpublished writers, 30% by authors who are new to the magazine. Receives 240 queries, 120 unsolicited mss yearly.

SUBMISSIONS

Query or send complete ms. Accepts hard copy. SAE/IRC. Response time varies.

- Articles: Word lengths vary. Informational articles; profiles; interviews; humor; and reviews. Topics include popular culture and entertainment, including television, movies, video games, books, music, and lifestyles.
- Depts/columns: Word lengths vary. Entertainment updates and gossip; nutrition briefs; nature briefs; and reviews of CDs, DVDs, books, video games, e-zines, and websites.

SAMPLE ISSUE

96 pages: 8 articles; 28 depts/columns. Sample copy, $3.95. Guidelines available with SAE/IRC or via email request to work@themagazine.ca.

- "Where In the World Is Wall-E?" Article profiles the robot character starring in the movie that could be Pixar's next big hit.
- "We Just Make Our Own Style." Article presents an interview with the popular Montreal-based band, Simple Plan.
- Sample dept/column: "Top Ten" lists the top shows within TV shows.

RIGHTS AND PAYMENT

All rights. Written material, payment rates vary. Payment policy varies.

❖EDITOR'S COMMENTS

In order to fulfill our mission to "empower youth to read and write," we are emphasizing youth and reader submissions, which means that other freelance opportunities are limited.

Magazine of Fantasy & Science Fiction

P.O. Box 3447
Hoboken, NJ 07030

Editor: Gordon Van Gelder

DESCRIPTION AND INTERESTS

This magazine boasts a collection of short stories, novellas, and novelettes that appeal to science fiction and fantasy readers. Its non-fiction element (book reviews and essays about the genre) are written by staff. Its stories have been enjoyed by fans of these genres for almost 60 years. Circ: 45,000.

Audience: YA–Adult
Frequency: Monthly
Website: www.sfsite.com/fsf

FREELANCE POTENTIAL

98% written by nonstaff writers. Publishes 60–90 freelance submissions yearly; 10% by unpublished writers, 20% by authors who are new to the magazine. Receives 6,000–8,400 unsolicited mss yearly.

SUBMISSIONS

Send complete ms. Accepts hard copy. No simultaneous submissions. SASE. Responds in 1 month.

- Fiction: 1,000–25,000 words. Short stories, novellas, and novelettes. Genres include science fiction, fantasy, and humor.
- Depts/columns: Staff written.

SAMPLE ISSUE

160 pages (1% advertising): 2 novelettes; 2 short stories; 1 novella; 7 depts/columns. Sample copy, $5. Guidelines available.

- "The Tamarisk Hunter." Story imagines a time in which severe drought has dried the American West, and bounty hunters earn a living killing water-sucking bushes and trees.
- "The Great White Bed." Story tells of the narrator's mysterious summer spent with his grandfather, a strange book, and an unexplainable assault.

RIGHTS AND PAYMENT

First world rights with option of anthology rights. Written material, $.06–$.09 per word. Pays on acceptance. Provides 2 contributor's copies.

❖EDITOR'S COMMENTS

We have no set formula for fiction. Want to really get our attention? Send us speculative science fiction and upbeat visions of the future. We see lots and lots of fantasy stories, but want to read more science fiction tales, however slight the science element is. And don't be afraid of using humor.

Mahoning Valley Parent

100 DeBartolo Place, Suite 210
Youngstown, OH 44512

Editor & Publisher: Amy Leigh Wilson

DESCRIPTION AND INTERESTS
Read by parents in Northeastern Ohio with children up to age 12, this regional magazine focuses on news, education, family activities, and entertainment. Circ: 50,000.

Audience: Parents
Frequency: Monthly
Website: www.forparentsonline.com

FREELANCE POTENTIAL
99% written by nonstaff writers. Publishes 100 freelance submissions yearly; 5% by unpublished writers, 20% by authors who are new to the magazine. Receives 500 unsolicited mss each year.

SUBMISSIONS
Send complete ms. Accepts hard copy and email submissions to editor@ mvparentmagazine.com. Retains all material on file for possible use; does not respond until publication. Include SASE if retaining ms is not acceptable.

- Articles: 1,000–1,800 words. Informational and how-to articles; profiles; and reviews. Topics include regional news, current events, parenting, the environment, nature, health, crafts, travel, recreation, hobbies, and ethnic and multicultural subjects.
- Depts/columns: Word lengths vary. Parenting issues, book reviews, events for kids.
- Artwork: B/W or color prints.
- Other: Submit seasonal material 3 months in advance.

SAMPLE ISSUE
42 pages (70% advertising): 5 articles; 6 depts/columns. Sample copy, free with 9x12 SASE. Guidelines and editorial calendar available.

- "Summer Snack Attack." Article provides tips for encouraging healthy eating habits during the active days of summer.
- Sample dept/column: "Successful Single Parenting" discourages using kids as messengers between parents.

RIGHTS AND PAYMENT
One-time rights. Articles, $20–$50. Pays on publication. Provides tearsheets.

☛EDITOR'S COMMENTS
We are interested in well-written and researched news articles and personality profiles. No first-person essays, please.

The Majellan
Champion of the Family

P.O. Box 43
Brighton, Victoria 3186
Australia

Editor: Father Paul Bird

DESCRIPTION AND INTERESTS
The Majellan, published by the Redemptorists of the Catholic Church, gives expression to the ideals of the League of St. Gerard. Its goals are to make St. Gerard known as the Mother's Saint, to maintain the sanctity of marriage, and to foster Christian family life. Circ: 23,000.

Audience: Parents
Frequency: Quarterly
Website: www.majellan.org.au

FREELANCE POTENTIAL
60% written by nonstaff writers. Publishes 60 freelance submissions yearly; 10% by unpublished writers, 20% by authors who are new to the magazine. Receives 24–36 unsolicited mss yearly.

SUBMISSIONS
Send complete ms. Accepts hard copy and email submissions to editor@majellan.org.au (Microsoft Word or RTF attachments). SAE/IRC. Response time varies.

- Articles: 750–1,500 words. Informational articles and personal experience pieces about marriage and Catholic family life.
- Depts/columns: Staff written.
- Other: Filler; reader prayers and photos.

SAMPLE ISSUE
48 pages (15% advertising): 10 articles; 3 depts/columns.

- "Family Rules." Article discusses how families have expectations for behavior of each of their members, and offers specific examples of situations that occur within families.
- "White Butterfly Day." Article tells how a woman was healed of her grief and guilt after she had two abortions.
- "Through the Eyes of a Loving Father." Article draws a parallel between our earthly father's love and acceptance, and that of our Heavenly Father.

RIGHTS AND PAYMENT
Rights vary. Written material, $50–$80 Australian. Pays on acceptance.

☛EDITOR'S COMMENTS
We seek articles that will provide guidance and inspiration to Catholic parents. Prospective authors are encouraged to thoroughly familiarize themselves with our magazine before submitting any material.

Maryland Family

10750 Little Patuxent Parkway
Columbia, MD 21044

Editor: Betsy Stein

DESCRIPTION AND INTERESTS

Distributed free throughout the Baltimore area, *Maryland Family* tackles the usual issues that concern parents: how to keep their children healthy, happy, and on the track to success. Like other regional magazines, it provides local resources for achieving these goals—including a calendar of events. Circ: 50,000.

Audience: Parents
Frequency: Monthly
Website: www.marylandfamilymagazine.com

FREELANCE POTENTIAL

75% written by nonstaff writers. Publishes 50 freelance submissions yearly; 10% by unpublished writers, 10% by authors who are new to the magazine. Receives 360–600 queries yearly.

SUBMISSIONS

Query describing area of expertise. Accepts hard copy. SASE. Responds in 1 month.

- Articles: 800–1,000 words. Informational and how-to articles; profiles; and personal experience pieces. Topics include family issues, parenting, education, recreation, travel, summer camp, sports, and health.
- Depts/columns: Word lengths vary. News briefs on timely, local subjects. Readers' photos. "Family Matters," 100–400 words.
- Artwork: Color prints and transparencies.
- Other: Submit seasonal material about holidays and events 2–3 months in advance.

SAMPLE ISSUE

46 pages (50% advertising): 7 articles; 2 depts/columns. Sample copy, free with 9x12 SASE.

- "The Match Game." Article offers advice to help parents find the right after-school activities for their children.
- Sample dept/column: "Live & Learn" tells how to decide whether twins should be in the same classroom.

RIGHTS AND PAYMENT

First and electronic rights. Written material, payment rates vary. Pays on publication. Provides 1 contributor's copy.

➥EDITOR'S COMMENTS

Editorial submissions are welcome. Please note that we reserve the right to edit, reject, or comment on all material submitted. We cannot be responsible for the return of unsolicited material.

Massive Online Gamer

4635 McEwen Drive
Dallas, TX 75244

Editor: Douglas Kale

DESCRIPTION AND INTERESTS

For fans of MMO (massively multiplayer online) games, *Massive Online Gamer* is the source for gaming news and information. It features articles on games, new releases, game strategies, and game reviews. Circ: 100,000.

Audience: YA–Adult
Frequency: 6 times each year
Website: www.beckettmog.com

FREELANCE POTENTIAL

50% written by nonstaff writers. Publishes 60 freelance submissions yearly; 50% by unpublished writers, 90% by authors who are new to the magazine.

SUBMISSIONS

Query with writing sample and list of MMO experience. Prefers email queries to dkale@beckett.com; will accept hard copy. SASE. Response time varies.

- Articles: Word lengths vary. Informational and how-to articles; personal experience pieces; and interviews. Topics include MMO game descriptions, strategies, and gaming techniques.
- Depts/columns: Word lengths vary. MMO etiquette, technology, contests, and news.

SAMPLE ISSUE

88 pages: 27 articles; 4 depts/columns. Sample copy and guidelines available upon email request to mog@beckett.com.

- "The High Elves and Dark Elves." Article explains three new classes of elves that have been revealed in Warhammer.
- "Pirates of the Burning Sea: Guide to the Economy." Article explains the basic structure of the Pirates of the Burning Sea game.
- Sample dept/column: "MOG News" reports the news that pitcher Curt Schilling intends to start an MMO company when he retires from baseball.

RIGHTS AND PAYMENT

All rights. Written material, $25–$150. Pays 45 days after publication.

➥EDITOR'S COMMENTS

We expect anyone who queries us to have extensive experience in a variety of MMO games; that's why we request your credentials when you send us your query. We love new insights about the games.

MetroFamily

306 South Bryant, Suite C-152
Edmond, OK 73034

Editor: Denise Springer

DESCRIPTION AND INTERESTS
Families in central Oklahoma use this magazine as a valuable resource for local information on parenting issues, products, services,
and activities. A comprehensive calendar is
included each month. The magazine is a member of the Parenting Publications of America.
Circ: 35,000.

Audience: Parents
Frequency: Monthly
Website: www.metrofamilymagazine.com

FREELANCE POTENTIAL
60% written by nonstaff writers. Publishes
50–60 freelance submissions yearly; 10% by
unpublished writers, 10% by authors who are
new to the magazine. Receives 1,000 queries
and unsolicited mss yearly.

SUBMISSIONS
Query or send complete ms. Accepts email to
editor@metrofamilymagazine.com. Responds to
queries in 3 weeks, to mss in 1 month.

- Articles: 300–600 words. Informational and
  how-to articles; profiles; and personal experience pieces. Topics include parenting, education, health and fitness, travel, and recreational activities.
- Depts/columns: Staff written.

SAMPLE ISSUE
58 pages: 4 articles; 16 depts/columns. Sample
copy and guidelines, free with 10x13 SASE.

- "Call a Truce on the Mommy Wars." Article
  discusses the ongoing debate between stay-
  at-home and working mothers.
- "Hot Products for Cool Moms." Article suggests a variety of mom-friendly products,
  from a day planner to flip flops.
- "Mothers Don't Always Start at the Maternity
  Ward." Article profiles a local woman who
  has fostered more than 20 foster children.

RIGHTS AND PAYMENT
First North American serial rights. Articles,
$25–$50. Kill fee, 100%. Pays on publication.
Provides 1 contributor's copy.

◆EDITOR'S COMMENTS
Our readers actively seek out our magazine
for the most up-to-date and useful family
information. We are committed to supporting
the best of what family and community life
has to offer throughout central Oklahoma.

MetroKids

4623 South Broad Street
Philadelphia, PA 19112

Executive Editor: Tom Livingston

DESCRIPTION AND INTERESTS
Read mainly by mothers in the greater
Philadelphia area—including southern New
Jersey and Delaware—this tabloid provides
practical information and regional resources
for parents. Circ: 130,000.

Audience: Parents
Frequency: Monthly
Website: www.metrokids.com

FREELANCE POTENTIAL
30% written by nonstaff writers. Publishes
25 freelance submissions yearly; 10% by
authors who are new to the magazine.
Receives 600 unsolicited mss yearly.

SUBMISSIONS
Send complete ms. Accepts email submissions
to editor@metrokids.com (Microsoft Word or
text attachments). Availability of artwork
improves chance of acceptance. Responds
only if interested.

- Articles: 800–1,100 words. Informational,
  how-to, and self-help articles. Topics include
  pregnancy, childbirth, parenting, pets, computers, education, health, fitness, nature, the
  environment, recreation, and social issues.
- Depts/columns: 550–700 words. School
  news, product recalls, health notes, opinions, book reviews, nature activities, special
  education information, and local events.
- Artwork: Color prints or transparencies.

SAMPLE ISSUE
68 pages: 4 articles; 1 special section; 8 depts/
columns. Sample copy, free with 9x12 SASE.
Guidelines available.

- "How Do Schools Decide If Kids Are Gifted?"
  Article provides an explanation of student
  identification criteria used in area schools.
- Sample dept/column: "Hot! Hot! Hot!" lists
  the most popular clothing, toys, and names
  for babies.

RIGHTS AND PAYMENT
One-time and electronic rights. Written material,
$35–$50. Artwork, payment rates vary. Pays on
publication. Provides 1 contributor's copy.

◆EDITOR'S COMMENTS
We would like to see fresh approaches to
common parenting issues. Please note that
we do not publish essays or personal experience pieces.

Metro Parent Magazine

22041 Woodward Avenue
Ferndale, MI 48220

Managing Editor: Julia Elliott

DESCRIPTION AND INTERESTS
Parents in the greater Detroit region turn to this magazine for the most up-to-date information on parenting concerns, child health and development issues, education, and relationships. Each issue also features local entertainment information. Circ: 80,000.

Audience: Parents
Frequency: Monthly
Website: www.metroparent.com

FREELANCE POTENTIAL
75% written by nonstaff writers. Publishes 250 freelance submissions yearly; 5% by unpublished writers, 35% by authors who are new to the magazine. Receives 960+ unsolicited mss each year.

SUBMISSIONS
Send complete ms. Accepts email submissions to jelliott@metroparent.com. Responds in 1–2 days.

- Articles: 1,500–2,500 words. Informational, self-help, and how-to articles; interviews; and personal experience pieces. Topics include parenting and family life, childbirth, education, social issues, child development, crafts, vacation travel, personal finance, fitness, health, and nature.
- Depts/columns: 850–900 words. Women's health, family fun, new product information, media reviews, crafts, and computers.

SAMPLE ISSUE
68 pages (60% advertising): 6 articles; 12 depts/columns. Sample copy, free. Guidelines available at website.

- "The Perks of Parenthood." Article celebrates Parenting Awareness Month by presenting 10 advantages to being a parent.
- "Stress." Article discusses various sources of stress in children's lives, and offers ways parents can reduce it.

RIGHTS AND PAYMENT
First rights. Articles, $150–$300. Depts/columns, $50–$100. Pays on publication. Provides 1 contributor's copy.

➡EDITOR'S COMMENTS
While we certainly welcome all articles on topics that are of interest to parents, we will give extra attention to articles that have a Detroit-area angle or resource.

Midwifery Today

P.O. Box 2672
Eugene, OR 97402

Managing Editor: Cheryl K. Smith

DESCRIPTION AND INTERESTS
Written for childbirth professionals and educators, this magazine provides information on midwifery, the natural birthing process, and childbirth trends. Circ: 4,000.

Audience: Childbirth practitioners
Frequency: Quarterly
Website: www.midwiferytoday.com

FREELANCE POTENTIAL
90% written by nonstaff writers. Publishes 80–100 freelance submissions yearly; 35% by unpublished writers, 50% by authors who are new to the magazine. Receives 120–144 queries, 96–108 unsolicited mss yearly.

SUBMISSIONS
Query with author background; or send complete ms. Accepts email submissions to editorial@midwiferytoday.com (Microsoft Word or RTF attachments). No simultaneous submissions. Responds in 1 month.

- Articles: 800–1,500 words. Informational and instructional articles; profiles; interviews; personal experience pieces; and media reviews. Topics include feminism, health, fitness, medical care and services, diet, nutrition, and multicultural and ethnic issues related to childbirth.
- Depts/columns: Staff written.
- Artwork: Digital images at 300 dpi.

SAMPLE ISSUE
72 pages (10% advertising): 20 articles; 5 depts/columns. Sample copy, $12.50. Guidelines and editorial calendar available.

- "Technology and Fear." Article raises questions about the use of technology and medicine during childbirth.
- "Birth Rape." Article tells the author's experience of a hospital birth and how it relates to why she became a midwife.

RIGHTS AND PAYMENT
Joint rights. Written material, no payment. Artwork, payment rates vary. Pays on publication. Provides 2 contributor's copies and a 1-year subscription for articles over 800 words.

➡EDITOR'S COMMENTS
We are interested in receiving more clinical pieces from writers with the appropriate credentials, as well as articles with international viewpoints.

Minnesota Conservation Volunteer

500 Lafayette Road
St. Paul, MN 55155-4046

Editorial Assistant: Michael A. Kallok

DESCRIPTION AND INTERESTS
Published by the Minnesota Department of Natural Resources, this magazine shares stories of wild and natural places close to home with its readers, who include Minnesota schoolchildren and their teachers. Circ: 150,000.

Audience: Middle-grade students and teachers
Frequency: 6 times each year
Website: www.mndnr.gov/magazine

FREELANCE POTENTIAL
50% written by nonstaff writers. Receives 40 queries yearly.

SUBMISSIONS
Query with synopsis for feature articles and "Field Notes." Send complete ms for essays. Accepts hard copy and email queries to mike.kallok@dnr.state.mn.us. SASE. Response time varies.

- Articles: 1,200–1,800 words. Informational articles and essays. Topics include the natural resources, wildlife, state parks, lakes, grasslands, groundwater, biofuels, fishing, and outdoor recreation of Minnesota.
- Depts/columns: "Field Notes," 200–500 words. "A Sense of Place" and "Close Encounters," 800–1,200 words. "Young Naturalists," conservationist Q&As, and wildlife profiles, word lengths vary.
- Other: Student activities and teacher guides.

SAMPLE ISSUE
82 pages: 5 articles; 7 depts/columns. Sample copy and guidelines available at website.

- "Vision for Vermilion." Article discusses the benefits of a proposed new state park on Minnesota's Lake Vermilion.
- Sample dept/column: "Young Naturalists" explains the process of metamorphosis to children.

RIGHTS AND PAYMENT
First North American serial rights with option to purchase electronic rights. Articles and essays, $.50 per word plus $100 for electronic rights. Payment policy varies.

❖ EDITOR'S COMMENTS
Please keep your tone casual, use quotes and anecdotes liberally, and feel free to use first-person narration as a device for pulling your reader into the story. We expect you to vouch for facts, names, and quotations.

Mission

223 Main Street
Ottawa, Ontario K1S 1C4
Canada

Editor: Peter Pandimakil

DESCRIPTION AND INTERESTS
Mission offers articles—in English and French—on interreligious and intercultural topics. Its audience includes individuals committed to mission work and evangelization. Circ: 400.

Audience: YA–Adult
Frequency: Twice each year
Website: www.ustpaul.ca

FREELANCE POTENTIAL
95% written by nonstaff writers. Publishes 8–10 freelance submissions yearly; 20% by unpublished writers, 20% by authors who are new to the magazine. Receives 12–24 unsolicited mss yearly.

SUBMISSIONS
Send complete ms with résumé. Accepts disk submissions (RTF files), email submissions to ppandimakil@ustpaul.ca, and simultaneous submissions if identified. SAE/IRC. Responds in 1–2 months.

- Articles: 8,000–10,000 words. Bilingual articles; reviews; and personal experience pieces. Topics include current events; history; religion; and multicultural, ethnic, and social issues.
- Fiction: Word lengths vary. Historical, multicultural, ethnic, and problem-solving stories.
- Artwork: 8x10 B/W and color prints.

SAMPLE ISSUE
150 pages (no advertising): 6 articles; 6 book reviews. Sample copy, $12 U.S. with 8x6 SAE/IRC. Guidelines available.

- "How the Church Has Grown in Her Understanding of Mission and Evangelization." Article presents a brief overview of how Catholics understand the terms mission and evangelization.
- "Magnificent Opportunity, Breathtaking Responsibility." Article focuses on Catholic and social trends in North America, and their implications for missions and evangelization.

RIGHTS AND PAYMENT
Rights vary. No payment. Provides 3 copies.

❖ EDITOR'S COMMENTS
We are interested in seeing more submissions of articles on the relevance and importance of religion, as well as new perspectives on missions. We are particularly looking for contemporary viewpoints.

Momentum

National Catholic Educational Association
1077 30th Street NW, Suite 100
Washington, DC 20007-3852

Editor: Brian Gray

DESCRIPTION AND INTERESTS
Catholic educators and school administrators
read *Momentum* for its well-researched articles
and innovative ideas. Circ: 23,000.

Audience: Teachers, school administrators,
and parish catechists
Frequency: Quarterly
Website: www.ncea.org

FREELANCE POTENTIAL
95% written by nonstaff writers. Publishes 90
freelance submissions yearly; 25% by unpublished writers, 80% by authors who are new to
the magazine. Receives 60 mss yearly.

SUBMISSIONS
Send complete ms with résumé and bibliography. Accepts hard copy, disk submissions
(Microsoft Word), and email submissions to
momentum@ncea.org. SASE. Responds in
1–3 months.

- Articles: 1,000–1,500 words. Informational
  and scholarly articles on education. Topics
  include teacher and in-service education,
  educational trends, technology, research,
  management, and public relations—all as they
  relate to Catholic schools.
- Depts/columns: Book reviews, 300 words.
  "Trends in Technology," 900 words. "From
  the Field," 700 words.

SAMPLE ISSUE
94 pages (20% advertising): 22 articles; 8
depts/columns. Sample copy, free with 9x12
SASE ($1.05 postage). Guidelines and editorial
calendar available.

- "Will Philanthropy Save Catholic Schools?"
  Article shows that Catholic school leaders
  must consider the opportunities and challenges of philanthropic gifts.
- Sample dept/column: "DRE Directions"
  offers a reflection on how a catechism class
  came to appreciate the needy they served.

RIGHTS AND PAYMENT
First rights. Articles, $75. Depts/columns, $50.
Pays on publication. Provides 2 copies.

➥EDITOR'S COMMENTS
We seek proven ideas for parish-based religious education programs. We always want
articles on creative ways to encourage voca-
tions, but not "inspirational" material. Writers
should check our online editorial calendar.

MOM Magazine

2397 NW Kings Boulevard, #105
Corvallis, OR 97330

Editors: Raeann Van Arsdall

DESCRIPTION AND INTERESTS
Mothers with young children who live in the
Pacific Northwest find plenty of humor and a
healthy dose of helpful information in this publication, subtitled "A Brilliant Little Magazine
for Moms." Circ: 20,000.

Audience: Mothers
Frequency: 6 times each year
Website: www.mommag.com

FREELANCE POTENTIAL
80% written by nonstaff writers. Publishes 25
freelance submissions yearly; 50% by unpublished writers, 50% by authors who are new to
the magazine. Receives 240 queries, 120
unsolicited mss yearly.

SUBMISSIONS
Query or send complete ms. Accepts email
submissions to editor@mommag.com
(Microsoft Word attachments). Availability of
artwork improves chance of acceptance.
Response time varies.

- Articles: 500 words. Informational articles;
  profiles; and personal experience pieces.
  Topics include parenting, family life, recreation and activities, pets, health, and fitness.
- Depts/columns: Word lengths vary. Family
  travel, fatherhood issues, book reviews.
- Artwork: 5x7 JPEG or TIFF images at 300
  dpi, with photo releases.
- Other: Call-outs, 25–50 words (statistics,
  resources, and recipes).

SAMPLE ISSUE
32 pages: 12 articles; 6 depts/columns.
Guidelines available at website.

- "How Does She Do It?" Article profiles rock
  singer Meredith Brooks, whose own motherhood inspires her career.
- Sample dept/column: "Gimme a Break"
  offers craft ideas for keeping children busy
  during the holidays.

RIGHTS AND PAYMENT
Reprint rights. No payment.

➥EDITOR'S COMMENTS
We are looking for fun, humorous, light-hearted
stories about being a mom. The tone and message should be friendly and appealing to the
everyday woman. Candid photos work best in
keeping with our realistic approach to motherhood and family.

MOMSense

2370 South Trenton Way
Denver, CO 80231-3822

Editor: Mary Darr

DESCRIPTION AND INTERESTS
Mothers of preschool children turn to this magazine to read informative and inspiring articles relating to motherhood and womanhood, all with a Christian perspective. Circ: 120,000.

Audience: Mothers
Frequency: 6 times each year
Website: www.MomSense.com

FREELANCE POTENTIAL
70% written by nonstaff writers. Publishes 24 freelance submissions yearly; 20% by unpublished writers, 20% by authors who are new to the magazine. Receives 360–480 unsolicited mss yearly.

SUBMISSIONS
Send complete ms. Prefers email submissions to MOMSense@mops.org (Microsoft Word attachments); will accept hard copy. Availability of artwork improves chance of acceptance. SASE. Response time varies.

- Articles: 500–1,000 words. Informational articles; profiles; and personal experience pieces. Topics include parenting, religion, and humor.
- Depts/columns: Word lengths vary. Parenting and family life.
- Artwork: B/W or color prints/transparencies.
- Other: Accepts seasonal material 6–12 months in advance.

SAMPLE ISSUE
32 pages; 8 articles; 6 depts/columns. Sample copy, free. Guidelines available.

- "Home Sweet Home." Article describes how moving 13 times helped a mother define what a real home is.
- Sample dept/column: "The View From Here" offers insight into in-law relationships.

RIGHTS AND PAYMENT
First rights. Written material, $.15–$.25 per word. Payment policy varies. Provides contributor's copies.

➦EDITOR'S COMMENTS
MOMSense seeks to meet the needs of every mom—urban, suburban, stay-at-home, working, teen, single, or married—who shares the desire to be the best she can be, both as an individual and as a parent. We are looking for more practical articles showing Mom as a mom and Mom as a woman.

Mom Writer's Literary Magazine

1006 Black Oak Drive
Liberty, MO 64068

Editor-in-Chief: Samantha Gianulis

DESCRIPTION AND INTERESTS
This journal features creative writing by mothers on the many facets of motherhood, as well as profiles of "mom writers," book reviews, and poetry. Circ: 10,000.

Audience: Mothers
Frequency: Quarterly
Website: www.momwriterslitmag.com

FREELANCE POTENTIAL
90% written by nonstaff writers. Publishes 64 freelance submissions yearly; 90% by unpublished writers, 95% by authors who are new to the magazine. Receives 600–1,200 queries and unsolicited mss yearly.

SUBMISSIONS
Query for author profiles. Send complete ms for creative nonfiction and book reviews. Accepts email to editor@momwriterslitmag.com (no attachments) and simultaneous submissions if identified. Responds in 2–4 weeks.

- Articles: Creative nonfiction and memoir excerpts, to 2,000 words. Profiles and Q&A interviews with mom writers, 800–2,000 words. Book reviews, 700–1,000 words. Topics include biological, adoptive, foster, and expectant mothers; stepmothers; grandmothers; and motherhood issues.
- Depts/columns: Staff written.
- Other: Poetry, line lengths vary.

SAMPLE ISSUE
52 pages: 8 essays; 5 profiles; 3 reviews; 11 depts/columns; 6 poems. Sample copy and guidelines available at website.

- "Stories on Strings." Article features an interview with singer-songwriter Lori McKenna, who tells how she put her dream on hold while she raised five children.
- "New Lives." Personal essay describes the author's struggle with empty-nest syndrome.

RIGHTS AND PAYMENT
One-time electronic rights. No payment.

➦EDITOR'S COMMENTS
Nonfiction pieces should read like fiction and center on a fresh topic or a surprising take on a global experience. We look for character development, an amusing narrative, and rich details with a distinctive personal voice. Book reviews should give a well-balanced, supported critique offered in a supportive tone.

Mothering

P.O. Box 1690
Santa Fe, NM 87504

Articles Editor: Candace Walsh

DESCRIPTION AND INTERESTS
With a strong focus on natural family living, *Mothering* magazine publishes articles that are meant to support and empower parents in their lifestyle choices. Circ: 100,000.

Audience: Parents
Frequency: 6 times each year
Website: www.mothering.com

FREELANCE POTENTIAL
90% written by nonstaff writers. Publishes 60 freelance submissions yearly; 20% by unpublished writers, 70% by authors who are new to the magazine. Receives 2,000 queries yearly.

SUBMISSIONS
Query with outline/synopsis. Prefers email queries to editorial@mothering.com; will accept hard copy. SASE. Responds in 2–4 weeks.

- Articles: 2,000 words. Informational and factual articles; profiles; and personal experience pieces. Topics include pregnancy, childbirth, midwifery, health, homeopathy, teen issues, and organic food.
- Depts/columns: Word lengths vary. Cooking, book and product reviews, health news, parenting updates, and inspirational pieces.
- Artwork: 5x7 B/W and color prints.
- Other: Children's activities and crafts. Poetry about motherhood and families. Submit seasonal material 6–8 months in advance.

SAMPLE ISSUE
104 pages (35% advertising): 3 articles; 9 depts/columns. Sample copy, $5.95 with 9x12 SASE. Guidelines available at website.

- "Real Girl vs. Real World." Article explores ways to raise self-confident daughters despite society's mixed messages.
- Sample dept/column: "Ways of Learning" offers support for parents who homeschool.

RIGHTS AND PAYMENT
First rights. Written material, $100+. Artwork, payment rates vary. Pays on publication. Provides 2 contributor's copies and a 1-year subscription.

☛EDITOR'S COMMENTS
We are more likely to publish your article if you are a *Mothering* reader familiar with the issues we discuss. The letters section is a good place to discover which topics are of interest to our readership.

M: The Magazine for Montessori Families

3 Werner Way
Lebanon, NJ 08833

Editor-in-Chief

DESCRIPTION AND INTERESTS
Parents of children enrolled in Montessori schools read this magazine for information about the Montessori teaching philosophy, news of Montessori schools, articles about education, and ways families can supplement education at home. Circ: Unavailable.

Audience: Parents
Frequency: Quarterly
Website: www.mthemagazine.com

FREELANCE POTENTIAL
25% written by nonstaff writers. Publishes 20 freelance submissions yearly.

SUBMISSIONS
Query with writing sample. Accepts hard copy. SASE. Response time varies.

- Articles: Word lengths vary. Informational articles; profiles; interviews; reviews; and personal experience pieces. Topics include Montessori teaching methods and benefits, history of Montessori, family issues, parenting, and student motivation.
- Depts/columns: Word lengths vary. Ethical issues, parenting advice, cooking with children, news, and reading resources.

SAMPLE ISSUE
40 pages: 7 articles; 6 depts/columns. Sample copy, $8.50.

- "The Lonely Prophet." Article explores the origins and development of Maria Montessori's peace philosophy.
- "Between Parent, Toddler, and Teacher." Article provides tips and techniques that parents can employ to ensure effective three-way communication with their child and the child's teacher.
- Sample dept/column: "Leadership" describes the philosophy of the Leadership Today training organization.

RIGHTS AND PAYMENT
First North American serial rights. Written material, payment rates vary. Pays on publication.

☛EDITOR'S COMMENTS
We believe that the Montessori philosophy is not just for the children, but for all of us—the school, the teachers, the children, and the families. We look for articles that support that belief and can provide information parents can use to enhance their child's education.

MultiCultural Review

194 Lenox Avenue
Albany, NY 12208

Editor: Lyn Miller-Lachman

DESCRIPTION AND INTERESTS
This journal for educators and librarians is dedicated to providing a better understanding of ethnic, racial, and religious diversity around the world through its comprehensive articles and book reviews. Circ: 3,500+.

Audience: Teachers and librarians
Frequency: Quarterly
Website: www.mcreview.com

FREELANCE POTENTIAL
80% written by nonstaff writers. Publishes 600 freelance submissions yearly; 10% by unpublished writers, 50% by authors who are new to the magazine. Receives 80 queries and unsolicited mss yearly.

SUBMISSIONS
Query for reviews. Send complete ms for articles. Accepts email to mcreview@aol.com (attach file). Responds in 3–4 months.

- Articles: 2,000–6,000 words. Informational and how-to articles; profiles; and opinion pieces. Topics include the arts; education; writing; and social, multicultural, and ethnic issues.
- Depts/columns: 1,500–2,000 words. News, reports on new book releases. Book and media reviews, 200–300 words.

SAMPLE ISSUE
98 pages (10% advertising): 4 articles; 4 depts/columns; 131 book reviews. Sample copy, $15. Writers' guidelines and theme list available.

- "The Librarians for Tomorrow Project: Preparing San Jose's New Generation of Librarians for Cultural Communities." Article explains a unique program aimed at increasing the number of librarians of color.
- "From Cenicienta to Caníbales: Books in Spanish for Children and Adolescents." Article lists several newly-published Spanish books arranged by age group.

RIGHTS AND PAYMENT
First serial rights. Articles, $50–$100. Depts/columns, $50. Reviews, no payment. Pays on publication. Provides 2 contributor's copies.

•◆EDITOR'S COMMENTS
We are always looking for reviewers of children's and adult books. Writing reviews is a great way to break into the market.

MultiMedia & Internet Schools

14508 NE 20th Avenue, Suite 102
Vancouver, WA 98686

Editor: David Hoffman

DESCRIPTION AND INTERESTS
This magazine provides educators of students in kindergarten through twelfth grade with the most current information on integrating technology with classroom learning. The Internet, computers, and other multi-media tools are covered. Circ: 12,000.

Audience: Librarians, teachers, and technology coordinators
Frequency: 6 times each year
Website: www.mmischools.com

FREELANCE POTENTIAL
90% written by nonstaff writers. Publishes 20–24 freelance submissions yearly; 20% by unpublished writers, 20% by authors who are new to the magazine. Receives 60 queries and unsolicited mss yearly.

SUBMISSIONS
Query or send complete ms. Accepts email submissions to hoffmand@infotoday.com. Availability of artwork improves chance of acceptance. Responds in 6–8 weeks.

- Articles: 1,500 words. Informational and how-to articles. Topics include K–12 education, the Internet, multimedia and electronic resources, technology-based tools, and curriculum integration.
- Depts/columns: Word lengths vary. Product news, reviews, and ideas from educators.
- Artwork: TIFF images at 300 dpi.

SAMPLE ISSUE
48 pages (15% advertising): 3 articles; 8 depts/columns. Sample copy and guidelines, $7.95 with 9x12 SASE.

- "Are Textbooks Becoming Extinct?" Article discusses the benefit of creating wikitexts to enable unlimited electronic access to curriculum topics.
- Sample dept/column: "The Pipeline" discusses how storyboarding is a great way to strengthen creative writing skills.

RIGHTS AND PAYMENT
First rights. Written material, $300–$500. Artwork, payment rates vary. Pays on publication. Provides 2 contributor's copies.

•◆EDITOR'S COMMENTS
Submissions are usually solicited from media specialists and teachers; however, we welcome ideas from other knowledgeable writers.

Muse

Carus Publishing
70 East Lake Street, Suite 300
Chicago, IL 60603

Editor

DESCRIPTION AND INTERESTS

With insightful articles on topics ranging from anthropology to zoology and everything in between, *Muse* calls itself "a guidebook for intellectual exploration." Its colorful, glossy format and humorous tone draws tweens and teens into the worlds of science, history, and the arts. Circ: 51,000.

Audience: 10+ years
Frequency: 9 times each year
Website: www.cricketmag.com

FREELANCE POTENTIAL

100% written by nonstaff writers. Of the freelance submissions published yearly, 20% are by authors who are new to the magazine.

SUBMISSIONS

All work is assigned. Send résumé and clips. Response time varies.

- Articles: To 2,500 words. Informational articles; interviews; and photo-essays. Topics include science, nature, the environment, history, culture, anthropology, sociology, technology, and the arts.
- Depts/columns: Word lengths vary. Science news, Q&As, and math problems.
- Other: Cartoons, contests, and activities.

SAMPLE ISSUE

42 pages (no advertising): 4 articles; 6 depts/columns. Sample copy and guidelines available at website.

- "Construction: Sight." Article explains the cognitive science of visual construction, or how we create what we see.
- "Oxygen: A Love Story." Article tells how the introduction of oxygen into Earth's atmosphere changed life on the planet.
- Sample dept/column: "Last Page" depicts the Swiss tradition of blowing up a cotton snowman to execute winter and usher in spring.

RIGHTS AND PAYMENT

Rights vary. Written material, payment rates vary. Payment policy varies.

➡ EDITOR'S COMMENTS

As a science and discovery magazine, we seek writers with subject expertise. Those interested in writing for us should submit a résumé and several writing samples. We are not accepting unsolicited manuscripts or queries at present.

Music Educators Journal

Music Educators National Conference
1806 Robert Fulton Drive
Reston, VA 20191

Submissions: Caroline Arlington

DESCRIPTION AND INTERESTS

This magazine publishes articles about all phases of music education in schools and communities, including practical instruction techniques and teaching philosophy. Its goal is to advance music education. Circ: 80,000.

Audience: Music teachers
Frequency: Quarterly
Website: www.menc.org

FREELANCE POTENTIAL

85% written by nonstaff writers. Publishes 30 freelance submissions yearly; 25% by unpublished writers. Receives 100 unsolicited mss yearly.

SUBMISSIONS

Send complete ms. Accepts email submissions to caroline@menc.org (Microsoft Word attachments) or 6 hard copies. No simultaneous submissions. SASE. Responds in 3 months.

- Articles: 1,800–3,500 words. Informational and instructional articles; and historical studies of music education. Topics include teaching methods and philosophy, and current trends in music teaching and learning.
- Depts/columns: Media reviews, teaching tips, technology updates, and association news.
- Artwork: High-contrast 5x7 or 8x10 prints; TIFF images at 300 dpi.
- Other: Submit seasonal material 8–12 months in advance.

SAMPLE ISSUE

62 pages (40% advertising): 5 articles; 7 depts/columns. Sample copy, $6 with 9x12 SASE ($2 postage). Guidelines available at website.

- "Mimes and Conductors: Silent Artists." Article states that music conductors can learn from mime technique how to convey more meaning through their gestures.
- "Successful Single-Sex Offerings in the Choral Department." Article tells of one high school's success with same-sex classes.
- Sample dept/column: "Take Note" includes news of the association's revamped website.

RIGHTS AND PAYMENT

All rights. Written material, no payment. Artwork, $10. Provides 2 contributor's copies.

➡ EDITOR'S COMMENTS

We welcome articles of value, assistance, or inspiration to practicing music teachers.

Muslim Girl

My Light Magazine

1179 King Street West, Suite 114
Toronto, Ontario M6K 3C5
Canada

Editor-in-Chief: Ausma Khan

Editor: Jennifer Gladen

DESCRIPTION AND INTERESTS
Muslim Girl provides an outlet for the voice of the Muslim girl—an outlet that has been missing from mainstream media in North America. It is like many lifestyle magazines for girls, except that it works within the parameters of Muslim values. Circ: Unavailable.

Audience: Girls, 12–19 years
Frequency: 6 times each year
Website: www.muslimgirlmagazine.com

FREELANCE POTENTIAL
40% written by nonstaff writers. Publishes 20 freelance submissions yearly.

SUBMISSIONS
Query with résumé and 2 clips. Accepts email to submissions@muslimgirlmagazine.com. Response time varies.

- Articles: Word lengths vary. Informational and self-help articles; profiles; interviews; and personal experience pieces. Topics include religion, social issues, multicultural and ethnic issues, family issues, relationships, and health.
- Depts/columns: Word lengths vary. Short profiles, Q&As, fashion, news.

SAMPLE ISSUE
112 pages: 14 articles; 14 depts/columns. Sample copy, $4.95 with 9x12 SASE. Guidelines available at website.

- "The Boy Next Door." Article provides a discussion with Muslim girls about their feelings about boys and dating.
- "Tasha & Tanika's Open Hearts." Article profiles twin girls, one of whom is Muslim and the other Christian.
- Sample dept/column: "Women to Watch" profiles the Muslim woman behind the first-ever sitcom on network TV that features characters who are Muslims.

RIGHTS AND PAYMENT
Rights vary. Written material, $150–$1,000. Payment policy varies.

➡ EDITOR'S COMMENTS
We want life-affirming stories of real North American Muslim teens. We mainly look for female writers of Muslim background who can relate to and understand the stories we are profiling. We do, however, include work by non-Muslim writers when there is a good fit.

DESCRIPTION AND INTERESTS
My Light is a Catholic-centered e-zine for children that seeks to help readers nurture a deep relationship with God. It features articles that help children understand God and Christianity, as well as fiction, prayers, poems, and activities. Hits per month: Unavailable.

Audience: 6–12 years
Frequency: Monthly
Website: www.mylightmagazine.com

FREELANCE POTENTIAL
95% written by nonstaff writers. Publishes 180 freelance submissions yearly; 50% by unpublished writers, 75% by authors who are new to the magazine.

SUBMISSIONS
Send complete ms with bibliography and brief author biography. Accepts email submissions to mylight@jennifergladen.com (Microsoft Word attachments). Response time varies.

- Articles: 300–800 words. Informational articles. Topics include prayer, Jesus's teachings and parables, the rosary, Mary, the saints, Holy Mass, God's creation, spiritual experiences, respect for parents and teachers, Catholic values, and living the Beatitudes.
- Fiction: 300–800 words. Stories that represent Christian lifestyles and morals.
- Depts/columns: 500 words. Activities, prayers, crafts.
- Artwork: Color line art.
- Other: Poetry, to 24 lines.

SAMPLE ISSUE
44 pages: 4 articles; 3 stories; 2 depts/columns; 1 prayer; 2 poems. Guidelines and theme list available at website.

- "Josiah and the Great Passover." Article is a retelling of the Bible story of Josiah, the king of Judah, and how he denounced worshipping any god but the true God.
- "God Throws Jimmy a Curveball." Story tells of a boy's realization that God answers our prayers in ways we may not expect.

RIGHTS AND PAYMENT
One-time electronic rights. No payment.

➡ EDITOR'S COMMENTS
We love articles that bring out Jesus's personality, and those involving Mary. All articles must be nonviolent and Christian in nature.

Nashville Parent Magazine

2270 Metro Center Boulevard
Nashville, TN 37228

Editor: Susan B. Day

DESCRIPTION AND INTERESTS
Serving parents in the Nashville region for more than a decade, *Nashville Parent* is a resource for parenting information, child health and development topics, and local entertainment. Circ: 85,000.

Audience: Parents
Frequency: Monthly
Website: www.parentworld.com

FREELANCE POTENTIAL
15–20% written by nonstaff writers. Publishes 400 freelance submissions yearly; 40% by authors who are new to the magazine. Receives 1,200 unsolicited mss yearly.

SUBMISSIONS
Send complete ms. Accepts hard copy, Macintosh disk submissions with hard copy, and email submissions to npinfo@ nashvilleparent.com. Availability of artwork improves chance of acceptance. SASE. Response time varies.

- Articles: 800–1,000 words. Informational and how-to articles; profiles; interviews; photo-essays; and personal experience pieces. Topics include parenting, family issues, current events, social issues, health, music, travel, recreation, religion, the arts, crafts, computers, and multicultural and ethnic issues.
- Depts/columns: Staff written.
- Artwork: B/W or color prints.
- Other: Submit Christmas, Easter, and Halloween material 2 months in advance.

SAMPLE ISSUE
116 pages (50% advertising): 7 articles; 9 depts/columns. Sample copy, free with 9x12 SASE. Guidelines available.

- "Coddling: When Overprotective Parenting Becomes a Hindrance." Article explains how coddling can be detrimental to a child.
- "Glad to Be a Dad." Article interviews mothers about what it takes to be a good father.

RIGHTS AND PAYMENT
One-time rights. Written material, $35. Pays on publication. Provides 3 copies upon request.

➡EDITOR'S COMMENTS
We look for articles that are written in a lively style, and that provide parents with a new parenting tool or idea.

National Geographic Kids

National Geographic Society
1145 17th Street NW
Washington, DC 20036-4688

Editor: Julie Agnone

DESCRIPTION AND INTERESTS
It's still the *National Geographic* that we all know and love, but this version is geared toward children. With the tagline "Dare to Explore," its articles are designed to instill a sense of wonder about the world, while teaching kids about the planet in a way that captures their interest. Circ: 1.3 million.

Audience: 6–14 years
Frequency: 10 times each year
Web: http://kids.nationalgeographic.com

FREELANCE POTENTIAL
85% written by nonstaff writers. Publishes 20–25 freelance submissions yearly; 1% by unpublished writers, 30% by authors who are new to the magazine. Receives 360 queries each year.

SUBMISSIONS
Query with clips. No unsolicited mss. Accepts hard copy. Responds if interested.

- Articles: Word lengths vary. Informational articles. Topics include geography, archaeology, paleontology, history, entertainment, and the environment.
- Depts/columns: Word lengths vary. News and trends, amazing animals, and fun facts.
- Other: Original games and jokes.

SAMPLE ISSUE
44 pages (20% advertising): 5 articles; 12 depts/columns; 2 activities. Sample copy, $3.95. Guidelines available.

- "Dolphins in Disguise." Article looks at some of the more unusual members of the dolphin family.
- "Leopard Rescue." Article tells how an orphaned, wounded cub was saved through the efforts of a park warden.
- Sample dept/column: "The Green List" offers simple things kids can do to help preserve the planet.

RIGHTS AND PAYMENT
All rights. Written material, payment rates vary. Artwork, $100–$600. Pays on acceptance. Provides 3–5 contributor's copies.

➡EDITOR'S COMMENTS
It's our mission to find fresh ways to entertain children while educating them about their world. We currently need ideas for our "Amazing Animals" section.

National Geographic Little Kids

National Geographic Society
1145 17th Street NW
Washington, DC 20036

Executive Editor: Julie Agnone

DESCRIPTION AND INTERESTS
Who says adults and older kids are the only ones who get to enjoy *National Geographic*? This version of the popular magazine is designed to appeal to young children. Its small size, short articles, and lots of great photos and activities make learning about the world fun. This magazine won a Parents' Choice Gold Award. Circ: Unavailable.

Audience: 3–6 years
Frequency: 6 times each year
**Web: http://littlekids.
      nationalgeographic.com**

FREELANCE POTENTIAL
10% written by nonstaff writers. Publishes 5 freelance submissions yearly.

SUBMISSIONS
Query with résumé. Accepts hard copy. SASE. Response time varies.

- Articles: Word lengths vary. Informational articles. Topics include nature, animals, science, geography, history, and multicultural subjects.
- Fiction: Word lengths vary. Rebus stories about animals and other cultures.
- Other: Original games, jokes, and activities.

SAMPLE ISSUE
24 pages (no advertising): 4 articles; 8 activities. Sample copy, $3.95.

- "Swimming Sea Otters." Article explains the habits and lifestyle of the sea otters.
- "Fruits Around the World." Article presents a plethora of fruits, and explains the origins of each one.
- "The Moon!" Article discusses the moon and presents photos of its various phases.

RIGHTS AND PAYMENT
All rights. Written material, payment rates vary. Pays on acceptance. Provides 3–5 contributor's copies.

➔EDITOR'S COMMENTS
We are proud of the fact that every issue is packed with teaching tools to help parents inspire a love of learning in their children. Obviously, given our readership, each piece is very short and highlighted by big, brilliant photographs. Each issue includes a host of activities and simple games and puzzles, and we welcome any and all ideas for them.

Natural Life

P.O. Box 112
Niagara Falls, NY 14304

Editor: Wendy Priesnitz

DESCRIPTION AND INTERESTS
This magazine, which recently merged with its former sister publications, *Life Learning* and *Natural Child*, is trusted by readers who want to create greener, healthier lifestyles for their families. Circ: 85,000.

Audience: Parents
Frequency: 6 times each year
Website: www.naturallifemagazine.com

FREELANCE POTENTIAL
50% written by nonstaff writers. Publishes 40 freelance submissions yearly; 20% by unpublished writers, 20–30% by authors who are new to the magazine. Receives 180 queries each year.

SUBMISSIONS
Query with detailed outline and 50- to 200-word synopsis. Accepts email queries to editor@ naturallifemagazine.com and simultaneous submissions if identified. Responds in 3–5 days.

- Articles: 2,500–3,500 words. Informational and how-to articles; profiles; interviews; and personal experience pieces. Topics include green living, eco-travel, natural parenting, unschooling, lifelong learning, self-directed learning, and social issues.
- Depts/columns: Staff written.
- Artwork: Color prints. High-resolution TIFF images at 300 dpi.

SAMPLE ISSUE
62 pages: 15 articles; 4 depts/columns. Sample copy, $6.95 with 9x12 SASE ($6 postage); also available at newsstands. Guidelines available at website.

- "Vindication of Trust." Article motivates parents to trust their instincts when it comes to alternative parenting styles.
- "Believe in Birth: Reflections of a Home Birth Midwife." Article explains why a home birth is also a green birth.

RIGHTS AND PAYMENT
One-time print and non-exclusive electronic rights. No payment. Provides contributor's copies.

➔EDITOR'S COMMENTS
Our readers are already quite knowledgeable about "green" issues and natural parenting, so articles need to dig deep into the subject at hand.

Nature Friend Magazine

4253 Woodcock Lane
Dayton, VA 22821

Editor: Kevin Shank

DESCRIPTION AND INTERESTS
Nature Friend Magazine is wild about nature, birds, plants, marine life, and astronomy—and it covers these topics through stories, puzzles, science experiments, and nature experiments. All of its material honors God as the Creator. Circ: 13,000.

Audience: 6–12 years
Frequency: Monthly
Website: www.naturefriendmagazine.com

FREELANCE POTENTIAL
90% written by nonstaff writers. Publishes 50 freelance submissions yearly; 5% by unpublished writers, 10% by authors who are new to the magazine. Receives 480–720 unsolicited mss yearly.

SUBMISSIONS
Send complete ms. Accepts hard copy. SASE. Response time varies.

- Articles: 250–900 words. Informational and how-to articles. Topics include science, nature, and wildlife.
- Fiction: 300–900 words. Genres include adventure stories and stories about the outdoors, wildlife, nature, and the environment.
- Depts/columns: 100–450 words. Seasonal, nature-related stories or activities.
- Artwork: High-resolution digital images with accompanying contact prints.
- Other: Nature-related puzzles and projects.

SAMPLE ISSUE
22 pages (no advertising): 6 articles; 2 depts/columns, 2 activities. Sample copy and writers' guidelines, $9.

- "Another Marvel: More Than Just Nuts." Article takes readers on an adventure in the Amazon jungle to discover the wonders of the Brazil nut.
- Sample dept/column: "Nature Trails" describes a Christmas bird-counting expedition in northern Ontario.

RIGHTS AND PAYMENT
One-time rights. Written material, $.05 per word. Artwork, $25–$75 per photo. Pays on publication. Provides 1 tearsheet.

➥EDITOR'S COMMENTS
We endeavor to create a sense of awe about nature's Creator and a respect for His creation. We'd like more hands-on activities.

The New Era

50 East North Temple Street, Room 2420
Salt Lake City, UT 84150-3220

Managing Editor: Richard M. Romney

DESCRIPTION AND INTERESTS
The New Era is read by Latter-day Saints youth, their parents, and youth leaders. Its contents show the connection between faith and the issues faced by young people. Circ: 230,000.

Audience: YA–Adults
Frequency: Monthly
Website: www.lds.org

FREELANCE POTENTIAL
20% written by nonstaff writers. Publishes 15–20 freelance submissions yearly; 5% by unpublished writers, 5% by authors who are new to the magazine. Receives 240 queries each year.

SUBMISSIONS
Query. Accepts hard copy and email queries to newera@ldschurch.org. SASE. Responds in 2 months.

- Articles: 200–500 words. Photo features and interviews, to 1,500 words. Informational and self-help articles; profiles; and personal experience pieces. Topics include Gospel messages, religion, social issues, family, marriage, and overcoming adversity.
- Depts/columns: Word lengths vary. News items, events, and youth Church activities.
- Artwork: Digital images at 300 dpi; color transparencies.
- Other: Poetry, to 30 lines.

SAMPLE ISSUE
50 pages (no advertising): 9 articles; 6 depts/columns; 1 poem. Sample copy, $1.50 with 9x12 SASE. Guidelines available.

- "Small Miracle." Article shows how the loss of an infant son strengthened a family.
- "Face Trials with Smiles." Article profiles a young man who excels in baseball despite having only one arm.
- Sample dept/column: "Questions and Answers" offers suggestions to teens on how to get along with their mothers.

RIGHTS AND PAYMENT
All rights. Written material, $.03–$.12 per word. Pays on acceptance. Provides 2 copies.

➥EDITOR'S COMMENTS
We need shorter pieces with sidebars. All submissions must be biblically sound. We urge prospective authors to read our magazine to determine if their work fits our style.

New Expression

Youth Communications
Columbia College
619 South Wabash, Suite 207
Chicago, IL 60605

Program Manager: Lurlene Brown

DESCRIPTION AND INTERESTS
This newsmagazine is published throughout the school year and distributed to 223 schools and youth centers in Chicago. It is also mailed to a subscriber list that includes college students, parents, journalists, educators, and corporate executives. Circ: 45,000.

Audience: YA
Frequency: Quarterly
Website: www.newexpression.org

FREELANCE POTENTIAL
10% written by nonstaff writers. Publishes 50 freelance submissions yearly; 5% by authors who are new to the magazine.

SUBMISSIONS
Send complete ms. Accepts hard copy and email submissions to lurlene@youthcommunications.org. Availability of artwork improves chance of acceptance. SASE. Response time varies.

- Articles: No word limit. Informational articles; reviews; and personal experience pieces. Topics include current events, music, popular culture, and social issues.
- Depts/columns: Word lengths vary. News briefs, media reviews, opinions, teen business, school issues, and sports.
- Artwork: B/W JPEG images.
- Other: Poetry.

SAMPLE ISSUE
36 pages (10% advertising): 11 articles; 12 depts/columns. Sample copy, $1. Guidelines and editorial calendar available.

- "Commuter Woes Over the CTA Cuts Continue into the New Year." Article describes how public transit is impacted by new budget.
- Sample dept/column: "News Briefs" reports on the discovery of Lincoln documents.

RIGHTS AND PAYMENT
All rights. No payment.

➥EDITOR'S COMMENTS
Freelance reporters comprise the majority of our writers, and we expect selected contributors to make a serious commitment to the publication. We are looking for more investigative news stories as well as more positive stories about teens and teen entrepreneurs. We continue to seek submissions on colleges and careers, entertainment, and health.

N.E.W. Kids

P.O. Box 12264
Green Bay, WI 54307

Publisher: Dr. Brookh Lyonns

DESCRIPTION AND INTERESTS
N.E.W. Kids is a recently redesigned parenting tabloid for families living in the Green Bay area. Its goal is to improve the lives of its readers in mind, body, and spirit. Circ: 40,000.

Audience: Parents
Frequency: Monthly
Website: www.newandfoxvalleykids.com

FREELANCE POTENTIAL
100% written by nonstaff writers. Publishes 12+ freelance submissions yearly; 5% by unpublished writers, 50% by authors who are new to the magazine. Receives 60–240 queries and unsolicited mss yearly.

SUBMISSIONS
Query or send complete ms. Accepts email submissions to brookh@bigchoicespublishing.com. Response time varies.

- Articles: To 750 words. Informational articles and humorous pieces. Topics include parenting, family issues, education, gifted and special education, regional and national news, crafts, hobbies, music, the arts, health, fitness, sports, animals, pets, travel, popular culture, and multicultural and ethnic issues.
- Depts/columns: To 750 words. Women's and children's health, school news, and essays.
- Other: Submit seasonal material 4 months in advance.

SAMPLE ISSUE
12 pages (50% advertising): 8 articles; 5 depts/columns. Sample copy, free with 9x12 SASE. Guidelines and editorial calendar available.

- "Dad's Annual Weekend Away." Article describes the rejuvenating power of a guys-only vacation.
- "Enjoy Safe Fun in the Sun on Spring Break." Article offers tips for keeping babies and toddlers protected from the sun's harmful rays.
- Sample dept/column: "The 411" describes a new website for 'green' schools information.

RIGHTS AND PAYMENT
Rights negotiable. All material, payment rates vary. Pays on publication. Provides 1 contributor's copy upon request.

➥EDITOR'S COMMENTS
We are currently interested in receiving more submissions of articles on proactive parenting and health. We always appreciate humor.

New Moon

2 West First Street, Suite 101
Duluth, MN 55802

Associate Editor

DESCRIPTION AND INTERESTS
This magazine "for girls and their dreams" is written and edited primarily by its target demographic. Content is meant to encourage girls to grow their spirits and share their creativity. Circ: 25,000.

Audience: 8–12 years
Frequency: 6 times each year
Website: www.newmoongirlmedia.com

FREELANCE POTENTIAL
90% written by nonstaff writers. Publishes 50 freelance submissions yearly; 85% by unpublished writers, 20% by authors who are new to the magazine. Receives 720 queries and unsolicited mss yearly.

SUBMISSIONS
Female authors only. Query or send complete ms. Prefers email submissions to girl@ newmoongirlmedia.com; will accept hard copy and simultaneous submissions if identified. Does not return mss. Responds in 4–6 months.

- Articles: 300–1,200 words. Profiles and interviews. Topics include activism, school, fitness, recreation, science, technology, and social and multicultural issues.
- Fiction: 900–1,600 words. Empowering stories about girls ages 8–12.
- Depts/columns: Word lengths vary. "Go Girl!" (activism and athletics), "Global Village," "Women's Work," "Herstory," "Body Language," and "Science Side Effects."
- Other: Poetry and artwork by girls ages 8–12.

SAMPLE ISSUE
52 pages (no advertising): 4 articles; 1 story; 18 depts/columns, 2 poems. Sample copy, $7. Guidelines and theme list available at website.

- "Great Dates with Mom." Article recommends special mother-daughter outings.
- Sample dept/column: "Global Village" tells of two girls' lives in Saudi Arabia.

RIGHTS AND PAYMENT
All rights. Written material, $.06–$.10 per word. Pays on publication. Provides 3 contributor's copies.

➥EDITOR'S COMMENTS
We will consider adult-written work for our "Herstory," "Women's Work," and fiction sections only. We will not publish adult work in any other departments.

New York Family

79 Madison Avenue, 16th Floor
New York, NY 10016

Editor: Eric Messinger

DESCRIPTION AND INTERESTS
New York Family reaches upper-income parents throughout Manhattan with articles on the topics they care about most: education, health, and child-rearing. It also offers a fair share of humor. Circ: 25,000.

Audience: Parents
Frequency: Monthly
Website: www.manhattanmedia.com

FREELANCE POTENTIAL
50% written by nonstaff writers. Publishes 40 freelance submissions yearly; 40% by authors who are new to the magazine. Receives 600 queries yearly.

SUBMISSIONS
Query with clips. Accepts hard copy. SASE. Response time varies.

- Articles: 800–1,200 words. Informational articles; profiles; interviews; photo-essays; and personal experience pieces. Topics include education, music, recreation, regional news, social issues, travel, and parenting advice and techniques.
- Depts/columns: 400–800 words. News and media reviews.

SAMPLE ISSUE
114 pages: 6 articles; 10 depts/columns. Sample copy, free with 9x12 SASE. Writers' guidelines available.

- "The Funniest Dad on the Block." Article profiles comedy writer and young father Ben Karlin, who shares his views on parenting.
- "Bringing Up Boys." Interview with a psychologist offers insights into how boys think and how to raise a son to be a great man.
- Sample dept/column: "Maternity" shares the thoughts of expectant fathers.

RIGHTS AND PAYMENT
First rights. Written material, $25–$300. Pays on publication. Provides 3 contributor's copies.

➥EDITOR'S COMMENTS
Our magazine caters to Manhattan's sophisticated, wealthy, and active parents who live in the finest doorman buildings and send their children to top private schools. We are looking for smart, provocative, and engaging articles that take an important parenting topic and offer expert advice in a lighthearted way.

New York Times Upfront

Scholastic Inc.
557 Broadway
New York, NY 10012-3999

Editor

DESCRIPTION AND INTERESTS

This newsmagazine created especially for teens makes it easy for teachers to connect current events to the high school curriculum. Content includes reports from *New York Times* correspondents around the world, history features on topics related to today's events, and essays by teens. It is available online as well as in print. Circ: 250,000.

Audience: 14–18 years
Frequency: 14 times each year
Website: www.upfrontmagazine.com

FREELANCE POTENTIAL

10% written by nonstaff writers. Publishes 2 freelance submissions yearly; 10% by authors who are new to the magazine. Receives 144 queries yearly.

SUBMISSIONS

Query with résumé and clips. Accepts hard copy. Availability of artwork improves chance of acceptance. SASE. Responds in 2–4 weeks if interested.

- Articles: 500–1,200 words. Informational articles; profiles; and interviews. Topics include current events, politics, history, media, technology, social issues, careers, the arts, the environment, and multicultural and ethnic issues.
- Depts/columns: Word lengths vary. Essays by teens, opinions, news briefs, and trends.
- Artwork: Color prints or transparencies.
- Other: Cartoons.

SAMPLE ISSUE

Sample copy and editorial calendar available at website. Guidelines available.

- "China's Olympic Challenge." Article discusses how the 2008 Summer Games in Beijing shined a spotlight on China's human-rights abuses.
- "Tree Shaker: The Story of Nelson Mandela." Article profiles the former South African president and gives a history of apartheid.

RIGHTS AND PAYMENT

All rights. All material, payment rates vary. Pays on publication.

❖EDITOR'S COMMENTS

Our publication brings together the reporting and analysis of *The New York Times* with the proven classroom experience of Scholastic.

The Next Step Magazine

86 West Main Street
Victor, NY 14564

Editor-in-Chief: Laura Jeanne Hammond

DESCRIPTION AND INTERESTS

Distributed to high school students across the country, this magazine offers objective and informative articles on college, career planning, and life skills. Topics include financial aid, choosing a major, career profiles, public speaking, and personal finance management. Circ: 800,000.

Audience: 14–21 years
Frequency: 5 times each year
Website: www.nextSTEPmag.com

FREELANCE POTENTIAL

90% written by nonstaff writers. Publishes 40 freelance submissions yearly.

SUBMISSIONS

Query. Accepts email queries to laura@ nextSTEPmag.com. Response time varies.

- Articles: 70–1,000 words. Informational, self-help, and how-to articles; profiles; interviews; personal experience and opinion pieces; humor; and essays. Topics include college planning, financial aid, campus tours, choosing a career, life skills, résumé writing, public speaking, personal finances, computers, multicultural and ethnic issues, sports, and special education.

SAMPLE ISSUE

54 pages: 18 articles. Sample copy available at website. Guidelines available.

- "Interview with a CEO." Article interviews the chairman of a mutual fund research company and reveals his tips for success.
- "What Does It Mean to Join ROTC?" Article profiles members of the Reserve Officers' Training Corps and discusses the benefits.
- "Careers in Criminal Justice." Article discusses the different careers paths available for someone who studies criminal justice.

RIGHTS AND PAYMENT

All rights. Articles, payment rates vary. Pays within 30 days of acceptance.

❖EDITOR'S COMMENTS

The best queries are specific, concise, entertaining, and well researched. If you have timely information that relates to college planning or starting a career, we'd like to hear from you. We are also interested in hearing from high school counselors who can address these topics.

North Star Family Matters

689 East Promontory Road
Shelton, WA 98584

Editor-in-Chief: Wendy Garrido

DESCRIPTION AND INTERESTS
Families who live in the Puget Sound region read this magazine, the only U.S. publication dedicated to the emotional and spiritual health of parents and children. Its articles seek to inspire kids to discover and contribute their unique gifts to the world, and feel confident in their interactions with others. Circ: 32,000.

Audience: Parents; children 1–15 years
Frequency: Monthly
Website: www.northstarfamilymatters.com

FREELANCE POTENTIAL
60% written by nonstaff writers. Publishes 50 freelance submissions yearly; 60% by unpublished writers, 80% by new authors.

SUBMISSIONS
Query. Accepts email queries to submit@ northstarfamilymatters.com. Response time varies.

- Articles: 300–1,500 words. Informational and how-to articles; personal experience pieces; and book excerpts. Topics include communication, empowerment, discipline, conflict resolution, anger management, emotional support, attention, and reading.
- Depts/columns: Staff written.
- Other: Puzzles, games, and activities.

SAMPLE ISSUE
32 pages: 5 articles; 5 depts/columns; 7 activities. Guidelines available.

- "That Fishing Feeling." Read-aloud story for parents and kids tells how a pre-teen learned to use meditation tactics when things in life seemed out of control.
- "Empowering Education: Rewired to Read." Article details how one family overcame their young daughter's struggle to learn to read.

RIGHTS AND PAYMENT
First and electronic rights. Written material, payment rates vary. Payment policy varies. Provides 1 contributor's copy.

⟞EDITOR'S COMMENTS
We are looking for articles that are relevant and interesting to our readers, whether they are children or grandparents, parents of newborns, or parents of teens. Submissions should provide an empowering philosophy and a positive focus, have consistent use of language, and a reader-friendly flow.

Northwest Baby & Child

4395 Rollinghill Road
Clinton, WA 98236

Editor: Betty Freeman

DESCRIPTION AND INTERESTS
A regional tabloid, *Northwest Baby & Child* provides information and support to parents in the greater Puget Sound community of Washington state. Circ: 35,000.

Audience: Parents
Frequency: Monthly
Website: www.nwbaby.com

FREELANCE POTENTIAL
60–75% written by nonstaff writers. Publishes 100 freelance submissions yearly; many by unpublished writers, 33% by authors who are new to the magazine.

SUBMISSIONS
Send complete ms. Accepts email submissions to editor@nwbaby.com (no attachments). Responds in 2–3 months if interested.

- Articles: 250–750 words. Informational and how-to articles; personal experience pieces; and interviews. Topics include pregnancy, childbirth, parenting, early education, party ideas, family life, travel, recreation, traditions, and the environment.
- Depts/columns: 250–750 words. Health and parenting tips, family activities, special events, and regional resources.
- Artwork: B/W and color TIFF or PDF images. No JPEGs.
- Other: Poetry.

SAMPLE ISSUE
12 pages (15% advertising): 3 articles; 9 depts/columns. Guidelines and editorial calendar available at website.

- "Becoming a Mother Takes Time." Article stresses the importance of taking time to reflect on pregnancy and parenthood, and recommends keeping a journal.
- Sample dept/column: "Healthy Families" recommends early screening for autism and lists warning signs to look for.

RIGHTS AND PAYMENT
First or reprint rights. Articles and depts/columns, $40. Poetry and artwork, $10. Pays on publication. Provides 1–2 contributor's copies.

⟞EDITOR'S COMMENTS
Many writers break into the business through publications like ours. If you have an article that would be appropriate for us, we would be happy to review it.

OC Family
The Newsmagazine for Parents

1451 Quail Street, Suite 201
Newport Beach, CA 92660

Editor: Lenée Harvey

DESCRIPTION AND INTERESTS
Parents in Orange County, California, turn to this magazine for information on topics and issues of interest to them; many of them localized to the region. Its articles on parenting, family, school, and recreation appeal to parents of children of all ages. Circ: 70,000.

Audience: Families
Frequency: Monthly
Website: www.ocfamily.com

FREELANCE POTENTIAL
82% written by nonstaff writers. Publishes 50 freelance submissions yearly; 1% by unpublished writers, 1% by authors who are new to the magazine. Receives 144 queries yearly.

SUBMISSIONS
Query. Accepts hard copy and email queries to lharvey@churmmedia.com. SASE. Responds in 1 month.

- Articles: 800–2,500 words. Informational articles and profiles. Topics include education, the Internet, family activities, fine arts, regional food and dining, consumer interest, parenting, grandparenting, and child development issues.
- Depts/columns: Word lengths vary. Family life, personal finances, book and software reviews, and women's health.
- Artwork: B/W and color prints.

SAMPLE ISSUE
204 pages (60% advertising): 3 articles; 20 depts/columns; 2 directories. Sample copy, free with 9x12 SASE. Editorial calendar available.

- "Happy, Happy Birthdays." Article provides ideas for age-appropriate themed birthday parties that can make a child's special day extra special.
- Sample dept/column: "Kid Fashion" looks at what's hot in shorts and other summer wear for the younger set.

RIGHTS AND PAYMENT
One-time rights. Written material, $100–$500. Artwork, $90. Kill fee, $50. Pays 45 days after publication. Provides 3 contributor's copies.

➥EDITOR'S COMMENTS
We are open to any idea that is helpful to parents or that they may find interesting. We are a regional publication; therefore, all articles must have local sources or angles.

Odyssey

Carus Publishing
30 Grove Street, Suite C
Peterborough, NH 03458

Senior Editor: Elizabeth E. Lindstrom

DESCRIPTION AND INTERESTS
Subtitled "Adventures in Science," *Odyssey* caters to the middle and high school students who are crazy about science, math, and technology. Its articles blend scientific accuracy with lively approaches to its subjects. Circ: 25,000.

Audience: 10–16 years
Frequency: 9 times each year
Website: www.odysseymagazine.com

FREELANCE POTENTIAL
70% written by nonstaff writers. Publishes 60 freelance submissions yearly; 2% by unpublished writers, 25% by authors who are new to the magazine. Receives 300 queries each year.

SUBMISSIONS
Query with outline, author biography, bibliography, and clips or writing samples. Accepts hard copy. Availability of artwork improves chance of acceptance. SASE. Responds in 5 months only if interested.

- Articles: 750–1,000 words. Informational articles; biographies; and interviews. Topics include math, science, and technology.
- Fiction: To 1,000 words. Science fiction and science-related stories.
- Depts/columns: Word lengths vary. Astronomy, animals, profiles, and science news.
- Artwork: B/W and color prints.
- Other: Activities, to 500 words. Seasonal material about space or astronomy events.

SAMPLE ISSUE
48 pages (no advertising): 7 articles; 8 depts/columns; 2 activities. Sample copy, $4.50 with 9x12 SASE (4 first-class stamps). Guidelines and theme list available.

- "Urban Geocaching." Article explains the growing popularity of this technical treasure hunt game.
- Sample dept/column: "Science Scoops" warns of the pitfalls of electronic multi-tasking.

RIGHTS AND PAYMENT
All rights. Written material, $.20–$.25 per word. Artwork, payment rates vary. Pays on publication. Provides 2 contributor's copies.

➥EDITOR'S COMMENTS
We look for cutting-edge science reporting written in a fun, youthful tone. Most material supports our planned themes.

The Old Schoolhouse

P.O. Box 8426
Gray, TN 37615

Editors: Paul & Gena Suarez

DESCRIPTION AND INTERESTS
One of the largest homeschooling magazines on the market, *The Old Schoolhouse* explores all facets of home-based education. It includes articles about family life, joyful parenting, and all aspects of homeschooling, many of them written from a Christian perspective. Circ: 50,000.
Audience: Homeschool families
Frequency: Quarterly
Web: www.thehomeschoolmagazine.com

FREELANCE POTENTIAL
98% written by nonstaff writers. Publishes 160 freelance submissions yearly; 30% by unpublished writers, 50% by authors who are new to the magazine. Receives 192 queries each year.

SUBMISSIONS
Query with outline, sample paragraphs, and brief author bio. Accepts queries submitted via website only (homeschoolblogger.com users should note their user name on query). No simultaneous submissions. Responds in 4–6 weeks.

- Articles: 1,000–2,000 words. Informational and how-to articles; and personal experience pieces. Topics include homeschooling, education, family life, art, music, spirituality, literature, child development, teen issues, science, history, and mathematics.
- Depts/columns: Word lengths vary. Short news items, teaching styles, opinion pieces, teaching special needs children, and humor.

SAMPLE ISSUE
200 pages (40% advertising): 3 articles; 21 depts/columns. Sample copy available. Guidelines available at website.

- "Homeschooling On a Shoestring." Article shows how homeschooling doesn't have to mean a big investment in books and supplies.
- "Homeschooling the Rebel." Article provides tips for working with a rebellious student.

RIGHTS AND PAYMENT
First rights. Written material, payment rates vary. Pays on publication. Provides 2 copies.

➡EDITOR'S COMMENTS
We appreciate queries that demonstrate an understanding of our magazine and our target audience.

Once Upon a Time . . .

553 Winston Court
St. Paul, MN 55118

Editor/Publisher: Audrey B. Baird

DESCRIPTION AND INTERESTS
This magazine, written for children's authors and illustrators, offers specialized articles to support, inform, instruct, inspire, and entertain them. Circ: 1,000.
Audience: Children's writers and illustrators
Frequency: Quarterly
Website: www.onceuponatimemag.com

FREELANCE POTENTIAL
34% written by nonstaff writers. Publishes 120 freelance submissions yearly; 50% by unpublished writers. Receives 80–175 unsolicited mss yearly.

SUBMISSIONS
Send complete ms. Accepts hard copy. SASE. Responds in 2 months.

- Articles: To 900 words. Informational, self-help, and how-to articles; and personal experience pieces. Topics include writing and illustrating for children.
- Depts/columns: Staff written
- Artwork: B/W line art.
- Other: Poetry, to 24 lines.

SAMPLE ISSUE
32 pages (2% advertising): 25 articles; 10 depts/columns; 6 poems. Sample copy, $5. Guidelines available.

- "You Can Write Every Day." Article offers tips for fitting writing time into a busy schedule.
- "Starting the Race at 65." Article shares the author's feelings about starting to write at age 65.
- "The Children's Publishing Merry-Go-Round." Article provides ten exercises to help picture book writers hang in there while trying to get their work published.

RIGHTS AND PAYMENT
One-time rights. No payment. Provides 2 contributor's copies.

➡EDITOR'S COMMENTS
We are always looking for fresh angles about the process of writing. How about an article on collaboration, including the problems and joys that come with working in tandem with another writer? Please note that we do not accept fiction, nor do we accept poetry for children. We do, however, publish poems for adults about writing for children.

On Course

General Council of the Assemblies of God
1445 North Boonville Avenue
Springfield, MO 65802-1894

Editor

DESCRIPTION AND INTERESTS
Each themed issue of this Christian teen magazine addresses topics and challenges that readers face on a daily basis through articles that provide Christian solutions and guidance. It offers fiction by young adults, and covers pop culture and entertainment. Circ: 160,000.

Audience: 12–18 years
Frequency: Quarterly
Website: www.oncourse.ag.org

FREELANCE POTENTIAL
95% written by nonstaff writers. Publishes 32 freelance submissions yearly; 30% by unpublished writers, 40% by authors who are new to the magazine.

SUBMISSIONS
All work is assigned. Send audition ms with résumé. Prefers hard copy and email submissions to oncourse@ag.org; will accept Macintosh-compatible CD submissions.

- Articles: To 800 words. Informational and how-to articles; profiles; humor; and personal experience pieces. Topics include social issues, music, health, religion, sports, careers, college, and multicultural subjects.
- Fiction: To 800 words. Genres include contemporary, humorous, multicultural, and sports-themed fiction.
- Depts/columns: Word lengths vary. Profiles and brief news items.

SAMPLE ISSUE
32 pages (33% advertising): 4 articles; 6 depts/columns. Sample copy, free. Guidelines available at website.

- "Restoring the Heart." Article tells the story of an abused young girl whose life was saved—literally—when she found God.
- "Superbad . . . Superhero?" Articles uses the popular film *Hancock* as a metaphor for understanding the consequences of bad behavior.

RIGHTS AND PAYMENT
First and electronic rights. Articles, $30–$80; $15 for reviews and sidebars. Depts/columns, $40. Fiction, $15. Payment policy varies. Provides 5 contributor's copies.

➥ EDITOR'S COMMENTS
We will evaluate your manuscript to determine if you, as a writer, fit our magazine. If you do, we will give you an assignment.

Organic Family Magazine

P.O. Box 1614
Wallingford, CT 06492-1214

Editor: Catherine Wong

DESCRIPTION AND INTERESTS
Organic Family Magazine does more than inform people about how to eat organically. It also inspires and educates families who want to move away from the consumer-based world and raise their families contentedly and confidently. Circ: 200.

Audience: Families
Frequency: Twice each year
Website: www.organicfamilymagazine.com

FREELANCE POTENTIAL
90% written by nonstaff writers. Publishes 40 freelance submissions yearly.

SUBMISSIONS
Query or send complete ms. Prefers email submissions to sciencelibrarian@hotmail.com; will accept hard copy. SASE. Response time varies.

- Articles: Word lengths vary. Informational articles; interviews; and personal experience pieces. Topics include nature, organic agriculture, conservation, parenting, natural pet care, herbs, organic gardening, nutrition, progressive politics, health, wellness, and environmental issues.
- Fiction: Word lengths vary. Stories about nature and the environment.
- Depts/columns: Word lengths vary. New product reviews, recipes, profiles of conservation organizations, and media reviews.
- Other: Poetry.

SAMPLE ISSUE
28 pages: 8 articles; 1 story; 11 depts/columns; 1 poem. Sample copy and guidelines available at website.

- "The Invisible Trust Fund: Organic Foods." Article touts the advantages of organic foods as a way to keep our children, lifestyles, the environment, and the economy healthier.
- "What Links Florida and Idaho? The Phosphate Risk." Article explains the long-term dangers of phosphate strip mining, common in both states.

RIGHTS AND PAYMENT
One-time rights. No payment. Provides 1 contributor's copy.

➥ EDITOR'S COMMENTS
We welcome a plethora of ideas and topics; but each article must be well-researched, with source lists provided.

Our Children

National PTA
541 North Fairbanks Court, Suite 1300
Chicago, IL 60611-3396

Editor: Ted Villaire

DESCRIPTION AND INTERESTS
Our Children is the official voice of the National PTA, so it is not surprising that it is concerned with issues relating to school and children. Its articles cover education issues and encourage parent involvement in their children's schools. Circ: 31,000.

Audience: Parents, teachers, administrators
Frequency: 5 times each year
Website: www.pta.org

FREELANCE POTENTIAL
50% written by nonstaff writers. Publishes 20–25 freelance submissions yearly; 75% by authors who are new to the magazine. Receives 180–240 queries and unsolicited mss yearly.

SUBMISSIONS
Query or send complete ms. Accepts email submissions to tvillaire@pta.org. No simultaneous submissions. Responds in 2 months.

- Articles: 600–1,100 words. Informational and how-to articles. Topics include running local PTA chapters, child welfare, education, and family life.
- Depts/columns: Word lengths vary. Short updates on parenting and education issues and member advice.
- Artwork: 3x5 or larger color prints and slides.
- Other: Submit seasonal material 3 months in advance.

SAMPLE ISSUE
24 pages (no advertising): 5 articles; 9 depts/columns. Sample copy, $2.50 with 9x12 SASE ($1 postage). Writers' guidelines and theme list available.

- "The 21st Century School Is a Green School." Article explains the educational advantages of an environmentally healthy school.
- "Can Your Children Hear Their Teachers?" Article explains the importance of acoustics in a classroom.

RIGHTS AND PAYMENT
First rights. No payment. Provides 3 contributor's copies.

➛EDITOR'S COMMENTS
If you have experience in running a successful PTA program, or have substantially increased parent involvement in your chapter, we want to hear your ideas.

Our Little Friend

Pacific Press Publishing
P.O. Box 5353
Nampa, ID 83653-5353

Editor: Aileen Andres Sox

DESCRIPTION AND INTERESTS
Children attending Bible school at Seventh-day Adventist churches receive this magazine, which has been published continuously since 1890. Bible lessons and stories that teach Christian values, beliefs, and practices are found in its pages. Circ: 35,000.

Audience: 1–6 years
Frequency: Weekly
Website: www.ourlittlefriend.com

FREELANCE POTENTIAL
20% written by nonstaff writers. Publishes 52 freelance submissions yearly; 10% by unpublished writers, 10% by authors who are new to the magazine. Receives 240 unsolicited mss each year.

SUBMISSIONS
Send complete ms. Accepts hard copy, email to ailsox@pacificpress.com, and simultaneous submissions if identified. SASE. Responds in 4 months.

- Articles: 500–650 words. Devotionals, Bible lessons, and true stories that teach Christian values. Topics include school and family.
- Fiction: 500–650 words. Short stories that portray God's love for children, personal faith, and contemporary issues.
- Artwork: Color photos. Line art.
- Other: Submit seasonal material 7 months in advance.

SAMPLE ISSUE
8 pages (no advertising): 6 stories. Sample copy, free with 9x12 SASE (2 first-class stamps). Guidelines available.

- "Fly Away Home." Story tells of a young girl and her mother freeing a trapped bird.
- "Visiting Grandma's Church." Story describes a girl's experience of helping out at church while visiting her grandmother.

RIGHTS AND PAYMENT
One-time rights. Written material, $25–$50. Pays on acceptance. Provides 3 contributor's copies.

➛EDITOR'S COMMENTS
We are looking for inspiring stories that are in an easy-to-read format for children. All material should be consistent with the Seventh-day Adventist beliefs and practices. Please check our website for editorial guidelines.

Pack-O-Fun

2400 East Devon, Suite 292
Des Plaines, IL 60018-4618

Editor: Annie Niemiec

DESCRIPTION AND INTERESTS
Pack-O-Fun is right! This magazine is packed with fun—from crafts that can be made from household objects, to children's group activities that encourage learning. It is read by parents, teachers, and others who work with children ages six to twelve. Circ: 15,000.

Audience: Parents, teachers
Frequency: 6 times each year
Website: www.pack-o-fun.com

FREELANCE POTENTIAL
50% written by nonstaff writers. Receives 504 unsolicited mss yearly.

SUBMISSIONS
Send complete ms with instructions, brief materials list, and photographs of project. Accepts hard copy and email submissions to aniemiec@amoscraft.com (attach files). SASE. Responds in 4–6 weeks.

- Articles: To 200 words. How-to articles; craft projects; and party ideas.
- Depts/columns: Word lengths vary. Art ideas, projects for children and adults to do together, vacation Bible school projects, pictures of readers' projects.
- Artwork: B/W and color prints; JPEG files. Line art.

SAMPLE ISSUE
66 pages (10% advertising): 39 crafts; 4 depts/columns. Sample copy, $4.99 with 9x12 SASE (2 first-class stamps). Writers' guidelines available at website.

- "Americana Tea Light Holders." Article explains how to make tea light holders decorated for the Fourth of July.
- "Take a Hike! Bottle Holder." Article explains how to make a bottle holder for use in hiking.

RIGHTS AND PAYMENT
All rights. All material, $25–$150. Pays 30 days after signed contract. Provides 3 contributor's copies.

⚙ EDITOR'S COMMENTS
We are looking for arts and crafts projects that are easy to make; made from recycled or low-cost materials; perfect for large groups such as camps or elementary classrooms; and gift ideas for parents, teachers, or friends. We'd love it even more if the project was based on a children's book or a period in history.

Pageantry

P.O. Box 160307
Alamonte Springs, FL 32716

Editor: Frank Abel

DESCRIPTION AND INTERESTS
Everything a girl needs to know to be crowned a beauty queen is covered in this magazine, including tips on fitness, modeling, makeup, and interview techniques. Circ: Unavailable.

Audience: YA–Adult
Frequency: Quarterly
Website: www.pageantrymagazine.com

FREELANCE POTENTIAL
10% written by nonstaff writers. Publishes 5 freelance submissions yearly.

SUBMISSIONS
Query. Accepts hard copy and email queries to editor@pageantrymagazine.com. SASE. Response time varies.

- Articles: Word lengths vary. Informational articles; profiles; interviews; and personal experience pieces. Topics include beauty pageants, celebrities, fitness, modeling, makeup tips, interviewing techniques, dance, winning psychology, judges' perspectives, etiquette, coaching, talent competitions, and fashion.
- Depts/columns: Word lengths vary. Jewelry, makeup, hairstyles, fitness, body shaping, modeling, personal advice, teen issues, winner profiles, etiquette, news, and opinions.

SAMPLE ISSUE
144 pages: 6 articles; 23 depts/columns. Sample copy, $4.95.

- "A New, Updated, and Relevant America—Miss America, That Is." Article profiles the latest Miss America and tells how she stands out from the crowd of past winners.
- "Tale of Two Miss Floridas." Article describes the friendship between pageant winners in the Sunshine State.
- Sample dept/column: "Goodie Bag" offers information on fashion accessories.

RIGHTS AND PAYMENT
First North American serial rights. Written material, payment rates vary. Payment policy varies. Provides 1 contributor's copy.

⚙ EDITOR'S COMMENTS
We are known as the "bible of the pageant industry." Our writers are esteemed experts in their fields, and our magazine is devoted to one thing: telling our readers everything they need to know to win a beauty competition.

Parent and Preschooler Newsletter

North Shore Child & Family Guidance Center
480 Old Westbury Road
Roslyn Heights, NY 11577-2215

Editor: Neala S. Schwartzberg, Ph.D.

DESCRIPTION AND INTERESTS
Parents, educators, and child-care professionals turn to this international newsletter for its trusted and supportive information on child development, discipline, and family issues and concerns. Written by respected psychologists, social workers, and other field experts, it is available in both English and Spanish editions. Circ: Unavailable.

Audience: Parents, caregivers, and early childhood professionals
Frequency: 6 times each year
Website: www.northshorechildguidance.org

FREELANCE POTENTIAL
90% written by nonstaff writers. Publishes 6 freelance submissions yearly; 50% by authors who are new to the magazine. Receives 12 queries yearly.

SUBMISSIONS
Query with outline. Accepts email queries to nealas@msn.com. Responds in 1 week.

- Articles: 2,000 words. Practical information and how-to articles. Topics include education, self-esteem, discipline, children's health, parenting skills, fostering cooperation through play, and coping with death.
- Depts/columns: Staff written.

SAMPLE ISSUE
8 pages (no advertising): 3 articles; 3 depts/columns. Sample copy, $3 with #10 SASE (1 first-class stamp). Guidelines and editorial calendar available.

- "Welcoming Daddyhood." Article recounts a father's first-time journey through pregnancy and childbirth.
- "Preschooler in the Kitchen." Article offers suggestions on ways to avoid food battles with kids.
- "Colic." Article discusses ways to cope with and help a colicky baby.

RIGHTS AND PAYMENT
First world rights. Articles, $200. Pays on publication. Provides 10 contributor's copies.

⚏EDITOR'S COMMENTS
When sending us a query, please remember that we have an international readership. Topics related to universal child development are always of interest to us. Information should be practical and positive.

Parentguide

419 Park Avenue South, 13th Floor
New York, NY 10016

Editor: Jenna Greditor

DESCRIPTION AND INTERESTS
Parents of elementary school-aged children living in the New York metropolitan area read *Parentguide* for articles on all aspects of child-rearing and family life. Circ: 280,000.

Audience: Parents
Frequency: Monthly
Website: www.parentguidenews.com

FREELANCE POTENTIAL
80% written by nonstaff writers. Publishes 100 freelance submissions yearly; 5% by unpublished writers, 50% by authors who are new to the magazine. Receives 12 queries and unsolicited mss yearly.

SUBMISSIONS
Query or send complete ms with résumé. Prefers email submissions to jenna@parentguidesnews.com (Microsoft Word attachment); will accept hard copy. SASE. Responds in 3–4 weeks.

- Articles: 750–1,500 words. Informational and self-help articles; humor; and personal experience pieces. Topics include parenting, family, education, regional and social issues, popular culture, and careers.
- Depts/columns: 500 words. Local news and events, health, travel, and reviews.

SAMPLE ISSUE
70 pages (39% advertising): 10 articles; 7 depts/columns. Sample copy, free with 10x13 SASE. Guidelines available.

- "Care to Spare." Article explains how visiting nurses help new moms and dads adjust to parenthood and provide valuable advice about newborn care.
- "Hot Under the Collar." Article offers tips and techniques for dealing with, and avoiding, aggression in young children.
- Sample dept/column: "Health" reviews a number of new products of interest to parents, from books to juice.

RIGHTS AND PAYMENT
Rights vary. No payment. Provides 1 contributor's copy.

⚏EDITOR'S COMMENTS
While we focus on regional events, news, and issues, we also publish articles that have a national appeal to parents of children ages 12 and under.

Parenting

135 West 50th Street, 3rd Floor
New York, NY 10026

Submissions Editor

DESCRIPTION AND INTERESTS
This popular magazine for parents of children through the age of 12 provides information and tips on all aspects of child-rearing. Articles cover health, behavior, relationships, child development, and pregnancy. Product reviews are also featured. Circ: 2 million+.

Audience: Parents
Frequency: 11 times each year
Website: www.parenting.com

FREELANCE POTENTIAL
80% written by nonstaff writers. Publishes 10–15 freelance submissions yearly; 5% by unpublished writers, 10% by authors who are new to the magazine. Receives 1,000 queries each year.

SUBMISSIONS
Query with clips. Accepts hard copy. SASE. Responds in 1–2 months.

- Articles: 1,000–2,500 words. Informational, how-to, and self-help articles; profiles; and personal experience pieces. Topics include child development, behavior, health, pregnancy, and family activities.
- Depts/columns: 100–1,000 words. Parenting tips and advice, child development by age range, work and family, and health and beauty advice for moms.

SAMPLE ISSUE
144 pages (50% advertising): 10 articles; 15 depts/columns. Sample copy, $5.95 (mark envelope Attn: Back Issues). Writers' guidelines available.

- "Don't Eat That!" Article discusses the rise in food allergies and how to determine if your child has them.
- Sample dept/column: "How to Keep Roughhousing From Being Too Rough" offers tips on keeping preschool play safe.

RIGHTS AND PAYMENT
First world rights with 2 months' exclusivity. Written material, payment rates vary. Pays on acceptance. Provides 1 contributor's copy.

➬ EDITOR'S COMMENTS
We are looking for informative, timely, and well-researched material that presents a fresh approach to topics of interests to parents and expectant mothers. Reality-tested advice and quick tips are always of interest to us.

Parenting New Hampshire

150 Dow Street
Manchester, NH 03101

Editor: Tracy Kittredge

DESCRIPTION AND INTERESTS
The pages of this regional publication are filled with resources, tips, and practical information on parenting topics such as childbirth, education, child development, and health. A monthly events calendar is also featured. Circ: 27,500.

Audience: Parents
Frequency: Monthly
Website: www.parentingnh.com

FREELANCE POTENTIAL
85% written by nonstaff writers. Publishes 25–35 freelance submissions yearly; 20% by unpublished writers, 50% by authors who are new to the magazine. Receives 1,200 queries, 240–360 unsolicited mss yearly.

SUBMISSIONS
Query or send complete ms. Accepts hard copy, disk submissions, and email submissions to news@parentingnh.com. SASE. Response time varies.

- Articles: Word lengths vary. Informational and how-to articles; profiles; and interviews. Topics include parenting, education, maternity, childbirth, special needs, gifted education, fathering, child development, summer fun, birthday parties, holidays, back-to-school issues, and health.
- Depts/columns: Word lengths vary. Child development, parenting issues, and health and wellness.
- Other: Submit seasonal material 3 months in advance.

SAMPLE ISSUE
54 pages (42% advertising): 5 articles; 12 depts/columns. Sample copy, free. Writers' guidelines available.

- "Stay-At-Home Dads Can Do the Job." Article discusses the state's increase in fathers becoming their children's primary caregiver.
- Sample dept/column: "House Calls" explains cerebral palsy and its effects on children.

RIGHTS AND PAYMENT
All rights. Articles, $30. Other material, payment rates vary. Pays on acceptance. Provides 3 contributor's copies.

➬ EDITOR'S COMMENTS
We are interested in seeing more articles that offer advice on parenting adolescents and teens. Articles must have a local angle.

ParentingUniverse.com

Best Parenting Resources, LLC
546 Charing Cross Drive
Marietta, GA 30066

Editor: Alicia Hagan

DESCRIPTION AND INTERESTS
This online magazine answers all the questions parents might ask, whether the topic is pregnancy or pacifiers, discipline or diapers. It also provides ideas for family activities. Hits per month: 2+ million.

Audience: Parents
Frequency: Daily
Website: www.parentinguniverse.com

FREELANCE POTENTIAL
90% written by nonstaff writers. Publishes 500+ freelance submissions yearly. Receives 5,000+ queries, 144 unsolicited mss yearly.

SUBMISSIONS
Query or send complete ms. Prefers submissions through website; will accept email submissions to alicia@parentinguniverse.com. Response time varies.

- Articles: Word lengths vary. Informational, how-to, and self-help articles; profiles; interviews; reviews; and personal experience pieces. Topics include pregnancy, childbirth, newborns, parenting, education, health, fitness, recreation, crafts, and family activities.
- Depts/columns: Word lengths vary. Parenting tips and guidelines.

SAMPLE ISSUE
Sample issue and writers' guidelines available at website.

- "Kick the Diaper Habit NOW!" Article offers pointers for toilet-training children over the age of 3.
- "10 Risk Factors That Every Pregnant Woman Should Be Aware Of." Article lists the most common pregnancy risk factors that can be controlled or influenced.
- Sample dept/column: "Health" tells how to prevent misbehavior, how to discipline when appropriate, and how to reduce stress.

RIGHTS AND PAYMENT
Electronic rights. Written material, payment rates vary. Pays on publication. Provides 2 contributor's copies.

➻EDITOR'S COMMENTS
Our readers consist of both expectant and experienced mothers. Please keep our audience in mind when formulating your query, and avoid irrelevant topics.

Parent Life

One LifeWay Plaza
Nashville, TN 37234-0172

Editor: Jodi Skulley

DESCRIPTION AND INTERESTS
This magazine is read by evangelical Christian parents of babies and children up to age 12. Its content is designed to encourage and equip parents with biblical solutions that will transform their family life. Circ: 76,000.

Audience: Parents
Frequency: Monthly
Website: www.lifeway.com/magazines

FREELANCE POTENTIAL
95% written by nonstaff writers. Publishes 12 freelance submissions yearly; 5% by unpublished writers, 5% by authors who are new to the magazine. Receives 300 queries and unsolicited mss yearly.

SUBMISSIONS
Query or send complete ms. Accepts hard copy and email submissions to jodi.skulley@lifeway.com. SASE. Response time varies.

- Articles: 500–1,500 words. Informational and how-to articles; and personal experience pieces. Topics include family issues, religion, education, health, and hobbies.
- Depts/columns: 500 words. Age-appropriate advice, fathers' perspectives, single parenting, working parents, expectant parents, and medical advice.
- Artwork: Color prints or transparencies.
- Other: Accepts seasonal material for Christmas and Thanksgiving.

SAMPLE ISSUE
50 pages: 6 articles; 15 depts/columns. Sample copy, $2.95 with 10x13 SASE. Writers' guidelines available.

- "The Best Thing About Easter." Article explains how families and churches can focus on the true meaning of Easter.
- Sample dept/column: "Single Parent Life" outlines some of the challenges particular to single parents.

RIGHTS AND PAYMENT
Non-exclusive rights. All material, payment rates vary. Pays on publication. Provides 1 contributor's copy.

➻EDITOR'S COMMENTS
We have redesigned our magazine to feature shorter articles on more topics, including fatherhood, single parents, adoptive families, and families with special needs.

Parents & Kids

2727 Old Canton Road, Suite 294
Jackson, MS 39216

Editor: Gretchen Cook

DESCRIPTION AND INTERESTS

Subtitled "Creative Solutions for Mississippi Families," this magazine provides parenting information and advice, along with local information on family activities. Circ: 35,000.

Audience: Parents
Frequency: 9 times each year
Website: www.parents-kids.com

FREELANCE POTENTIAL

80% written by nonstaff writers. Publishes 80 freelance submissions yearly; 50% by unpublished writers. Receives 396 unsolicited mss each year.

SUBMISSIONS

Send complete ms. Accepts email submissions to magazine@parents-kids.com (text in body of message and as Microsoft Word attachment). Responds in 6 weeks.

- Articles: 700 words. Informational, self-help, and how-to articles. Topics include the arts, computers, crafts and hobbies, health and fitness, multicultural and ethnic issues, recreation, regional news, social issues, special education, sports, and family travel.
- Depts/columns: 500 words. Travel, cooking, and computers.
- Artwork: Prefers digital images; will accept B/W prints and transparencies. Line art.
- Other: Submit seasonal material 3–6 months in advance.

SAMPLE ISSUE

48 pages (54% advertising): 11 articles; 7 depts/columns. Sample copy, free with 9x12 SASE ($1.06 postage). Writers' guidelines available at website.

- "Preventing Summer 'Brain Drain.'" Article offers tips for keeping kids' minds active during the summer months.
- Sample dept/column: "Family Matters" offers ideas for getting kids involved in cooking.

RIGHTS AND PAYMENT

One-time rights. Written material, $25. Pays on publication. Provides 1 tearsheet.

➥EDITOR'S COMMENTS

We are a regional publication covering central Mississippi. We tend to choose writers who live in the area, as they will be most able to relate local information to our readers. All articles must use local sources.

Parents' Press

1454 Sixth Street
Berkeley, CA 94710

Editor: Dixie M. Jordan

DESCRIPTION AND INTERESTS

Parents in the San Francisco Bay area turn to this tabloid for information on topics regarding healthy, happy families. It publishes articles on education, parenting, relationships, and health. Circ: 75,000.

Audience: Parents
Frequency: Monthly
Website: www.parentspress.com

FREELANCE POTENTIAL

30–40% written by nonstaff writers. Publishes 60 freelance submissions yearly; 15% by authors who are new to the magazine. Receives 720 unsolicited mss yearly.

SUBMISSIONS

Send complete ms. Accepts hard copy and email submissions to parentsprs@aol.com. SASE. Responds in 2 months.

- Articles: To 1,500 words. Informational and how-to articles. Topics include child development, education, health, safety, party planning, and local family events and activities.
- Depts/columns: Staff written.
- Artwork: B/W prints and transparencies. Line art.
- Other: Submit seasonal material 2 months in advance.

SAMPLE ISSUE

32 pages (63% advertising): 7 articles; 4 depts/columns. Sample copy, $3 with 9x12 SASE ($1.93 postage). Writers' guidelines available at website.

- "Certified Green." Article explains the ways in which Bay Area schools are adopting eco-friendly policies.
- "Humor and Tears: A New Mom Copes with Breast Cancer." Article profiles a writer and new mom who struggled through cancer.

RIGHTS AND PAYMENT

All or second rights. Articles, $50–$500. Pays 45 days after publication.

➥EDITOR'S COMMENTS

Our focus is on practical, down-to-earth articles. We do not accept political material, fiction, poetry, or childbirth stories. If you have a parenting article, be forewarned that we are not interested in your opinions or experiences. Rather, your article must be well-researched and use local sources.

Parent:Wise Austin

5501 A Balcones Drive, Suite 102
Austin, TX 78731

Editor/Publisher: Kim Pleticha

DESCRIPTION AND INTERESTS

Read by parents living in central Texas, this magazine aims to depict different aspects of the parenting journey through its essays, informational articles, and humor columns. Calendars, resource lists, and book and restaurant reviews appear in each issue. Circ: 32,000.

Audience: Parents
Frequency: Monthly
Website: www.parentwiseaustin.com

FREELANCE POTENTIAL

25% written by nonstaff writers. Publishes 6 freelance submissions yearly; 33% by authors who are new to the magazine.

SUBMISSIONS

Query with 4 clips for articles. Send complete ms for "Essays" and "My Life as a Parent" (monthly humor article). Accepts email submissions to storyideas@parentwiseaustin.com. Response time varies.

- Articles: 2,000–2,500 words for cover stories; to 1,000 words for news articles. Informational articles and profiles. Topics include parenting, family life, education, regional news, and people in the Austin community who work to make life better for others.
- Depts/columns: Word lengths vary. Restaurant and music reviews, humor about parenting and family life, and medical advice.
- Other: Poetry about parenting, children, or families; line lengths vary.

SAMPLE ISSUE

40 pages: 3 articles; 6 depts/columns; 3 calendars. Sample copy and writers' guidelines available at website.

- "Color Blind." Essay portrays a mother's gratification when she realizes her son is color blind to skin color.
- Sample dept/column: "Book Reviews" presents four new children's books.

RIGHTS AND PAYMENT

First North American serial and Internet rights. Cover stories, $200. Other written material, payment rates vary. Payment policy varies.

➔ EDITOR'S COMMENTS

All articles must focus on the Austin and central Texas areas and be well researched. Local writers are preferred, but we will consider new writers.

Partners

Christian Light Publications
P.O. Box 1212
Harrisonburg, VA 22803-1212

Editor: Etta G. Martin

DESCRIPTION AND INTERESTS

Written as the Sunday school take-home story paper for children ages 9 to 14, *Partners* promotes the principles of the Mennonite faith. Included in its pages are short stories, poems, puzzles, and activities. Circ: 6,707.

Audience: 9–14 years
Frequency: Monthly
Website: www.clp.org

FREELANCE POTENTIAL

98% written by nonstaff writers. Publishes 200–500 freelance submissions yearly; 5% by unpublished writers, 5% by authors who are new to the magazine. Receives 300–400 unsolicited mss yearly.

SUBMISSIONS

Send complete ms. Prefers email submissions to partners@clp.org; will accept hard copy, disk submissions, and simultaneous submissions if identified. SASE. Responds in 6 weeks.

- Articles: 200–800 words. Informational articles. Topics include nature, customs, history, and teachings.
- Fiction: 400–1,600 words. Stories emphasizing Mennonite beliefs.
- Other: Puzzles and activities with Christian themes. Poetry, no line limits. Submit seasonal material 6 months in advance.

SAMPLE ISSUE

16 pages (no advertising): 4 articles; 4 stories; 8 activities; 4 poems. Sample copy, free with 9x12 SASE ($.97 postage). Guidelines and theme list available.

- "Peace-Making Day." Article discusses how children can promote peace within their own lives on Veteran's Day.
- "Race on Parson's Hill." Story shares a boy's experience of befriending a lonely child while sledding.

RIGHTS AND PAYMENT

First, reprint, or multiple-use rights. Articles and stories, $.03–$.05 per word. Poetry, $.35–$.75 per line. Other material, payment rates vary. Pays on acceptance. Provides 1 contributor's copy.

➔ EDITOR'S COMMENTS

Stories we publish must foster reverence to God, present His Word as truth, and build conviction and Christian character.

Passport

WordAction Publishing
2923 Troost Avenue
Kansas City, MO 64109

Assistant Editor: Kimberly Adams

DESCRIPTION AND INTERESTS
This colorful, upbeat Sunday School take-home supplement for pre-teens contains articles and activities that emphasize spirituality and holy living. Circ: 55,000.
Audience: 11–12 years
Frequency: Weekly
Website: www.wordaction.com

FREELANCE POTENTIAL
90% written by nonstaff writers. Publishes 30 freelance submissions yearly; 20% by unpublished writers, 20% by authors who are new to the magazine. Receives 240 queries and unsolicited mss yearly.

SUBMISSIONS
Query with author biography; or send complete ms. Accepts hard copy and email submissions to kdadams@wordaction.com. SASE. Responds in 4–6 weeks.

- Articles: Staff written.
- Depts/columns: "Survival Guide," 400–500 words; hot topics with spiritual applications. "Curiosity Island," 200–300 words; hobbies, fun activities, tips, and career ideas.
- Other: Puzzles, activities, and cartoons.

SAMPLE ISSUE
8 pages (no advertising): 1 article; 2 depts/columns; 2 activities; 1 cartoon. Sample copy, free with 5x7 SASE. Guidelines available.

- Sample dept/column: "Survival Guide" encourages children to use the gifts God has given them.
- Sample dept/column: "Curiosity Island" describes the history and wonder of the Pacific Ocean's Marianas Trench.

RIGHTS AND PAYMENT
Multi-use rights. "Survival Guide," $25. "Curiosity Island," spot cartoons, and puzzles, $15. Pays on publication. Provides 1 contributor's copy.

➻EDITOR'S COMMENTS
Submissions to our "Survival Guide" column do not have to relate directly to the curriculum's lesson of the day, but they should conform to the doctrine of the Church of the Nazarene. Submissions to our "Curiosity Island" column may be about any miscellaneous area of interest to pre-teens and are not required to be spiritual in nature.

Pediatrics for Parents

35 Starknaught Heights
Gloucester, MA 01930

Editor: Richard J. Sagall, M.D.

DESCRIPTION AND INTERESTS
With much of its content written by medical professionals, this newsletter serves as an informed and practical guide to child health and well-being, with an emphasis on preventive care. Circ: 140,000.
Audience: Parents
Frequency: Monthly
Website: www.pedsforparents.com

FREELANCE POTENTIAL
50% written by nonstaff writers. Publishes 30 freelance submissions yearly; 50% by unpublished writers, 50% by authors who are new to the magazine. Receives 50 queries and unsolicited mss yearly.

SUBMISSIONS
Query or send complete ms. Accepts hard copy and email submissions to articles@pedsforparents.com. SASE. Response time varies.

- Articles: 750–1,500 words. Informational articles. Topics include prevention, fitness, medical advances, new treatment options, wellness, and pregnancy.
- Depts/columns: Word lengths vary. New product information, and article reprints.

SAMPLE ISSUE
16 pages (no advertising): 12 articles; 2 depts/columns. Sample copy, $5. Writers' guidelines available.

- "Strep Throat." Article details the six main approaches to the diagnosis and treatment of this common childhood ailment.
- "Hiccups in Babies and Children." Article explains the causes, special concerns, and treatment of hiccups in youngsters.
- "Children and Food Allergies." Article discusses how food allergies develop, and how to cope with them.

RIGHTS AND PAYMENT
First rights. Written material, to $25. Pays on publication. Provides 3 contributor's copies and a 1-year subscription.

➻EDITOR'S COMMENTS
We want to provide parents with patient-friendly, medically sound information. Prospective writers should share the editorial goal to help make parents active and informed partners in their child's healthcare.

Piedmont Parent

P.O. Box 11740
Winston-Salem, NC 27116

Editor: Eve White

DESCRIPTION AND INTERESTS
Parents living in the Greensboro, Winston-Salem, and High Point triad of North Carolina read this "family resource," which offers regional coverage of child-related issues, events, and activities. Circ: 39,000.

Audience: Parents
Frequency: Monthly
Website: www.piedmontparent.com

FREELANCE POTENTIAL
50% written by nonstaff writers. Publishes 36–40 freelance submissions yearly; 25% by unpublished writers, 50% by authors who are new to the magazine. Receives 1,000+ queries and unsolicited mss yearly.

SUBMISSIONS
Query or send complete ms. Accepts email to editor@piedmontparent.com (Microsoft Word attachments) and simultaneous submissions if identified. Responds in 1–2 months.

- Articles: 500–1,200 words. Informational and how-to articles; interviews; and personal experience pieces. Topics include child development, day care, summer camps, gifted and special education, local and regional news, science, social issues, sports, popular culture, health, and travel.
- Depts/columns: 600–900 words. Family health and parenting news.
- Other: Family games and activities.

SAMPLE ISSUE
40 pages (47% advertising): 8 articles; 5 depts/columns. Sample copy, free with 9x12 SASE ($1.50 postage). Writers' guidelines and theme list available.

- "A Life-Changing Experience." Article tells of the author's volunteer work on behalf of abused and neglected children.
- Sample dept/column: "Food" features recipes for outdoor cooking.

RIGHTS AND PAYMENT
One-time rights. Written material, payment rates vary. Pays on publication. Provides 1 tearsheet.

☛EDITOR'S COMMENTS
Submissions from freelance writers are welcome, as long as they have local relevancy. While we demand exclusivity within our region, reprints of articles from publications outside our region are also considered.

Pikes Peak Parent

30 South Prospect Street
Colorado Springs, CO 80903

Editor: Lisa Carpenter

DESCRIPTION AND INTERESTS
Parents living in the Colorado Springs area pick up this free tabloid each month for pointers on raising, disciplining, feeding, educating, and—most importantly—entertaining their children. Circ: 30,000.

Audience: Parents
Frequency: Monthly
Website: www.pikespeakparent.com

FREELANCE POTENTIAL
5% written by nonstaff writers. Publishes 4 freelance submissions yearly; 2% by authors who are new to the magazine. Receives 60 queries yearly.

SUBMISSIONS
Query with writing samples. Accepts hard copy and email queries to parent@gazette.com. SASE. Response time varies.

- Articles: 800–1,500 words. Informational and how-to articles. Topics include regional news and resources, parenting issues, family life, travel, health, safety, sports, social issues, and recreation.
- Depts/columns: Word lengths vary. News, opinions, grandparenting, health, family issues, profiles, and events.

SAMPLE ISSUE
24 pages (50% advertising): 8 articles; 5 depts/columns. Sample copy, free with 9x12 SASE.

- "Fido, Meet Baby." Article tells how to help pets prepare for a newborn's arrival.
- "Star Gazing." Article recommends taking children to watch the U.S. Senior Open golf championship in Colorado Springs.
- Sample dept/column: "Tidbits" includes instructions for carving a turtle out of a watermelon, then serving fruit salad in it.

RIGHTS AND PAYMENT
All rights on assigned pieces; second rights on reprints and unsolicited pieces. Written material, payment rates vary. Pays on publication. Provides 1 contributor's copy.

☛EDITOR'S COMMENTS
We pride ourselves on providing relevant, reliable, and wide-ranging resources for families living in our area. Articles on general parenting subjects must have a Colorado angle, such as interviews with local professionals or a sidebar listing area services.

Pittsburgh Parent

Playground Magazine

P.O. Box 374
Bakerstown, PA 15007

Editor: Pat Poshard

DESCRIPTION AND INTERESTS

This regional parenting magazine aims to educate its readers with informative and upbeat articles on a variety of family-oriented topics. Circ: 50,000+.

Audience: Parents
Frequency: Monthly
Website: www.pittsburghparent.com

FREELANCE POTENTIAL

100% written by nonstaff writers. Publishes 150 freelance submissions yearly; 20% by authors who are new to the magazine. Receives 1,500 queries and mss yearly.

SUBMISSIONS

Query or send complete ms. Accepts hard copy, email submissions to editor@ pittsburghparent.com, and simultaneous submissions if identified. SASE. Response time varies.

- Articles: Cover story; 2,500–2,750 words. Other material; 400–900 words. Informational articles; profiles; and interviews. Topics include family issues, parenting, education, science, fitness, health, nature, college, computers, and multicultural subjects.
- Fiction: 1,000 words. Genres include mystery, adventure, and historical and multicultural fiction.
- Depts/columns: Word lengths vary. Education, teen issues, book reviews, and humor.
- Other: Submit seasonal material 3 months in advance.

SAMPLE ISSUE

44 pages (65% advertising): 10 articles; 5 depts/columns. Sample copy, free. Guidelines and editorial calendar available.

- "Are You Parenting the Next President?" Article discusses how parents can teach their children to become good leaders.
- "The Ups and Downs of Backyard Playgrounds." Article compares the different playground categories.

RIGHTS AND PAYMENT

First serial rights. Written material, payment rates vary. Pays 45 days after publication. Provides 1 tearsheet.

⇢ EDITOR'S COMMENTS

Submissions must have a local angle and include local resource information.

360 B Street
Idaho Falls, ID 83402

Editor: Shannon Amy Stockwell

DESCRIPTION AND INTERESTS

Previously listed as *Today's Playground Magazine*, this publication is directed toward those who install, own, or manage playgrounds. It covers regulations, structural issues, and materials of playscapes, skate parks, water parks, and others. It also includes information about playground safety and the importance of play. Circ: 35,000.

Audience: Adults
Frequency: 7 times each year
Website: www.playgroundmag.com

FREELANCE POTENTIAL

40% written by nonstaff writers. Publishes 14–20 freelance submissions yearly; 30% by authors who are new to the magazine.

SUBMISSIONS

Query or send ms. Accepts hard copy and email to shannon@playgroundmag.com. SASE. Responds in 1–2 months.

- Articles: 800–1,200 words. Informational and how-to articles. Topics include various types of play structures.
- Depts/columns: Word lengths vary. Legal issues, news, industry updates, company profiles, landscaping, and design.

SAMPLE ISSUE

46 pages: 3 articles; 9 depts/columns. Sample copy, $5. Guidelines available.

- "Wet and Wondrous! Enter the Aquatic Zone." Article profiles companies that create water parks, and provides some examples of their work.
- "Rebuilding Wildcat Park." Article reports on one community's efforts to replace its aging playground, incorporating the ideas of its children.
- Sample dept/column: "Spotlight" profiles Gummiwerk Kraiburg, a company that manufactures playground safety surfaces.

RIGHTS AND PAYMENT

First serial rights. Articles, $100–$300. Depts/columns, $50–$175. Payment policy varies.

⇢ EDITOR'S COMMENTS

We are interested in hearing from people with professional experience with playgrounds, whether it be through a parks and rec department, school, or playground company. We are always looking for new ideas for playgrounds.

Plays

The Drama Magazine for Young People

P.O. Box 600160
Newton, MA 02460

Editor: Elizabeth Preston

DESCRIPTION AND INTERESTS

Plays—one-acts, monologues, and skits—all appear in this magazine designed to provide material for school or children's theater groups. Plays for various age levels are included in each issue. Circ: 5,300.

Audience: 6–17 years
Frequency: 7 times each year
Website: www.playsmag.com

FREELANCE POTENTIAL

100% written by nonstaff writers. Publishes 75 freelance submissions yearly; 25% by unpublished writers, 50% by authors who are new to the magazine. Receives 250 queries and unsolicited mss yearly.

SUBMISSIONS

Query for adaptations of classics and folktales. Send complete ms for other material. Accepts hard copy. SASE. Responds to queries in 2 weeks, to mss in 1 month.

- Fiction: One-act plays for high school, to 5,000 words; for middle school, to 3,750 words; for elementary school, to 2,500 words. Also publishes skits, monologues, puppet plays, and dramatized classics. Genres include patriotic, historical, and biographical drama; mystery; melodrama; fairy tales and folktales; comedy; and farce.
- Other: Submit seasonal material 4 months in advance.

SAMPLE ISSUE

64 pages (5% advertising): 5 plays; 1 dramatized classic; 1 skit. Sample copy, free with 6x9 SASE ($.87 postage). Guidelines available.

- "Sibling Switch." One-act play for the middle and lower grades provides some interesting lessons in friendship, doled out by a modern fairy godmother.
- "Winning Combination." Play for junior and senior high schools shows the class clown finding his true talent when he judges a beauty contest.

RIGHTS AND PAYMENT

All rights. Written material, payment rates vary. Pays on acceptance. Provides 1 author's copy.

❖EDITOR'S COMMENTS

You must supply us with all production notes, list of characters, and information about the age group for which the play is written.

Pockets

The Upper Room
1908 Grand Avenue
P.O. Box 340004
Nashville, TN 37203-0004

Editor: Lynn W. Gilliam

DESCRIPTION AND INTERESTS

Each themed issue of this Christian magazine includes stories, articles, and activities that reinforce the truth of a loving God, and show young readers how to live their faith each day. Circ: 98,000.

Audience: 6–11 years
Frequency: 11 times each year
Website: www.pockets.org

FREELANCE POTENTIAL

97% written by nonstaff writers. Publishes 220 freelance submissions yearly; 20% by unpublished writers, 30% by authors who are new to the magazine. Receives 2,000 mss yearly.

SUBMISSIONS

Send complete ms. Accepts hard copy. SASE. Responds in 2 months.

- Articles: 400–1,000 words. Informational articles; profiles; and personal experience pieces. Topics include multicultural and community issues, and persons whose lives reflect their Christian commitment.
- Fiction: 600–1,400 words. Stories that demonstrate Christian values.
- Depts/columns: Word lengths vary. Scripture readings and lessons; recipes.
- Artwork: Color prints; digital images to 300 dpi.
- Other: Poetry. Puzzles, activities, games.

SAMPLE ISSUE

48 pages (no advertising): 1 article; 6 stories; 6 depts/columns; 10 activities; 3 poems. Sample copy, free with 9x12 SASE (4 first-class stamps). Guidelines and theme list available at website.

- "Nothing Special." Story shows how a girl learns to appreciate the many blessings in her life.
- "Boredom Busters." Story tells how a young boy changes his ungrateful attitude.

RIGHTS AND PAYMENT

First and second rights. Written material, $.14 per word. Poetry, $2 per line. Games, $25–$50. Pays on acceptance. Provides 3–5 copies.

❖EDITOR'S COMMENTS

We need more 600-word stories written for children ages five to seven. Because each issue revolves around a theme, writers should always consult our theme list.

Pocono Parent Magazine

P.O. Box 291
Analomink, PA 18320

Editor: Teri O'Brien

DESCRIPTION AND INTERESTS
Each issue of this regional publication distributed in the Pocono Mountains is packed with information on raising healthy, happy children. As "Monroe County's family resource," it presents reader-friendly feature stories on such topics as education and family activities, as well as a comprehensive calendar of events. Circ: 10,000.
Audience: Parents
Frequency: 6 times each year
Website: www.poconoparent.com

FREELANCE POTENTIAL
25% written by nonstaff writers. Publishes 20 freelance submissions yearly.

SUBMISSIONS
Send complete ms. Accepts email submissions to editor@poconoparent.com (no attachments). Response time varies.

- Articles: 750–1,000 words. Informational and how-to articles; profiles; interviews; reviews; photo-essays; and personal experience pieces. Topics include parenting, family issues, child care, education, social issues, current events, health, fitness, nature, the environment, recreation, and regional news.
- Depts/columns: Word lengths vary. Family finances, teen and tween issues, restaurant reviews, pets, and family fitness.

SAMPLE ISSUE
40 pages: 6 articles; 13 depts/columns. Sample copy and guidelines available at website.

- "Beyond Being a Father." Essay recognizes the men who have eagerly embraced the role of grandfather.
- "Summer Is a Hot Time to Practice Writing Skills." Article recommends encouraging children to record summer memories.
- Sample dept/column: "Teen & Tween" features the opinions of two girls, ages 17 and 12, on the subject of understanding boys.

RIGHTS AND PAYMENT
Rights vary. Written material, payment rates vary. Payment policy varies. Provides 2 contributor's copies.

➥EDITOR'S COMMENTS
Being a parent makes you an automatic expert on parenting issues. If you have other expertise, you're a step ahead of other freelancers.

Pointe

110 William Street, 23rd Floor
New York, NY 10038

Managing Editor: Jocelyn Anderson

DESCRIPTION AND INTERESTS
Pointe provides practical and timely information about the world of ballet. It is read by professional dancers, students, artistic directors, and choreographers, among others. Circ: 40,000.
Audience: All ages
Frequency: 6 times each year
Website: www.pointemagazine.com

FREELANCE POTENTIAL
75% written by nonstaff writers. Publishes 1–2 freelance submissions yearly; 10% by unpublished writers, 25% by authors who are new to the magazine. Receives 12 queries each year.

SUBMISSIONS
Query. Accepts hard copy. SASE. Responds in 2 months.

- Articles: 1,200 words. Informational articles; profiles; interviews; personal experience pieces; and photo essays. Topics include ballet companies, dancers, choreographers, news, trends, festivals and events, premieres, and auditions.
- Depts/columns: 800–1,000 words. Premieres, news, interviews with directors, profiles of dancers and companies, reviews, advice, and tips on technique.
- Artwork: B/W and color prints and transparencies; digital photos. Line art.

SAMPLE ISSUE
80 pages (50% advertising): 5 articles; 14 depts/columns. Sample copy and guidelines available.

- "Oklahoma Rising." Article describes the high standards and great repertoire of the Tulsa Ballet.
- Sample dept/column: "Dancer Profile" explains how Kathleen Thielhelm found early success at The Joffrey Ballet.

RIGHTS AND PAYMENT
All rights. Written material, payment rates vary. Pays on acceptance. Provides 2 author's copies.

➥EDITOR'S COMMENTS
We are always looking for writers with a strong background in and knowledge of ballet. Currently, we are seeking informative articles on new ideas that would be of interest to our teenage readers.

Popcorn Magazine for Children

8320 Brookfield Road
Richmond, VA 23227

Editor: Charlene Warner Coleman

DESCRIPTION AND INTERESTS
This magazine for children is available in many forms—as a print magazine, an interactive website, and as the *Popcorn Magazine* television program. It is filled with informative articles, fiction, poetry, and activities of interest to its target age group. Circ: 250,000.
Audience: 5–13 years
Frequency: 6 times each year
Website: www.popcornville.com

FREELANCE POTENTIAL
75% written by nonstaff writers. Publishes 500+ freelance submissions yearly. Receives many unsolicited mss yearly.

SUBMISSIONS
Send complete ms. Accepts hard copy and email submissions to cwcoleman1@ comcast.net. SASE. Response time varies.

- Articles: Word lengths vary. Informational and how-to articles; and profiles. Topics include sports, art, science, travel, fashion, cooking, technology, nature, and animals.
- Fiction: To 800 words. Genres include humor, fantasy, and adventure.
- Depts/columns: 500 words. Arts and crafts, fashion, cooking, building projects, product reviews, and book and movie reviews.
- Other: Puzzles and games. Poetry.

SAMPLE ISSUE
30 pages: 4 articles; 3 stories; 6 depts/ columns; 2 poems. Guidelines available.

- "Hammer & Nail." Article explains how to build a dog house that will truly elevate a dog's standing in the canine community.
- Sample dept/column: "Book, CD, and Video Game Reviews" offers a review of *Misty of Chincoteague*, which tells the story of a special colt caught up in the Pony Penning Day on Chincoteague, Virginia.

RIGHTS AND PAYMENT
Rights vary. No payment.

◆ EDITOR'S COMMENTS
Our number-one priority when it comes to submissions is that they be engaging. We will publish material on a variety of topics, but it must resonate with our readership's age group and be able to translate well into other mediums. If you're still in doubt, ask yourself, "Is this piece fun?"

Prehistoric Times

145 Bayline Circle
Folsom, CA 95630-8077

Editor: Mike Fredericks

DESCRIPTION AND INTERESTS
Enjoyed by teen and adult dinosaur enthusiasts, *Prehistoric Times* is full of informational and scientific articles on paleontology and other topics related to prehistoric life. Updates on collectibles and artwork are also found in its pages. Circ: Unavailable.
Audience: YA–Adult
Frequency: Quarterly
Website: www.prehistorictimes.com

FREELANCE POTENTIAL
30% written by nonstaff writers. Publishes 20+ freelance submissions yearly; 75% by unpublished writers, 75% by authors who are new to the magazine. Receives 24+ unsolicited mss each year.

SUBMISSIONS
Send complete ms. Accepts email submissions to pretimes@comcast.net (attach file). Response time varies.

- Articles: 1,500–2,000 words. Informational articles. Topics include dinosaurs, paleontology, prehistoric life, drawing dinosaurs, and dinosaur-related collectibles.
- Depts/columns: Word lengths vary. Field news, dinosaur models, media reviews, interviews, and in-depth descriptions of dinosaurs and other prehistoric species.

SAMPLE ISSUE
60 pages (30% advertising): 8 articles; 15 depts/columns. Sample copy, $7. Guidelines available via email to pretimes@comcast.net.

- "Professor Earl Douglass." Article profiles an early paleontologist and his accomplishments.
- "Edmontosaurus." Article offers an in-depth look at this dinosaur with details on fossil discovery and research.
- Sample dept/column: "Painting Thylacoleo" describes how to paint a dinosaur model.

RIGHTS AND PAYMENT
All rights. Written material, payment rates and policy vary. Provides contributor's copies.

◆ EDITOR'S COMMENTS
Writers with a strong knowledge of dinosaurs and prehistoric life are invited to submit material on the latest information in the field. We are always in need of interviews with scientists, artists, and makers and collectors of dinosaur models.

Prep Traveler

621 Plainfield Road, Suite 406
Willowbrook, IL 60527

Editorial Coordinator: Randy Mink

DESCRIPTION AND INTERESTS
Designed to meet the needs of the youth travel
market, *Prep Traveler* focuses on youth sports
and performance travel groups. It is filled with
information they need to know to plan their
trips, such as where to stay, where to perform
(or play), and destination ideas. Circ: 26,000.
Audience: Youth travel planners
Frequency: Twice each year
Website: www.preptraveler.com

FREELANCE POTENTIAL
10% written by nonstaff writers. Publishes 2–3
freelance submissions yearly; 50% by unpublished writers. Receives 50 queries yearly.

SUBMISSIONS
Query with résumé and field experience.
Accepts email queries to randy@ptmgroups.com
(no attachments). Responds in 1 week.

- Articles: Word lengths vary. Informational
  and how-to articles; and profiles. Topics
  include travel destinations, music, health,
  and fitness.
- Depts/columns: 700–2,000 words.
  Information related to youth travel planning.

SAMPLE ISSUE
46 pages (40% advertising): 2 articles; 8
depts/columns. Writers' guidelines and theme
list available.

- "Volunteers: A Key Ingredient for Success."
  Article discusses getting volunteers involved
  in helping host a large activity or competition.
- "Where to Perform: The Art of Getting There."
  Article tackles the problem many school and
  youth groups face of rising travel and performance fees.
- Sample dept/column: "Regional Resources:
  Southeast" describes lodging and recreation
  resources in various parts of the southeast-
  ern U.S.

RIGHTS AND PAYMENT
All rights. Written material, payment rates vary.
Pays on publication. Provides author's copies.

➼EDITOR'S COMMENTS
We welcome fresh ideas from people who
have experience in planning travel for youth
groups. Whether it's a marching band, travel
soccer team, or drama group—if you've done
it successfully and can write about it in an
informative way, we'd like to hear from you.

Preschool Playhouse

Urban Ministries
1551 Regency Court
Calumet City, IL 60409

Senior Editor: Dr. Rosa Sailes

DESCRIPTION AND INTERESTS
Published by Urban Ministries, this take-home
story paper augments church school lessons. It
features Christ-centered and Bible-based stories and activities designed to teach African
American preschoolers that Jesus loves them,
and it teaches them how Bible lessons can
apply to their lives. Circ: 50,000.
Audience: 2–5 years
Frequency: Quarterly, with weekly take-
   home sections
Website: www.urbanministries.com

FREELANCE POTENTIAL
25% written by nonstaff writers. Publishes 12
freelance submissions yearly; 10% by unpublished writers, 25% by authors who are new to
the magazine.

SUBMISSIONS
All material is written on assignment. Send
résumé with clips or writing samples. Accepts
hard copy. SASE. Response time varies.

- Articles: Word lengths vary. Informational,
  life-application stories for children; and
  Bible stories.
- Other: Bible-based activities.

SAMPLE ISSUE
4 pages (no advertising): 1 contemporary story;
1 Bible story; 2 activities; 2 memory verses.
Sample copy, free. Guidelines available.

- "Two Brothers." Bible story tells of brothers
  Cain and Abel to illustrate that we must
  show our love for God as well as for our
  brothers and sisters.
- "Growing." Activity encourages children to
  do something for their brother or sister to
  show their love.

RIGHTS AND PAYMENT
All rights. Written material, payment rates
vary. Pays on publication. Provides 1 copy.

➼EDITOR'S COMMENTS
We look for writers who know the Bible well
and who have experience writing curriculum
or teaching, either in a church or a school.
Experience with this age group is preferred,
but not required. Although all of the material
is written on assignment, we encourage you
to contact us with your credentials if you
have the ability to present Bible concepts in
age-appropriate ways.

PresenTense Magazine

214 Sullivan Street, Suite 2A
New York, NY 10012

Editor: Ariel Beery

DESCRIPTION AND INTERESTS
PresenTense Magazine calls itself "a grassroots, volunteer effort by hundreds of young Jews spread across four continents." Its editorial mission is to define what it means to be Jewish, and to share how being Jewish adds value to readers' lives. Circ: 30,000.

Audience: YA–Adult
Frequency: 3 times each year
Website: www.presentensemagazine.org

FREELANCE POTENTIAL
80% written by nonstaff writers. Publishes 40 freelance submissions yearly; 40% by unpublished writers, 70% by new authors.

SUBMISSIONS
Query or send complete ms. Accepts email to editor@presentensemagazine.org (include submission type, title, and word count in subject line, and attach document) and simultaneous submissions if identified. Responds to queries in 2 weeks, to mss in 2 months.

- Articles: Features, 800–1,200 words. Profiles, 600–700 words. Sidebars, 50–250 words. Informational articles and profiles. Topics include Judaism, the Diaspora, Zionism, Israel, activism, community, education, the environment, ethnic issues, relationships, religion, and health.
- Fiction: 1,000–3,000 words. Genres vary.
- Depts/columns: "Paradigm Shift" essay, 1,200–2,000 words. "Around the World," 600–800 words. News and pop-culture briefs, 400–500 words. Reviews, 300–800 words.
- Artwork: Digital files of photos, line art, paintings, and cartoons at 200–300 dpi.
- Other: Photo-essays; 12 photos, 600 words. Poetry, to 300 words.

SAMPLE ISSUE
64 pages: 8 articles; 28 depts/columns. Sample copy and guidelines available at website.

- "Zionism A–Z." Article lists terms that have made a palpable impact on today's Zionism.
- Sample dept/column: "Around the World" describes Jewish matchmaking in India.

RIGHTS AND PAYMENT
First rights. No payment. Provides 3 copies.

☛EDITOR'S COMMENTS
Please see our guidelines for a list of department editors to pitch.

Primary Street

Urban Ministries
1551 Regency Court
Calumet City, IL 60409

Senior Editor: Dr. Rosa Sailes

DESCRIPTION AND INTERESTS
Scripture lessons, Bible stories, and Jesus-based activities are all found in this take-home story paper. It is written for African American primary school students as a complement to their regular Sunday school lessons. Its goal is to teach Christian values and a love of Jesus Christ. A teacher's guide is included. Circ: 50,000.

Audience: 6–8 years
Frequency: Quarterly
Website: www.urbanministries.com

FREELANCE POTENTIAL
25% written by nonstaff writers. Publishes 15 freelance submissions yearly; 25% by unpublished writers, 25% by authors who are new to the magazine. Receives 180 queries yearly.

SUBMISSIONS
All material is assigned. Query with résumé and writing samples. Accepts hard copy. Response time varies.

- Articles: Word lengths vary. Informational and how-to articles; personal experience pieces; photo-essays; and Bible stories. Topics include religion, Christian values, nature, the environment, animals, pets, crafts, hobbies, African history, multicultural subjects, regional news, and social issues.
- Other: Bible verses, activities, puzzles, and games.

SAMPLE ISSUE
4 pages (no advertising): 1 article; 1 story; 1 memory verse; 2 activities. Sample copy, free. Guidelines available.

- "God's Throne Room." Story tells of the apostle John's visit to heaven, during which he saw the angels bowing to God as He sat on His throne, worthy of their worship.
- Sample dept/column: "Good Attitude" is an activity that has students using words from a memory verse to complete a cartoon.

RIGHTS AND PAYMENT
All rights. Written material, payment rates vary. Pays on publication. Provides 1 author's copy.

☛EDITOR'S COMMENTS
We need writers who know the Bible and have experience teaching in a church or school setting. We may assign you a story if you have to ability to present Bible concepts in a way that will appeal to this age group.

Primary Treasure

Pacific Press Publishing
P.O. Box 5353
Nampa, ID 83653-5353

Editor: Aileen Andres Sox

DESCRIPTION AND INTERESTS
Primary Treasure is written for elementary-age students attending Sabbath school at the Seventh-day Adventist Church. It is filled with true-to-life stories that teach Christian values, beliefs, and practices. Circ: 250,000.

Audience: 6–9 years
Frequency: Weekly
Website: www.primarytreasure.com

FREELANCE POTENTIAL
10% written by nonstaff writers. Publishes 52 freelance submissions yearly; 10% by unpublished writers, 30% by authors who are new to the magazine. Receives 240 mss yearly.

SUBMISSIONS
Query for serials. Send complete ms for other submissions. Accepts hard copy, email submissions to ailsox@pacificpress.com, and simultaneous submissions if identified. SASE. Responds in 4 months.

- Articles: 600–1,000 words. True stories about children in Christian settings; true, problem-solving pieces that help children learn about themselves in relation to God and others. Material must be consistent with Seventh-day Adventist beliefs and practices.
- Other: Submit seasonal material 7 months in advance.

SAMPLE ISSUE
16 pages (no advertising): 4 articles; 1 Bible lesson; 1 puzzle. Sample copy, free with 9x12 SASE (2 first-class stamps). Writers' guidelines available.

- "Catching Bowling Balls with Your Teeth." Story tells of a boy who learns not to give in to peer pressure.
- "Christopher Visits Grandpa." Story describes a young boy learning the joys of spending time with his grandfather.
- "A 'Home Run' for Paul." Story describes how a class accepts a special-needs student.

RIGHTS AND PAYMENT
One-time rights. Written material, $25–$50. Pays on acceptance. Provides 3 copies.

➤EDITOR'S COMMENTS
We are interested in stories that are age-appropriate with topics of interest to kids. Material must help children understand and develop a relationship with God.

Principal

1615 Duke Street
Alexandria, VA 22314

Editor: Raven Padgett

DESCRIPTION AND INTERESTS
The National Association of Elementary School Principals publishes this magazine for educators interested in the latest news concerning school administration policies and educational issues. Circ: 36,000.

Audience: K–8 school administrators
Frequency: 5 times each year
Website: www.naesp.org

FREELANCE POTENTIAL
90% written by nonstaff writers. Publishes 20 freelance submissions yearly; 80% by authors who are new to the magazine. Receives 84–96 unsolicited mss yearly.

SUBMISSIONS
Send complete ms. Accepts hard copy, PC-compatible disk submissions, and email submissions to publications@naesp.org. No simultaneous submissions. SASE. Responds in 6 weeks.

- Articles: 1,000–2,500 words. Informational and instructional articles; profiles; and opinion and personal experience pieces. Topics include elementary education, gifted and special education, parenting, mentoring, and technology.
- Depts/columns: 750–1,500 words. "Parents & Schools," "It's the Law," "Practitioner's Corner," "Tech Support," "A Touch of Humor," "Principal's Bookshelf," "Ten to Teen," "The Reflective Principal," and "Speaking Out."

SAMPLE ISSUE
72 pages (25% advertising): 9 articles; 9 depts/columns. Sample copy, $8. Guidelines and theme list available.

- "Kindergarten Readiness: A Challenge." Article suggests new guidelines for assessing students entering kindergarten.
- Sample dept/column: "Practitioner's Corner" details a one-of-a-kind school-business alliance that pairs students with runners participating in the Boston Marathon.

RIGHTS AND PAYMENT
All North American serial rights. No payment. Provides 3 contributor's copies.

➤EDITOR'S COMMENTS
Keep in mind that we are not a journal; please avoid jargon and clarify all technical terms.

Queens Parent

1040 Avenue of the Americas, 4th Floor
New York, NY 10018

Editor-in-Chief: Helen Freedman

DESCRIPTION AND INTERESTS
Filled with articles on parenting, childhood
health, and education issues, *Queens Parent*
also provides readers with information on local
recreation and family activities. It covers local
current events, and provides a comprehensive
calendar of events. Circ: 285,000.
Audience: Parents
Frequency: Monthly
Website: www.nymetroparents.com

FREELANCE POTENTIAL
50% written by nonstaff writers. Publishes
300 freelance submissions yearly. Receives
12 queries, 72 unsolicited mss yearly.

SUBMISSIONS
Query or send complete ms. Accepts hard
copy and email submissions to hfreedman@
davlermedia.com. SASE. Responds in 1 week.

- Articles: 800–1,000 words. Informational
  articles; profiles; interviews; and personal
  experience pieces. Topics include family
  issues, health, nutrition, fitness, current
  events, and regional news.
- Depts/columns: Staff written
- Other: Submit seasonal material 4 months
  in advance.

SAMPLE ISSUE
78 pages: 12 articles; 9 depts/columns.
Sample copy, free with 10x13 SASE.

- "Camp '08: What's New." Article provides
  tips and advice for parents choosing a camp
  program for their children.
- "Ski Bunny Bulletin." Article tells where to
  find the best deals and the newest resort
  additions in the region's ski areas.
- "Get Milk!" Article discusses recent studies
  that promote the health benefits of drinking
  milk instead of soda or fruit juice.

RIGHTS AND PAYMENT
First New York area rights. No payment.

➽EDITOR'S COMMENTS
We generally do not pay for freelance work.
We will, occasionally, pay for the few pieces
that we assign. If you live in Queens and
have experience with children's or parenting
issues, we'd like to hear your ideas. We're
open to working with new writers; getting
published in our pages provides a great way
to build up your clip file.

Rainbow Kids

P.O. Box 202
Harvey, LA 70059

Editor: Martha Osborne

DESCRIPTION AND INTERESTS
This e-zine is an adoption advocacy website
that contains information and resource material
for families involved in any step of the adop-
tion process. Hits per month: 1.5 million.
Audience: Adoptive families
Frequency: Updated monthly
Website: www.rainbowkids.com

FREELANCE POTENTIAL
80% written by nonstaff writers. Publishes 40
freelance submissions yearly; 50% by authors
who are new to the magazine.

SUBMISSIONS
Query or send complete ms. Accepts hard
copy and email submissions to martha@
rainbowkids.com (attach file). SASE. Responds
in 2–3 days.

- Articles: Word lengths vary. Informational
  articles; and personal experience pieces.
  Topics include all matters related to adoption and adoptive families, both domestic
  and foreign. Also publishes adoption guidelines, adoption events, and photo listings.

SAMPLE ISSUE
Sample issue and writers' guidelines available
at website.

- "Adopting from a Disruption." Article shares
  the experience of a single woman who chose
  to adopt an older girl from Bulgaria.
- "Digging Up Roots." Article explains the pros
  and cons of families searching for the foreign birth families of their adoptive children.
- "Perfect Adoptive Parents." Article provides
  advice for adoptive parents who are faced
  with the challenge of broaching the subject
  of adoption and differing races and cultures
  within the adoptive family.

RIGHTS AND PAYMENT
Electronic rights. No payment.

➽EDITOR'S COMMENTS
We serve as a central resource for people
looking for information on the adoption
process, as well as a support system for
people who are just beginning, are in the
middle of, or have recently completed adoption. As such, we require that writers have
personal experience with adoption, either as
a parent, a child, or as a professional working with adoptive families.

Rainbow Rumpus

P.O. Box 6881
Minneapolis, MN 55406

Editor-in-Chief: Beth Wallace

DESCRIPTION AND INTERESTS
Rainbow Rumpus is an online magazine for youth with parents who are lesbian, gay, bisexual, or transgender. It features articles, fiction, and poetry celebrating the diversity of such families and their communities. Hits per month: 9,000.
Audience: 4–18 years
Frequency: Monthly
Website: www.rainbowrumpus.org

FREELANCE POTENTIAL
60% written by nonstaff writers. Publishes 24 freelance submissions yearly. Receives 12 unsolicited mss yearly.

SUBMISSIONS
Send complete ms. Accepts email submissions to fictionandpoetry@rainbowrumpus.org (Microsoft Word or RTF attachments) with "Submission" in subject line. Responds in 2 weeks.

- Articles: Staff written.
- Fiction: 800–2,500 words for children 4–12 years. To 5,000 words for 12–18 years. Publishes most genres.
- Depts/columns: Staff written.
- Other: Poetry, no line limit.

SAMPLE ISSUE
Sample copy and writers' guidelines available at website.

- "Big Deal." Story tells of a boy who, although too small to win a rope-climbing competition, finds meaning as a team manager.
- "The Great Burkes." Story follows a boy whose clairvoyant mother packs him, his senile grandmother, and a clothes-wearing dog into a circus caravan after falling in love with a performer.

RIGHTS AND PAYMENT
First North American electronic rights and anthology rights. Fiction, $75. Poetry, payment rates vary. Pays on publication.

➡EDITOR'S COMMENTS
Stories should be written from the point of view of children with LGBT parents or other family members. The primary plot point of the story should *not* be family structure, or children being teased about their families. Stories featuring families of color, bisexual parents, transgender parents, and mixed-race families are particularly welcome.

Raising Arizona Kids

7000 East Shea Boulevard, Suite 1470
Scottsdale, AZ 85254-5257

Assistant Editor: Mary Holden

DESCRIPTION AND INTERESTS
Geared toward Arizona parents of children from birth through high school, this magazine provides a multitude of ideas about parenting, childcare, health, nutrition, and family lifestyles. Circ: Unavailable.
Audience: Parents
Frequency: Monthly
Website: www.raisingarizonakids.com

FREELANCE POTENTIAL
65% written by nonstaff writers. Publishes 12 freelance submissions yearly; 1% by unpublished writers, 1% by authors who are new to the magazine. Receives 44 queries yearly.

SUBMISSIONS
Query with clips. Accepts hard copy and email queries to maryh@raisingarizonakids.com. SASE. Response time varies.

- Articles: 1,000–2,000 words. Informational and how-to articles; profiles; interviews; personal experience pieces; and photo-essays. Topics include parenting issues, children, health and fitness, college and careers, current events, education, social issues, travel, and recreation.
- Depts/columns: Word lengths vary. News, parenting issues, family matters, profiles, sports, food, nutrition.
- Other: Journal articles, 500 words.

SAMPLE ISSUE
46 pages: 2 articles; 6 depts/columns; 1 journal article. Sample copy and guidelines available at website.

- "Emergency! Big Help for Little People." Article profiles the pediatric services of a local hospital, and offers tips for parents about being prepared for an emergency.
- Sample dept/column: "Ages & Stages" explains how a cooperative preschool works.

RIGHTS AND PAYMENT
Rights vary. Articles, $125. Depts/columns, payment rates vary. Journal articles, $100. Pays 30 days after publication.

➡EDITOR'S COMMENTS
Our magazine is intended to foster networking and support among Arizona parents and professionals who are dedicated to quality parenting. Articles should quote Arizona-based sources.

Ranger Rick

National Wildlife Federation
1100 Wildlife Center Drive
Reston, VA 20190-5362

Editor: Mary Dalheim

DESCRIPTION AND INTERESTS
Published by the National Wildlife Federation, *Ranger Rick* has been inspiring a love of nature in children for generations. It includes articles, stories, and activities to engage young minds. Circ: 560,000.

Audience: 7–12 years
Frequency: Monthly
Website: www.nwf.org/gowild

FREELANCE POTENTIAL
10% written by nonstaff writers. Publishes 1–2 freelance submissions yearly; 1% by authors who are new to the magazine.

SUBMISSIONS
Send résumé and clips. No queries or unsolicited mss.

- Articles: 900 words. Informational articles. Topics include nature, animals, insects, dinosaurs, oceanography, the environment, outdoor adventure, and crafts.
- Fiction: 900 words. Genres include mystery, adventure, fantasy, fables, and stories about the natural world.
- Depts/columns: Staff written.
- Other: Puzzles and activities with nature themes.

SAMPLE ISSUE
40 pages (no advertising): 4 articles; 2 stories; 10 activities. Sample copy and guidelines available at website.

- "A Great American Bike Ride." Article follows a group of kids on a bike ride across Iowa.
- "Fun on Two Wheels." Article suggests ways for kids to pursue adventure on their bikes.
- "A Berry Good Time." Article provides tips for berry-picking.
- "Ladybug Lore." Article shares five fun facts about everyone's favorite insect.

RIGHTS AND PAYMENT
Rights vary. Written material, payment rates vary. Pays on acceptance. Provides 2 copies.

➥EDITOR'S COMMENTS
Due to the bulk of editorial material we have on hand, we are not reviewing queries or unsolicited manuscripts at this time. Please check our guidelines, which are posted at our website, for the most current information concerning freelance submissions.

The Reading Teacher

International Reading Association
800 Barksdale Road
P.O. Box 8139
Newark, DE 19714-8139

Executive Editor: Sandra Patterson

DESCRIPTION AND INTERESTS
A professional journal serving reading teachers and other professionals involved in literacy, *The Reading Teacher* is a research-based information pipeline that features solutions to transform reading-challenged students and thereby transform their lives. Circ: 57,500.

Audience: Literacy educators
Frequency: 8 times each year
Website: www.reading.org

FREELANCE POTENTIAL
95% written by nonstaff writers. Publishes 50 freelance submissions yearly; 20% by unpublished writers, 30% by authors who are new to the magazine. Receives 300 unsolicited mss each year.

SUBMISSIONS
Send complete ms. Accepts online submissions via www.mc.manuscriptcentral.com/rt. Responds in 1–2 months.

- Articles: To 6,000 words. Informational and how-to articles; profiles; and personal experience pieces. Topics include literacy, reading education, instructional techniques, classroom strategies, reading research, and educational technology.
- Depts/columns: 1,500–2,500 words. Reviews of children's books, teaching tips, and material on cultural diversity.

SAMPLE ISSUE
86 pages (17% advertising): 5 articles; 5 depts/columns. Sample copy, $10. Writers' guidelines available.

- "The Shape of Direct Quotation." Article explains how understanding of direct quotation fits into the comprehension of the story.
- Sample dept/column: "Teaching Tips" presents the idea of using a descriptive video as a tool to enhance students' writing.

RIGHTS AND PAYMENT
All rights. No payment. Provides 5 contributor's copies for articles, 2 copies for depts/columns.

➥EDITOR'S COMMENTS
All articles in our magazine are edited by leading professionals in the field, peer reviewed, and produced by a professional publication staff. Authors will receive careful evaluation of their work and detailed advice regarding how it can be best presented.

Reading Today

International Reading Association (IRA)
800 Barksdale Road
P.O. Box 8139
Newark, DE 19714-8139

Editor-in-Chief: John Micklos, Jr.

DESCRIPTION AND INTERESTS
A tabloid, *Reading Today* addresses the needs and interests of reading educators and specialists at all grade levels. It covers new programs, people in the field, and reading-related conferences. Circ: 82,000.

Audience: IRA members
Frequency: 6 times each year
Website: www.reading.org

FREELANCE POTENTIAL
50% written by nonstaff writers. Publishes 30 freelance submissions yearly; 10% by unpublished writers, 25% by authors who are new to the magazine. Receives 300 queries, 180 unsolicited mss yearly.

SUBMISSIONS
Prefers query; will accept complete ms. Prefers email submissions to readingtoday@reading.org; will accept hard copy and simultaneous submissions if identified. SASE. Responds in 1 month.

- Articles: 500–1,500 words. Informational and factual articles; and interviews. Topics include reading and reading education, community programs, staffing, assessment, program funding, and censorship.
- Depts/columns: To 750 words. News, book reviews, education policy information, ideas for administrators.
- Artwork: B/W and color prints. Line art.

SAMPLE ISSUE
48 pages (30% advertising): 9 articles; 11 depts/columns. Sample copy, $6. Writers' guidelines available.

- "Response to Intervention: Hot, Getting Hotter." Article analyzes the growing acceptance of RTI and its implementation in early intervention strategies.
- Sample dept/column: "Ideas for Administrators" showcases successful methods schools have used to engage student reading.

RIGHTS AND PAYMENT
All rights. Written material, $.20–$.30 per word. Pays on acceptance. Provides 3 copies.

⇢ EDITOR'S COMMENTS
We are in need of features on innovative reading promotion programs at various school levels.

Relate Magazine

1601 Parkway Drive
Findlay, OH 45840

Submissions: Mary Bowman

DESCRIPTION AND INTERESTS
Relate Magazine inspires teen girls to pursue their dreams, and teaches them to be an example for others by their life, love, faith, and purity. Its content has a Christian perspective, but never preaches. Circ: 10,000.

Audience: Girls, 15–19 years
Frequency: Quarterly
Website: www.relatemag.com

FREELANCE POTENTIAL
80% written by nonstaff writers. Publishes 60 freelance submissions yearly; 10% by unpublished writers, 80% by authors who are new to the magazine. Receives 360 queries yearly.

SUBMISSIONS
Query with outline and opening paragraph. Accepts email queries to mary@relatemag.com. Response time varies.

- Articles: 800–1,800 words. Informational articles; profiles; interviews; and personal experience pieces. Topics include college, careers, self-help, personal development, faith, social issues, design, fashion, beauty, and entertainment.
- Depts/columns: 200–800 words. Entertainment, college, careers, faith, relationships, and trends.

SAMPLE ISSUE
42 pages; 1 article; 10 depts/columns. Guidelines available at website.

- "Trendy." Article offers a look at the latest back-to-school fashions for teen girls in a photo-essay format.
- Sample dept/column: "You Can Do Anything" relates the author's passion for writing and how her confidence keeps her going.
- Sample dept/column: "The Ins and Outs of Cliques" advises readers to focus on their schoolwork rather than popularity.

RIGHTS AND PAYMENT
Rights vary. Written material, $50–$700. Payment policy varies.

⇢ EDITOR'S COMMENTS
We are interested in publishing more full-length articles, and encourage submissions from freelance writers. We would especially like to see stories about real teens who are designers, volunteers, and entrepreneurs, as well as more articles on interior design.

Research in Middle Level Education Online

Portland State University
Graduate School of Education
615 SW Harrison
Portland, OR 97201

Editor: Micki M. Caskey

DESCRIPTION AND INTERESTS
This international, peer-reviewed research journal is published by the National Middle School Association. It contains quantitative and qualitative studies, case studies, action research studies, research syntheses, integrative reviews, and interpretations of research literature—all pertaining to middle-grade education. Hits per month: 30,000.

Audience: Educators
Frequency: Twice each year
Website: www.nmsa.org

FREELANCE POTENTIAL
100% written by nonstaff writers. Publishes 23–25 freelance submissions yearly. Receives 72 unsolicited mss yearly.

SUBMISSIONS
Send complete ms with 150- to 200-word abstract. Accepts email submissions to caskeym@pdx.edu (Microsoft Word attachments). Responds in 1 week.

- Articles: 7,000–12,000 words. Informational articles; quantitative and qualitative studies; case studies; action research studies; research syntheses; integrative reviews; and interpretations of research literature—all pertaining to middle-grade education.

SAMPLE ISSUE
Sample articles and writers' guidelines available at website.

- "Orienting to the Public Good: Developing a Moral Self in the Middle Grades." Article describes a study of the extent to which schools prepare young adolescents to commit themselves to serve the public interest.
- "Factors Associated with Perceived Parental Academic Monitoring in a Population of Low-Income, African American Young Adolescents." Study of incoming urban sixth-graders helps explain why some urban youth succeed academically while others do not.

RIGHTS AND PAYMENT
All rights. No payment. Provides 1 copy.

⇢EDITOR'S COMMENTS
Aside from a separate title page with author name, affiliation, and contact information, manuscripts should have no reference to the author(s) to ensure a blind review. Please follow APA style guidelines.

Reunions Magazine

P.O. Box 11727
Milwaukee, WI 53211-0727

Editor: Edith Wagner

DESCRIPTION AND INTERESTS
This magazine is a valuable source of information and inspiration for individuals planning reunions of all kinds. It also features accounts of successful reunions. Circ: 20,000.

Audience: Adults
Frequency: 5 times each year
Website: www.reunionsmag.com

FREELANCE POTENTIAL
75% written by nonstaff writers. Publishes 100 freelance submissions yearly; 60% by unpublished writers, 80% by authors who are new to the magazine.

SUBMISSIONS
Query with outline; or send complete ms. Prefers email submissions to reunions@execpc.com (Microsoft Word attachments); will accept hard copy. SASE. Responds in 12–18 months.

- Articles: Word lengths vary. Informational, factual, and how-to articles; profiles; and personal experience pieces. Topics include organizing reunions, choosing locations, entertainment, activities, and genealogy.
- Depts/columns: 250–1,000 words. Opinion and personal experience pieces; and resource information.
- Artwork: Color digital images at 300 dpi or higher.
- Other: Recipes, games, cartoons, and filler.

SAMPLE ISSUE
54 pages (45% advertising): 6 articles; 10 depts/columns. Sample copy, $3. Guidelines and editorial calendar available.

- "Nashville: A Reunion Kind of Town." Article explains why the Music City is a great place to hold a reunion.
- Sample dept/column: "Alum & I" profiles the Alumni Association of an Oregon high school.

RIGHTS AND PAYMENT
One-time and electronic rights. Written material, payment rates vary. Payment policy varies. Provides 2 contributor's copies.

⇢EDITOR'S COMMENTS
As a reader-driven publication, we seek submissions that answer our readers' questions about reunion locations, activities, and materials. We welcome field-tested advice and the most up-to-date information.

Richmond Parents Monthly

5511 Staples Mill Road, Suite 103
Richmond, VA 23228

Editor: Angela Lehman-Rios

DESCRIPTION AND INTERESTS
Providing parenting and family life information
and resources to parents in the Richmond
area, this magazine also includes local recre-
ation and entertainment news. Circ: 30,000.
Audience: Parents
Frequency: Monthly
Website: www.richmondparents.com

FREELANCE POTENTIAL
75% written by nonstaff writers. Publishes
50–60 freelance submissions yearly; 5% by
authors who are new to the magazine.
Receives 600 queries yearly.

SUBMISSIONS
Query. Accepts email queries to mail@
richmondpublishing.com. No simultaneous
submissions. Availability of artwork improves
chance of acceptance. Responds in 1–3 weeks.

- Articles: 600–1,000 words. Informational
  and self-help articles. Topics include the
  arts, camps for children, pets, home and
  garden, birthday parties, school, education,
  children's health, and holidays.
- Depts/columns: Word lengths vary. Family-
  related news, media reviews, technology.
- Artwork: Color prints and transparencies.

SAMPLE ISSUE
30 pages (15% advertising): 4 articles; 8
depts/columns; 1 calendar; 1 contest. Sample
copy, free. Writers' guidelines and editorial
calendar available.

- "You Are Getting Verrry Sleeepy." Article discusses children's sleep needs and offers tips
  for solving bedtime and sleep problems.
- "When Parenting Secrets Get Revealed."
  Article shares a woman's sleep-deprived
  journey through parenting.
- Sample dept/column: "Family Connection"
  profiles a local couple's fostering and adoption experience.

RIGHTS AND PAYMENT
One-time rights. Written material, $52–$295
based on page length. Pays on publication.

➡EDITOR'S COMMENTS
We will accept family travel pieces from
writers outside of our local area. Our goal for
the magazine is to facilitate a shared sense
of community among parents and offer a
forum for dialogue.

Sacramento Parent

457 Grass Valley Highway, Suite 5
Auburn, CA 95603

Editor-in-Chief: Shelly Bokman

DESCRIPTION AND INTERESTS
Published by a group of mothers who hold
master's degrees in child development,
Sacramento Parent offers sound, practical par-
enting advice made simple and entertaining. It
is read by parents and grandparents in greater
Sacramento and the Sierra foothills of
California. Circ: 50,000.
Audience: Parents
Frequency: Monthly
Website: www.sacramentoparent.com

FREELANCE POTENTIAL
75% written by nonstaff writers. Publishes 50
freelance submissions yearly; 10% by unpublished writers, 25% by authors who are new to
the magazine. Receives 780 queries yearly.

SUBMISSIONS
Query with writing samples. Accepts email
queries to shelly@sacramentoparent.com.
Response time varies.

- Articles: 700–1,000 words. Informational and
  how-to articles; personal experience pieces;
  and humor. Topics include parenting, health,
  fitness, finance, family travel, education,
  grandparenting, adoption, sports, recreation,
  learning disabilities, and regional news.
- Depts/columns: 300–500 words. Child development, opinions, and hometown highlights.
- Other: Submit seasonal or themed material
  3 months in advance.

SAMPLE ISSUE
50 pages (50% advertising): 7 articles; 6 depts/
columns. Sample copy, free with 9x12 SASE
($1.29 postage). Guidelines and theme list
available at website.

- "Ten Ways to Raise College-Bound Kids."
  Article provides tips for keeping teens moti-
  vated to make the most of high school.
- Sample dept/column: "Let's Go" recommends taking a "Fauxcation" to a nearby
  hotel for the weekend.

RIGHTS AND PAYMENT
Second rights. Articles, $50. Depts/columns,
$25–$40. Pays on publication. Provides contributor's copies.

➡EDITOR'S COMMENTS
Story ideas, outlines, or excerpts from articles may be submitted, as well as photos
that may accompany a manuscript.

San Diego Family Magazine

P.O. Box 23960
San Diego, CA 92193

Publisher/Editor-in-Chief: Sharon Bay

DESCRIPTION AND INTERESTS

This comprehensive regional publication covers a myriad of things parents need to know about raising a family in the San Diego area. Circ: 120,000.

Audience: Parents
Frequency: Monthly
Website: www.sandiegofamily.com

FREELANCE POTENTIAL

90% written by nonstaff writers. Publishes 50 freelance submissions yearly; 50% by unpublished writers. Receives 360–600 queries and unsolicited mss yearly.

SUBMISSIONS

Query or send complete ms. Accepts hard copy. SASE. Responds in 1 month.

- Articles: 800–1,200 words. Informational, self-help, and how-to articles. Topics include parenting, pregnancy, childbirth, child care, education, summer camp, health, safety, nutrition, gardening, dining out, recreation, travel, sports, family finance, local events, and multicultural issues. Also publishes humorous essays.
- Depts/columns: Word lengths vary. News briefs, tips, trends, restaurant reviews, book reviews, cooking, and gardening.
- Artwork: 3x5 or 5x7 four-color glossy prints.

SAMPLE ISSUE

178 pages (60% advertising): 24 articles; 16 depts/columns. Sample copy, $4.50 with 9x12 SASE ($1 postage). Guidelines available at website.

- "Meeting Your Summer Childcare Needs." Article provides guidance in deciding between day camp or home care.
- Sample dept/column: "Family Dining" reviews a local vegetarian restaurant.

RIGHTS AND PAYMENT

First or second rights and all regional rights. Written material, $1.25 per column inch. Pays on publication. Provides 1 contributor's copy.

➦ EDITOR'S COMMENTS

Always give your story a headline, and localize it within the first couple of paragraphs. When mentioning parents or professionals, always give the city and state where they live or work. Strive for interviews and resources with a San Diego County focus.

Santa Barbara Family Life

P.O. Box 4867
Santa Barbara, CA 93140

Editor: Nansie Chapman

DESCRIPTION AND INTERESTS

Residents of Santa Barbara and its surrounding counties turn to this magazine for the latest local information on recreation, arts, health, and education. Circ: 60,000.

Audience: Parents
Frequency: Monthly
Website: www.sbfamilylife.com

FREELANCE POTENTIAL

30% written by nonstaff writers. Publishes 5–15 freelance submissions yearly; 5% by unpublished writers, 10% by authors who are new to the magazine. Receives 240 queries each year.

SUBMISSIONS

Query or send complete ms. Accepts email to nansie@sbfamilylife.com. Responds only if interested.

- Articles: 500–1,200 words. Informational articles; profiles; photo-essays; and personal experience pieces. Topics include regional events and activities, parenting, family life, education, recreation, crafts, hobbies, and current events.
- Depts/columns: Word lengths vary. Love and relationships, arts and entertainment, and health issues.
- Other: Puzzles and activities.

SAMPLE ISSUE

40 pages: 6 articles; 8 depts/columns; 3 puzzles. Sample copy and writers' guidelines available at website.

- "Making a Difference Through Mentoring." Article describes a regional mentoring program and includes how to get involved.
- Sample dept/column: "Health Watch" discusses ways for parents to become involved in what their children eat at school.

RIGHTS AND PAYMENT

Rights vary. Written material, $25–$35. Payment policy varies.

➦ EDITOR'S COMMENTS

Our goal is to act as a resource that educates and entertains our readers, while connecting families to members, events, activities, and businesses within our community. We are always in need of articles pertaining to family health issues, education, and camps for children.

Scholastic Choices

Scholastic Inc.
557 Broadway
New York, NY 10012-3999

Editor: Bob Hugel

DESCRIPTION AND INTERESTS
This "life skills magazine" for young adults is distributed through schools as a tool for discussing the challenges faced by today's teens. It tackles such issues as peer pressure, personal responsibility, substance abuse, self-esteem, and nutrition. Circ: 200,000.

Audience: 12–18 years
Frequency: 6 times each year
Website: www.scholastic.com

FREELANCE POTENTIAL
90% written by nonstaff writers. Publishes 30–40 freelance submissions yearly; 10% by unpublished writers. Receives 60 queries, 60 unsolicited mss yearly.

SUBMISSIONS
Query or send complete ms. Accepts hard copy and email submissions to choicesmag@scholastic.com. SASE. Responds to queries in 2 months, to mss in 3 months.

- Articles: 500–1,000 words. Informational and self-help articles; profiles; and personal experience pieces. Topics include health, nutrition, fitness, sports, personal development, personal responsibility, family issues, relationships, safety, social issues, conservation, the environment, popular culture, careers, and substance abuse prevention.
- Depts/columns: Staff written.
- Other: Quizzes, word games, and recipes.

SAMPLE ISSUE
32 pages (20% advertising): 5 articles; 7 depts/columns. Sample copy, free with 9x12 SASE. Guidelines and editorial calendar available.

- "I Live with My Dad." Article describes the experience of a 16-year-old girl who lives with her divorced father.
- "Messed Up On Meth." Article emphasizes the dangers of methamphetamine.
- "Color Code." Article recommends choosing fruits and vegetables by their colors to achieve a well-balanced diet.

RIGHTS AND PAYMENT
All rights. Written material, payment rates vary. Pays on publication. Provides 10 copies.

⊶EDITOR'S COMMENTS
We seek articles that will help young adults learn essential living skills and deal successfully with life's challenges.

Scholastic DynaMath

Scholastic Inc.
557 Broadway, Room 4052
New York, NY 10012-3999

Editor: Matt Friedman

DESCRIPTION AND INTERESTS
Committed to helping teachers introduce math concepts to students in grades three through six, *Scholastic DynaMath* is filled with creative stories and activities. Each issue is designed to entertain while illustrating how math fits into everyday life. Circ: 200,000.

Audience: 8–11 years
Frequency: 8 times each year
Website: www.scholastic.com/dynamath

FREELANCE POTENTIAL
10% written by nonstaff writers. Publishes 5 freelance submissions yearly; 25% by unpublished writers, 25% by authors who are new to the magazine. Receives 48 queries and unsolicited mss yearly.

SUBMISSIONS
Query with outline/synopsis; or send complete ms. Accepts hard copy and simultaneous submissions. SASE. Responds in 1–2 months.

- Articles: To 600 words. Informational articles about math skills. Topics include critical thinking, chart and graph reading, measurement, addition, subtraction, fractions, division, problem solving, interdisciplinary issues, sports, and consumer awareness.
- Other: Filler, puzzles, games, and jokes. Submit holiday material 4–6 months in advance.

SAMPLE ISSUE
16 pages (15% advertising): 7 articles; 5 activities. Sample copy, $4 with 9x12 SASE. Guidelines and editorial calendar available.

- "Math On the Job: Finding Averages." Article explains how TV meteorologist Janice Huff figures averages for her forecasts.
- "A Look Back at MLK." Article works backward to look at the accomplishments of Dr. Martin Luther King, Jr.

RIGHTS AND PAYMENT
All rights. Articles, $250–$400. Puzzles, $25–$50. Pays on acceptance. Provides 3 copies.

⊶EDITOR'S COMMENTS
We strive to motivate students to learn curricular math, prepare them for standardized testing, *and* make them laugh. If you have an idea for a feature that uses math in a real-world application—particularly a humorous one—we'd like to see it.

Scholastic Math Magazine

Scholastic Inc.
557 Broadway
New York, NY 10012-3999

Editor: Jack Silbert

DESCRIPTION AND INTERESTS

This take-home paper for math students in grades six through nine shows how math is connected to real life and can be—gulp!—fun. It blends students' interest in pop culture, celebrities, TV, and music with math problems. Circ: 200,000.

Audience: 11–15 years
Frequency: Monthly
Website: www.scholastic.com/classmags

FREELANCE POTENTIAL

30% written by nonstaff writers. Publishes 10 freelance submissions yearly; 10% by unpublished writers. Receives 24 queries yearly.

SUBMISSIONS

Query. Accepts hard copy. SASE. Responds in 2–3 months.

- Articles: 600 words. Informational articles. Topics include real-world math, consumer math, math-related news, teen issues, sports, celebrities, TV, music, movies, and current events.
- Depts/columns: 140 words. Skill-building exercises, quizzes, and practice tests.
- Other: Puzzles, activities, comic strips, Q&As, and mystery photos.

SAMPLE ISSUE

16 pages (no advertising): 6 articles; 2 depts/columns; 1 comic strip. Sample copy, free with 9x12 SASE (3 first-class stamps). Guidelines and editorial calendar available.

- "America's Doggy Dough." Article explains the cost of owning a dog, and provides problems for readers to calculate expenditures.
- "Kick Off the Season." Article explains the statistics that are used to calculate football kickers' total points.
- Sample dept/column: "Fast Math" presents a math problem that asks readers to determine the birthday of one of the Jonas Brothers.

RIGHTS AND PAYMENT

All rights. Articles, $300+. Depts/columns, $35. Pays on publication.

➔EDITOR'S COMMENTS

All articles must present math problems to be solved. The problems may be mixed throughout the article or may be listed at end, but the problems must be related to the subject of the article.

Scholastic Parent & Child

Scholastic Inc.
557 Broadway
New York, NY 10012-3999

Assistant Editor: Samantha Brody

DESCRIPTION AND INTERESTS

The mission of Scholastic Parent & Child is to help parents keep pace with their children's developmental growth both at home and at school. Its articles help parents understand their children's various stages, as well as provide some fun and imaginative parenting advice. Circ: 1.2 million.

Audience: Parents
Frequency: 8 times each year
Website: www.parentandchildonline.com

FREELANCE POTENTIAL

90% written by nonstaff writers. Publishes 20 freelance submissions yearly; 90% by unpublished writers, 10% by authors who are new to the magazine. Receives 144 queries and unsolicited mss yearly.

SUBMISSIONS

Query or send complete ms. Accepts hard copy. SASE. Responds to queries in 3 months, to mss in 2 months.

- Articles: 500–1,000 words. Informational articles and interviews. Topics include child development, education, and parenting.
- Depts/columns: Word lengths vary. Literacy, health, parent/teacher relationships, arts and crafts, child development, product reviews, travel, cooking, and family issues.

SAMPLE ISSUE

92 pages (33% advertising): 6 articles; 15 depts/columns. Sample copy, $2.95. Writers' guidelines available.

- "Get In the Game." Article explains the right way to encourage and support a young athlete without demanding too much.
- "Into the Wild." Article presents the many developmental advantages to outside play, and provides tips for fostering children's natural love of nature.
- Sample dept/column: "First Steps" explains why babies bite and offers tips for getting them to stop.

RIGHTS AND PAYMENT

All rights. Written material, payment rates vary. Pays on publication. Provides author's copies.

➔EDITOR'S COMMENTS

We accept submissions on a variety of topics. Each article, however, must be fun and lively as well as informative.

Scholastic Scope

Scholastic Inc.
557 Broadway
New York, NY 10012-3999

Associate Editor: Lisa Feder-Feitel

DESCRIPTION AND INTERESTS

Subtitled "Reading, Writing, and Reality for Teens," this publication is dedicated to instilling in teenagers a love of writing and the written word. It contains articles on writing as well as classroom plays, interviews, and pop culture articles, all designed to increase reading and writing skills. Circ: 550,000.

Audience: 12–18 years
Frequency: 17 times each year
Website: www.scholastic.com/scope

FREELANCE POTENTIAL

30% written by nonstaff writers. Of the freelance submissions published yearly, 2% is by unpublished writers, 10% is by authors who are new to the magazine. Receives 200–300 queries yearly.

SUBMISSIONS

Query with résumé, outline/synopsis, and clips. Accepts hard copy. SASE. Response time varies.

- Articles: 1,000 words. News and features that appeal to teens; and profiles of young adults who have overcome obstacles, performed heroic acts, or had interesting experiences.
- Fiction: 1,500 words. Contemporary, realistic stories about relationships and family problems, school issues, and other teen concerns; and science fiction.
- Depts/columns: Staff written.
- Other: Puzzles and activities. Submit seasonal material 4 months in advance.

SAMPLE ISSUE

22 pages (8% advertising): 4 articles; 2 stories; 1 puzzle. Sample copy, $1.75 with 9x12 SASE (2 first-class stamps); also available at website.

- "Interview: Spielberg Speaks." Article provides an interview with director Steven Spielberg about *Indiana Jones* and filmmaking.
- "Readers Theater Play." Classroom play is based on *The Chronicles of Narnia*.

RIGHTS AND PAYMENT

Rights vary. Written material, $100+. Pays on acceptance. Provides contributor's copies upon request.

❖EDITOR'S COMMENTS

Everything we publish is designed to increase reading and comprehension skills in addition to capturing teens' attention.

School Arts

Davis Publications
2223 Parkside Drive
Denton, TX 76201

Editor: Nancy Walkup

DESCRIPTION AND INTERESTS

Art teachers read this magazine for the project ideas, teaching strategies, and classroom inspiration it provides. Each issue is divided according to grade level to better provide grade-appropriate material. It also publishes articles on teacher development topics. Circ: 20,000.

Audience: Art teachers, grades K–12
Frequency: 9 times each year
Website: www.schoolartsonline.com

FREELANCE POTENTIAL

75% written by nonstaff writers. Publishes 200 freelance submissions yearly; 60% by unpublished writers, 60% by authors who are new to the magazine. Receives 300 unsolicited mss each year.

SUBMISSIONS

Send complete ms with artwork. Prefers disk submissions with images; will accept email submissions to nwalkup@verizon.net (attach files). Responds in 1–2 months.

- Articles: 300–800 words. Informational, how-to, and self-help articles. Topics include teaching art; techniques; art history; projects and activities; curriculum development; and art programs for the gifted, handicapped, and learning-disabled student.
- Depts/columns: 500–1,200 words. Crafts, new product reviews, and opinion pieces.
- Artwork: B/W and color prints and slides; high-quality digital images. Line art.

SAMPLE ISSUE

84 pages (40% advertising): 18 articles; 6 depts/columns. Sample copy, $5. Guidelines and editorial calendar available at website.

- "Printmaking on a Budget." Article explains how to make print repeat cards using homemade stamps.
- Sample dept/column: "Meeting Individual Needs" offers ideas for teaching weaving to students at varying developmental stages.

RIGHTS AND PAYMENT

First serial rights. Written material, $25–$150. Artwork, payment rates vary. Pays on publication. Provides 6 contributor's copies.

❖EDITOR'S COMMENTS

Check our website for a list of upcoming themes. Most of our writers are, or were, art teachers themselves.

The School Librarian's Workshop

1 Deerfield Court
Basking Ridge, NJ 07920

Editor: Ruth Toor

DESCRIPTION AND INTERESTS
School librarians and media specialists read this professional journal for its discussions of fiction, nonfiction, and poetry books. It also provides literacy units and lessons for use in elementary through high school, and articles on professional development subjects. Circ: 7,000.

Audience: School librarians
Frequency: 6 times each year
Web: www.school-librarians-workshop.com

FREELANCE POTENTIAL
10% written by nonstaff writers. Publishes 20 freelance submissions yearly; 5–10% by unpublished writers. Receives 24 unsolicited mss yearly.

SUBMISSIONS
Send 2 copies of complete ms. Prefers disk submissions (Microsoft Word); will accept hard copy. SASE. Responds in 3 weeks.

- Articles: To 1,000 words. Informational, how-to, and practical application articles; profiles; and interviews. Topics include librarianship, literature, special education, ethnic studies, computers, technology, social and multicultural issues, and the environment.
- Artwork: Line art.
- Other: Submit seasonal material 8 months in advance.

SAMPLE ISSUE
24 pages (no advertising): 15 articles. Sample copy, free with 9x12 SASE. Guidelines and theme list available at website.

- "Winds of Change." Article recommends several books about learning and education that librarians should read and discuss with their administrators.
- "Dreaming Big." Article explains how to set up a literacy unit for Black History Month.
- "Ubiquitous Computing." Article reports on a new initiative known as "one-to-one laptops," which will enhance the ability of technology to advance learning.

RIGHTS AND PAYMENT
First rights. No payment. Provides 3 copies.

➥EDITOR'S COMMENTS
Although we are a professional periodical, we do not like articles that are laden with industry jargon. Present practical ideas in an easy, conversational tone.

School Library Journal

360 Park Avenue South
New York, NY 10010

Executive Editor: Rick Margolis

DESCRIPTION AND INTERESTS
Providing librarians with up-to-date information on technology and trends, this journal features multi-media and book reviews in each issue. Articles cover library services and management, and issues pertaining to children's and young adult literature. Circ: 34,500.

Audience: School librarians
Frequency: Monthly
Website: www.slj.com

FREELANCE POTENTIAL
80% written by nonstaff writers. Publishes 25 freelance submissions yearly; 60% by unpublished writers, 60% by authors who are new to the magazine. Receives 48–72 mss yearly.

SUBMISSIONS
Query or send complete ms. Accepts disk submissions (ASCII or Microsoft Word) and email submissions to rmargolis@reedbusiness.com. SASE. Responds to queries in 1 month, to mss in 3 months.

- Articles: 1,500–2,500 words. Informational articles and interviews. Topics include children's and young adult literature, school library management, and library careers.
- Depts/columns: 1,500–2,500 words. Book and media reviews, descriptions of successful library programs, and opinion pieces.
- Artwork: Color prints. Color tables, charts, and cartoons.

SAMPLE ISSUE
188 pages (25% advertising): 5 articles; 15 depts/columns; 3 review sections. Sample copy, $6.75. Guidelines available at website.

- "Face the Facts." Article contends that the subject of slavery is taught incorrectly to young people.
- "Test Drive: Neuros OSD." Article reviews this open-source media recorder/player.

RIGHTS AND PAYMENT
First rights. Articles, $400. Depts/columns, $100–$200. Pays on publication. Provides 4 contributor's copies.

➥EDITOR'S COMMENTS
We are the world's largest reviewer of books and multi-media technology for children and young adults. Our goal is to provide librarians with the most up-to-date information on the industry and book selections.

School Library Media Activities Monthly

3520 South 35th Street
Lincoln, NE 68506

Managing Editor: Deborah Levitov

DESCRIPTION AND INTERESTS
This magazine provides school librarians and media specialists with ideas for developing and reinforcing information literacy skills, teaching research strategies, and introducing literature to children. It also includes complete, collaborative lessons and reproducible pages in all subject areas. Circ: 12,000.

Audience: School library and media specialists, grades K–8
Frequency: 10 times each year
Website: www.schoollibrarymedia.com

FREELANCE POTENTIAL
90% written by nonstaff writers. Publishes 30 freelance submissions yearly; 20% by unpublished writers, 30% by authors who are new to the magazine. Receives 36 queries, 36 unsolicited mss yearly.

SUBMISSIONS
Query or send complete ms with bibliographic citations and brief author vita. Accepts email to deborah.levitov@lu.com (Microsoft Word attachments). Responds in 2 months.

- Articles: 1,000–1,500 words. Informational and factual articles. Topics include media education and promotion, information technology, integration of curriculum materials, and library management.
- Depts/columns: Word lengths vary. Activities, lesson plans, tips for professional growth.
- Artwork: B/W prints. Line drawings.

SAMPLE ISSUE
58 pages: 4 articles; 16 depts/columns. Sample copy and guidelines available at website.

- "One Indivisible Day." Article summarizes a symposium dedicated to teaching social justice to children through literature.
- Sample dept/column: "Notes From the Field" explores how chess clubs can be a valuable part of library services.

RIGHTS AND PAYMENT
All rights. Written material, payment rates vary. Pays on publication. Provides 3+ copies.

➤EDITOR'S COMMENTS
We would like to receive more lesson plans for kindergarten through grade 12. We welcome unsolicited manuscripts and contributions to all columns and departments, as well as unsolicited feature articles.

The School Magazine

Private Bag 3
Ryde, New South Wales 2112
Australia

Editor: Suzanne Eggins

DESCRIPTION AND INTERESTS
This magazine combines poetry, fiction, and factual articles for children in each of its four editions (each one targeting a different reading and comprehension level). Designed for the Australian school system, the editions are also available to individual subscribers across the globe. Circ: 150,000.

Audience: 8–12 years
Frequency: 10 times each year
Website: www.curriculumsupport. education.nsw.gov.au/services/ schoolmagazine

FREELANCE POTENTIAL
80% written by nonstaff writers. Publishes 100 freelance submissions yearly; 10% by unpublished writers, 10% by authors who arc new to the magazine. Receives 240 queries and unsolicited mss each year.

SUBMISSIONS
Query for nonfiction. Send complete ms for fiction. Accepts hard copy. SAE/IRC. Responds in 6–8 weeks.

- Articles: 800–2,000 words. Informational articles. Topics include nature, pets, the environment, history, biography, science, technology, and multicultural and ethnic issues.
- Fiction: 800–2,000 words. Genres include adventure; humor; fantasy; science fiction; horror; mystery; folktales; and contemporary, multicultural, and historical fiction.
- Depts/columns: Staff written.

SAMPLE ISSUE
34 pages (no advertising): 1 article; 4 stories; 3 depts/columns; 4 poems. Sample copy, free with IRC ($2 postage). Writers' guidelines available at website.

- "Pearl: Treasure of the Mollusc." Article explains the facts about oysters and pearls.
- "My New Pet." Story is a heartwarming tale of a farm boy and a new turkey chick.

RIGHTS AND PAYMENT
One-time serial rights. Written material, $270 (Australian) per 1,000 words. Poetry, payment rates vary. Pays on acceptance. Provides 2 contributor's copies.

➤EDITOR'S COMMENTS
We would like to receive plays that children can read and perform at school.

Science Activities

Heldref Publications
1319 18th Street NW
Washington, DC 20036-1802

Managing Editor

DESCRIPTION AND INTERESTS
Science Activities is a source of experiments, projects, and curriculum ideas for today's classroom science teacher. The material is appropriate for elementary to advanced high school students. Circ: 1,286.

Audience: Teachers, grades K–12
Frequency: Quarterly
Website: www.heldref.org

FREELANCE POTENTIAL
95% written by nonstaff writers. Publishes 25 freelance submissions yearly; 25% by unpublished writers, 50% by authors who are new to the magazine. Receives 60 queries, 48 unsolicited mss yearly.

SUBMISSIONS
Query or send complete ms. Accepts email to SA@heldref.org. Responds in 3 months.

- Articles: Word lengths vary. Informational and how-to articles. Topics include behavioral, biological, chemical, Earth, environmental, physical, and technological science.
- Depts/columns: Word lengths vary. News, "Classroom Aids," and book reviews.
- Artwork: B/W prints and slides. Line art and diagrams.

SAMPLE ISSUE
80 pages (1% advertising): 6 articles; 2 depts/columns. Sample copy available via email request to SA@heldref.org. Guidelines available.

- "How to Create and Use Analogies Effectively in the Teaching of Science Concepts." Article explains how science teachers use analogies to organize knowledge in student memories.
- "Estimating Cloud Cover." Article provides an activity to help students understand percentage of cloud cover.
- Sample dept/column: "Classroom Aids" recommends a new series of science games.

RIGHTS AND PAYMENT
All rights. No payment. Provides contributors with free access to articles online.

➔ EDITOR'S COMMENTS
Most of the accepted submissions are from high school science teachers and professors of education, although we do consider innovative, classroom-tested ideas from other writers. All material must be original, current, and adequately referenced.

Science and Children

National Science Teachers Association
1840 Wilson Boulevard
Arlington, VA 22201-3000

Managing Editor: Monica Zerry

DESCRIPTION AND INTERESTS
Science and Children serves as a forum of shared ideas for elementary-level science teachers from around the country. Peer-reviewed articles showcasing examples of activities, partnerships, and programs in today's science education are found in each issue. Circ: 18,000.

Audience: Science teachers, preK–grade 8
Frequency: 9 times each year
Website: www.nsta.org/elementaryschool

FREELANCE POTENTIAL
99% written by nonstaff writers. Publishes 25 freelance submissions yearly; 95% by unpublished writers, 50% by authors who are new to the magazine. Receives 360 unsolicited mss each year.

SUBMISSIONS
Accepts submissions from practicing educators only. Send complete ms. Accepts email submissions to msrs.nsta.org. Responds in 6 months.

- Articles: To 1,500 words. Informational and how-to articles; profiles; interviews; personal experience pieces; and reviews. Topics include science education, teacher training and techniques, staff development, classroom activities, astronomy, biology, chemistry, physics, and Earth science.
- Depts/columns: To 1,500 words. "Helpful Hints" and "In the Schools," to 500 words.
- Other: Submit seasonal material 1 year in advance.

SAMPLE ISSUE
76 pages (2% advertising): 7 articles; 8 depts/columns; 2 calendars. Sample copy, free. Guidelines available at website.

- "A Garden of Learning." Article explains a fourth-grade project of cultivating a Native Plant Learning Garden.
- "Going Places with Books." Article offers summer reading suggestions for teachers' professional development.

RIGHTS AND PAYMENT
All rights. No payment. Provides 5 copies.

➔ EDITOR'S COMMENTS
We are looking for original and creative ideas. Articles should include complete details of the experience you are sharing.

The Science Teacher

National Science Teacher Association
1840 Wilson Boulevard
Arlington, VA 22201-3000

Managing Editor: Meghan Sullivan

DESCRIPTION AND INTERESTS
High school science teachers look to this magazine for an exchange of ideas on classroom activities, teaching techniques, and scientific research. It is published by the National Science Teacher Association. Circ: 29,000.

Audience: Science educators, grades 7–12
Frequency: 9 times each year
Website: www.nsta.org/highschool

FREELANCE POTENTIAL
100% written by nonstaff writers. Of the freelance submissions published yearly, 70% are by unpublished writers and 50% are by authors who are new to the magazine. Receives 360 unsolicited mss yearly.

SUBMISSIONS
Send complete ms. Accepts email submissions to msullivan@nsta.org. Responds in 1 month.

- Articles: 2,000 words. Informational articles; classroom projects; and experiments. Topics include science education, biology, Earth science, computers, social issues, space, technology, and sports medicine.
- Depts/columns: 500 words. Science updates, association news, science careers.
- Artwork: 5x7 or larger B/W glossy prints. Line art, tables, and diagrams.

SAMPLE ISSUE
84 pages (40% advertising): 6 articles; 6 depts/columns. Sample copy, $4.25. Guidelines available.

- "Celebrate Summer with Reading." Article suggests a list of summer-reading books for personal enrichment and enjoyment.
- "Supporting Beginning Science Teachers." Article discusses strategies used to assist new science teachers.
- Sample dept/column: "Idea Bank" describes an effective technique that helps students learn science vocabulary.

RIGHTS AND PAYMENT
First rights. No payment. Provides copies.

➻EDITOR'S COMMENTS
We are always in need of material that describes new and creative ideas for the secondary science classroom. Your submission should provide worthwhile ideas and practical help for teachers. The magazine's "Call for Papers" section includes our themes.

Science Weekly

2141 Industrial Parkway, Suite 202
Silver Spring, MD 20904

Publisher: Dr. Claude Mayberry

DESCRIPTION AND INTERESTS
This magazine provides hands-on activities and workbook exercises to make learning science fun. *Science Weekly* is published in six different reading levels. Circ: 200,000.

Audience: Grades K–6
Frequency: 14 times each year
Website: www.scienceweekly.com

FREELANCE POTENTIAL
100% written by nonstaff writers. Of the freelance submissions published yearly, 5% is by authors who are new to the magazine. Receives 12 queries yearly.

SUBMISSIONS
Send résumé only. All work is assigned to writers in the District of Columbia, Maryland, or Virginia. Accepts hard copy. SASE. Response time varies.

- Articles: Word lengths vary. Informational articles. Topics include space exploration, ecology, the environment, nature, biology, the human body, meteorology, ocean science, navigation, nutrition, photography, physical science, and secret codes.
- Other: Theme-related puzzles, games, and activities.

SAMPLE ISSUE
4 pages (no advertising): 1 article; 6 activities. Sample copy, theme list, and writers' guidelines available.

- "Sharks." Article explains the nature of sharks, what they eat, where they live, and whether or not they are endangered.
- "Weekly Lab." Hands-on activity uses tomato juice and water to illustrate the varying odor levels human can detect.
- "Where in the World?" Workbook activity tests students' knowledge of geography as they attempt to pinpoint regions prone to shark attacks.

RIGHTS AND PAYMENT
All rights. All material, payment rates vary. Pays on publication.

➻EDITOR'S COMMENTS
All writing for us is done on assignment only. We work from a list of contributing writers who live in the greater-Washington, DC, area. Send us your résumé if you would like to be considered.

Scouting

Boy Scouts of America
1325 West Walnut Hill Lane
P.O. Box 152079
Irving, TX 75015-2079

Managing Editor: Scott Daniels

DESCRIPTION AND INTERESTS
Volunteer and professional Boy Scout leaders read this magazine for information on successful programs, activities, and leadership training techniques. It also features profiles of outstanding leaders and material aimed at strengthening families. Circ: 1 million.
Audience: Scout leaders and parents
Frequency: 6 times each year
Website: www.scoutingmagazine.org

FREELANCE POTENTIAL
80% written by nonstaff writers. Publishes 4–6 freelance submissions yearly; 5–10% by authors who are new to the magazine. Receives 150 queries yearly.

SUBMISSIONS
Query with outline. Accepts hard copy. SASE. Responds in 3 weeks.

- Articles: 500–1,200 words. Informational and how-to articles; personal experience pieces; profiles; and humor. Topics include scout programs, leadership, volunteering, nature, social issues and trends, and history.
- Depts/columns: 500–700 words. Family activities, outdoor activities, short profiles, scouting news.
- Other: Quizzes, puzzles, and games.

SAMPLE ISSUE
56 pages (33% advertising): 6 articles; 8 depts/columns. Sample copy, $2.50 with 9x12 SASE. Guidelines available at website.

- "Tracking Trinkets and Treasure." Article explains geocaching, a GPS-based hide-and-seek adventure that takes searchers through the woods and beyond.
- Sample dept/column: "Outdoor Smarts" offers basic information for keeping safe in the sun.

RIGHTS AND PAYMENT
First North American serial rights. Written material, $300–$800. Pays on acceptance. Provides 2 contributor's copies.

⟶EDITOR'S COMMENTS
We look for articles that are a mixture of information, instruction, and inspiration, designed to strengthen readers' abilities to perform their leadership roles in Scouting and also to assist them as parents in strengthening families.

Seattle's Child

511 Second Avenue West
Seattle, WA 98119

Managing Editor: Liz Gillespie

DESCRIPTION AND INTERESTS
Every article in *Seattle's Child* has a local angle, whether it be parenting advice from area experts or coverage of family-friendly events in the region. Founded by four young mothers in the late 1970s, the magazine is now part of the Washington Post Company— yet it retains its dedication to the Seattle community. Circ: 80,000.
Audience: Parents
Frequency: Monthly
Website: www.seattleschild.com

FREELANCE POTENTIAL
80% written by nonstaff writers. Publishes 30 freelance submissions yearly; 10% by unpublished writers, 25% by authors who are new to the magazine. Receives 120+ queries yearly.

SUBMISSIONS
Query with outline. Accepts hard copy, email queries to editor@seattleschild.com, and simultaneous submissions if identified. SASE. Responds in 1 month.

- Articles: Word lengths vary. Informational and how-to articles; and personal experience pieces. Topics include family, parenting, and social issues; health, fitness, and nutrition; regional news; family travel; and recreation.
- Depts/columns: Word lengths vary. Profiles, cooking, and media reviews.

SAMPLE ISSUE
50 pages (30% advertising): 3 articles; 3 depts/columns. Sample copy, $3 with 9x12 SASE. Guidelines and theme list available.

- "Protect Children's Eyes This Summer." Article recommends safety eyewear for children who play sports, and wide-brimmed hats and UV-filtering sunglasses for all children.
- Sample dept/column: "The Full Plate" provides information on backyard gardening and Community Supported Agriculture.

RIGHTS AND PAYMENT
Rights vary. Written material, $100–$450. Pays 30 days after publication. Provides 2 contributor's copies.

⟶EDITOR'S COMMENTS
We pride ourselves on a tradition of good journalism, community involvement, and service. Articles should inspire, inform, enlighten, and entertain our readers.

Seek

Standard Publishing Company
8805 Governor's Hill Drive, Suite 400
Cincinnati, OH 45249

Editor: Margaret K. Williams

DESCRIPTION AND INTERESTS
A pamphlet for use in Sunday school, *Seek* presents inspirational stories and Bible lessons in an effort to provide Christian-based guidance for young adults living in a modern world. Circ: 27,000.

Audience: YA–Adult
Frequency: Weekly
Website: www.standardpub.com

FREELANCE POTENTIAL
85% written by nonstaff writers. Publishes 150–200 freelance submissions yearly; 80% by authors who are new to the magazine.

SUBMISSIONS
Send complete ms. Prefers email to seek@standardpub.com; will accept hard copy. No simultaneous submissions. SASE. Responds in 3–6 months.

- Articles: 500–1,200 words. Inspirational, devotional, and personal experience pieces. Topics include religious and contemporary issues, Christian living, coping with moral and ethical dilemmas, and controversial subjects.
- Fiction: 500–1,200 words. Stories about Christian living, moral and ethical problems, controversial topics, and dealing with contemporary life challenges.
- Other: Submit seasonal material 1 year in advance.

SAMPLE ISSUE
8 pages (no advertising): 3 articles; 1 Bible lesson. Sample copy, free with 6x9 SASE. Guidelines and theme list available.

- "The Promise Keeper." Article explains that God is a promise maker and a promise keeper, and examines how we can keep our promises in our lives.
- "Blood Pressure and God." Articles compares seeking medical treatment to seeking out God to help right our lives.

RIGHTS AND PAYMENT
First and second rights. Written material, $.05–$.07 per word. Pays on acceptance. Provides 5 contributor's copies.

⇢EDITOR'S COMMENTS
Writers wishing to send material would be well-served to check our theme list as outlined on our website, and submit accordingly. All submissions must relate to a theme.

Seventeen

300 West 57th Street, 17th Floor
New York, NY 10019

Editor

DESCRIPTION AND INTERESTS
This popular consumer magazine contains information—and lots of it—on topics that are of interest to teen girls. It features articles on fashion, beauty, fitness, relationships, boys, celebrities, and popular culture. Circ: 2 million.

Audience: 13–21 years
Frequency: Monthly
Website: www.seventeen.com

FREELANCE POTENTIAL
20% written by nonstaff writers. Publishes 20 freelance submissions yearly; 5% by unpublished writers, 40% by authors who are new to the magazine. Receives 46 queries, 200 unsolicited mss yearly.

SUBMISSIONS
Query with outline and clips or writing samples for nonfiction. Send complete ms for fiction. Accepts hard copy and simultaneous submissions if identified. SASE. Responds time varies.

- Articles: 650–3,000 words. Informational and self-help articles; profiles; and personal experience pieces. Topics include relationships, dating, family issues, current events, social concerns, friendship, and pop culture.
- Fiction: 1,000–3,000 words. Stories that feature female teenage experiences.
- Depts/columns: 500–1,000 words. Fashion, beauty, health, and fitness.
- Other: Submit seasonal material 6 months in advance.

SAMPLE ISSUE
166 pages (50% advertising): 7 articles; 24 depts/columns. Sample copy, $2.99 at newsstands. Guidelines available.

- "Miley Cyrus." Article features an interview with the actress/singer about her career, her love life, and how she feels about being a Disney superstar.
- "Sample dept/column: "Hair+Skin+Makeup" offers makeup and hairstyling tips, and reviews several new perfumes.

RIGHTS AND PAYMENT
First rights. Written material, $1–$1.50 per word. Pays on acceptance.

⇢EDITOR'S COMMENTS
When submitting your query, check our list of editors to determine the most appropriate person to receive your pitch.

Sharing the Victory

Fellowship of Christian Athletes
8701 Leeds Road
Kansas City, MO 64129

Editorial Assistant: Ashley Burns

DESCRIPTION AND INTERESTS
This publication features articles on faith and sports that inspire and inform the coaches and players of the Fellowship of Christian Athletes. Circ: 80,000.

Audience: Athletes and coaches, grades 7 and up
Frequency: 9 times each year
Website: www.sharingthevictory.com

FREELANCE POTENTIAL
40% written by nonstaff writers. Publishes 20 freelance submissions yearly; 25% by unpublished writers, 10% by authors who are new to the magazine. Receives 48 queries and unsolicited mss yearly.

SUBMISSIONS
Query with outline/synopsis and clips or writing samples; or send complete ms. Accepts hard copy. SASE. Availability of artwork improves chance of acceptance. Response time varies.

- Articles: To 1,200 words. Informational articles; profiles; interviews; and personal experience pieces. Topics include sports, competition, training, and Christian education.
- Depts/columns: Staff written.
- Artwork: Color prints.
- Other: Submit seasonal material 3–4 months in advance.

SAMPLE ISSUE
38 pages (30% advertising): 5 articles; 10 depts/columns. Sample copy, $1 with 9x12 SASE (3 first-class stamps). Guidelines available.

- "Get Focused." Article relays an interview with NBA athlete and FCA camp coach Anthony Parker.
- "Coach's Profile: Tony Bennett." Article profiles college basketball coach Bennett, whose faith is an inspiration to others.

RIGHTS AND PAYMENT
First serial rights. Articles, $150–$400. Pays on publication.

➡️EDITOR'S COMMENTS
We need more submissions of stories, including profiles and interviews, of Christian professional athletes. We continue to be very interested in stories of the female athletes and coaches in the FCA. All stories must have an FCA connection.

Shine Brightly

P.O. Box 87334
Canton, MI 48187

Editor: Sara Lynne Hilton

DESCRIPTION AND INTERESTS
This Christian magazine for girls has a goal of bringing its readers into a dynamic relationship with Jesus Christ. Its articles cover contemporary topics, yet show readers how to apply their Christian beliefs to become empowered and help others. Circ: 15,500.

Audience: Girls, 9–14 years
Frequency: 9 times each year
Website: www.gemsgc.org

FREELANCE POTENTIAL
30% written by nonstaff writers. Publishes 20 freelance submissions yearly; 15% by unpublished writers, 90% by new authors. Receives 500 unsolicited mss yearly.

SUBMISSIONS
Send complete ms. Accepts hard copy and simultaneous submissions if identified. SASE. Responds in 1 month.

- Articles: 50–500 words. Informational and how-to articles; profiles; humor; and personal experience pieces. Topics include community service, stewardship, contemporary social issues, family and friend relationships, and peer pressure.
- Fiction: 400–900 words. Genres include contemporary fiction, romance, mystery, science fiction, and adventure. Also publishes stories about nature, animals, and sports.
- Depts/columns: Staff written.
- Artwork: 5x7 or larger B/W and color prints.
- Other: Puzzles, activities, and cartoons.

SAMPLE ISSUE
22 pages (no advertising): 4 articles; 1 story; 3 depts/columns; 2 activities. Sample copy, $1 with 9x12 SASE ($.75 postage). Guidelines available.

- "Lavender and Vitamins." Story tells of a girl's new-found relationship with an elderly aunt.
- "Let's Make a Difference." Article offers an idea of donating shoes to children of Zambia.

RIGHTS AND PAYMENT
First, second, and simultaneous rights. Articles and fiction, $.02–$.05 per word. Other material, payment rates vary. Pays on publication. Provides 2 contributor's copies.

➡️EDITOR'S COMMENTS
We seek articles that address contemporary issues in a positive, uplifting way.

Single Mother

National Organization of Single Mothers
P.O. Box 68
Midland, NC 28107-0068

Editor: Andrea Engber

DESCRIPTION AND INTERESTS
Single Mother is the official publication of the National Organization of Single Mothers. As such, it contains informative and inspirational articles relating to the many challenges women face while raising children alone. Hits per month: 3,000–5,000.

Audience: Single mothers
Frequency: Monthly
Website: www.singlemothers.org

FREELANCE POTENTIAL
10% written by nonstaff writers. Publishes 6 freelance submissions yearly. Receives 12–24 queries and unsolicited mss yearly.

SUBMISSIONS
Query or send complete ms. Accepts hard copy. SASE. Response time varies.

- Articles: Word lengths vary. Informational articles; essays; and opinion and personal experience pieces. Topics include parenting, money and time management, absent dads, dating, handling ex-families, death, pregnancy and childbirth, adoption, donor insemination, child support, paternity, custody, and visitation rights.
- Depts/columns: Word lengths vary. News, book reviews, advice.

SAMPLE ISSUE
Sample issue available at website.

- "Unconditional Love." Essays tackles the question of when a mother's love starts and when or if it should ever end.
- "Kids' 'Firsts' Come at all Ages." Essay explains the author's excitement at having all of her children reunited in her first post-divorce house.

RIGHTS AND PAYMENT
Rights vary. Written material, payment rates vary. Payment policy varies.

⇝EDITOR'S COMMENTS
We welcome all articles, thoughts, and essays from single mothers. Whether you are a MOM (a mother outside of marriage) or a DIVA (divorced or separated), your experiences are valuable to our readers. We are interested in articles that tackle the practical issues of single mothering, such as juggling job responsibilities with home life, finances, child support, and other legal issues.

Sisterhood Agenda

524 Ridge Street
Newark, NJ 07104

Editor: Angela D. Coleman

DESCRIPTION AND INTERESTS
As the voice of a nonprofit organization by the same name, *Sisterhood Agenda* seeks to uplift women and girls of African descent. It appears in print as well as online. Circ: 500,000.

Audience: 13–32 years
Frequency: Quarterly
Website: www.sisterhoodagenda.com

FREELANCE POTENTIAL
80% written by nonstaff writers. Publishes 40 freelance submissions yearly; 75% by unpublished writers, 90% by authors who are new to the magazine.

SUBMISSIONS
Send complete ms. Accepts email submissions to acoleman@sisterhoodagenda.com (Microsoft Word attachments). Availability of artwork improves chance of acceptance. Response time varies.

- Articles: To 500 words. Informational articles; profiles; interviews; and photo-essays. Topics include Africa, ancestry, heritage, history, current events, health, nutrition, music, fashion, beauty, self-esteem, fitness, technology, life skills, celebrities, and community service.
- Fiction: Word lengths vary. Multicultural, ethnic, and inspirational fiction.
- Depts/columns: First-person essays, to 300 words. News briefs, book reviews, hair tips, and affirmations, word lengths vary.
- Artwork: Digital images at 300 dpi.
- Other: Poetry, to 15 lines.

SAMPLE ISSUE
50 pages: 6 articles; 23 depts/columns. Sample copy, guidelines, and editorial calendar available at website.

- "Super Sister Tyra Banks." Article profiles the model-turned-media-mogul.
- Sample dept/column: "Sister Connections" summarizes a documentary about French women who trace their roots back to Mali.

RIGHTS AND PAYMENT
One-time rights. No payment.

⇝EDITOR'S COMMENTS
We look for thoughtful, informative, well-written articles, essays, reviews, and poetry. Whenever possible, please submit high-resolution photographs or graphics with your submission, even for fiction.

Six78th

P.O. Box 450
Newark, CA 94560

Editorial Director: Carol S. Rothchild

DESCRIPTION AND INTERESTS

This junior high school lifestyle magazine strives to be as intelligent, trendy, and constantly changing as the girls who read it. *Six78th* presents age-appropriate content that piques girls' interests, but still protects them from growing up too soon. Circ: Unavailable.

Audience: Girls, 10–14 years
Frequency: 6 times each year
Website: www.six78th.com

FREELANCE POTENTIAL

12% written by nonstaff writers. Publishes 5–10 freelance submissions yearly; 5% by unpublished writers, 5% by authors who are new to the magazine. Receives 60–120 queries each year.

SUBMISSIONS

Query. Accepts email queries to carol@six78th.com. Responds in 2 months.

- Articles: Word lengths vary. Informational and self-help articles; profiles; interviews; and personal experience pieces. Topics include school, current events, health and fitness, music, popular culture, sports, celebrities, and social issues.
- Depts/columns: Word lengths vary. Reviews, healthy eating, fashion, friendships.

SAMPLE ISSUE

56 pages: 10 articles; 25 depts/columns. Sample copy, $4.50 with 9x12 SASE. Writers' guidelines available.

- "Amanda Bynes." Article presents an interview with the actress, singer, and new fashion designer.
- Sample dept/column: "Random" reviews four websites that cater to tween girls.

RIGHTS AND PAYMENT

Rights vary. Written material, payment rates vary. Pays on publication. Provides 5 contributor's copies.

➥EDITOR'S COMMENTS

Now in our third year, we have established our voice, and contributing writers will need to be in tune with it. We are currently interested in receiving ideas for articles regarding relationship, scholastic, and self-esteem boosting topics. We pride ourselves on tackling serious issues for young girls, as well as offering lighter, fun material.

Skating

United States Figure Skating Association
29 First Street
Colorado Springs, CO 80906

Director of Publications: Troy Schwindt

DESCRIPTION AND INTERESTS

This member publication for the U.S. Figure Skating Association reports on the sport's programs, events, news, and trends. Personalities are also regularly covered. Circ: 45,000.

Audience: 5 years–Adult
Frequency: 10 times each year
Website: www.usfigureskating.org

FREELANCE POTENTIAL

70% written by nonstaff writers. Publishes 15 freelance submissions yearly; 10% by unpublished writers, 20% by authors who are new to the magazine. Receives 72 queries and unsolicited mss yearly.

SUBMISSIONS

Query with résumé, clips or writing samples, and photo ideas; or send complete ms. Accepts hard copy, MacIntosh Zip disk submissions, and email to skatingmagazine@usfigureskating.org. SASE. Responds in 1 month.

- Articles: 750–2,000 words. Informational articles; profiles; and interviews. Topics include association news, competitions, techniques, personalities, and training.
- Depts/columns: 600–800 words. Competition results; profiles of skaters and coaches; sports medicine; fitness; and technique tips.
- Artwork: B/W and color prints, slides, or transparencies; digital images at 300 dpi.

SAMPLE ISSUE

56 pages: 5 articles; 11 depts/columns. Sample copy, $3 with 9x12 SASE. Writers' guidelines available.

- "Coming Of Age." Article profiles the recent winner of the junior title at the U.S. Figure Skating Championships.
- Sample dept/column: "InSynch" discusses the best way to select skaters for a synchronized team.

RIGHTS AND PAYMENT

First serial rights. Articles, $75–$150. Depts/columns, $75. Artwork, payment rates vary. Pays on publication. Provides 5–10 copies.

➥EDITOR'S COMMENTS

We welcome competition reviews; features on skaters, coaches, and judges; and news that would fit in our regular columns.

Skipping Stones
A Multicultural Magazine

P.O. Box 3939
Eugene, OR 97403-0939

Editor: Arun N. Toké

DESCRIPTION AND INTERESTS
The winner of the 2007 NAME Award for Children's Publishing, *Skipping Stones* features a variety of articles that share cultural experiences by youths as well as adults. Circ: 2,500.

Audience: 7–17 years
Frequency: 5 times each year
Website: www.skippingstones.org

FREELANCE POTENTIAL
90% written by nonstaff writers. Publishes 175–200 freelance submissions yearly; 60% by unpublished writers, 75% by authors new to the magazine. Receives 2,500+ mss yearly.

SUBMISSIONS
Send complete ms with author biography. Accepts hard copy, Macintosh disk submissions, and email submissions to editor@ skippingstones.org (Microsoft Word attachments). SASE. Responds in 4–6 months.

- Articles: 750–1,000 words. Informational articles; photo-essays; personal experience pieces; profiles; and humor. Topics include cultural and religious celebrations, architecture, living abroad, family, careers, disabilities, sustainable living, nature, technology, parenting, and activism.
- Fiction: To 1,000 words. Genres include multicultural fiction and folktales.
- Depts/columns: 100–200 words. Health issues, book reviews, school topics, Q&As.
- Other: Puzzles and games. Poetry by children, to 30 lines. Submit seasonal material 3–4 months in advance.

SAMPLE ISSUE
36 pages (no advertising): 11 articles; 1 story; 7 depts/columns; 16 poems. Sample copy, $5 with 9x12 SASE ($1 postage). Guidelines and editorial calendar available.

- "The Country I'm From, the Heritage I Have." Article shares the story of the author's family.
- Sample dept/column: "Health Rocks" explains a vegan diet.

RIGHTS AND PAYMENT
First and non-exclusive reprint rights. No payment. Provides author's copies and discounts.

➤EDITOR'S COMMENTS
We invite your creative submissions on any of our upcoming planned themes. See our website for a full list.

Slap

High Speed Productions
1303 Underwood Avenue
San Francisco, CA 94124

Editor: Mark Whiteley

DESCRIPTION AND INTERESTS
Written in a language only true "skaters" can understand, this magazine of "Skateboarding, Life, Art, and Progression" covers everything from techniques to terrain. Circ: 130,000.

Audience: YA
Frequency: Monthly
Website: www.slapmagazine.com

FREELANCE POTENTIAL
40% written by nonstaff writers. Publishes 24 freelance submissions yearly; 20% by unpublished writers.

SUBMISSIONS
Send complete ms. Accepts hard copy, disk submissions, and simultaneous submissions if identified. Availability of artwork improves chance of acceptance. SASE. Responds in 2 months.

- Articles: Word lengths vary. Informational and how-to articles; profiles; interviews; and personal experience pieces. Topics include skateboarding techniques, equipment and competition; music; art; and pop culture.
- Depts/columns: Word lengths vary. Media reviews, interviews, gear, gossip, and skateboard tricks.
- Artwork: 35mm B/W negatives; color prints and transparencies. Line art.
- Other: Photo-essays, cartoons, and contests.

SAMPLE ISSUE
146 pages (40% advertising): 6 articles; 11 depts/columns. Sample copy, free with 9x12 SASE ($1.95 postage). Guidelines and editorial calendar available.

- "Sean Malto." Interview with the up-and-coming skater discusses everything from his haircut to his Tampa Am slam.
- "AZ-DC." Photo-essay captures the Am Vets and Flow Trash skaters as they hit the AZ meat grinder . . . dude!
- Sample dept/column: "Tuner" reviews new CD releases from obscure bands like Monotonix, Human Bell, and the Silver Jews.

RIGHTS AND PAYMENT
First rights. All material, payment rates vary. Pays on publication. Provides 1 author's copy.

➤EDITOR'S COMMENTS
You must be intimately familiar with the skate world to write for us.

Softball Youth

Dugout Media Inc.
P.O. Box 983
Morehead, KY 40351

Managing Editor: Nathan Clinkenbeard

DESCRIPTION AND INTERESTS
Dubbed "The Nation's Softball Magazine for Girls," *Softball Youth* provides girls with all the information and inspiration they need to continue to grow in the sport. It features profiles of players and programs, as well as articles about playing technique, strategies, and gear. Circ: Unavailable.

Audience: Girls
Frequency: Quarterly
Website: www.softballyouth.com

FREELANCE POTENTIAL
50% written by nonstaff writers. Publishes 10–20 freelance submissions yearly.

SUBMISSIONS
Query or send complete ms. Accepts hard copy and email submissions to nathan@dugoutmedia.com. SASE. Response time varies.

- Articles: Word lengths vary. Informational and how-to articles; profiles; and personal experience pieces. Topics include techniques, training, and softball personalities.
- Depts/columns: Word lengths vary. Gear, health and fitness, beauty tips, youth ballpark profiles, product reviews, college profiles.
- Other: Contests, horoscopes, quizzes, puzzles, crafts, and activities.

SAMPLE ISSUE
24 pages: 3 articles; 5 depts/columns. Guidelines available.

- "Catch Her If You Can." Article profiles Natasha Watley, a professional softball player on the PFX tour and a member of the Team USA softball team.
- Sample dept/column: "College Profile" offers a profile of the UCLA Bruins women's softball program.

RIGHTS AND PAYMENT
Exclusive rights. Written material, payment rates vary. Pays on publication.

⚫◆EDITOR'S COMMENTS
We are always interested in articles that profile players on professional softball teams. Our favorites are human interest stories on professional players or on youth players who have overcome adversity to play. Both topics provide what we most like to offer our readers—inspiration to continue playing and to take their game to a higher level.

South Florida Parenting

6501 Nob Hill Road
Tamarac, FL 33321

Managing Editor: Vicki McCash Brennan

DESCRIPTION AND INTERESTS
With a mission to be the most valuable source of parenting information and local resources for families in South Florida, this magazine puts a local perspective on parenting, family, children's health and development, and education. Circ: 110,000.

Audience: Parents
Frequency: Monthly
Website: www.sfparenting.com

FREELANCE POTENTIAL
85% written by nonstaff writers. Publishes 90 freelance submissions yearly; 10% by authors who are new to the magazine. Receives 996 unsolicited mss yearly.

SUBMISSIONS
Prefers complete ms; will accept query. Accepts hard copy and email submissions to vmcash@sfparenting.com. SASE. Responds in 2–3 months.

- Articles: 800–2,000 words. Informational and how-to articles; profiles; interviews; and personal experience pieces. Topics include family life, travel, parenting, education, leisure, music, health, and regional events and activities.
- Depts/columns: To 750 words. Family finances, health, nutrition, infant care, and pre-teen issues.

SAMPLE ISSUE
142 pages (60% advertising): 3 articles; 13 depts/columns. Guidelines available.

- "Pacifiers: Peacemakers or Breakers?" Article explores the pros and cons of allowing babies to have pacifiers.
- "Kids and Sports." Article explains how parents can help balance sports and school.

RIGHTS AND PAYMENT
One-time regional rights. Written material, $100–$300. Pays on publication. Provides contributor's copies upon request.

⚫◆EDITOR'S COMMENTS
We are committed to enhancing the lives of families by maintaining excellence in editorial content, highlighting high-quality events, and encouraging community awareness. Our features require careful research, independent reporting, and well-developed interviews with South Florida sources.

South Jersey Mom

P.O. Box 2413
Vineland, NJ 08362-2413

Editor: Adrienne Richardson

DESCRIPTION AND INTERESTS
South Jersey Mom does not mimic what national parenting magazines publish. Instead, it provides a plethora of information, opinion, and inspiration related to the joys and challenges of parenting. Circ: 35,000.
Audience: Parents
Frequency: Monthly
Website: www.southjerseymom.com

FREELANCE POTENTIAL
100% written by nonstaff writers. Publishes 200 freelance submissions yearly; 20% by unpublished writers, 60% by authors who are new to the magazine.

SUBMISSIONS
Query with two writing samples. Accepts email queries to adrienne@southjerseymom.com. Response time varies.

- Articles: Word lengths vary. Informational articles; profiles; and personal experience pieces. Topics include parenting, trends, family issues, pregnancy, technology, education, exercise, safety, sports, and recreation.
- Depts/columns: Word lengths vary. Health topics, gear, technology.

SAMPLE ISSUE
32 pages: 7 articles; 13 depts/columns. Sample copy available at website. Writers' guidelines available.

- "Get On the Green Team." Article reports on eco-friendly school supplies.
- "From Baby Talk to the Boardroom." Article provides information and inspiration for women who are thinking about returning to the workforce after having a child.
- Sample dept/column: "Learning 2 Learn" tells parents how to prepare for a smooth transition back to school after the long summer vacation.

RIGHTS AND PAYMENT
Rights vary. No payment.

➤EDITOR'S COMMENTS
All of our articles are written by local freelancers using sources and settings in southern New Jersey. Our mission is to empower mothers and recognize all they do for their families. If you think you can help us in that endeavor, send us a query. Be sure to enclose writing samples or clips.

Spaceports & Spidersilk

Editor: Marcie Lynn Tentchoff

DESCRIPTION AND INTERESTS
Formerly known as *KidVisions*, this online magazine for young readers features short stories, poems, interviews, and essays on science and the environment. It looks for fantasy, science fiction, and mildly spooky "shadow stories" that make readers say, "Ewwww, gross!" Hits per month: Unavailable.
Audience: 8–17 years
Frequency: Updated continually
Website: www.samsdotpublishing.com/
         spacesilk/main.htm

FREELANCE POTENTIAL
99% written by nonstaff writers. Publishes 50 freelance submissions yearly.

SUBMISSIONS
Send complete ms. Accepts email submissions to spacesilk@yahoo.com (fiction and nonfiction as RTF attachments, poetry in body of email; include "Submission: Title" in subject line). Response time varies.

- Articles: To 800 words. Interviews and essays. Topics include science and the environment.
- Fiction: To 2,000 words. Genres include fantasy, science fiction, and mild horror.
- Artwork: B/W and color line art.
- Other: Quizzes. Poetry, to 25 lines.

SAMPLE ISSUE
5 stories; 1 article; 6 poems; 1 serial. Sample copy and guidelines available at website.

- "Give Me Your Lunch Money." Story tells of a junior high school boy who uses sorcery to exact revenge on a bully.
- "When the Tides Bloom." Article explains the phenomenon of red tide.

RIGHTS AND PAYMENT
First or reprint and electronic rights. Written material, $2 for original, $1 for reprints. Cover artwork, $10. Pays on acceptance.

➤EDITOR'S COMMENTS
We are interested in adventurous, humorous, or mildly spooky fiction and poetry, with a strong speculative bent. More stories and poems set in space would be nice, though fantasy is always needed as well. Please note that we are not interested in stories involving sex or bad language, nor do we wish to see pieces that encourage sexist behavior.

Sparkle

P.O. Box 7259
Grand Rapids, MI 49510

Senior Editor: Sara Lynne Hilton

DESCRIPTION AND INTERESTS
Sparkle's mission is to prepare young Christian girls to live out their faith and become world changers. In short, its service-oriented articles are designed to help girls make a difference in the world. Circ: 5,065.

Audience: Girls, 6–9 years
Frequency: 6 times each year
Website: www.gemsgc.org

FREELANCE POTENTIAL
80% written by nonstaff writers. Publishes 20 freelance submissions yearly; 80% by unpublished writers, 90% by authors who are new to the magazine. Receives 100 unsolicited mss each year.

SUBMISSIONS
Send complete ms. Accepts hard copy. SASE. Responds in 4–6 weeks.

- Articles: 100–400 words. Informational articles. Topics include animals, sports, music, musicians, famous people, interaction with family and friends, service projects, and dealing with school work.
- Fiction: 100–400 words. Genres include adventure, mystery, and contemporary fiction.
- Other: Puzzles, games, recipes, party ideas, short humorous pieces, cartoons, and inexpensive craft projects.

SAMPLE ISSUE
16 pages (no advertising): 3 articles; 2 stories; 2 activities; 2 Bible lessons. Sample copy, $1 with 9x12 SASE. Writers' guidelines and theme list available.

- "The Best Ice Cream." Story finds a little girl waiting at her window, watching for the ice cream man.
- "What Does God Say About Being Joyful?" Bible lesson tells about Miriam and her brother, Moses.

RIGHTS AND PAYMENT
Rights vary. Articles, $20. Other material, payment rates vary. Pays on publication. Provides 2 contributor's copies.

☛EDITOR'S COMMENTS
Please remember to check our writers' guidelines before submitting. We are currently in need of articles and stories about living a green lifestyle, and articles and stories about Zambia, Africa.

Spider
The Magazine for Children

Cricket Magazine Group
70 East Lake Street, Suite 300
Chicago, IL 60601

Submissions Editor

DESCRIPTION AND INTERESTS
Original stories, poetry, articles, and illustrations fill the pages of this fun-filled children's magazine. An activity pullout section is also found in each issue. Circ: 60,000.

Audience: 6–9 years
Frequency: 9 times each year
Website: www.cricketmag.com

FREELANCE POTENTIAL
97% written by nonstaff writers. Publishes 50 freelance submissions yearly; 30% by unpublished writers, 50% by authors who are new to the magazine. Receives 3,600 unsolicited mss yearly.

SUBMISSIONS
Send complete ms; include bibliography for nonfiction. Accepts hard copy and simultaneous submissions if identified. SASE. Responds in 6 months.

- Articles: 300–800 words. Informational and how-to articles; profiles; and interviews. Topics include nature, animals, science, technology, history, multicultural issues, foreign cultures, and the environment.
- Fiction: 300–1,000 words. Easy-to-read stories. Genres include humor; fantasy; fairy tales; folktales; and realistic, historical, and science fiction.
- Other: Recipes, crafts, puzzles, games, brainteasers, and math and word activities. Poetry, to 20 lines.

SAMPLE ISSUE
34 pages (no advertising): 1 article; 3 stories; 3 poems; 5 activities. Sample copy, $5 with 9x12 SASE. Guidelines available at website.

- "Doodlebug & Dandelion Tracking Sasquatch." Story tells of the adventure two boys have during a family camping trip.
- "Creeping, Crawling Nightlights." Article describes several of the world's most interesting night-time insects.

RIGHTS AND PAYMENT
All rights. Articles and fiction, $.25 per word. Poetry, to $3 per line. Other material, rates vary. Pays on publication. Provides 2 copies.

☛EDITOR'S COMMENTS
We are always looking for thought-provoking articles and stories that will foster a love of reading and discovery.

SportingKid

2050 Vista Parkway
West Palm Beach, FL 33411

Managing Editor: Greg Bach

DESCRIPTION AND INTERESTS
SportingKid is the official member publication
of the National Alliance for Youth Sports. Its
articles help parents, volunteer coaches, and
other adults who work with children in sports
programs to understand their roles and responsibilities. Circ: 300,000.
Audience: Parents; coaches; officials
Frequency: Quarterly
Website: www.nays.org

FREELANCE POTENTIAL
20% written by nonstaff writers. Publishes 10
freelance submissions yearly; 10% by unpublished writers. Receives 180 queries, 36 unsolicited mss yearly.

SUBMISSIONS
Query or send complete ms. Accepts email
submissions to sportingkid@nays.org.
Responds in 1 month.

- Articles: To 1,000 words. Informational and
  how-to articles; and profiles. Topics include
  youth sports, coaching, parenting, officiating,
  health, and safety.
- Depts/columns: 750 words. New product
  information, the culture of youth sports, and
  coaching and parenting tips.

SAMPLE ISSUE
32 pages: 8 articles; 4 depts/columns.

- "Survey Says." Article reports on the national
  survey conducted by NAYS that reveals the
  troubling trend of violence in youth sports.
- "Commander Turned Coach." Article profiles
  a former U.S. Navy commander who is now
  a volunteer coach stressing respect and fun.
- Sample dept/column: "Product Locker"
  offers reviews of products from a pitching
  machine to an ice pack for nosebleeds.

RIGHTS AND PAYMENT
First and electronic rights. Written material,
payment rates vary. Pays on publication.

☞ EDITOR'S COMMENTS
We'd like to see more articles that provide
tips for making sports safe and fun. In particular, we seek submissions on creative coaching techniques, and advice for parents
regarding the various issues associated with
having children involved in sports. We're
always looking for profiles of individuals, but
we're not interested in first-person stories.

Sports Illustrated Kids

Time & Life Building
1271 Avenue of the Americas, 32nd Floor
New York, NY 10020

Managing Editor: Bob Der

DESCRIPTION AND INTERESTS
This offspring of *Sports Illustrated* delivers the
excitement, passion, and fun of sports to kids
in an action-oriented and interactive manner.
Its pages include profiles of young athletes,
sports trivia, and games, among other features.
Circ: 1.1 million.
Audience: 8–14 years
Frequency: Monthly
Website: www.sikids.com

FREELANCE POTENTIAL
3–5% written by nonstaff writers. Publishes
3–5 freelance submissions yearly. Receives
1–2 queries yearly.

SUBMISSIONS
Query or send complete ms. Accepts hard
copy. SASE. Responds in 2 months.

- Articles: Lead articles and profiles, 500–700
  words. Short features, 500–600 words.
  Informational articles; profiles; and interviews. Topics include professional and aspiring athletes, various sports, fitness, health,
  safety, hobbies, and technology.
- Depts/columns: Word lengths vary. Events
  coverage, team and player profiles, news,
  pro tips, humor, and video-game reviews.
- Other: Puzzles, games, trivia, comics, and
  sports cards. Poetry and artwork by children.

SAMPLE ISSUE
56 pages (24% advertising): 6 articles; 7 depts/
columns; 1 comic; 1 puzzle; 1 quiz. Sample
copy and editorial calendar available at website. Guidelines available.

- "Ryan Sheckler." Interview with the celebrity
  skateboarder discusses his career, including
  his goals for the upcoming X Games.
- Sample dept/column: "Warmup" includes a
  brief about 9-year-old basketball prodigy
  Mark Walker's charitable work.

RIGHTS AND PAYMENT
All rights. Articles, $100–$1,500. Depts/
columns, payment rates vary. Pays on acceptance. Provides contributor's copies.

☞ EDITOR'S COMMENTS
We are the authority on kids and sports, and
our contents reflect the interests and humor
of our audience. Writers are advised to check
our website to determine which departments
to submit to.

Start

EduGuide: Partnership for Learning
321 North Pine
Lansing, MI 48933

Editor: Rebecca Kavanaugh

DESCRIPTION AND INTERESTS

This publication from EduGuide: Partnership for Learning is a free resource for parents on children's development and education from birth to the preschool years. Articles, written to inspire and inform, typically include a parent's and an expert's point of view. Circ: 90,000.

Audience: Parents
Frequency: Unavailable
Website: www.eduguide.org
www.partnershipforlearning.org

FREELANCE POTENTIAL

100% written by nonstaff writers. Publishes 60 freelance submissions yearly.

SUBMISSIONS

Query. Accepts hard copy and email queries to info@partnershipforlearning.org. SASE. Response time varies.

- Articles: 150–800 words. Informational and how-to articles; and personal experience pieces. Topics include parenting, health and safety, literacy, and child development.

SAMPLE ISSUE

8 pages: 3 articles. Sample copy, free with 9x12 SASE. Guidelines available at website.

- "Slowing Down to Read the Signs." Article profiles a young boy and his family's journey from a devastating diagnosis of autism to intervention and success.
- "Playful Ways to Recognize Delays." Article lists different developmental milestones and offers tips on how to recognize delays.
- "Hayden's Story." Article details a toddler's diagnosis of Fragile X Syndrome; includes interviews with two experts about the next steps his parents should take and treatment options available.

RIGHTS AND PAYMENT

Rights vary. Articles, 150–800 words; $75–$150. Pays on publication.

⟢EDITOR'S COMMENTS

We're not reviewing unsolicited manuscripts at this time, as we have been using the work of writers who have published with us before. The freelance pieces we do use are written by experienced, previously published authors. Features most open to freelancers include skill-building activities and journal-style stories that share lessons learned as a parent.

Stone Soup
The Magazine by Young Writers and Artists

P.O. Box 83
Santa Cruz, CA 95063

Editor: Gerry Mandel

DESCRIPTION AND INTERESTS

This fully illustrated literary magazine publishes work by writers and artists under 14 years of age only. Circ: 20,000.

Audience: 8–14 years
Frequency: 6 times each year
Website: www.stonesoup.com

FREELANCE POTENTIAL

100% written by nonstaff writers. Publishes 50 freelance submissions yearly; 85% by authors who are new to the magazine. Receives 9,600 unsolicited mss yearly.

SUBMISSIONS

Send complete ms. Accepts submissions from writers under 14 years of age only. Accepts hard copy. No simultaneous submissions. SASE. Responds in 6 weeks if interested.

- Fiction: To 2,500 words. Genres include adventure; mystery; suspense; science fiction; and multicultural, ethnic, and historical fiction.
- Depts/columns: Word lengths vary. Book reviews.
- Artwork: B/W and color drawings.
- Other: Poetry, line lengths vary.

SAMPLE ISSUE

48 pages (no advertising): 7 stories; 2 book reviews; 3 poems. Sample copy and guidelines available at website.

- "Trek to the Peak." Story follows a father and son on a climb that ends in an avalanche accident.
- "The Black Cat." Story tells of an elderly woman who reminisces about overcoming her fears when she helped a brave doctor rescue runaway slaves.
- Sample dept/column: "Book Review" offers a brief synopsis of *Dark Water Rising*, a novel about a 1920 storm in Texas.

RIGHTS AND PAYMENT

All rights. Written material, $40. Artwork, $25. Pays on publication. Provides 2 contributor's copies.

⟢EDITOR'S COMMENTS

We urge prospective writers to review our online guidelines carefully in order to ensure that they are submitting work that meets our expectations and requirements. We respond only to submissions that we can use.

Stories for Children Magazine

54 East 490 South
Ivins, UT 84738

Editor: V. S. Grenier

DESCRIPTION AND INTERESTS
This free e-zine seeks to inspire the imaginations of children through its short stories, articles, coloring pages, poems, crafts, puzzles, reviews of children's books, and interviews with children's book authors and illustrators. Hits per month: 5,000+.

Audience: 3–12 years
Frequency: Monthly
Web: http://storiesforchildrenmagazine.org

FREELANCE POTENTIAL
95% written by nonstaff writers. Publishes 400 freelance submissions yearly; 25–30% by unpublished writers, 65–70% by authors who are new to the magazine.

SUBMISSIONS
Prefers complete ms for fiction and nonfiction; will accept query for nonfiction. Accepts email submissions (Microsoft Word attachment); see website for proper editor and email address. Response time varies.

- Articles: 150–1,200 words. Informational and how-to articles; and personal experience pieces. Topics include animals, history, nature, the environment, science, technology, and multicultural and ethnic issues.
- Fiction: 150–1,200 words. Genres include fantasy, light horror, humor, mystery, suspense, and science fiction.
- Other: Puzzles, games, crafts, recipes. Poetry, to 16 lines.

SAMPLE ISSUE
4 articles; 19 stories; 8 poems; 5 activities. Guidelines available at website.

- "Melody's Smelly Guest." Story brings a smelly creature to young Melody's door, and shares Melody's frustrations in trying to get it to take a bath.
- "Little Leapin' Lizzie the Lizard." Story follows the adventures of animal friends as they search for a girl's lost doll.

RIGHTS AND PAYMENT
One-time non-exclusive rights. No payment (except for anthology inclusion).

✎ EDITOR'S COMMENTS
We welcome submissions from adult writers as well as young writers. We like to see engaging articles that read more like stories or have a "wow" factor.

Story Mates

Christian Light Publications
P.O. Box 1212
Harrisonburg, VA 22803-1212

Editor: Crystal Shank

DESCRIPTION AND INTERESTS
Used in Mennonite Sunday school classrooms, *Story Mates* offers inspirational stories, poems, and Bible-based activities and crafts that show children how to live in a way that pleases God. Each issue is theme-based. Circ: 6,340.

Audience: 4–8 years
Frequency: Monthly
Website: www.clp.org

FREELANCE POTENTIAL
90% written by nonstaff writers. Publishes 200 freelance submissions yearly. Receives 600 unsolicited mss yearly.

SUBMISSIONS
Send complete ms. Accepts hard copy and email submissions to storymates@clp.org. SASE. Responds in 6 weeks.

- Fiction: Stories related to Sunday school lessons and true-to-life stories, to 800 words. Picture stories, 120–150 words.
- Other: Bible puzzles, crafts, and activities. Poetry, word lengths vary. Submit seasonal material 6 months in advance.

SAMPLE ISSUE
20 pages (no advertising): 4 stories; 4 poems; 8 puzzles. Sample copy, free with 9x12 SASE ($1.20 postage). Writers' guidelines and theme list available.

- "Elliot's Excuses." Story revolves around a young boy who learns the consequences of not taking care of his responsibilities.
- "Hunter and the Straying Cat." Story tells of a boy who learns about stealing.
- "Baby Ezra Comes Home." Story tells of a little girl's anticipation as she awaits her adopted brother's arrival.

RIGHTS AND PAYMENT
First, reprint, or multiple-use rights. Fiction, $.03–$.05 per word. Poetry, $.35–$.75 per line. Other material, payment rates vary. Pays on acceptance. Provides 1 contributor's copy.

✎ EDITOR'S COMMENTS
Because we target young Mennonite children, references to popular culture must be avoided. Stories should have a clear spiritual lesson that is age appropriate and follows the King James Version of the Bible. All material follows a Sunday school lesson theme for a given date.

Student Assistance Journal

1270 Rankin Drive, Suite F
Troy, MI 48083

Editor: Julie Lofquist

DESCRIPTION AND INTERESTS

This professional journal is dedicated to people who work with troubled kids in preschool through grade 12. It provides information about student assistance programs and initiatives that deal with alcohol and drugs, school violence, and behavioral health issues. Circ: 5,000.

Audience: Student assistance professionals
Frequency: Quarterly
Website: www.prponline.net

FREELANCE POTENTIAL

90% written by nonstaff writers. Publishes 12 freelance submissions yearly; 50% by unpublished writers. Receives 36 queries yearly.

SUBMISSIONS

Professionals should send complete ms. Accepts hard copy, IBM disk submissions, and simultaneous submissions if identified. SASE. Accepts email queries from those outside the field to editorial@prponline.net. Responds only if interested.

- Articles: 1,500 words. Informational and how-to articles; and personal experience pieces. Topics include high-risk students, special education, drug testing, substance abuse prevention, school violence, legal issues, federal funding, and staff development.
- Depts/columns: 750–800 words. Book reviews, events, commentaries, news briefs, legal issues, media resources, and research.

SAMPLE ISSUE

34 pages (20% advertising): 4 articles; 2 depts/columns. Sample copy, free. Writers' guidelines available.

- "Sex, Drugs, and Alcohol . . . Online?" Article explains how teens are using the Internet to trade information on how to use illicit drugs without getting caught.
- "At Risk: Our Communities, Schools, and Youth." Article explores the increase in gang activity, and explains a new program in Los Angeles designed to fight it.

RIGHTS AND PAYMENT

First rights. No payment. Provides 5 contributor's copies.

➙EDITOR'S COMMENTS

We pride ourselves on our practical, solution-focused editorial. All articles must provide information that other professionals can use.

SuperScience

Scholastic Inc.
557 Broadway
New York, NY 10012-3999

Editor: Britt Norlander

DESCRIPTION AND INTERESTS

SuperScience, distributed in elementary school classrooms, gets kids interested in science through upbeat articles, quirky facts, and hands-on activities. Circ: 200,000.

Audience: Grades 3–6
Frequency: 8 times each year
Website: www.scholastic.com/superscience

FREELANCE POTENTIAL

80% written by nonstaff writers. Publishes 50 freelance submissions yearly; 15% by authors who are new to the magazine. Receives 60–120 queries yearly.

SUBMISSIONS

Query with résumé and clips. Accepts hard copy. SASE. Response time varies.

- Articles: 100–600 words. Informational and how-to articles; profiles; interviews; and personal experience pieces. Topics include Earth, physical, and life sciences; health; technology; chemistry; nature; and the environment.
- Depts/columns: Word lengths vary. Science news and experiments.
- Artwork: 8x10 B/W and color prints. Line art.
- Other: Puzzles and activities.

SAMPLE ISSUE

16 pages (no advertising): 4 articles; 4 depts/columns. Sample copy, free with 9x12 SASE. Sample copy available at website. Guidelines and editorial calendar available.

- "Behind the Scenes of *Nim's Island*." Article reveals the real-life science behind the movie.
- "Turkey Power." Article describes how electricity can be created from turkey droppings.
- Sample dept/column: "Hands-On" features an electrical experiment using lemons.

RIGHTS AND PAYMENT

First rights. Articles, $75–$650. Other material, payment rates vary. Pays on acceptance. Provides 2 contributor's copies.

➙EDITOR'S COMMENTS

We look for fascinating science news stories that teach Earth, life, and physical science concepts; cross-curricular activities that build reading comprehension, math, critical thinking, writing, and vocabulary skills; hands-on activities and experiments with the fun of real science discoveries; and profiles of real-life scientists.

SW Florida Parent & Child

2422 Dr. Martin Luther King Jr. Boulevard
Fort Myers, FL 33901

Editor: Pamela Smith Hayford

DESCRIPTION AND INTERESTS
Dedicated to bringing information about parenting and family issues to parents living in the counties of southwest Florida, this magazine is filled with ideas, advice, and activities of interest to families. Most of the material has a regional focus. Circ: 23,000.
Audience: Parents
Frequency: Monthly
Website: www.gulfcoast.momslikeme.com

FREELANCE POTENTIAL
75% written by nonstaff writers. Publishes 160 freelance submissions yearly; 5% by unpublished writers, 25% by authors who are new to the magazine. Receives 275 queries and unsolicited mss yearly.

SUBMISSIONS
Query or send complete ms. Accepts email submissions to pamela@swflparentchild.com. Response time varies.

- Articles: To 500 words. Informational articles; profiles; and personal experience pieces. Topics include family issues, parenting, education, travel, sports, health, fitness, computers, and social and regional issues.
- Depts/columns: To 500 words. Dining, travel, parenting, education, and nutrition.

SAMPLE ISSUE
84 pages (50% advertising): 4 articles; 21 depts/columns. Guidelines available.

- "Swimming Program Missed in Bonita Springs." Article reports on the cancellation of a children's swim team.
- "Tropicalia Kids: Summer Reading Promoted." Article reviews the various summer reading programs going on in libraries throughout the region.
- Sample dept/column: "Tots to Teen" offers ideas for talking to young kids about drugs.

RIGHTS AND PAYMENT
All rights. Written material, $25–$200. Pays on publication.

➥EDITOR'S COMMENTS
We are always interested in hearing from moms and dads who can provide helpful information or tips to other parents *and* do so in an entertaining style. If it's not general parenting advice you offer, your article should be about something local.

Swimming World and Junior Swimmer

90 Bell Rock Plaza, Suite 200
Sedona, AZ 86351

Editor: Jayson Marsteller

DESCRIPTION AND INTERESTS
Packed with information on training techniques and reports from competitions throughout the country, this magazine is read by competitive swimmers of all ages and skill levels. It features profiles of athletes and coaches, and includes a special section for junior swimmers. Circ: 59,000.
Audience: All ages
Frequency: Monthly
Website: www.swimmingworldmagazine.com

FREELANCE POTENTIAL
60% written by nonstaff writers. Publishes 100 freelance submissions yearly; 5% by unpublished writers. Receives 192+ queries yearly.

SUBMISSIONS
Query. Accepts hard copy and email queries to jaysonm@swimmingworldmagazine.com. SASE. Responds in 1 month.

- Articles: 500–3,500 words. Informational and how-to articles; profiles; and personal experience pieces. Topics include swimming, training, competition, medical advice, swim drills, nutrition, dry land exercise, exercise physiology, and fitness.
- Depts/columns: 500–750 words. Swimming news, product reviews, and nutrition advice.
- Artwork: Color prints and transparencies. Line art.
- Other: Activities, games, and jokes. Submit seasonal material 1–2 months in advance.

SAMPLE ISSUE
62 pages (30% advertising): 24 articles; 4 depts/columns. Sample copy, $4.50 with 9x12 SASE ($1.80 postage). Guidelines available.

- "One Last Chance." Article reports on the final U.S. qualifying event for the Olympics, and how one swimmer made the most of a lucky break.
- "The Workout Card." Article explains how hypoxic breathing can be an integral part of a successful swimming regime.

RIGHTS AND PAYMENT
All rights. Written material, $.12 per word. Artwork, payment rates vary. Pays on publication. Provides 2–5 contributor's copies.

➥EDITOR'S COMMENTS
We are looking for writers who understand the complexities of competitive swimming.

Syracuse Parent

5910 Firestone Drive
Syracuse, NY 13206

Editor: Brittany Jerred

DESCRIPTION AND INTERESTS
Syracuse Parent is central New York's go-to guide for family news and events. It covers parenting and education issues from a local perspective. Circ: 26,500.
Audience: Parents
Frequency: Monthly
Website: www.syracuseparent.net

FREELANCE POTENTIAL
40% written by nonstaff writers. Publishes 15 freelance submissions yearly; 25% by unpublished writers, 10% by authors who are new to the magazine. Receives 95 queries yearly.

SUBMISSIONS
Query. Accepts hard copy. SASE. Responds in 4–6 weeks.

- Articles: 800–1,000 words. Informational and how-to articles; personal experience and practical application pieces; profiles; interviews; and humor. Topics include parenting, family issues, pets, education, health, current events, regional news, social issues, nature, the environment, technology, music, travel, recreation, and sports.
- Depts/columns: Staff written.
- Other: Submit seasonal material 3–4 months in advance.

SAMPLE ISSUE
24 pages (50% advertising): 6 articles; 7 depts/columns. Sample copy, $1 with 9x12 SASE. Guidelines and editorial calendar available.

- "Grandparent and Grandchild Fishing Days Set at Carpenter's Brook." Article describes an annual fishing derby hosted by an area fish hatchery.
- "Research: Mom's Encouragement Key to Dad's Involvement in Child Care." Article summarizes a recent study on the sharing of parenting duties.

RIGHTS AND PAYMENT
First North American serial rights. Articles, $25–$30. Pays on publication.

⇥EDITOR'S COMMENTS
We like personality profiles of Syracuse-area people who touch the lives of local parents and children. We can also always use well-written articles on birthday parties, sibling issues, baby care, back-to-school, and holidays, among other popular topics.

Take Five Plus

General Council of the Assemblies of God
1445 North Boonville Avenue
Springfield, MO 65802-1894

Director of Editorial Services: Paul Smith

DESCRIPTION AND INTERESTS
This is not your mother's devotional guide. *Take Five Plus*, a pocket-sized guide just for teens, is filled with devotionals and Bible lessons that use modern, real-life situations as motivating or cautionary tales. The publication is designed to encourage teens to spend consistent time with God's Word. Circ: 20,000.
Audience: 12–19 years
Frequency: Quarterly
Website: www.gospelpublishing.com

FREELANCE POTENTIAL
98% written by nonstaff writers. Of the freelance submissions published yearly, 10% are by authors who are new to the magazine.

SUBMISSIONS
All material is assigned. Send letter of introduction with résumé, church background, and clips or writing samples. Accepts hard copy. SASE. Responds in 3 months.

- Articles: 200–235 words. Daily devotionals based on Scripture readings.
- Artwork: Accepts material from teens only. 8x10 B/W prints and 35mm color slides. 8x10 or smaller line art.
- Other: Poetry by teens, to 20 lines.

SAMPLE ISSUE
104 pages (no advertising): 89 devotionals; 4 poems. Sample copy and writers' guidelines available.

- "Idols." Devotional teaches that idols are not often what they seem, and following them can derail us from our partnership with God.
- "Did God Really Say?" Devotional warns that modern media and even friends can try to discount the messages of the Bible.

RIGHTS AND PAYMENT
First rights. Written material, $.05 per word. Artwork, payment rates vary. Pays on publication. Provides 2 contributor's copies.

⇥EDITOR'S COMMENTS
All of our devotionals are written on assignment. Interested freelancers are welcome to request a sample devotional assignment, which will provide us with a sense of their style. We have a rather detailed set of guidelines that spell out exactly what we are looking for in devotionals. Teens may submit poetry and artwork any time.

tandT News

419 Park Avenue South, 13th Floor
New York, NY 10016

Editor: Jenna Greditor

DESCRIPTION AND INTERESTS
This regional tabloid for middle-grade and high school students seeks to "take the turbulence out of adolescence" with its informative articles on friends, drugs, alcohol, sex, and self-esteem. It also features a calendar of events happening in the greater New York City area. Circ: 280,000.

Audience: 8–18 years
Frequency: Monthly
Website: www.tweensandteensnews.com

FREELANCE POTENTIAL
80% written by nonstaff writers. Publishes 100 freelance submissions yearly; 5% by unpublished writers, 50% by authors who are new to the magazine.

SUBMISSIONS
Query or send complete ms. Accepts email submissions to jenna@teensandtweensnews.com. Response time varies.

- Articles: 750–1,500 words. Informational articles; profiles; interviews; and personal experience pieces. Topics include college, careers, computers, technology, crafts and hobbies, current events, health and fitness, music, culture, the environment, recreation, self-esteem, social issues, sex, drugs, alcohol, education, family matters, sports, and pets.
- Depts/columns: 500 words. Things to see and do, media reviews, most embarrassing moments, advice, and events calendars.

SAMPLE ISSUE
28 pages: 9 articles; 6 depts/columns. Sample copy available at website. Guidelines available.

- "Keeping Brains Active." Article provides ways to rest, recharge, play, and learn during summer break.
- Sample dept/column: "Don't Miss" includes information on the grand opening of the Sports Museum of America.

RIGHTS AND PAYMENT
Rights negotiable. No payment. Provides 1+ contributor's copies.

➷EDITOR'S COMMENTS
Our articles are written by parents, teachers, psychologists, and other experts who contribute their wisdom, making the adolescent years easier to navigate.

TAP: The Autism Perspective

10153½ Riverside Drive, Suite 243
Toluca, CA 91602

Publisher: Nicki Fischer

DESCRIPTION AND INTERESTS
This magazine was founded on the belief that those living with and treating autism should have a resource that presents all options and points of view, including new therapies, cutting-edge research, and inspirational stories. Circ: Unavailable.

Audience: Adults
Frequency: Quarterly
Website: www.theautismperspective.org

FREELANCE POTENTIAL
100% written by nonstaff writers. Publishes 120–140 freelance submissions yearly. Receives 360 queries yearly.

SUBMISSIONS
Query. Accepts hard copy and email queries to submissions@theautismperspective.org. SASE. Response time varies.

- Articles: 800–2,000 words. Informational articles and personal experience pieces. Topics include autism spectrum disorders and related developmental disabilities, treatments, therapies, intervention, and research.
- Depts/columns: "Understanding Autism" personal accounts of living with ASD, 800–1,200 words. Grandparent and sibling perspectives, to 550 words.
- Artwork: B/W or color prints. Line art.

SAMPLE ISSUE
100 pages: 16 articles; 12 depts/columns. Sample copy and guidelines available at website.

- "Follow the Heartline." Article offers advice to help married couples with autistic children avoid strain on their relationships.
- "The Power of Therapy Dogs." Article describes the program for autistic children offered by Georgia Canines for Independence.
- "I Am Sue Rubin." Personal essay by an autistic woman tells how Facilitated Communication is her only means of expression.

RIGHTS AND PAYMENT
All rights. All material, payment rates vary. Payment policy varies. Provides 1 copy.

➷EDITOR'S COMMENTS
Write from your own perspective about what you believe to be true about autism. Submissions giving opinions or negative comments about other people or other therapies will not be accepted.

Tar Heel Junior Historian

North Carolina Museum of History
4650 Mail Service Center
Raleigh, NC 27699-4650

Editor: Lisa Coston Hall

DESCRIPTION AND INTERESTS
Tar Heel Junior Historian was created by the North Carolina Museum of History to present the history of the state to students through a combination of scholarly articles, photographs, and illustrations. Circ: 9,000.
Audience: 9–18 years
Frequency: Twice each year
Website: www.ncmuseumofhistory.org

FREELANCE POTENTIAL
50% written by nonstaff writers. Publishes 16 freelance submissions yearly; 20% by unpublished writers, 50% by authors who are new to the magazine.

SUBMISSIONS
Query. Accepts hard copy and email queries to lisa.hall@ncmall.net. SASE. Response time varies.

- Articles: 700–1,000 words. Informational articles; profiles; interviews; and personal experience pieces. Topics include regional history; geography; government; and social, multicultural, and ethnic issues.
- Fiction: Word lengths vary. Genres include historical, ethnic, and multicultural fiction; folklore; and folktales.
- Artwork: B/W or color prints or transparencies. Line art.
- Other: Puzzles, activities, and word games.

SAMPLE ISSUE
38 pages (no advertising): 14 articles; 1 activities page. Sample copy, $4 with 9x12 SASE ($2 postage). Writers' guidelines and theme list available.

- "When World War II Was Fought Off North Carolina's Beaches." Article reports on the many German U-boat attacks on U.S. ships off North Carolina's coast.
- "Women Step Up to Serve." Article discusses the military service of North Carolina women during World War II.

RIGHTS AND PAYMENT
All rights. No payment. Provides 10 contributor's copies.

➡️EDITOR'S COMMENTS
We typically solicit authors for our articles. You may query us with your qualifications, and we may assign you an article if you have the experience and scholarship that we seek.

TC Magazine

915 East Market, Box 10750
Searcy, AR 72149

Managing Editor: Laura Kaiser

DESCRIPTION AND INTERESTS
Formerly known as *Teenage Christian Magazine*, this publication takes a positive, faith-based approach to the issues teens face and the culture in which they live. Its goal is to be a relevant, challenging, and interesting magazine to teens from all sorts of backgrounds. Circ: 7,000+.
Audience: 13–19 years
Frequency: Quarterly
Website: www.tcmagazine.org

FREELANCE POTENTIAL
35% written by nonstaff writers. Publishes 5–10 freelance submissions yearly; 20% by unpublished writers, 30% by authors who are new to the magazine. Receives 60 queries and unsolicited mss yearly.

SUBMISSIONS
Query or send complete ms. Prefers email to editor@tcmagazine.org (Microsoft Word attachments); will accept hard copy. SASE. Response time varies.

- Articles: 450–700 words. Informational articles; profiles; interviews; and personal experience pieces. Topics include faith, youth groups, sports, health, humor, music, film, fashion, art, and relationships.
- Depts/columns: Word lengths vary. College, style, entertainment, media reviews, humor, personal essays, and Q&A interviews.

SAMPLE ISSUE
48 pages: 5 articles; 9 depts/columns. Sample copy, $3.95. Guidelines available at website.

- "You Say Goodbye, But I Say Hello." Personal essay discusses the author's mixed feelings about graduating from high school.
- Sample dept/column: "Ten Minutes With . . ." is a Q&A with 16-year-old Olympic gymnast Shawn Johnson.

RIGHTS AND PAYMENT
All rights. Written material, payment rates vary. Payment policy varies.

➡️EDITOR'S COMMENTS
Because we want teens to feel they have an ownership in *TC*, we welcome submissions from teen writers. Articles that talk down to teens or contain excessive or misused Christian lingo will not be considered. Non-Christians should be able to read any article.

Teacher Librarian

4501 Forbes Boulevard, Suite 200
Lanham, MD 20706

Managing Editor: Corinne O. Burton

DESCRIPTION AND INTERESTS
Serving school library professionals since 1973, this magazine identifies and responds to the challenges encountered by school librarians every day. It covers library advocacy, library services, library management issues, and literacy programs. Circ: 10,000.

Audience: School librarians
Frequency: 5 times each year
Website: www.teacherlibrarian.com

FREELANCE POTENTIAL
60% written by nonstaff writers. Publishes 10 freelance submissions yearly; 25% by unpublished writers, 5% by authors who are new to the magazine. Receives 6 queries and unsolicited mss yearly.

SUBMISSIONS
Query or send complete ms with résumé, abstract, or bibliography. Accepts hard copy, disk submissions, and email submissions to editor@teacherlibrarian.com. SASE. Responds in 2 months.

- Articles: 2,000+ words. Informational and analytical articles; and profiles. Topics include library funding, technology, leadership, library management, audio/visual material, cooperative teaching, and young adult library services.
- Depts/columns: Staff written.
- Artwork: B/W or color prints. Cartoons and line art.

SAMPLE ISSUE
78 pages (20% advertising): 7 articles; 16 depts/columns; 17 reviews. Guidelines and editorial calendar available.

- "Everyone Wins: Differentiation in the School Library." Article addresses ways to diversify the path to learning.
- "Position Yourself at the Center: Co-Teaching Reading Comprehension Strategies." Article suggests ways that librarians can make themselves a larger part of literacy programs.

RIGHTS AND PAYMENT
All rights. Written material, $100. Pays on publication. Provides 2 contributor's copies.

❖EDITOR'S COMMENTS
All articles are reviewed by at least two members of our advisory board, all of whom are scholars or recognized professionals.

Teachers of Vision

227 North Magnolia Avenue, Suite 2
Anaheim, CA 92801

Managing Editor: Denise Trippett

DESCRIPTION AND INTERESTS
A publication of Christian Educators Association International, this magazine is written for Christian teachers in both public and private educational settings. Circ: 8,000.

Audience: Christian educators
Frequency: Quarterly
Website: www.ceai.org

FREELANCE POTENTIAL
75% written by nonstaff writers. Publishes 55–70 freelance submissions yearly; few by unpublished writers, 35–40% by new authors. Receives 132 unsolicited mss yearly.

SUBMISSIONS
Send complete ms with brief biography. Prefers email submissions to tov@ceai.org; will accept hard copy. SASE. Responds in 2–3 months.

- Articles: Word lengths vary. How-to articles, personal experience pieces, and documented reports; 800–1,000 words. Topics include education issues, educational philosophy, and methodology. Interviews with noted Christian educators; 500–800 words. Teaching techniques, news, and special event reports; 400–500 words.
- Depts/columns: 100–200 words. Reviews of books, videos, curricula, games, and other curricula resources for K–12 teachers.
- Other: Submit seasonal material 4 months in advance.

SAMPLE ISSUE
24 pages (1% advertising): 11 articles; 10 depts/columns. Sample copy, free with 9x12 SASE (5 first-class stamps). Guidelines available at website.

- "The ABCs of Dealing with Defiance." Article outlines strategies teachers can employ to deal with defiant students.
- Sample dept/column: "Disappointed, But Not Discouraged" offers encouragement for teachers faced with daily disappointments.

RIGHTS AND PAYMENT
First and electronic rights. Articles, $20–$40. Reviews, $5. Pays on publication. Provides 2 contributor's copies.

❖EDITOR'S COMMENTS
Keep in mind that our articles need a distinctly Christian viewpoint. However, please avoid even the appearance of preaching.

Teaching Theatre

2343 Auburn Avenue
Cincinnati, OH 45219

Editor: James Palmarini

DESCRIPTION AND INTERESTS
Middle school and high school drama teachers turn to this magazine for informative articles on curricula, teaching strategies and techniques, and ideas they can use in their classrooms. Circ: 4,500.

Audience: Theater teachers
Frequency: Quarterly
Website: www.edta.org

FREELANCE POTENTIAL
70% written by nonstaff writers. Publishes 15 freelance submissions yearly; 30% by unpublished writers, 50% by authors who are new to the magazine. Receives 75 queries yearly.

SUBMISSIONS
Query with outline. Accepts hard copy. SASE. Responds in 1 month.

- Articles: 1,000–3,000 words. Informational articles and personal experience pieces. Topics include theater education, the arts, and curricula.
- Depts/columns: Word lengths vary. Classroom exercises, ideas, technical advice, and textbook or play suggestions.

SAMPLE ISSUE
30 pages (3–5% advertising): 3 articles; 2 depts/columns. Sample copy, $2 with 9x12 SASE ($2 postage). Guidelines available.

- "Serving All." Article explains how to incorporate students with disabilities into a theater class, and promotes the experience as being beneficial to teacher and class.
- "The Language of Movement." Article touts mime as a fundamental performance technique that can be a useful building block in student actors' skills.
- Sample dept/column: "Promptbook" explains how to create effective press releases and programs for theater productions.

RIGHTS AND PAYMENT
One-time rights. Written material, payment rates vary. Pays on publication. Provides 3 contributor's copies.

✎ EDITOR'S COMMENTS
Our readers are experienced theater teachers and artists who are looking to us for fresh, insightful ideas that they can use to advance their students' skills. We welcome writers who can provide that for us.

Teach Kids! Essentials

P.O. Box 348
Warrenton, MO 63383-0348

Editor

DESCRIPTION AND INTERESTS
Formerly listed as *Teach Kids!*, this newsletter offers Christian educators curriculum ideas, provides programming ideas, and inspires them to get the most out of their special ministry of educating children. It is geared toward teachers of children ages four to eleven. Circ: Unavailable.

Audience: Christian educators
Frequency: Monthly
Web: www.teachkidsforum.com/essentials

FREELANCE POTENTIAL
75% written by nonstaff writers. Publishes 50 freelance submissions yearly.

SUBMISSIONS
Query with outline; or send complete ms. Accepts hard copy and email submissions to TKeditor@cefpress.com. SASE. Response time varies.

- Articles: Word lengths vary. Informational and how-to articles; and personal experience pieces. Topics include religion, education, teaching techniques, classroom management, discipleship, curriculum, and programming and activity ideas.
- Depts/columns: Word lengths vary. Teaching tips, short news items, Bible lessons.

SAMPLE ISSUE
4 pages: 7 articles; 2 depts/columns. Sample copy available.

- "Including Kids with Special Needs." Article provides tips that teachers can employ to include students with special needs in classroom lessons.
- Sample dept/column: "Teach Kids! 9-1-1" offers ideas for bringing snack time into the Bible lessons.

RIGHTS AND PAYMENT
All, first, one-time, or electronic rights. Written material, payment rates vary. Payment policy varies.

✎ EDITOR'S COMMENTS
Writers should note that we have replaced our *Teach Kids!* magazine with this monthly newsletter. It has shorter articles and quick teaching tips designed to provide insights and ideas to spark teachers' creativity. We welcome all ideas from writers who have experience in Christian education.

Tech Directions

Prakken Publications
832 Phoenix Drive
P.O. Box 8623
Ann Arbor, MI 48107

Managing Editor: Susanne Peckham

DESCRIPTION AND INTERESTS
This magazine designed for technical and vocational educators features articles on today's technology and technical career development. Circ: 43,000.

Audience: Teachers and administrators
Frequency: 10 times each year
Website: www.techdirections.com

FREELANCE POTENTIAL
80% written by nonstaff writers. Publishes 40 freelance submissions yearly; 50% by unpublished writers, 50% by authors who are new to the magazine. Receives 192 mss yearly.

SUBMISSIONS
Query or send complete ms. Accepts hard copy and email submissions to susanne@ techdirections.com. SASE. Availability of artwork improves chance of acceptance. Responds to queries in 1 week, to mss in 1 month.

- Articles: To 3,000 words. Informational and how-to articles. Topics include teaching techniques and unusual projects in the fields of automotives, building construction, computers, drafting, electronics, graphics, hydraulics, industrial arts, lasers, manufacturing, radio and TV, robotics, software, welding, woodworking, and other vocational education.
- Depts/columns: Word lengths vary. Legislation updates; technology news and history; media reviews; and product information.
- Artwork: Color prints, slides, and transparencies; B/W prints. B/W line art. CAD plots.
- Other: Puzzles, games, and quizzes.

SAMPLE ISSUE
32 pages (40% advertising): 6 articles; 7 depts/columns. Sample copy, $5 with 9x12 SASE (2 first-class stamps). Guidelines available.

- "How Engineers Engineer." Article gives teachers insight into the engineering process.
- Sample dept/column: "Technology Today" explains wind tunnels.

RIGHTS AND PAYMENT
All rights. Articles, $50+. Depts/columns, to $25. Pays on publication. Provides 3 copies.

➡️EDITOR'S COMMENTS
We need hands-on projects for technology education in any field. Please be sure to include step-by-step instructions.

Techniques

ACTE
1410 King Street
Alexandria, VA 22314

Managing Editor: Susan Emeagwali

DESCRIPTION AND INTERESTS
Published by the Association for Career and Technical Education, *Techniques* offers success stories, solutions to problems, case studies, and profiles of interesting people within the career and technical education community. Circ: 35,000.

Audience: Educators
Frequency: 8 times each year
Website: www.acteonline.org

FREELANCE POTENTIAL
50% written by nonstaff writers. Publishes 10–20 freelance submissions yearly; 15% by unpublished writers, 30% by authors who are new to the magazine. Receives 96 queries and unsolicited mss yearly.

SUBMISSIONS
Query or send complete ms. Accepts email submissions to semeagwali@acteonline.org (Microsoft Word attachments). Availability of artwork improves chance of acceptance. Responds in 4 months.

- Articles: 1,000–2,500 words. Informational and how-to articles; case studies; and profiles. Topics include career and technical education programming, practices, and policy; integrating career education and academics; technology; and environmental issues.
- Depts/columns: "Research Report" to 2,500 words.
- Artwork: Digital images at 300 dpi. Charts and graphs.

SAMPLE ISSUE
62 pages (30% advertising): 10 articles; 11 depts/columns. Sample copy, guidelines, and editorial calendar available at website.

- "The World of Sports Medicine." Article describes several sports-related careers.
- Sample dept/column: "Research Report" discusses the application of teaching and learning principles.

RIGHTS AND PAYMENT
All rights. No payment.

➡️EDITOR'S COMMENTS
Our style is journalistic. This basically means "write what you need to write." Most of our features begin with the issue/concern/ highlight, move into the solution/options/ reason, and close with the results.

Teen

3000 Ocean Park Boulevard, Suite 3048
Santa Monica, CA 90405

Editor: Jane Fort

DESCRIPTION AND INTERESTS
Everything a tween and young teen girl is interested in is covered in this magazine—including boys, celebrities, TV shows and their stars, fashion, beauty, fitness, relationship advice, pop music, and popular culture. Circ: 650,000.

Audience: Girls, 12–16 years
Frequency: Quarterly
Website: www.teenmag.com

FREELANCE POTENTIAL
60% written by nonstaff writers. Of the freelance submissions published yearly, 5% is by authors who are new to the magazine. Receives 500 queries yearly.

SUBMISSIONS
Query for nonfiction. Send complete ms for fiction. Accepts hard copy and simultaneous submissions if identified. SASE. Responds in 2 months.

- Articles: 800 words. Informational and how-to articles; profiles; interviews; and personal experience pieces. Topics include relationships, beauty, fashion, music, popular culture, recreation, the arts, crafts, current events, and social issues.
- Fiction: 1,000 words. Genres include romance and inspirational fiction.
- Depts/columns: Word lengths vary. Advice.
- Other: Games, puzzles, and quizzes.

SAMPLE ISSUE
112 pages (10% advertising): 5 articles; 22 depts/columns; 5 quizzes and puzzles. Sample copy, $3.99 at newsstands.

- "Boys of Summer." Article profiles the Jonas Brothers as they enjoy the perks of teen superstardom.
- Sample dept/column: "Health" offers easy, at-home exercise tips from a professional fitness trainer.

RIGHTS AND PAYMENT
All rights. Written material, payment rates vary. Pays on publication. Provides 2 author's copies.

↪EDITOR'S COMMENTS
We look for ideas that are new and different, and presented from a fresh angle. We welcome a change from the same old articles about teen issues. Of course, writers must be able to present material in a way that appeals to our demographic.

Teen Graffiti

P.O. Box 452721
Garland, TX 75045-2721

Publisher: Sharon Jones-Scaife

DESCRIPTION AND INTERESTS
The pages of this magazine become the walls on which teenagers can communicate with each other. It features the concerns, opinions, and poetry of teenagers, and offers informative articles on serious topics such as education, relationships, money management, and individualism. Circ: 10,000.

Audience: 12–19 years
Frequency: 6 times each year
Website: www.teengraffiti.com

FREELANCE POTENTIAL
70% written by nonstaff writers. Publishes 30–40 freelance submissions yearly. Receives many queries and unsolicited mss yearly.

SUBMISSIONS
Query or send complete ms. Prefers email submissions to editor@teengraffiti.com; will accept hard copy. SASE. Response time varies.

- Articles: 250 words. Informational articles; personal experience and opinion pieces; and essays. Topics include college, careers, current events, popular culture, sex, health, and social issues.
- Depts/columns: 100–200 words. Advice and resources from teachers; teen-to-teen advice; and book, movie, and music reviews.
- Artwork: B/W and color prints from teens.
- Other: Poetry written by teens.

SAMPLE ISSUE
32 pages (3% advertising): 4 articles; 4 depts/columns; 6 poems. Sample copy, $2.75 with 9x12 SASE. Guidelines available.

- "Health: You Make the Choice." Article discusses the dangers of obesity and offers ideas readers can use to formulate a weight-loss plan.
- Sample dept/column: "Advice" offers ideas for managing money.

RIGHTS AND PAYMENT
One-time rights. No payment.

↪EDITOR'S COMMENTS
We are always interested in hearing good ideas for articles and columns, but they must be able to resonate with our audience. Our goal is to help teens express themselves not by defacing property, but by sharing their poems, thoughts, and experiences on our pages. If your idea can help, let's hear it!

Teen Tribute

71 Barber Greene Road
Toronto, Ontario M3C 2A2
Canada

Editor: Toni-Marie Ippolito

DESCRIPTION AND INTERESTS
Teen Tribute delivers the buzz on entertainment and teen culture, from the latest movies, TV shows, celebrities, video games, and music, to relationship advice and fashion tips. Circ: 300,000.

Audience: 14–18 years
Frequency: Quarterly
Website: www.teentribute.ca

FREELANCE POTENTIAL
10% written by nonstaff writers. Publishes 5–10 freelance submissions yearly; 1% by authors who are new to the magazine. Receives 24 queries yearly.

SUBMISSIONS
Query with clips or writing samples. Accepts hard copy and email queries to tippolito@tribute.ca. Availability of artwork improves chance of acceptance. SAE/IRC. Responds in 1–2 months.

- Articles: 400–500 words. Informational articles; profiles; interviews; and personal experience pieces. Topics include movies, the film industry, music, the arts, entertainment, popular culture, and social issues.
- Depts/columns: Word lengths vary. Media reviews, fashion and beauty tips, new product reviews, relationship advice, gear and gadgets, and horoscopes.
- Artwork: Color prints or transparencies.

SAMPLE ISSUE
38 pages (50% advertising): 9 articles; 10 depts/columns. Sample copy, $1.95 Canadian with 9x12 SAE/IRC ($.86 postage).

- "Sharkwater." Article previews a true-life adventure film about shark poaching.
- Sample dept/column: "Sound Bytes" offers reviews of CDs from The Shins, Bloc Party, and other recording artists.

RIGHTS AND PAYMENT
First serial rights. Written material, $100–$400 Canadian. Artwork, payment rates vary. Pays on publication. Provides 1 contributor's copy.

➤EDITOR'S COMMENTS
We cover everything that teens are talking about, especially new movies and the people who make them. We are always on the lookout for celebrity interviews that tie in with the latest film or music releases.

Teen Voices

P.O. Box 120-027
Boston, MA 02112-0027

Managing Editor: Rebecca Steinitz

DESCRIPTION AND INTERESTS
Teen Voices features articles on topics that specifically concern teen girls and young women, and challenges the stereotypes prevalent in mainstream magazines. Circ: 55,000.

Audience: YA
Frequency: Monthly (website);
Twice each year (print)
Website: www.teenvoices.com

FREELANCE POTENTIAL
95% written by nonstaff writers. Publishes 100 freelance submissions yearly; 95% by unpublished writers, 95% by authors who are new to the magazine. Receives 2,000 unsolicited mss each year.

SUBMISSIONS
Accepts mss written by girls ages 13–19 only. Send complete ms. SASE. Response time varies.

- Articles: Word lengths vary. Informational and self-help articles; interviews; and profiles. Topics include ethnic and religious traditions, the Internet, multicultural issues, surviving sexual assault, family relationships, teen motherhood, international life, disability, health, nutrition, cooking, the arts, the media, and activism.
- Fiction: Word lengths vary. Humorous, inspirational, contemporary, ethnic, and multicultural fiction.
- Depts/columns: Word lengths vary. "Read It," "Hear It," "Watch It," "Girl Talk," "Top 10," and "Say What?!"
- Other: Poetry. Comic strips.

SAMPLE ISSUE
58 pages (8% advertising): 7 articles; 7 depts/columns; 18 poems. Guidelines and editorial calendar available.

- "Got the Knowledge to Go to College?" Article provides a guide to help girls through the college admissions process.
- Sample dept/column: "Say What?!" criticizes ads for movies featuring large black women.

RIGHTS AND PAYMENT
First or one-time rights. No payment. Provides 5 contributor's copies.

➤EDITOR'S COMMENTS
We will consider any submission that fits our editorial guidelines and our mission: to create social change for girls and young women.

Teenwire.com

Texas Child Care Quarterly

434 West 33rd Street
New York, NY 10001

Editor: Amy Bryant

DESCRIPTION AND INTERESTS
Teenwire is an online magazine aimed at helping young adults make responsible choices regarding sex, drugs, alcohol, and education. It is published by the Planned Parenthood Federation of America. Hits per month: 500,000.
Audience: 13–21 years
Frequency: Daily
Website: www.teenwire.com

FREELANCE POTENTIAL
25% written by nonstaff writers. Publishes 100–156 freelance submissions yearly; 10% by unpublished writers, 25% by new authors.

SUBMISSIONS
Query with biography and 2–3 clips or writing samples. Accepts email queries to tw.staff@ppfa.org (Microsoft Word or URL attachments; include "Write for Teenwire" in subject line). Responds in 1 week.

- Articles: 600 words. Informational articles; profiles; and interviews. Topics include teen relationships; sexual health and activity; birth control; pregnancy; sexually transmitted diseases; lesbian, gay, and transgender issues; teen activism; international youth issues; college; careers; current events; the arts; popular culture; recreation; substance abuse; social concerns; and multicultural and ethnic issues.
- Other: Puzzles, games, and quizzes.

SAMPLE ISSUE
Sample copy and writers' guidelines available at website.

- "Dealing with Dating Abuse." Article describes the health effects of abusive relationships and a program that helps tweens and teens avoid them.
- "Not Cooties, But UTIs!" Article discusses the symptoms, treatment, and prevention of urinary tract infections in girls.
- "Surviving Puberty: Moods and Emotions." Article guides adolescents through the roller-coaster ride of pubescent feelings.

RIGHTS AND PAYMENT
All rights. Articles, $300. Pays on acceptance.

➡EDITOR'S COMMENTS
Start your query with a sample lede. Grab our attention and show us that you can write in a teen-friendly tone.

P.O. Box 162881
Austin, TX 78716-2881

Editor: Louise Parks

DESCRIPTION AND INTERESTS
This magazine offers practical and research-based articles about the development, care, and education of children from birth to age eight. Most of its readers are teachers and in-home child-care providers. Circ: 32,000.
Audience: Teachers and child-care workers
Frequency: Quarterly
Website: www.childcarequarterly.com

FREELANCE POTENTIAL
50% written by nonstaff writers. Publishes 12–15 freelance submissions yearly; 10% by unpublished writers, 50% by authors who are new to the magazine. Receives 24–36 unsolicited mss yearly.

SUBMISSIONS
Query with outline; or send complete ms. Accepts hard copy with disk and email to editor@childcarequarterly.com. No simultaneous submissions. SASE. Responds in 3 weeks.

- Articles: 2,500 words. Informational and how-to articles. Topics include child growth and development, school-family communication, health and safety, program administration, professional development, and hands-on activities.
- Depts/columns: Staff written.

SAMPLE ISSUE
44 pages (no advertising): 6 articles; 5 depts/columns. Sample copy, $6.25. Guidelines available at website.

- "Supporting Problem Solving in the Early Childhood Classroom." Article tells how to help young children think out and solve problems on their own.
- "The Best Toys—and Teaching Materials—Are (Almost) Free." Article gives an alphabetical list of household items that can be used to foster creativity and discovery in children.

RIGHTS AND PAYMENT
All rights. No payment. Provides 3 contributor's copies and a 1-year subscription.

➡EDITOR'S COMMENTS
We look for well-developed, active learning themes that promote children's social, cognitive, emotional, and physical growth. Activities must be age-appropriate, inexpensive to duplicate, support cooperative learning, value creative expression, and enhance self-esteem.

Thrasher

1303 Underwood Avenue
San Francisco, CA 94121

Managing Editor: Ryan Henry

DESCRIPTION AND INTERESTS
Thrasher offers comprehensive coverage of skateboarding and snowboarding in a way that appeals to teenage boys who are fanatical about these sports. Circ: 200,000.
Audience: Boys, 12–20 years
Frequency: Monthly
Website: www.thrashermagazine.com

FREELANCE POTENTIAL
20% written by nonstaff writers. Publishes 20 freelance submissions yearly; 100% by unpublished writers. Receives 72–120 unsolicited mss yearly.

SUBMISSIONS
Send complete ms. Prefers email submissions to info@thrashermagazine.com (Microsoft Word attachments); will accept Macintosh disk submissions, hard copy, and simultaneous submissions if identified. Availability of artwork improves chance of acceptance. SASE. Responds in 1 month.

- Articles: To 1,500 words. Informational articles; profiles; and interviews. Topics include skateboarding, snowboarding, and music.
- Fiction: To 2,500 words. Stories with skateboarding and snowboarding themes.
- Depts/columns: 750–1,000 words. News, tips, art, 'zines, and profiles.
- Artwork: Color prints or transparencies; 35mm B/W negatives. Line art.

SAMPLE ISSUE
220 pages (45% advertising): 12 articles; 10 depts/columns. Sample copy, $3.99. Guidelines available.

- "Rune Glifberg." Article profiles the pioneering vertical skater.
- Sample dept/column: "Trash" contains a brief about a commemorative skateboard.

RIGHTS AND PAYMENT
First North American serial rights. Written material, $.15 per word. Artwork, payment rates vary. Pays on publication. Provides 2 copies.

➥EDITOR'S COMMENTS
If you know what the title means, then you're halfway to qualifying as a contributor to this magazine. Don't even try to fake it. If you can't read our magazine, then you definitely can't write for it. Our readers can spot posers a mile away.

Tiger Beat

330 North Brand, Suite 1150
Glendale, CA 91203

Editor: Leesa Coble

DESCRIPTION AND INTERESTS
From heartthrobs to *Hannah Montana*, this colorful magazine promotes a "personal" relationship between tweens and their favorite young celebrities through interview excerpts, behind-the-scenes coverage, personality quizzes, and exclusive photos. Circ: 200,000.
Audience: 10–16 years
Frequency: Monthly
Website: www.tigerbeatmag.com

FREELANCE POTENTIAL
1% written by nonstaff writers. Publishes 2 freelance submissions yearly; 50% by authors who are new to the magazine. Receives 20 queries yearly.

SUBMISSIONS
Query with résumé and clips for celebrity angles only. Accepts hard copy and simultaneous submissions if identified. SASE. Responds in 3 months.

- Articles: To 700 words. Interviews and profiles. Topics include young celebrities in the film, television, and recording industries.
- Depts/columns: Staff written.
- Artwork: Color digital images.
- Other: Quizzes, contests, and posters. Submit seasonal material 3 months in advance.

SAMPLE ISSUE
108 pages (5% advertising): 16 articles; 10 depts/columns; 5 quizzes; 5 contests. Sample copy, $3.99 at newsstands.

- "Is Nick Ready for Love?" Interview with Nick Jonas of the Jonas Brothers discusses his approach to dating after his breakup with Miley Cyrus; includes a contest element.
- "My Sweet 16." Article tells how actress Selena Gomez would like to celebrate her milestone birthday.
- "Do Stars Get Jealous?" Article features quotes from celebrity friends on the topic.

RIGHTS AND PAYMENT
All rights. Written material, payment rates vary. Pays on publication. Provides 2 author's copies.

➥EDITOR'S COMMENTS
We are always looking for transcripts of interviews with pre-teen and teen celebrities, as well as coverage of the Hollywood events they attend. If you have access to these stars, we want to hear from you.

Time for Kids

Time-Life Building
1271 Avenue of the Americas
New York, NY 10020

Editor: Martha Pickerill

DESCRIPTION AND INTERESTS

It's like *Time* magazine, except it's written for kids. This publication, designed to be used as a teaching tool in the classroom, provides a great way for children to learn about the world around them. Circ: 4.1 million.

Audience: 5–12 years
Frequency: Weekly
Website: www.timeforkids.com

FREELANCE POTENTIAL

4% written by nonstaff writers. Publishes 4 freelance submissions yearly.

SUBMISSIONS

All work is assigned. Send résumé only. No queries or unsolicited mss. Accepts hard copy. Responds if interested.

- Articles: Word lengths vary. Informational and biographical articles. Topics include world news, current events, animals, education, health, fitness, science, technology, math, social studies, geography, multicultural and ethnic issues, music, popular culture, recreation, regional news, sports, travel, and social issues.
- Depts/columns: Word lengths vary. Profiles and short news items.
- Artwork: Color prints and transparencies.
- Other: Theme-related activities.

SAMPLE ISSUE

8 pages (no advertising): 3 articles; 4 depts/columns. Sample copy, $3.95.

- "Good News for Gorillas." Article reports that wildlife workers have discovered 125,000 more lowland gorillas than were previously thought to exist.
- "The Games Get Going." Article reports on events at the 2008 Summer Olympics in Beijing and explains what is at stake for the athletes.

RIGHTS AND PAYMENT

All rights. Written material, payment rates vary. Pays on publication.

➥EDITOR'S COMMENTS

We assign articles on a contract basis only, so don't send us manuscripts. Your résumé will give us an idea of your qualifications and if we want to pursue an arrangement with you. We are known for our engaging style, which draws students into the magazine.

Today's Catholic Teacher

2621 Dryden Road, Suite 300
Dayton, OH 45439

Editor-in-Chief: Mary Noschang

DESCRIPTION AND INTERESTS

Catholic educators turn to this magazine for information not found elsewhere, including developments in curriculum, testing, and technology; and coverage of national issues and trends. Circ: 50,000.

Audience: Teachers, grades K–12
Frequency: 6 times each year
Website: www.peterli.com

FREELANCE POTENTIAL

95% written by nonstaff writers. Publishes 20 freelance submissions yearly; 50% by authors who are new to the magazine. Receives 190+ queries and unsolicited mss yearly.

SUBMISSIONS

Query or send complete ms. Accepts hard copy, disk submissions with hard copy, email submissions to mnoschang@peterli.com, and simultaneous submissions. SASE. Responds to queries in 1 month, to mss in 3 months.

- Articles: 600–1,500 words. Informational, self-help, and how-to articles. Topics include technology, fundraising, classroom management, curriculum development, administration, and educational issues and trends.
- Depts/columns: Word lengths vary. Opinions, news, software, character development, curricula, teaching tools, and school profiles.
- Artwork: 8x10 color prints, slides, or transparencies.
- Other: Classroom-ready reproducible activity pages.

SAMPLE ISSUE

62 pages (45% advertising): 7 articles; 8 depts/columns. Sample copy, $3. Writer's guidelines available.

- "Fostering Moral Development." Article explores school-wide conditions that facilitate student moral development.
- Sample dept/column: "When Students Ask" explains workers' rights from the Catholic viewpoint.

RIGHTS AND PAYMENT

All rights. Written material, $100–$250. Pays on publication. Provides contributor's copies.

➥EDITOR'S COMMENTS

Query letters are encouraged. We always seek practical information that can be used by classroom teachers.

Today's Christian Woman

465 Gundersen Drive
Carol Stream, IL 60188

Assistant Editor: Andrea Bianchi

DESCRIPTION AND INTERESTS
The articles in *Today's Christian Woman* are designed to encourage women to engage their culture and world with God's grace and redeeming power. It covers practical topics as well as contemporary issues faced by women. Circ: 210,000.

Audience: Women
Frequency: 6 times each year
Website: www.todayschristianwoman.com

FREELANCE POTENTIAL
85% written by nonstaff writers. Publishes 65 freelance submissions yearly; 5% by unpublished writers. Receives 1,200 queries yearly.

SUBMISSIONS
Query with résumé and author credentials. Accepts hard copy. No simultaneous submissions. SASE. Responds in 2–3 months.

- Articles: 1,000–2,000 words. Informational and self-help articles; personal experience pieces; and humor. Topics include parenting, family issues, relationships, spiritual living, contemporary women's concerns, and turning points in life.
- Depts/columns: 100–300 words. First-person narratives, reviews, and pieces on parenting and faith.

SAMPLE ISSUE
56 pages (25% advertising): 12 articles; 6 depts/columns. Sample copy, $5 with 9x12 SASE ($3.19 postage). Guidelines available.

- "Married in the Middle." Personal experience piece details the challenges and triumphs of marrying midway through life.
- Sample dept/column: "Great Question" offers answers to whether or not the book *The Secret* contains biblical truth.

RIGHTS AND PAYMENT
First rights. Written material, $.20 per word. Pays on acceptance. Provides 2 author's copies.

⁌EDITOR'S COMMENTS
We don't shy away from covering the tough issues real women face. Life is messy, and we seek to acknowledge that reality by avoiding pat answers and formulaic conclusions. We look for authors who do the same. Your query should clearly state the purpose of your article, and all articles must feature an evangelical perspective.

Today's Parent

1 Mount Pleasant Road, 8th Floor
Toronto, Ontario M4Y 2Y5
Canada

Managing Editor

DESCRIPTION AND INTERESTS
This Canadian parenting magazine strikes a balance between the lighthearted and the investigative, with an overall emphasis on the positive. Its goal is to support its readers in making good parenting choices. Circ: 215,000.

Audience: Parents
Frequency: Monthly
Website: www.todaysparent.com

FREELANCE POTENTIAL
Of the freelance submissions published yearly, many are by unpublished writers and authors who are new to the magazine. Receives many queries yearly.

SUBMISSIONS
Query with clips and projected word count. Accepts hard copy, faxes to 416-764-2801, and email queries to queries@tpg.rogers.com. SAE/IRC. Responds in 6 weeks if interested.

- Articles: 1,500–2,500 words. Informational, how-to, and self-help articles. Topics include parenting, discipline, child development, health, nutrition, pregnancy, and childbirth.
- Depts/columns: "Your Turn," 650 words. "Mom Time," "Education," "Health," and "Behaviour," 1,200 words.

SAMPLE ISSUE
146 pages (15% advertising): 6 articles; 13 depts/columns. Sample copy, $4.50 Canadian at newsstands. Guidelines available.

- "Separation Anxiety Survival Guide." Article tells why separation anxiety occurs and how to cope with it.
- "The Babysitter Diaries." Article shares what babysitters across Canada love and hate about caring for kids.
- Sample dept/column: "Mom Time" tells how motherhood makes women stronger.

RIGHTS AND PAYMENT
All North American serial rights. Articles, $1,500–$2,500. "Your Turn," $200. Other depts/columns, $1,200. Pays 1 month after acceptance. Provides 2 contributor's copies.

⁌EDITOR'S COMMENTS
Because we promote ourselves as a Canadian magazine, we favor Canadian writers. Queries should have a specific hook, not just a subject area, and should have wide appeal for a national audience.

Toledo Area Parent News

1120 Adams Street
Toledo, OH 43624

Managing Editor: Jason Webber

DESCRIPTION AND INTERESTS
Informative and inspirational articles on parenting topics and family issues fill this magazine for parents living in the Toledo region of Ohio. It also presents local information on services, events, and entertainment. Circ: 81,000.

Audience: Parents
Frequency: Monthly
Website: www.toledoparent.com

FREELANCE POTENTIAL
75% written by nonstaff writers. Publishes 12 freelance submissions yearly; 10% by unpublished writers, 20% by authors who are new to the magazine. Receives 48 queries and unsolicited mss yearly.

SUBMISSIONS
Query with clips; or send complete ms. Prefers email to jwebber@toledoparent.com; will accept hard copy. SASE. Responds in 1 month.

- Articles: 700–2,000 words. Informational articles; profiles; and interviews. Topics include family issues, parenting, teen issues, education, social issues, health, and fitness.
- Depts/columns: Word lengths vary. Restaurant reviews; brief news items related to family issues.

SAMPLE ISSUE
48 pages (60% advertising): 1 article; 9 depts/columns; 1 calendar. Sample copy available via email request to kdevol@toledocitypaper.com. Guidelines available. Editorial calendar available at website.

- "School's Back!" Article offers tips parents can use to make their children's transition back to school run smoothly.
- Sample dept/column: "Parent Profile" spotlights the executive director of "Mom's House," an organization that supports young single mothers.
- "Proud Parvin." Article profiles the Sylvania Children's Center and its innovative director.

RIGHTS AND PAYMENT
All North American serial rights. Written material, $30–$200. Pays on publication.

•♦EDITOR'S COMMENTS
We welcome articles that bring something new to the parenting discussion, or that can answer a question or solve a problem of a Toledo parent.

Transitions Abroad

P.O. Box 745
Bennington, VT 05201

Editor/Publisher: Sherry Schwarz

DESCRIPTION AND INTERESTS
This e-zine is all about cultural-immersion travel. It caters to travelers who want to extend their time abroad through work, study, or low-cost travel. Hits per month: 12,000.

Audience: YA–Adult
Frequency: 6 times each year
Website: www.transitionsabroad.com

FREELANCE POTENTIAL
90% written by nonstaff writers. Publishes 250 freelance submissions yearly; 70% by unpublished writers, 90% by new authors. Receives 300–420 queries and unsolicited mss yearly.

SUBMISSIONS
Prefers query with outline; will accept complete ms with bibliography. Prefers email submissions to editor@transitionsabroad.com; will accept hard copy. Availability of artwork improves chance of acceptance. SASE. Responds in 1–2 months.

- Articles: To 1,500 words. Informational and how-to articles. Topics include living, working, and studying abroad; immersion travel by region; and cultural travel trends.
- Depts/columns: Word lengths vary. Budget travel, responsible travel, volunteer travel, adventure travel, and senior travel; interviews; food; music; and book reviews.
- Artwork: Color slides or prints. High-resolution JPEG or TIFF images.

SAMPLE ISSUE
Sample copy, guidelines, and editorial calendar available at website.

- "Safe Solo Travel for Women: Ten Rules." Article provides tips and advice for women who travel alone.
- "A Walking Tour Through Tuscia, Italy." Article reveals how best to explore this undiscovered corner of Italy.

RIGHTS AND PAYMENT
First rights. Written material, $2 per column inch; minimum $25. Artwork, payment rates vary. Pays on publication.

•♦EDITOR'S COMMENTS
Our writers must provide the "nuts and bolts" that readers need to make their immersion travel experience a success. We are looking for usable, first-hand experience from writers who themselves travel abroad.

Treasure Valley Family

13191 West Scotfield Street
Boise, ID 83713-0899

Publisher: Liz Buckingham

DESCRIPTION AND INTERESTS

A free publication for Idaho parents, *Treasure Valley Family* covers local activities as well as the general themes of health, education, summer camp, sports, and travel. Circ: 20,000.

Audience: Parents
Frequency: 10 times each year
Website: www.treasurevalleyfamily.com

FREELANCE POTENTIAL

95% written by nonstaff writers. Publishes 20 freelance submissions yearly; 1–2% by authors who are new to the magazine. Receives 500 queries and unsolicited mss yearly.

SUBMISSIONS

Prefers query with clips; will accept complete ms with 1-sentence author bio. Accepts hard copy and email submissions to magazine@ treasurevalleyfamily.com (Microsoft Word attachments). SASE. Responds in 2–3 months.

- Articles: 1,000–1,300 words. Informational and how-to articles. Topics include health, preschool, child care, education, summer camp, sports, travel, recreation, teenage issues, college, party planning, crafts, hobbies, and the arts.
- Depts/columns: 700–900 words. Events, activities, advice, news, product reviews, age-specific issues, book reviews, women's health, profiles of local agencies, and interviews with area families.

SAMPLE ISSUE

58 pages (45% advertising): 3 articles; 10 depts/columns. Sample copy, free with 9x12 SASE ($1.50 postage). Guidelines available.

- "Why, When and How to Set Limits." Article provides guidance on discipline and explains the effects of different parenting styles.
- "Lake Hikes for Summer Fun." Article gives an overview of several Idaho parks.
- Sample dept/column: "Bits & Pieces" offers sun safety tips for children of various ages.

RIGHTS AND PAYMENT

First North American serial rights. All material, payment rates vary. Pays 10% more for Web rights. Pays on publication. Provides 2 contributor's copies.

➥EDITOR'S COMMENTS

In assigning or purchasing a story, we understand that we will receive market exclusivity.

Tulsa Kids Magazine

1820 South Boulder Avenue, Suite 400
Tulsa, OK 74119-4409

Editor: Betty Casey

DESCRIPTION AND INTERESTS

This award-winning regional publication takes pride in "serving children, informing parents, and enriching families." It is read by parents in the greater Tulsa area who wish to stay apprised of child-related trends and events. Circ: 20,000.

Audience: Parents
Frequency: Monthly
Website: www.tulsakids.com

FREELANCE POTENTIAL

99% written by nonstaff writers. Publishes 100+ freelance submissions yearly; 5% by unpublished writers, 1% by authors who are new to the magazine. Receives 1,200 unsolicited mss yearly.

SUBMISSIONS

Send complete ms. Accepts hard copy, disk submissions, and simultaneous submissions if identified. SASE. Responds in 2–3 months.

- Articles: 500–800 words. Informational articles; profiles; interviews; humor; and personal experience pieces. Topics include family life, education, parenting, recreation, entertainment, college, health, fitness, careers, crafts, and social issues.
- Depts/columns: 100–300 words. News, book reviews, safety, and family cooking.

SAMPLE ISSUE

52 pages (50% advertising): 2 articles; 11 depts/columns. Sample copy, free with 10x13 SASE ($.75 postage). Guidelines available.

- "So Much to Do, So Little Time Between Naps." Article suggests 10 fun things to do with toddlers in tow.
- "OK Mozart: A Family Classic." Article describes the OK Mozart Festival held each year in Bartlesville, Oklahoma.
- Sample dept/column: "Get Cooking" provides recipes for Fourth of July treats.

RIGHTS AND PAYMENT

One-time rights. Written material, $25–$100. Payment policy varies. Provides 1 author's copy.

➥EDITOR'S COMMENTS

We would like to receive more short, punchy travel articles, and cooking articles with four or five recipes included. Please do not send us personal essays or very long articles! We can't accommodate them.

Turtle

Children's Better Health Institute
1100 Waterway Boulevard
P.O. Box 567
Indianapolis, IN 46206-0567

Editor: Terry Harshman

DESCRIPTION AND INTERESTS
Designed for preschoolers, *Turtle* magazine is filled with colorfully illustrated read-aloud stories, poems, and rebuses, as well as games and activities that keep little minds thinking. Circ: 382,000.

Audience: 2–5 years
Frequency: 6 times each year
Website: www.turtlemag.org

FREELANCE POTENTIAL
20% written by nonstaff writers. Publishes 40 freelance submissions yearly.

SUBMISSIONS
Send complete ms. Accepts hard copy. SASE. Responds in 3 months.

- Articles: To 500 words. Informational articles and book reviews. Topics include health, fitness, nutrition, nature, the environment, science, hobbies, and crafts.
- Fiction: 100–150 words for rebus stories. Genres include mystery; adventure; fantasy; humor; problem-solving stories; and contemporary, ethnic, and multicultural fiction.
- Other: Puzzles, activities, and games. Poetry, 4–8 lines. Submit seasonal material 8 months in advance.

SAMPLE ISSUE
34 pages (6% advertising): 2 articles; 3 stories; 13 activities. Sample copy, $1.75 with 9x12 SASE. Guidelines available.

- "Ella's Castle Mystery." Story tells of a little girl who tries to solve the mystery of what happened to her sandcastle.
- "Friends." Story shows how friends come in all shapes and sizes, and that each one has its own value.
- "Measuring Fun." Activity teaches volume measurement using cups, rice, and measuring spoons.

RIGHTS AND PAYMENT
All rights. Articles and fiction, to $.32 per word. Other material, payment rates vary. Pays on publication. Provides up to 10 author's copies.

❖EDITOR'S COMMENTS
At this time, we are accepting poems, short rebuses, easy recipes, and experiments only. We have a sizeable backlog of fiction and nonfiction items. Our website will alert you if this current policy changes.

Twins

5748 South College Avenue, Unit D
Fort Collins, CO 80525

Editor-in-Chief: Christa D. Reed

DESCRIPTION AND INTERESTS
Parents of twins, triplets, and other multiples read this magazine for informative articles on baby care, child development, nutrition, and travel. New product information is also included. It has been published since 1984. Circ: 40,000.

Audience: Parents
Frequency: 6 times each year
Website: www.twinsmagazine.com

FREELANCE POTENTIAL
50% written by nonstaff writers. Publishes 10–12 freelance submissions yearly. Receives 252 queries yearly.

SUBMISSIONS
Query. Accepts email queries to editor@ businessword.com. Responds in 3 months.

- Articles: 800–1,300 words. Informational and how-to articles; profiles; and personal experience pieces. Topics include parenting, family life, health, fitness, education, music, the arts, house and home, nutrition, diet, sports, social issues, crafts, and hobbies.
- Depts/columns: To 800 words. News, new product information, opinion pieces, and short items on child development.

SAMPLE ISSUE
56 pages (30% advertising): 9 articles; 15 depts/columns. Sample copy, $5.50. Guidelines available at website.

- "To Maui and Back with Twins." Article details a family trip and includes tips for a smooth vacation.
- "Hypotonia: A Moving Story." Article shares the story of a girl diagnosed with hypotonia and explains how her twin helped with her treatment and progress.
- Sample dept/column: "Family Fare" shares a healthy take-along family meal.

RIGHTS AND PAYMENT
All rights. Written material, $50–$100. Pays on publication. Provides 2 contributor's copies.

❖EDITOR'S COMMENTS
We are always looking for stories that relate to the joys and challenges of parenting twins and multiples. Articles should be written in a friendly, conversational tone with a happy ending. Please do not send material that does not pertain to our audience.

U Mag

United Services Automotive Association
9800 Fredericksburg Road
San Antonio, TX 78288-0264

Editor

DESCRIPTION AND INTERESTS
Targeting the children of United Services Automotive Association members (USAA), U Mag seeks to educate its readers about relevant topics while preparing them to become responsible adults. Many of its readers are current or former military dependents. Circ: 600,000.

Audience: 8–12 years
Frequency: Quarterly
Website: www.usaa.com/umag

FREELANCE POTENTIAL
90% written by nonstaff writers. Publishes 10 freelance submissions yearly; 1% by unpublished writers, 5% by authors who are new to the magazine. Receives 20 queries yearly.

SUBMISSIONS
Query with résumé and clips. Accepts hard copy. SASE. Responds in 6–8 weeks if interested.

- Articles: Word lengths vary. Informational, self-help, and how-to articles; profiles; interviews; and personal experience pieces. Topics include hobbies, history, math, music, popular culture, current events, science, technology, social issues, the environment, the arts, travel, finance, and safety issues.
- Other: Puzzles, games, activities, and jokes.

SAMPLE ISSUE
15 pages (no advertising): 4 articles; 3 activities. Sample copy, free with 9x12 SASE ($2 postage). Guidelines and theme list available.

- "Don't Mope, Cope." Article offers suggestions for children coping with the absence of a parent.
- "Best Foot Forward." Profile features a girl and her passion for ballet.
- "Gadget Guy." Article profiles a man and his many inventions.

RIGHTS AND PAYMENT
All rights. Written material, payment rates vary. Pays on acceptance. Provides 3–5 copies.

➥EDITOR'S COMMENTS
We are more interested in concepts and ideas from writers. Please refer to our theme list when sending submissions. If you are submitting a profile query, we would prefer that it be about a USAA member. Of particular interest are age-appropriate activity ideas for boys and girls.

The Universe in the Classroom

Astronomical Society of the Pacific
390 Ashton Avenue
San Francisco, CA 94112

Editor: Anna Hurst

DESCRIPTION AND INTERESTS
An electronic educational newsletter, The Universe in the Classroom is for teachers who want to help students of all ages learn more about the wonders of the universe through astronomy. Each issue covers a particular topic of current astronomical interest, complete with classroom activities and resource links. Hits per month: 10,000.

Audience: Teachers
Frequency: Quarterly
Website: www.astrosociety.org

FREELANCE POTENTIAL
75% written by nonstaff writers. Publishes 8 freelance submissions yearly; 10% by unpublished writers, 75% by authors who are new to the magazine. Receives 12 queries yearly.

SUBMISSIONS
Query. Accepts email queries to astroed@astrosociety.org. Availability of artwork improves chance of acceptance. Responds in 1 month.

- Articles: 3,000 words. Informational and factual articles; and classroom activities. Topics include astronomy and astrobiology.
- Artwork: Color prints and transparencies.
- Other: Hands-on classroom activities.

SAMPLE ISSUE
Sample copy available at website.

- "A Silent Cry for Dark Skies." Article discusses the effects of light pollution on humans, plants, and animals.
- "Invisible Galaxies." Article explains the concept of dark matter using simple language and a playful anecdote.
- "How Fast Are You Moving When You Are Sitting Still?" Article provides an in-depth look at the movement of the Earth, sun, galaxy, and universe.

RIGHTS AND PAYMENT
One-time rights. No payment.

➥EDITOR'S COMMENTS
You can help teachers all over the world bring astronomy to their students by writing an article or translating the newsletter into another language. Each issue is designed for teachers who want to learn more about astronomy themselves while also learning how to bring it alive for their students.

USA Gymnastics

Pan American Plaza
201 South Capitol Avenue, Suite 300
Indianapolis, IN 46225

Publication Director: Luan Peszek

DESCRIPTION AND INTERESTS
As the official publication of USA Gymnastics, the governing body for the sport of gymnastics in the United States, this magazine covers events, meets, and training techniques, and offers profiles of gymnasts. Circ: 100,000.

Audience: 10+ years
Frequency: 6 times each year
Website: www.usa-gymnastics.org

FREELANCE POTENTIAL
5% written by nonstaff writers. Publishes 5–10 freelance submissions yearly. Receives 6 queries and unsolicited mss yearly.

SUBMISSIONS
Query or send complete ms. Accepts email submissions to publications@usa-gymnastics.org. Responds in 1–2 months.

- Articles: To 1,000 words. Informational articles; profiles; and personal experience pieces. Topics include all issues related to gymnastics; sports psychology; and nutrition.
- Depts/columns: Word lengths vary. Event schedules and results, gym updates.

SAMPLE ISSUE
50 pages (35% advertising): 3 articles; 4 depts/columns. Sample copy, $3.95. Guidelines available.

- "Bhavsar Added to 2008 U.S. Olympic Team." Article reports on the newest member of the national men's gymnastics team, replacing another gymnast who withdrew.
- "USA Gymnastics Finalizes 2008 U.S. Olympic Team for Women's Gymnastics." Article reports on the six women and three replacement athletes that comprised the U.S. women's team at the Beijing Olympics.

RIGHTS AND PAYMENT
All rights. Written material, payment rates vary. Pays on publication.

➡EDITOR'S COMMENTS
More than 110,000 athletes and professionals are members of USA Gymnastics—more than 90,000 athletes are registered in competitive programs, and more than 20,000 are professional instructors and club members. We tell you this to emphasize that all of our readers are serious about gymnastics; therefore, our writers must be thoroughly familiar and personally involved with this sport.

U-Turn Magazine

United Services Automotive Association
9800 Fredericksburg Road
San Antonio, TX 78288-0264

Editor

DESCRIPTION AND INTERESTS
Current and former military dependents are the audience for this free publication of the United Services Automotive Association (USAA). Its goal is to strengthen the character and knowledge of its young members by providing valuable, positive information and advice. Circ: 500,000.

Audience: 13–17 years
Frequency: Quarterly
Website: www.usaa.com/uturn

FREELANCE POTENTIAL
80% written by nonstaff writers. Publishes 5 freelance submissions yearly; 5% by unpublished writers, 10% by authors who are new to the magazine. Receives 20 queries yearly.

SUBMISSIONS
Query with clips. Accepts hard copy and email queries to uturn@usaa.com. SASE. Responds in 6–8 weeks if interested.

- Articles: Word lengths vary. Informational articles; profiles; and personal experience pieces. Topics include the arts, college, careers, hobbies, current events, history, music, popular culture, recreation, science, technology, social issues, sports, travel, teen driving, money, safety, and relationships.
- Other: Puzzles, games, activities, jokes, and quizzes.

SAMPLE ISSUE
32 pages (no advertising): 3 articles; 4 depts/columns. Sample copy, free with 9x12 SASE ($2 postage). Writer's guidelines and theme list available.

- "Redo My Room." Article follows two teens as they shop for supplies and make over rooms in their homes.
- Sample dept/column: "College Knowledge" suggests unusual scholarships to investigate.

RIGHTS AND PAYMENT
All rights. Written material, payment rates vary. Pays on acceptance. Provides 3–5 copies.

➡EDITOR'S COMMENTS
Please read previous issues of our magazine to understand the focus and tone of our articles. We are interested in the following topics: earning, saving, and spending money; driving; preparing for college; military life; and any other matters relevant to teens.

U.25

United Services Automotive Association
9800 Fredericksburg Road
San Antonio, TX 78288

Editor: Carol Barnes

DESCRIPTION AND INTERESTS
Education, career advice, financial planning, and lifestyle choices are among the topics covered in this publication for teens and young adults. Distributed to members of the United Services Automotive Association (USAA), its purpose is to prepare its readers for adulthood. Circ: 500,000.

Audience: 18–24 years
Frequency: Quarterly
Website: www.usaa.com

FREELANCE POTENTIAL
90% written by nonstaff writers. Publishes 10–15 freelance submissions yearly. Receives 60 queries yearly.

SUBMISSIONS
Query with résumé and clips for feature articles. Send complete ms for shorter pieces. Accepts hard copy. Mss are not returned. Responds in 6–8 weeks.

- Articles: 1,000 words. Shorter pieces, 300 words. Informational and how-to articles; profiles; interviews; and personal experience pieces. Topics include college, careers, saving and investing money, driving, and lifestyle issues.
- Depts/columns: Word lengths vary. News items, money advice, and career guidance.
- Other: Activities, games.

SAMPLE ISSUE
30 pages: 4 articles; 6 depts/columns. Sample copy, free with 9x12 SASE ($1.80 postage). Guidelines available.

- "Drive Like A Pro." Article offers driving tips from several professionals who spend time on the road.
- "Recycling By Design." Article provides ideas for reusing common household items.
- Sample dept/column: "Career Coach" answers questions about work life.

RIGHTS AND PAYMENT
All rights. Feature articles, $500–$1,000. Other material, payment rates vary. Kill fee varies. Pays on acceptance.

➥EDITOR'S COMMENTS
Our articles focus on ideas that help our readers be successful at life and at work. Please familiarize yourself with the tone and style of our writing.

Vancouver Family Magazine

P.O. Box 820264
Vancouver, WA 98682

Editor: Nikki Klock

DESCRIPTION AND INTERESTS
Families living in the Clark County region of Washington turn to this magazine for informational articles on parenting, education, and childcare issues, as well as recreational events in the area. It specializes in presenting local information about places to go, things to do, and where to have family fun. Circ: 7,000.

Audience: Families
Frequency: Monthly
Web: www.vancouverfamilymagazine.com

FREELANCE POTENTIAL
100% written by nonstaff writers. Publishes 12 freelance submissions yearly; 30% by unpublished writers, 40% by authors who are new to the magazine. Receives 600 queries each year.

SUBMISSIONS
Query. Accepts hard copy. SASE. Response time varies.

- Articles: Word lengths vary. Informational articles. Topics include parenting, family-related issues, health and fitness, relationships, and recreation.
- Depts/columns: Word lengths vary. Parenting and family issues, local family-related businesses, local news.

SAMPLE ISSUE
32 pages (50% advertising): 5 articles; 3 depts/columns. Editorial calendar available.

- "Rock 'N' Roll Mama." Article profiles a local mother who moonlights as a disc jockey.
- "A Place to Be Me." Article spotlights a center that enables specially challenged children to play and make music.
- Sample dept/column: "Local Business Spotlight" features a local restaurant.

RIGHTS AND PAYMENT
Rights vary. Assigned articles, $.15 per word. Payment policy varies.

➥EDITOR'S COMMENTS
We give preference to freelance writers who live in the Vancouver region. Local issues, goods, and services; local events; and features spotlighting local businesses are the most important parts of our editorial content. We prefer our writers to have first-hand knowledge of the area. New medical research is always of interest to us.

VegFamily

4920 Silk Oak Drive
Sarasota, FL 34232

Editor: Cynthia Mosher

DESCRIPTION AND INTERESTS
This online "magazine of vegan family living" provides not only meat- and dairy-free recipes, but also expert advice on nutrition, parenting, animal rights, and environmental concerns. Hits per month: 10,000.
Audience: YA–Adult
Frequency: Monthly
Website: www.vegfamily.com

FREELANCE POTENTIAL
50% written by nonstaff writers. Publishes 150+ freelance submissions yearly; 90% by unpublished writers, 50% by authors who are new to the magazine. Receives 60 queries yearly.

SUBMISSIONS
Query. Accepts email queries to cynthia@ vegfamily.com. Responds in 2 weeks.

- Articles: 700+ words. Informational, self-help, and how-to articles; profiles; and personal experience pieces. Topics include vegan pregnancy and health, vegan cooking, natural parenting, animal rights, the environment, and green living.
- Depts/columns: 700+ words. Nutrition advice, recipes, cooking tips, health issues, vegan news, opinions, family profiles, and parenting issues by age group.
- Artwork: JPEG and GIF files.
- Other: Activities. Submit seasonal material 2 months in advance.

SAMPLE ISSUE
Sample copy available at website.

- "Recession-Proof Your Diet: Go Vegan." Article tells how maintaining a vegan diet nullifies the high cost of animal products.
- "Camp Cooking: Thinking Outside the Recipe in the Great Outdoors." Article suggests ways to prepare nourishing foods over a campfire.
- Sample dept/column: "Vegan Children" provides swimming safety tips.

RIGHTS AND PAYMENT
First and electronic rights. Articles, $10–$30. Pays on publication.

✦EDITOR'S COMMENTS
We accept submissions of any material dealing with the vegan lifestyle or green living, as long as it is a minimum of 700 words long. We are solely a digital magazine, available only online. Please see our homepage for sample content.

VerveGirl

401 Richmond Street West, Suite 245
Toronto, Ontario M5V 1X5
Canada

Editor-in-Chief: Sara Graham

DESCRIPTION AND INTERESTS
VerveGirl offers all the stuff that teenage girls obsess over while also recognizing that they have brains. Therefore, it features the usual fashion, beauty, and entertainment coverage alongside thoughtful articles on such topics as education and social service. Each issue is printed in two language editions. Circ: 150,000 English; 30,000 French.
Audience: Girls, 14–18 years
Frequency: 8 times each year
Website: www.vervegirl.com

FREELANCE POTENTIAL
75% written by nonstaff writers. Publishes 10–20 freelance submissions yearly. Receives 200 queries yearly.

SUBMISSIONS
Query. Accepts hard copy. SAE/IRC. Response time varies.

- Articles: Word lengths vary. Informational and self-help articles; profiles; interviews; and personal experience pieces. Topics include health, nutrition, fitness, fashion, beauty, social issues, current events, the environment, education, careers, and music.
- Depts/columns: Word lengths vary. Entertainment, fashion and beauty, and health and wellness.
- Other: Quizzes.

SAMPLE ISSUE
50 pages: 5 articles; 5 depts/columns. Sample copy available at website.

- "Building Hope By Building Homes." Article describes how two 19-year-old women became involved with Habitat for Humanity.
- "Classroom Culture." Article examines how immigrant students are treated in Canadian high schools.
- Sample dept/column: "Fashion & Beauty" includes a description of the new anti-blister shoe from MBT.

RIGHTS AND PAYMENT
Rights vary. Written material, payment rates vary. Pays on publication.

✦EDITOR'S COMMENTS
We are especially interested in stories of Canadian girls' involvement with social and environmental causes, including those that take them abroad.

Voice of Youth Advocates

4501 Forbes Boulevard, Suite 200
Lanham, MD 20706

Editor: Stacy L. Creel

DESCRIPTION AND INTERESTS
Voice of Youth Advocates is a professional journal devoted exclusively to the informational needs of teenagers. It is read by librarians, educators, and other professionals who work with teens, and deals primarily with library services, intellectual freedom issues, and youth advocacy. Circ: 7,000.
Audience: Professionals who work with youth
Frequency: 6 times each year
Website: www.voya.com

FREELANCE POTENTIAL
95% written by nonstaff writers. Publishes 50 freelance submissions yearly; 5% by unpublished writers, 60% by authors who are new to the magazine. Receives 60 queries yearly.

SUBMISSIONS
Query with résumé, synopsis, and market analysis. Accepts hard copy and email queries to editors@voya.com. Availability of artwork improves chance of acceptance. SASE. Responds in 2–4 months.

- Articles: 750–3,500 words. Informational and how-to articles; book reviews; and book lists. Topics include young adult literature, contemporary authors, and library programs.
- Depts/Columns: Staff written.
- Artwork: B/W and color photos. Illustrations.
- Other: Submit seasonal material 1 year in advance.

SAMPLE ISSUE
96 pages (20% advertising): 4 articles; 4 depts/columns; 180 reviews. Sample copy, free with 9x12 SASE. Guidelines available at website.

- "Tune in Online." Article extols the virtues of graphic novels and webcomics as ways to get teens excited about stories.
- "Go Where the Teens Are." Article describes innovative methods for marketing library services online to teens.

RIGHTS AND PAYMENT
All rights. Written material, $50–$125. Pays on publication. Provides 3 contributor's copies.

➦EDITOR'S COMMENTS
Most of our writers have experience working with teens in reading-related programs. We are interested in articles about the use of technology in the library, and about ways to get teens involved in their libraries.

Voices from the Middle

University of Texas at San Antonio
Dept. of Interdisciplinary Learning & Teaching
One UTSA Circle
San Antonio, TX 78249

Submissions

DESCRIPTION AND INTERESTS
Voices from the Middle devotes each issue to one concept related to literacy and learning at the middle school level. Authentic classroom practices, reviews of adolescent literature, technology information, and reviews of professional resources are included in each issue. Circ: 9,000.
Audience: Teachers
Frequency: Quarterly
Website: www.ncte.org/pubs/journals/vm

FREELANCE POTENTIAL
70% written by nonstaff writers. Publishes 12 freelance submissions yearly; 60% by unpublished writers, 85% by authors who are new to the magazine. Receives 150 unsolicited mss each year.

SUBMISSIONS
Send 3 copies of complete ms. Accepts hard copy and email to voices@utsa.edu (Microsoft Word attachments; include issue for which you are submitting in the subject line). SASE. Responds in 3–5 months.

- Articles: 2,500–4,000 words. Educational articles and personal experience pieces related to the issue's theme. Topics include middle school language arts and English instruction.
- Depts/columns: Staff written.

SAMPLE ISSUE
64 pages (9% advertising): 8 articles; 12 depts/columns. Sample copy, $6. Guidelines and theme list available at website.

- "Connecting Literacy and Learning Through Collaborative Action Research." Article explores research from a recent study on using argument during literature discussions.
- "Bumps in the Road: Textese." Article describes the quandry faced by teachers and parents about the use of textese vs standard spelling.

RIGHTS AND PAYMENT
First and second rights. No payment. Provides 2 contributor's copies.

➦EDITOR'S COMMENTS
Submissions should include specific classroom practices that are grounded in research and explore the connection between theory and practice.

Washington Family Magazine

485 Spring Park Place, Suite 550
Herndon, VA 20170

Managing Editor: Marae Leggs

DESCRIPTION AND INTERESTS
Parents living in and around Washington, D.C., pick up this free magazine to read about all aspects of family life in the area. Circ: 100,000.

Audience: Parents
Frequency: Monthly
Website: www.washingtonfamily.com

FREELANCE POTENTIAL
75% written by nonstaff writers. Publishes 90 freelance submissions yearly; 50% by unpublished writers, 50% by authors who are new to the magazine. Receives 1,200 queries yearly.

SUBMISSIONS
Query with outline. Accepts email queries to editor@thefamilymagazine.com (Microsoft Word attachments). No simultaneous submissions. Response time varies.

- Articles: 800–900 words. Informational, how-to, and self-help articles; and personal experience pieces. Topics include parenting, family life, relationships, fitness, crafts, hobbies, the arts, education, music, multicultural and ethnic issues, social issues, and travel.
- Depts/columns: Word lengths vary. News, tips, and trends; family fun guide; children's health; home decorating; cooking; and travel.
- Artwork: B/W prints or transparencies. Line art.
- Other: Submit seasonal material at least 3 months in advance.

SAMPLE ISSUE
126 pages (50% advertising): 14 articles; 9 depts/columns. Sample copy, $4. Guidelines and editorial calendar available at website.

- "Dads In Delivery." Article helps men decide whether they want to be present for the birth of their children.
- Sample dept/column: "Healthy Kids" provides nutritional guidance for athletic children.

RIGHTS AND PAYMENT
Exclusive regional and Web rights. Articles, $50. Depts/columns and artwork, payment rates vary. Pays on publication. Provides 1 tearsheet.

➡️EDITOR'S COMMENTS
Most of our writers are local professionals who share their expertise with our readers, but we accept with enthusiasm work from freelance writers. We do not accept manuscripts that have been simultaneously submitted to other publications within our region.

Washington Parent

4701 Sangamore Road, Suite N720
Bethesda, MD 20186

Editor: Margaret Hut

DESCRIPTION AND INTERESTS
Washington Parent offers DC-area parents topical articles on family issues, including health, education, and recreation. Each issue also features special sections that spotlight activities in the region. Circ: 75,000.

Audience: Families
Frequency: Monthly
Website: www.washingtonparent.com

FREELANCE POTENTIAL
90% written by nonstaff writers. Publishes 20 freelance submissions yearly. Receives 1,000 queries yearly.

SUBMISSIONS
Query. Accepts email queries to contactus@washingtonparent.net (Microsoft Word or WordPerfect attachments). SASE. Response time varies.

- Articles: 1,000–1,200 words. Informational and how-to articles. Topics include regional news and events, parenting and family issues, entertainment, gifted and special education, child development, health, fitness, the environment, and multicultural and ethnic issues.
- Depts/columns: Word lengths vary. Family travel, book and media reviews, education, topics relating to children with special needs, and short news items.

SAMPLE ISSUE
122 pages (63% advertising): 10 articles; 8 depts/columns; 2 special resource sections. Sample copy, writers' guidelines, and editorial calendar available.

- "The Whole Tooth and Nothing But the Tooth." Article offers suggestions for preventing fear of the dentist.
- "Getting Started." Article explains effective ways to toilet train autistic children.
- Sample dept/column: "Book Reviews" suggests some titles worth reading.

RIGHTS AND PAYMENT
First rights. Written material, payment rates vary. Provides 3 contributor's copies.

➡️EDITOR'S COMMENTS
We prefer local writers, but will consider all well-written and carefully researched submissions that offer information and support to families living in the Washington area.

Weatherwise

1319 Eighteenth Street NW
Washington, DC 20036

Managing Editor: Margaret Benner

DESCRIPTION AND INTERESTS
Engaging features and breathtaking photography relating to weather make up the bulk of this magazine. Its articles take a creative look at everyday occurrences and are accurate, authoritative, and easily understood by a non-technical readership. Circ: 5,800.

Audience: YA–Adult
Frequency: 6 times each year
Website: www.weatherwise.org

FREELANCE POTENTIAL
50% written by nonstaff writers. Publishes 36 freelance submissions yearly; 30% by authors who are new to the magazine. Receives 100 queries yearly.

SUBMISSIONS
Query with outline, résumé, and clips. Accepts email queries to ww@heldref.org. No simultaneous submissions. Availability of artwork improves chance of acceptance. Responds in 2 months.

- Articles: 1,500–2,000 words. Informational articles; photo-essays; and reviews. Topics include storms, storm tracking, safety issues, and other topics related to the weather—all with a scientific basis.
- Depts/columns: 800–1,500 words. Weather Q&As; book and software reviews; weather highlights; forecasts; and historical pieces.
- Artwork: Color prints or transparencies; digital files.

SAMPLE ISSUE
82 pages: 6 articles; 9 depts/columns. Guidelines available at website.

- "Cloud Watching from 35,000 Feet." Article explains how looking at clouds from above can help in identifying them and their causal weather systems.
- Sample dept/column: "WeatherWatch" offers weather information over a two-month period for various regions of the country.

RIGHTS AND PAYMENT
All rights. All material, payment rates vary. Pays on publication.

➥EDITOR'S COMMENTS
Don't query us proposing a broad-brush article, such as global warming or the ozone hole. Give us the story within the story, and emphasize the human element.

Westchester Family

7 Purdy Street, Suite 201
Harrison, NY 10528

Editor: Jean Sheff

DESCRIPTION AND INTERESTS
This regional family magazine has it all, from parenting advice and child health and safety information, to where to go in New York's Westchester County for entertainment and recreation. Circ: 59,000.

Audience: Parents
Frequency: Monthly
Website: www.westchesterfamily.com

FREELANCE POTENTIAL
80% written by nonstaff writers. Publishes 40 freelance submissions yearly; 10% by unpublished writers, 30% by authors who are new to the magazine. Receives 600 queries yearly.

SUBMISSIONS
Query with clips. Accepts hard copy. SASE. Response time varies.

- Articles: 800–1,200 words. Informational articles; profiles; interviews; photo-essays; and personal experience pieces. Topics include education, music, recreation, regional news, social issues, special and gifted education, travel, and women's issues.
- Depts/columns: 400–800 words. News and media reviews.

SAMPLE ISSUE
74 pages (52% advertising): 3 articles; 8 depts/columns; 1 activities calendar. Sample copy, free with 9x12 SASE. Writers' guidelines available.

- "When Kids Fly Solo." Article presents tips for making a child's solo airplane flight a successful one.
- "27 Products That Make Parenting Easier." Article presents the winners of the National Parenting Publications Awards, each one a product designed to help parents.
- Sample dept/column: "Family FYI" reports on the pros and cons of co-sleeping.

RIGHTS AND PAYMENT
First rights. Written material, $25–$200. Pays on publication. Provides 1 contributor's copy.

➥EDITOR'S COMMENTS
We are always excited to receive a query that presents something new and fresh—a topic we haven't thought of before, a new twist to an age-old issue, or even a personal insight that will help enlighten other parents. If you have that, we'd like to see it.

Westchester Parent

1040 Avenue of the Americas, 4th Floor
New York, NY 10018

Editor-in-Chief: Helen Freedman

DESCRIPTION AND INTERESTS
Parents in Westchester County, New York, read
this magazine for articles on parenting trends
and news and current events as they relate
to parenting. It also provides information on
local recreation, education, entertainment,
and services. Circ: 285,000.

Audience: Parents
Frequency: Monthly
Website: www.nymetroparents.com

FREELANCE POTENTIAL
50% written by nonstaff writers. Publishes
300 freelance submissions yearly. Receives
12 queries, 72 unsolicited mss yearly.

SUBMISSIONS
Query or send complete ms. Accepts hard copy
and email to hfreedman@davlermedia.com.
SASE. Responds in 1 week.

- Articles: 800–1,000 words. Informational
  articles; profiles; and personal experience
  pieces. Topics include family issues, health,
  nutrition, fitness, and regional news.
- Depts/columns: Staff written.
- Other: Submit seasonal material 4 months
  in advance.

SAMPLE ISSUE
94 pages: 17 articles; 8 depts/columns.
Sample copy, free with 10x13 SASE.

- "The Back-to-Basics Baby." Article explores
  the question of whether fancy products and
  gadgets give babies a leg up on learning.
- "The Well-Functioning Kid." Article provides
  expert advice about limiting kids' use of
  electronic games.
- "As Your Child Grows, Consider a Health
  Care Proxy." Article encourages parents to
  set up health care proxies.

RIGHTS AND PAYMENT
First New York area rights. No payment.

➥EDITOR'S COMMENTS
In general, we do not pay freelancers for articles. We do pay for a few assigned pieces.
We are a great market for writers trying to get
their first article published or to build their
clip file. We would like to hear your ideas or
see your stories about subjects that would be
of interest to parents in the Westchester area.
We're looking for news on parenting trends
and current affairs.

West Coast Families

13988 Maycrest Way, Unit 140
Richmond, British Columbia V6V 3C3
Canada

Editor: Michelle Froese

DESCRIPTION AND INTERESTS
This guide to family life in British Columbia
helps parents plan everything from birthday
parties to college savings. Circ: 50,000.

Audience: Parents
Frequency: 8 times each year
Website: www.westcoastfamilies.com

FREELANCE POTENTIAL
75% written by nonstaff writers. Publishes 40
freelance submissions yearly; 25% by authors
who are new to the magazine. Receives 300
queries yearly.

SUBMISSIONS
Query. Accepts hard copy and email queries
to editor@westcoastfamilies.com. SAE/IRC.
Response time varies.

- Articles: 600–800 words. Informational, self-
  help, and how-to articles; profiles; interviews; and personal experience pieces.
  Topics include family life, parenting, recreation, travel, religion, current events, health,
  fitness, finance, education, sports, hobbies,
  science, technology, nature, and pets.
- Depts/columns: Staff written.
- Other: Puzzles, activities, and jokes. Submit
  seasonal material 3 months in advance.

SAMPLE ISSUE
52 pages (8% advertising): 10 articles; 12
depts/columns. Sample copy, free with 9x12
SAE/IRC ($1.45 postage); also available for
download at website. Guidelines and editorial
calendar available.

- "Girly Girlz Wanted!" Article describes the
  Girly Girlz Club, a local party venue where
  children can play dress-up.
- "Daddy's Little Girl." Article features an inter-
  view with *Global Television* star Steve
  Darling about recently becoming a father.
- "A New Contraceptive Device for Men."
  Article describes the Intra Vas Device (IVD).

RIGHTS AND PAYMENT
One-time and electronic rights. Written material,
payment rates vary. Pays on publication.
Provides contributor's copies upon request.

➥EDITOR'S COMMENTS
As a regional publication, we insist on printing articles of local relevancy for our readers.
While you don't have to be Canadian to contribute, you must take a Canadian angle.

Western New York Family Magazine

3147 Delaware Avenue, Suite B
Buffalo, NY 14217

Editor: Michele Miller

DESCRIPTION AND INTERESTS
As its name suggests, this publication approaches parenting issues from a regional perspective, emphasizing where and how to find family-oriented events, goods, and services in the greater Buffalo area. Circ: 25,000.

Audience: Parents
Frequency: Monthly
Website: www.wnyfamilymagazine.com

FREELANCE POTENTIAL
90% written by nonstaff writers. Publishes 125–150 freelance submissions yearly; 30% by unpublished writers, 30% by authors who are new to the magazine, 40% by experts. Receives 1,200 unsolicited mss yearly.

SUBMISSIONS
Local writers receive preference. Send complete ms with 2-sentence bio. Prefers email submissions to michele@wnyfamilymagazine.com (paste text into body of email and attach Microsoft Word document); will accept hard copy and simultaneous submissions if identified and out of market. SASE. Responds upon publication.

- Articles: "UpFront" articles, 2,000–2,500 words. Other features, 750–1,500 words. Informational, how-to, and self-help articles; creative nonfiction; humor; and personal experience pieces. Topics include parenting, education, and special needs children.
- Depts/columns: Word lengths vary. News briefs, reviews, family travel, recipes, fatherhood, and single parenting.

SAMPLE ISSUE
72 pages (40% advertising): 14 articles; 12 depts/columns. Sample copy, $2.50 with 9x12 SASE ($1.79 postage). Guidelines and editorial calendar available at website.

- "It's a Balancing Act." Article provides new mothers with time management techniques.
- Sample dept/column: "Simple Pleasures" celebrates family board games.

RIGHTS AND PAYMENT
First or second rights. Articles, $35–$200. Pays on publication. Provides 1 contributor's copy.

➥ EDITOR'S COMMENTS
Articles written by fathers on fatherhood issues are of great interest to us, as are features about family travel, education, eldercare, and children with special needs.

What If?

19 Lynwood Place
Guelph, Ontario N1G 2V9
Canada

Managing Editor: Mike Leslie

DESCRIPTION AND INTERESTS
"Canada's Creative Magazine for Teens" publishes the writing and artwork of young adults. Its goal is to inspire young Canadian readers, writers, and artists by exposing them to the work of their peers. Circ: 3,000.

Audience: 12–19 years
Frequency: Quarterly
Website: www.whatifmagazine.com

FREELANCE POTENTIAL
95% written by nonstaff writers. Publishes 100 freelance submissions yearly; 95% by unpublished writers, 90% by authors who are new to the magazine. Receives 3,000 unsolicited mss each year.

SUBMISSIONS
Send complete ms. Accepts hard copy, email submissions to editor@whatifmagazine.com (Microsoft Word attachments), and simultaneous submissions if identified. Availability of artwork improves chance of acceptance. SAE/IRC. Responds in 3 months.

- Articles: To 1,500 words. Opinion pieces and personal essays. Topics vary.
- Fiction: To 3,000 words. Genres include mystery; suspense; fantasy; humor; science fiction; and contemporary, realistic, and inspirational fiction.
- Depts/columns: Word lengths vary. Reviews and interviews.
- Artwork: Color prints. Line art.
- Other: Poetry, to 20 lines.

SAMPLE ISSUE
48 pages (3% advertising): 8 stories; 3 essays; 11 depts/columns; 7 poems. Sample copy, $8 with 9x12 SAE. Guidelines available.

- "Wilderness Excursion." Essay tells how a six-day canoe trip on the Magnetawan River transformed the author's outlook.
- Sample dept/column: "What If? Interviews" is a Q&A with the editors of *girlSpoken*.

RIGHTS AND PAYMENT
First rights. No payment. Provides 3 copies.

➥ EDITOR'S COMMENTS
We are not the typical teen fashion and dating magazine, so we're not packed with advertising but with teen creativity. We rely on the fantastic talent of young writers and artists from across Canada.

What's Up Kids Family Magazine

496 Metler Road
Ridgeville, Ontario L0S 1M0
Canada

Editor-in-Chief: Paul Baswick

DESCRIPTION AND INTERESTS

This Canadian family magazine offers editorial content for both parents and their children. Parents turn to the informative and positive articles on everything from child-rearing and health to travel and entertainment while the "Kids' Fun section" is filled with puzzles, stories, and hands-on activities. Circ: 200,000.

Audience: Families
Frequency: 6 times each year
Website: www.whatsupkids.com

FREELANCE POTENTIAL

80% written by nonstaff writers. Publishes 30 freelance submissions yearly; 60% by authors who are new to the magazine. Receives 348 queries yearly.

SUBMISSIONS

Canadian authors only. Query. Accepts email queries only to paul@whatsupkids.com. Response time varies.

- Articles: Word lengths vary. Informational articles; profiles; and interviews. Topics include education, family issues, travel, fitness, nutrition, health, the arts, and entertainment.
- Depts/columns: Word lengths vary. "Mom Time," "Health Matters," "Family Fit," "Family Tips," "Family Finances," "Baby Steps," "From the Kitchen," "Learning Curves," "Family Travel," "What's Up with Dad?," "Kids' Space," "Cool Careers," "Kid Craft," and other children's activities.

SAMPLE ISSUE

98 pages (15% advertising): 3 articles; 13 depts/columns; 6 activities.

- "Beat the Morning Bell." Article provides several strategies for getting children to school organized and on time.
- Sample dept/column: "Understanding Families" offers parents tips on successfully coaching their child's sports team.

RIGHTS AND PAYMENT

All rights. Written material, payment rates vary. Payment policy varies. Provides author's copies.

➥EDITOR'S COMMENTS

We work with Canadian writers only. We are in need of celebrity profiles and up-to-the-minute news of interest to families. Activities should cater to children up to age 13.

Winner

55 West Oak Ridge Drive
Hagerstown, MD 21740

Editor: Jan Schleifer

DESCRIPTION AND INTERESTS

Winner conveys the dangers of drug use to students in grades four through six while encouraging the development of life skills and real-world problem solving. Circ: 4,500.

Audience: 9–13 years
Frequency: 9 times each year
Website: www.winnermagazine.org

FREELANCE POTENTIAL

25% written by nonstaff writers. Publishes 15–18 freelance submissions yearly; 5% by unpublished writers, 20% by authors who are new to the magazine. Receives 100 unsolicited mss yearly.

SUBMISSIONS

Send complete ms. Accepts email submissions to jschleifer@rhpa.org. Responds in 1–2 months.

- Articles: 600–650 words. Informational, self-help, and how-to articles; and profiles. Topics include life skills, problem solving, positive role models, hobbies, and activities.
- Fiction: 600–650 words. Stories depicting smart choices and positive behavior.
- Artwork: Color prints and transparencies. Line art.
- Other: Puzzles, games, quizzes, and comics.

SAMPLE ISSUE

16 pages (no advertising): 2 articles; 2 stories; 7 activities; 1 comic strip. Sample copy, $2 with 9x12 SASE (3 first-class stamps). Guidelines available at website.

- "Xochil Garcia: Street-Smart Superkid." Article profiles an 11-year-old girl who escaped a would-be kidnapper by using her wits.
- "The Real Winner." Story tells of a boy who gives his wrestling medal to a teammate.
- "Monster on the School Bus." Story of bullying allows readers to choose the ending.

RIGHTS AND PAYMENT

First rights. Articles, $80. Artwork, payment rates vary. Pays on acceptance. Provides 3 contributor's copies.

➥EDITOR'S COMMENTS

Our theme list/editorial calendar includes only the specific drug each month that we educate against the use of. Those articles are always assigned. Articles about other appropriate, helpful, middle-school topics can be submitted at any time.

Wire Tap Magazine

Independent Media Institute
77 Federal Road
San Francisco, CA 94107

Associate Editor

DESCRIPTION AND INTERESTS
This award-winning national news and culture e-zine is written by and for "a new generation of progressives." Its articles, most of which are written by readers, cover overlooked social justice issues, promote social justice, inspire action, and provide young people with a voice in the media. Hits per month: 60,000.

Audience: 18–30 years
Frequency: Updated daily
Website: www.wiretapmag.org

FREELANCE POTENTIAL
95% written by nonstaff writers. Publishes 120 freelance submissions yearly. Receives 300 queries yearly.

SUBMISSIONS
Query. Accepts email queries to submissions@wiretapmag.org (no attachments). Response time varies.

- Articles: Word lengths vary. Informational articles; profiles; interviews; and personal experience pieces. Topics include social issues, politics, contemporary culture, current events, the environment, immigration, sex and relationships, peace, education, and youth activism.
- Depts/columns: Word lengths vary. Reviews, politics, and news.
- Other: Poetry.

SAMPLE ISSUE
Sample copy and writers' guidelines available at website.

- "Money Changers: Youth Tackle a Tough Economy." Article chronicles the multiple economic challenges faced by the Millennial Generation.
- "Stargazing in Kabul." Article tells the story of a young woman in Afghanistan who is bold enough to petition for a divorce.

RIGHTS AND PAYMENT
Electronic rights. Written material, $50–$400 for assigned pieces. No payment for unsolicited submissions. Payment policy varies.

➤EDITOR'S COMMENTS
We are looking for smart, passionate, and socially engaged young people who can add a unique style of writing to our mix. While we publish articles from all generations, we favor work from young adults.

Writers' Journal

Val-Tech Media
P.O. Box 394
Perham, MN 56573-0394

Editor: Leon Ogroske

DESCRIPTION AND INTERESTS
Covering all types of writing for all types of writers, this journal strives to be "the complete writer's magazine." It sponsors several writing contests each year. Circ: 23,000.

Audience: YA–Adult
Frequency: 6 times each year
Website: www.writersjournal.com

FREELANCE POTENTIAL
50% written by nonstaff writers. Publishes 40 freelance submissions yearly; 20% by unpublished writers, 80% by authors who are new to the magazine. Receives 200 queries and unsolicited mss yearly.

SUBMISSIONS
Query with clips; or send complete ms. Accepts hard copy and email submissions to writersjournal@writersjournal.com (text only). SASE. Responds in 2–6 months.

- Articles: 1,200–2,200 words. Informational and how-to articles; profiles; and interviews. Topics include fiction writing, travel writing, technical writing, business writing, screenwriting, journalism, poetry, writing skills and styles, punctuation, interviewing techniques, research, record-keeping, income venues, finance, and self-publishing.
- Depts/columns: Staff written.
- Other: Poetry, to 15 lines.

SAMPLE ISSUE
64 pages (10% advertising): 8 articles; 5 stories; 11 depts/columns. Sample copy, $6 with 9x12 SASE. Guidelines available at website.

- "Selling Smart Quizzes That Teach." Article tells how to create and market puzzles for older students.
- "Finding the 'Where.'" Article stresses the importance of setting.

RIGHTS AND PAYMENT
First North American rights. Articles, $30. Poetry, $5 per poem. Pays on publication. Provides 2 contributor's copies upon request and a 1-year subscription.

➤EDITOR'S COMMENTS
We are constantly seeking articles about unique markets for freelance writers. Please note that we do not publish fiction, except for the winning entries of our various contests. See our website for details.

Yes Mag

501-3960 Quadra Street
Victoria, British Columbia V8X 4A3
Canada

Managing Editor: Jude Isabella

DESCRIPTION AND INTERESTS
Each themed issue of *Yes Mag* offers informative articles about the wonders of science in a way that makes learning fun and entertaining for middle-grade readers. It also features book and other media reviews. Circ: 23,000.

Audience: 9–14 years
Frequency: 6 times each year
Website: www.yesmag.ca

FREELANCE POTENTIAL
70% written by nonstaff writers. Publishes 30 freelance submissions yearly; 5% by unpublished writers, 15% by authors who are new to the magazine. Receives 300 queries yearly.

SUBMISSIONS
Query. Accepts email queries to editor@yesmag.ca. Response time varies.

- Articles: Features, 300–800 words. Short, theme-related articles, 300–600 words. Informational articles, 250 words. Topics include astronomy, engineering, math, science, technology, nature, and the environment.
- Depts/columns: 250 words. Science and technology news, entomology, world records, environmental updates, book and product reviews, experiments to try at home.

SAMPLE ISSUE
32 pages (no advertising): 5 articles; 6 depts/columns. Sample copy, $4.50 with SAE/IRC. Guidelines and theme list available.

- "Lost!" Article provides do's and don'ts as guidelines for survival if readers ever find themselves lost in the woods.
- "Falling Food." Article profiles a German restaurant that uses a robotic serving device in the form of spiralling metal rails.
- Sample dept/column: "Sci & Tech Watch" introduces readers to a breakthrough in chewing gum.

RIGHTS AND PAYMENT
First and one-time Korean rights. "Sci & Tech Watch," $100. Articles, $145 per page. Pays on publication. Provides 1 contributor's copy.

➥EDITOR'S COMMENTS
We encourage new writers to approach us by submitting items to our "Sci & Tech Watch" department. Please keep in mind, however, that we feature mainly Canadian scientists and science topics.

Young Adult Today

1551 Regency Court
Calumet City, IL 60409

Editor: LaTonya Taylor

DESCRIPTION AND INTERESTS
Young Adult Today is a Christian education magazine published by Urban Ministries. Its articles and Bible studies help readers apply Scripture to modern life. Circ: 25,000.

Audience: 18–24 years
Frequency: Quarterly
Website: www.youngadulttoday.net

FREELANCE POTENTIAL
90% written by nonstaff writers. Publishes 52 freelance submissions yearly; 50% by unpublished writers, 50% by authors who are new to the magazine. Receives 240 queries yearly.

SUBMISSIONS
All articles and Bible lessons are assigned. Query with résumé. No unsolicited mss. Accepts email to ltaylor@urbanministries.com. Responds in 2 months.

- Articles: To 400 words. Lessons consist of discussion pieces, questions, devotional readings, and Bible study guides that explain how Scripture lessons can be applied to modern life.

SAMPLE ISSUE
80 pages (4% advertising): 1 article; 13 teaching plans; 13 corresponding Bible study guides. Sample copy, $2.25 with 9x12 SASE ($.87 postage). Writers' guidelines provided on assignment.

- "Five Ways to Look Like Jesus." Article explains that serving others, avoiding sin, and practicing forgiveness are some of the ways to be more Christlike.
- "Getting Help." Teaching plan shows, through a brief true-to-life example, how reaching out for help brings us closer to God and each other.
- "Christ as Intercessor." Bible study guide explores the Scriptural references for a lesson in how Jesus intercedes on our behalf, how we can be modern-day intercessors, and how to approach God through prayer.

RIGHTS AND PAYMENT
Rights negotiable. Written material, $150 per lesson. Pays on publication.

➥EDITOR'S COMMENTS
We are always looking for feature articles and lesson plans that will effectively enlighten and enrich our young, urban readership.

Young Adult Today Leader

1551 Regency Court
Calumet City, IL 60409

Editor: LaTonya Taylor

DESCRIPTION AND INTERESTS
Young Adult Today Leader is published in conjunction with the Christian student magazine *Young Adult Today*. It contains the Bible teaching plans and study guides to be used by teachers to support the lessons in *Young Adult Today*. It also offers thought-provoking articles to help teachers in their ministry. Circ: 15,000.

Audience: Religious educators
Frequency: Quarterly
Website: www.youngadulttoday.net

FREELANCE POTENTIAL
90% written by nonstaff writers. Publishes 52 freelance submissions yearly; 50% by unpublished writers, 50% by authors who are new to the magazine. Receives 240 queries yearly.

SUBMISSIONS
All work is assigned. Query with résumé. No unsolicited mss. Accepts email to ltaylor@urbanministries.com. Responds in 2 months.

- Articles: Devotionals, 400 words. Topics include current events and social issues as they relate to Christianity and the Bible.

SAMPLE ISSUE
96 pages (no advertising): 1 article; 13 teaching plans; 13 Bible study guides. Sample copy, $2.25 with 9x12 SASE ($.87 postage). Writers' guidelines available.

- "Five Ways to Look Like Jesus." Article explains that making time for others, speaking out for justice, and practicing forgiveness are some of the ways to be more like Christ.
- "Love and Parties." Teaching plan shows that discipline is a way that parents show they love their children, as it is with God and us.
- "Christ as Leader." Bible study guide explores Scripture for a lesson in leadership, including why Jesus was a good leader and why we should follow Jesus' example of godly discipline.

RIGHTS AND PAYMENT
Rights negotiable. Written material, $150. Pays on publication.

➥EDITOR'S COMMENTS
We continue to be interested in submissions from experienced Christian leaders and educators regarding how to teach young adults to live according to Scripture in the modern urban setting.

Young People's Press

374 Fraser Street
North Bay, Ontario P1B 2W7
Canada

Chief Executive Officer: Don Curry

DESCRIPTION AND INTERESTS
Teen and young adult writers use this website as an outlet for their true-life, journalistic, profile, and opinion pieces based on current social issues. Articles cover current events, politics, the arts, and culture, and also appear in both U.S. and Canadian newspapers. Hits per month: 4,000.

Audience: YA
Frequency: Updated on an ongoing basis
Website: www.ypp.net

FREELANCE POTENTIAL
100% written by nonstaff writers. Publishes 300 freelance submissions yearly; 70% by unpublished writers, 100% by authors who are new to the magazine. Receives 600 queries, 600 unsolicited mss yearly.

SUBMISSIONS
Send complete ms. Accepts email submissions to doncurry@ontera.net (Microsoft Word attachments). Availability of artwork improves chance of acceptance. Responds in 1 week.

- Articles: 400–1,000 words. Informational and self-help articles; profiles; reviews; and personal experience pieces. Topics include music, the arts, current events, multicultural and social issues, and popular culture.
- Depts/columns: Word lengths vary. Music, movie, and website reviews.
- Artwork: Electronic images only.
- Other: Submit seasonal material 1 month in advance.

SAMPLE ISSUE
Sample copy and writers' guidelines available at website.

- "World Should Boycott Olympics." Article explains author's view on why the 2008 Beijing Olympics should be boycotted because of the situation in Tibet.
- "Wal-Mart Draws Me Like a Magnet." Article examines the struggle of consciousness between saving money and supporting overseas labor.

RIGHTS AND PAYMENT
Rights vary. No payment.

➥EDITOR'S COMMENTS
Writers are free to publish their work elsewhere after it is published on our website. Send us your original work.

Young Rider

P.O. Box 8237
Lexington, KY 40533

Editor: Lesley Ward

DESCRIPTION AND INTERESTS
This photo-filled magazine offers a balanced mix of entertaining horse stories and practical information for young horse riders and horse lovers. Circ: 92,000.
Audience: 6–14 years
Frequency: 6 times each year
Website: www.youngrider.com

FREELANCE POTENTIAL
20% written by nonstaff writers. Publishes 20 freelance submissions yearly; 10% by unpublished writers, 10% by authors who are new to the magazine. Receives 60 queries yearly.

SUBMISSIONS
Query. Accepts email queries to yreditor@bowtieinc.com (Microsoft Word attachments). Responds in 2 weeks.

- Articles: Word lengths vary. Informational and how-to articles; and profiles. Topics include horseback riding, training, careers, English and Western riding techniques, and general horse care.
- Fiction: 1,200 words. Stories that feature horses, ponies, and youth themes.
- Artwork: Color prints, transparencies, and high-resolution digital images.

SAMPLE ISSUE
74 pages (28% advertising): 8 articles; 9 depts/columns. Sample copy, $3.99 with 9x12 SASE ($1 postage). Guidelines and editorial calendar available.

- "Brianne Goutal." Article interviews this talented young showjumper who has competed in and won many international shows.
- "Blonde Bombshells." Article profiles an unusual-looking breed of equine known as a Haflinger Pony.
- Sample dept/column: "Horse and Pony Problems" describes a piece of English-riding equipment called garters.

RIGHTS AND PAYMENT
First serial rights. Written material, $.10 per word. Artwork, payment rates vary. Pays on publication. Provides 2 contributor's copies.

•❖EDITOR'S COMMENTS
Freelance writers are encouraged to submit horse stories with humor, a bit of conflict, realistic plots, and characters our young readers can relate to.

Young Salvationist

The Salvation Army
615 Slaters Lane
Alexandria, VA 22314

Editor-in-Chief: Major Ed Forster

DESCRIPTION AND INTERESTS
This publication of the Salvation Army is devoted to meeting the challenges of adolescence, while moving toward Christian maturity, by offering inspirational and practical articles and stories. Circ: 48,000.
Audience: 13–21 years
Frequency: 10 times each year
Website: www.salpubs.com

FREELANCE POTENTIAL
50% written by nonstaff writers. Publishes 20 freelance submissions yearly; 5% by unpublished writers, 10% by authors who are new to the magazine. Receives 60 unsolicited mss each year.

SUBMISSIONS
Send complete ms. Accepts hard copy, email submissions to ys@usn.salvationarmy.org, and simultaneous submissions if identified. SASE. Responds in 4–6 weeks.

- Articles: To 1,000 words. How-to, inspirational, and personal experience articles; profiles; interviews; and humor. Topics include religion and issues of relevance to teens.
- Fiction: 500–1,000 words. Genres include contemporary and realistic fiction.
- Depts/columns: Word lengths vary. "Blog of the Month," "My Word on His Word," and advice.
- Other: Submit seasonal material 6 months in advance.

SAMPLE ISSUE
24 pages (no advertising): 5 articles; 1 story; 5 depts/columns. Sample copy, guidelines, and theme list available at website.

- "Just Hanging Around." Article recounts the author's introduction to true fellowship in the Salvation Army.
- Sample dept/column: "My Word on His Word" is an essay about showing love to all.

RIGHTS AND PAYMENT
First and second rights. Written material, $.15 per word for first rights; $.10 per word for reprints. Pays on acceptance. Provides 4 copies.

•❖EDITOR'S COMMENTS
We publish very little material from writers outside the Salvation Army. A good way to break in is to send a short inspirational piece based on living the Gospel.

Youth & Christian Education Leadership

1080 Montgomery Avenue
Cleveland, TN 37311

Editor: Wanda Griffith

DESCRIPTION AND INTERESTS

This magazine is designed to equip and inspire men and women who work in the ministries of Christian education. It features articles on Sunday school programming and creating excitement in youth groups. Circ: 10,000.

Audience: Adults
Frequency: Quarterly
Website: www.pathwaypress.org

FREELANCE POTENTIAL

10% written by nonstaff writers. Publishes 10 freelance submissions yearly; 90% by unpublished writers, 10% by authors who are new to the magazine. Receives 30–35 queries, 20–25 unsolicited mss yearly.

SUBMISSIONS

Prefers complete ms with author biography; will accept query. Accepts disk submissions (Microsoft Word or WordPerfect), and email submissions to wanda_griffith@pathwaypress.org. SASE. Responds in 3 weeks.

- Articles: 500–1,000 words. Informational and how-to articles; profiles; interviews; humor; and personal experience pieces. Topics include current events, music, religion, social issues, psychology, parenting, and multicultural and ethnic subjects.
- Depts/columns: Staff written.

SAMPLE ISSUE

30 pages (2% advertising): 11 articles; 6 depts/columns. Sample copy, $1 with 9x12 SASE (2 first-class stamps). Guidelines available at website.

- "Don't Exclude the Married 20-Somethings." Article explains the importance of creating ways to draw young married couples into church programs.
- "Backyard Campout." Personal experience piece explains how a group of boys found fellowship during a rainstorm in a minister's backyard.

RIGHTS AND PAYMENT

First rights. Written material, $25–$50. Kill fee, 50%. Pays on publication. Provides 1–10 contributor's copies.

⟶EDITOR'S COMMENTS

We love informational and instructional articles on building youth ministries, as well as success stories that can inspire others.

Youth Today

1200 17th Street NW, 4th Floor
Washington, DC 20036

Editor: Patrick Boyle

DESCRIPTION AND INTERESTS

Youth Today is a national trade newspaper for people who work with children and youth. It covers issues involving youth and those who work on their behalf—direct-case services, health, justice, government policies, funding for youth programs, and youth development. Circ: 12,000.

Audience: Youth workers
Frequency: 10 times each year
Website: www.youthtoday.org

FREELANCE POTENTIAL

50% written by nonstaff writers. Publishes 25 freelance submissions yearly; 10% by authors who are new to the magazine. Receives 36 queries yearly.

SUBMISSIONS

Query with résumé and clips. Accepts hard copy and email queries to pboyle@youthtoday.org. SASE. Responds in 3 months.

- Articles: 1,000–2,500 words. Informational articles; news and research reports; profiles of youth workers and youth programs; and business features. Topics include foster care, child abuse, program management, violence, adolescent health, juvenile justice, job training, school-to-work programs, after-school programs, mentoring, and other social issues related to youth development.
- Depts/columns: Word lengths vary. Book and video reviews, news briefs, opinion pieces, and people in the news.

SAMPLE ISSUE

36 pages (50% advertising): 42 articles; 5 depts/columns. Sample copy, $5. Writers' guidelines available.

- "It's Almost Easy Being Green." Article profiles a new Family Services headquarters built with an environmentally friendly design.
- Sample dept/column: "News Briefs" discusses a rent hike for a Boy Scout headquarters.

RIGHTS AND PAYMENT

First and Internet rights. Written material, $.50–$.75 per word. Pays on acceptance. Provides 2 contributor's copies.

⟶EDITOR'S COMMENTS

Most stories are written on assignment. Freelancers should have extensive experience writing for newspapers or magazines.

YouthWorker Journal

104 Woodmont Boulevard, Suite 300
Nashville, TN 37205

Editor: Steve Rabey

DESCRIPTION AND INTERESTS

Read by Christian youth ministers, *YouthWorker Journal* offers themed issues that provide solutions to everyday challenges while addressing today's important theological issues. Its goal is to help their readers make an impact on children's lives. Circ: 15,000.

Audience: Adults who work with youth
Frequency: 6 times each year
Website: www.youthworker.com

FREELANCE POTENTIAL

95% written by nonstaff writers. Publishes 50+ freelance submissions yearly; 15% by unpublished writers, 25% by authors who are new to the magazine. Receives 720 queries yearly.

SUBMISSIONS

Query with short biography. Prefers email queries to steve@youthworker.com (include "Query" in subject line); will accept hard copy and faxes to 615-385-4412. SASE. Responds in 6–8 weeks.

- Articles: Word lengths vary. Informational and practical application articles; personal experience pieces; and reviews. Topics include youth ministry, theology, spreading Christ's word, student worship, family ministry, education, family issues, popular culture, the media, and volunteering.
- Depts/columns: Word lengths vary. National and regional trends; youth workers' quotes.

SAMPLE ISSUE

72 pages (30% advertising): 8 articles; 12 depts/columns. Sample copy, $8. Guidelines and theme list available at website.

- "Transcending 'Young' and 'Old.'" Article discusses strategies to making intergenerational ministry work.
- Sample dept/column: "Soul Care" explains the importance of taking it easy on the Sabbath and setting aside time to worship.

RIGHTS AND PAYMENT

All rights. Written material, $15–$300. Pays on publication. Provides 1 contributor's copy.

➡ EDITOR'S COMMENTS

Over the years we have worked with the leading voices in the field as well as with unpublished writers. Our submission guidelines include information on the departments most accessible to new writers.

Zamoof!

644 Spruceview Place South
Kelowna, British Columbia V1V 2P7
Canada

Editor/Publisher: TeLeni Koochin

DESCRIPTION AND INTERESTS

This highly creative magazine is packed with comics, games, stories, movie previews, interviews, and activities. Its editorial content also educates youth about how to make safe and healthy choices. Circ: 5,000–7,000.

Audience: 7–12 years
Frequency: 6 times each year
Website: www.zamoofmag.com

FREELANCE POTENTIAL

7% written by nonstaff writers. Publishes 9 freelance submissions yearly; 50% by unpublished writers, 100% by authors who are new to the magazine. Receives 60 queries, 36 unsolicited mss yearly.

SUBMISSIONS

Canadian authors only. Query or send complete ms. Prefers email queries to mail@zamoofmag.com; will accept hard copy. SASE. Responds in 2–4 weeks.

- Articles: Staff written.
- Fiction: Currently closed to submissions.
- Depts/columns: 375 words. "Feet Up" written for parents. True stories about parenting and family life that can be humorous, serious, or thought-provoking.

SAMPLE ISSUE

82 pages: 11 articles; 3 stories; 6 depts/columns; 4 comics. Sample copy, $3.99 with 6x9 SASE ($2.01 postage) or at website. Guidelines available.

- "Stranded." Story tells of a group of kids who get left alone on an island while on a cruise to the Bahamas.
- Sample dept/column: "Safe Communities" looks at the consequences of boredom and offers tips for avoiding it.

RIGHTS AND PAYMENT

Writer retains rights. Written material, $.20 per word. Pays on publication. Provides 3 contributor's copies.

➡ EDITOR'S COMMENTS

As part of our qualification requirements for Canadian grants, we can publish work from Canadian authors only. Note that we are currently not accepting fiction submissions due to a backlog, but we do need submissions for "Feet Up." This new column features short narrative pieces about parenthood.

Additional Listings

We have selected the following magazines to offer you additional publishing opportunities. Many of these magazines range from general interest publications to women's magazines to craft and hobby magazines. While children, young adults, parents, or teachers are not their primary target audience, these publications do publish a limited amount of material related to or of interest to children.

As you review the listings that follow, use the Description and Interests section as your guide to the particular needs of each magazine. This section offers general information about the magazine and its readers' interests, as well as the type of material it usually publishes. The Freelance Potential section will provide information about the publication's receptivity to freelance manuscripts.

After you survey the listings to determine if your work meets the magazine's specifications, be sure to read a recent sample copy and the current writers' guidelines before submitting your material.

Action Pursuit Games

P.O. Box 417
Licking, MO 65542

Editor: Daniel Reeves

DESCRIPTION AND INTERESTS: Paintball enthusiasts read this monthly for information on technique, strategy, and products. It also offers a small amount of fiction. Circ: 80,000.
Website: www.actionpursuitgames.com

FREELANCE POTENTIAL: 60% written by non-staff writers. Publishes 150+ freelance submissions yearly; 20% by unpublished writers, 30% by authors who are new to the magazine. Receives 480 unsolicited mss yearly.

SUBMISSIONS AND PAYMENT: Sample copy, $4.99 with 9x12 SASE (15 first-class stamps). Send complete ms with artwork (digital images). Accepts disk submissions and email submissions to editor@actionpursuitgames.com. SASE. Responds in 1 month. All rights. Articles, 300–2,000 words. Depts/columns, 300–500 words. Fiction, 300–600 words. Written material, payment rates vary. Pays on publication. Provides 1 contributor's copy.

Amazing Kids!

3224 East Yorba Linda Boulevard, #442
Fullerton, CA 92831

Editor: Alyse Rome

DESCRIPTION AND INTERESTS: The goal of this e-zine, part of a nonprofit educational organization, is to uncover the potential of every child and inspire them to pursue excellence. A mix of articles, stories, interviews, and reviews are featured, many of them written by children and young adults. Adult volunteers are always needed to write for the "Kids of the Month" section. Hits per month: 640,000.
Website: www.amazing-kids.org

FREELANCE POTENTIAL: 40% written by non-staff writers. Publishes 70 freelance submissions yearly; 90% by unpublished writers, 70% by authors who are new to the magazine. Receives 3,000 queries and unsolicited mss yearly.

SUBMISSIONS AND PAYMENT: Sample copy available at website. Query or send complete ms. Accepts email to info@amazing-kids.org. Response time varies. All rights. Articles, word lengths vary. No payment.

The ALAN Review

College of Liberal Arts & Sciences
Arizona State University, Department of English
P.O. Box 870302
Tempe, AZ 85287

Editor: Dr. James Blasingame

DESCRIPTION AND INTERESTS: Focusing on young adult literature, this magazine offers reviews as well as author interviews. Read by members of the National Council of Teachers of English, it is published three times each year. Circ: 2,500.
Website: www.alan.ya.org

FREELANCE POTENTIAL: 84% written by non-staff writers. Publishes 38 freelance submissions yearly; 5% by unpublished writers, 65% by new authors. Receives 90 unsolicited mss yearly.

SUBMISSIONS AND PAYMENT: Guidelines available in magazine. Sample copy, free. Send 3 copies of complete ms with disk (ASCII or Microsoft Word 5.1 or higher). Accepts simultaneous submissions if identified. Availability of artwork improves chance of acceptance. SASE. Responds in 2 months. All rights. Articles, to 3,000 words. Depts/columns, word lengths vary. No payment. Provides 2 contributor's copies.

American History

Weider History Magazine Group
741 Miller Drive SE, Suite D2
Leesburg, VA 20175

Editorial Director: Roger Vance

DESCRIPTION AND INTERESTS: Written for anyone with an interest in history, *American History* magazine offers thoroughly researched and entertaining articles that reveal the people and events that have influenced American history. Published six times each year, it looks for material that is well researched but highly readable. Circ: 100,000.
Website: www.thehistorynet.com/ahi

FREELANCE POTENTIAL: 80% written by non-staff writers. Publishes 30 freelance submissions yearly; 50% by authors who are new to the magazine. Receives 1,200 queries yearly.

SUBMISSIONS AND PAYMENT: Sample copy and guidelines, $6 with return label. Query with 1- to 2-page proposal. Accepts hard copy. SASE. Responds in 10 weeks. All rights. Articles, 2,000–4,000 words; $.20 per word. Depts/columns, word lengths vary; $75. Pays on acceptance. Provides 5 contributor's copies.

American School Board Journal

1680 Duke Street
Alexandria, VA 22314

Editor-in-Chief: Glenn Cook

DESCRIPTION AND INTERESTS: Published monthly by the National School Boards Association, this journal strives to chronicle change, interpret issues, and offer advice on a range of educational topics. Read by teachers, school board members, and administrators, it covers school management, policy making, and student achievement. Circ: 50,200.
Website: www.asbj.com

FREELANCE POTENTIAL: 50% written by nonstaff writers. Publishes 35 freelance submissions yearly. Receives 360 queries yearly.

SUBMISSIONS AND PAYMENT: Sample copy, $5. Prefers query with clips; will accept complete ms. Accepts hard copy. SASE. Responds in 2 months. All rights. Articles, 2,200–2,500 words. Depts/columns, 1,000–1,200 words. Solicited articles, $800. Unsolicited articles and depts/columns, no payment. Pays on publication. Provides 3 contributor's copies.

The Apprentice Writer

Susquehanna University, Box GG
Selinsgrove, PA 17870-1001

Writers' Institute Director: Gary Fincke

DESCRIPTION AND INTERESTS: Published annually in September, this literary tabloid showcases the writings and artwork of high school students. Distributed to high schools and colleges across the U.S., it accepts fiction, poetry, artwork, and photography. Circ: 10,500.
Website: www.susqu.edu/writers (click on High School Students)

FREELANCE POTENTIAL: 100% written by nonstaff writers. Publishes 80 freelance submissions yearly; 95% by unpublished writers, 95% by authors who are new to the magazine. Receives 5,000 unsolicited mss yearly.

SUBMISSIONS AND PAYMENT: Sample copy, $3 with 9x12 SASE ($1.17 postage). Send complete ms by March 1. Accepts hard copy and simultaneous submissions if identified. SASE. Responds during the month of May. First rights. Articles and fiction, 7,000 words. Poetry, no line limits. No payment. Provides 2 author's copies.

AMomsLove.com

1308 Midland Beaver Road
Industry, PA 15052

Editor: Caroline G. Shaw

DESCRIPTION AND INTERESTS: Written for the stay-at-home, work-from-home, and working mother, this online magazine provides inspiration and support through information, advice, and tips. A variety of topics are covered, including health and fitness, pregnancy, home and garden, parenting issues, and entertainment. Hits per month: 30,000.
Website: www.amomslove.com

FREELANCE POTENTIAL: 75% written by nonstaff writers. Publishes 60 freelance submissions yearly; 20% by authors who are new to the magazine.

SUBMISSIONS AND PAYMENT: Sample copy available at website. Send complete ms with short bio. Accepts email submissions to mom@amomslove.com (prefers HTML attachments; will accept Microsoft Word attachments). Response time varies. First rights. Articles, 700–1,100 words. No payment.

Art Jewelry

21027 Crossroads Circle
Waukesha, WI 53187

Editorial Assistant: Katie Streeter

DESCRIPTION AND INTERESTS: Individuals who make jewelry, either as professionals or as hobbyists, read *Art Jewelry* for informative articles on materials and methods. Appearing six times each year, the magazine presents content aimed at all skill levels, and also offers project ideas and tips. Circ: 40,000.
Website: www.artjewelry.com

FREELANCE POTENTIAL: 50% written by nonstaff writers. Publishes 54 freelance submissions yearly; 30% by unpublished writers, 50% by authors who are new to the magazine. Receives 350 queries yearly.

SUBMISSIONS AND PAYMENT: Sample copy and guidelines, $6.95. Query with jewelry samples or photos. Accepts hard copy and email queries with JPEG images to editor@jewelrymag.com. SASE. Responds in 1–2 months. All rights. Written material, word lengths and payment rates vary. Pays on acceptance. Provides 2 author's copies.

Athens Parent

P.O. Box 1251
Athens, GA 30603

Editor-in-Chief: Shannon Walsh Howell

DESCRIPTION AND INTERESTS: This local resource for families in the Athens area of Georgia features original articles on topics of interest to parents, grandparents, educators, and others interested in the well-being of children and families. It also offers a calendar of family-oriented events and information on schools, camps, and birthday parties. It is published eight times each year. Circ: Unavailable.
Website: www.athensparent.com

FREELANCE POTENTIAL: 85% written by non-staff writers. Publishes 40 freelance submissions yearly. Receives 500 queries yearly.

SUBMISSIONS AND PAYMENT: Guidelines and theme list available at website. Query. Accepts hard copy and email queries to editor@athensparent.com. SASE. Response time varies. First rights. Articles and depts/columns, word lengths and payment rates vary. Payment policy varies.

Austin Family

P.O. Box 7559
Round Rock, TX 78683-7559

Editor: Dianna Dworin

DESCRIPTION AND INTERESTS: Families in the Austin area of Texas find local event information and tips on today's parenting and family life in each issue of this journal, which is published monthly. Circ: 35,000.
Website: www.austinfamily.com

FREELANCE POTENTIAL: 70% written by non-staff writers. Publishes 18 freelance submissions yearly; 10% by unpublished writers, 50% by authors who are new to the magazine. Receives 1,200 queries and unsolicited mss yearly.

SUBMISSIONS AND PAYMENT: Sample copy, free. Query or send complete ms. Accepts email submissions to editor2003@austinfamily.com and simultaneous submissions if identified. Availability of artwork improves chance of acceptance. Responds in 3–6 months. First and second serial rights. Articles, 800–1,200 words. Depts/columns, 800 words. All material, payment rates vary. Pays on publication.

Baton Rouge Parents Magazine

11831 Wentling Avenue
Baton Rouge, LA 70816-6055

Editor: Amy Foreman-Plaisance

DESCRIPTION AND INTERESTS: This monthly magazine is dedicated to providing timely and meaningful regional information to families in Baton Rouge and the surrounding areas. Family life articles and local events and resources are found in each issue. Circ: 55,000.
Website: www.brparents.com

FREELANCE POTENTIAL: 95% written by non-staff writers. Publishes 50+ freelance submissions yearly; 15% by unpublished writers, 30% by authors who are new to the magazine.

SUBMISSIONS AND PAYMENT: Guidelines available via email request to brpm@brparents.com. Query with outline, source list, brief author biography, and 2 writing samples. Accepts hard copy and email queries to brpm@brparents.com. SASE. Response time varies. First North American serial rights. Written material, word lengths vary; $25-$70. Kill fee, $10. Pays on publication. Provides 2 copies.

Bay Area Baby

1660 South Amphlett Boulevard, Suite 335
San Mateo, CA 94402

Special Sections Editor

DESCRIPTION AND INTERESTS: *Bay Area Baby* is written for expectant and new parents living in the San Francisco Bay area. Comprehensive articles are featured on all issues related to pregnancy, childbirth, and infant care and development. In addition, regional information on moms' groups and baby activities, as well as relevant resources, are included. All articles are written on assignment at this time; therefore, queries and unsolicited manuscripts are not reviewed. Circ: 80,000.
Website: www.bayareaparent.com

FREELANCE POTENTIAL: 30% written by nonstaff writers. Publishes 21 freelance submissions yearly; 30% by authors who are new to the magazine.

SUBMISSIONS AND PAYMENT: Sample copy, guidelines, and theme list, free with 9x12 SASE (5 first-class stamps). All work is written on assignment at this time.

Beckett Plushie Pals

4635 McEwen Road
Dallas, TX 75244

Editor: Doug Kale

DESCRIPTION AND INTERESTS: Plush toys and their virtual worlds are the focus of this magazine from Beckett Media. It targets children ages seven and up who are interested in Webkinz, Club Penguin, Ty Girlz, Shining Stars, or Beanie Babies. New product information and how-to articles for use with the online games are always of interest. Circ: Unavailable.
Website: www.beckett.com

FREELANCE POTENTIAL: 50% written by nonstaff writers. Publishes 20 freelance submissions yearly; 50% by authors who are new to the magazine. Receives 50–100 queries yearly.

SUBMISSIONS AND PAYMENT: Sample copy and guidelines available. Query with 3–5 article ideas. Accepts email queries to plushiepals@beckett.com. Responds in 1–2 months. First rights. Written material, word lengths vary; $25–$100. Pays on publication. Provides 2 contributor's copies.

Biography Today

Omnigraphics Inc.
P.O. Box 625
Holmes, PA 19043

Managing Editor

DESCRIPTION AND INTERESTS: Children ages nine and up are the target audience of this magazine, which features a compilation of profiles on a range of personalities. These include authors, musicians, political leaders, sports figures, actors, cartoonists, TV personalities, and scientists. *Biography Today* appears six times each year. Profiles are extensively researched and are written in a consistent and easy-to-read format. Circ: 9,000.
Website: www.biographytoday.com

FREELANCE POTENTIAL: 50% written by nonstaff writers. Publishes several freelance submissions yearly. Receives 12 queries yearly.

SUBMISSIONS AND PAYMENT: Sample copy and writers' guidelines available with 9x12 SASE. Query with résumé. Accepts hard copy. SASE. Responds in 2 months. All rights. Articles, 2,000–5,000 words; payment rates vary. Provides 2 contributor's copies.

Beta Journal

National Beta Club
151 Beta Club Way
Spartanburg, SC 29306-3012

Editor: Lori Guthrie

DESCRIPTION AND INTERESTS: Members of the National Beta Club in elementary and high school read this magazine for member news as well as information on community service, academics, and leadership. Published five times each year, it regularly features student contributions. Circ: 400,000.
Website: www.betaclub.org

FREELANCE POTENTIAL: 10% written by nonstaff writers. Publishes 2–4 freelance submissions yearly; 80% by unpublished writers. Receives 12 unsolicited mss yearly.

SUBMISSIONS AND PAYMENT: Send complete ms. Accepts hard copy and email submissions to lguthrie@betaclub.org. Availability of artwork improves chance of acceptance. SASE. Responds in 2 months. Rights vary. Articles, 700–1,000 words; $25–$50. B/W prints or transparencies and line art; payment rates vary. Pays on publication. Provides 10 contributor's copies.

Bird Times

4642 West Market Street, #368
Greensboro, NC 27407

Editor: Rita Davis

DESCRIPTION AND INTERESTS: Bird owners find a wealth of information and advice on caring for their pets in this magazine, which is published six times each year. Health, breeding, new products, training, and basic care are all covered. Articles related to bird care, as well as new information on bird gadgets and food, are always welcome. Circ: 20,000.
Website: www.birdtimes.com

FREELANCE POTENTIAL: 90% written by nonstaff writers. Publishes 30–40 freelance submissions yearly; 10% by unpublished writers, 50% by authors who are new to the magazine.

SUBMISSIONS AND PAYMENT: Sample copy, $5 with 9x12 SASE (4 first-class stamps). Query or send complete ms. Accepts hard copy. SASE. Responds in 1 month. All rights. Articles, 1,200–2,000 words. Depts/columns, 600–800 words. Written material, $.10 per word. Pays on publication. Provides 1 contributor's copy.

Black Woman and Child

P.O. Box 47045, 300 Borough Drive
Toronto, Ontario M1P 4P0
Canada

Editor: Nicole Osbourne James

DESCRIPTION AND INTERESTS: This quarterly magazine is for African American women who are pregnant, plan to become pregnant, or have a child under age seven. Articles focus on the cultural aspects of these experiences and cover birth, nutrition, health and wellness, spirituality, discipline, and safety. Circ: Unavailable.
Website: www.blackwomanandchild.com

FREELANCE POTENTIAL: 75% written by non-staff writers. Publishes 25–40 freelance submissions yearly. Receives 75–100 queries, 30 unsolicited mss yearly.

SUBMISSIONS AND PAYMENT: Guidelines available. Query or send complete ms. Accepts hard copy and email to bwac@nubeing.com (text files). No simultaneous submissions. SAE/IRC. Response time varies. Rights vary. Articles, 750–1,500 words. Depts/columns, word lengths vary. Written material, payment rates vary. Pays on publication.

Brass

P.O. Box 1220
Corvallis, OR 97339

Editor

DESCRIPTION AND INTERESTS: This lifestyle magazine for young adults ages 16 to 25 offers information on money, investments, and savings. It is interested in well-written articles and tips that will help its audience understand how young adult lifestyles and money interact. Interested writers should visit the website for updates on current article assignrments. Circ: Unavailable.
Website: www.brasscu.com

FREELANCE POTENTIAL: 50% written by non-staff writers. Publishes 20 freelance submissions each year.

SUBMISSIONS AND PAYMENT: Guidelines available at website. Prefers query; will accept complete ms. Accepts hard copy and email submissions to BrassService@brassmedia.com. SASE. Response time varies. First rights. Articles, 400–800 words. Short snippets, 100–300 words. No payment. Provides 2 contributor's copies.

Brain, Child

P.O. Box 714
Lexington, VA 24450

Editors: Jennifer Niesslein & Stephanie Wilkinson

DESCRIPTION AND INTERESTS: Since 2000, *Brain, Child* has been dedicated to women who have both. More than just a parenting magazine, *Brain, Child* serves as a platform for mothers to share their stories and thoughts, often with humor. It is specifically looking for essays about parenting teenagers. Circ: 30,000.
Website: www.brainchildmag.com

FREELANCE POTENTIAL: 90% written by non-staff writers. Publishes 40 freelance submissions yearly; 15% by unpublished writers, 60% by authors who are new to the magazine. Receives 300 queries, 2,400 unsolicited mss yearly.

SUBMISSIONS AND PAYMENT: Sample copy and guidelines, $5. Query or send complete ms. Accepts email to editor@brainchildmag.com. Responds in 2 months. Electronic rights. Articles, 3,000 words. Personal essays, 800–4,500 words. Fiction, 1,500–4,500 words. Written material, payment rates vary. Pays on publication.

Caledonia Times

Box 278
Prince Rupert, British Columbia V8J 3P6
Canada

Editor: Debby Shaw

DESCRIPTION AND INTERESTS: Christian youth and adults living in Canada are the target audience for this small newsletter that combines opinion and personal experience pieces, biographies, and inspirational fiction and poetry. News from each parish is presented along with articles on music, nature, religion, and social and multi-cultural issues. It is published 10 times each year. Freelance writers are encouraged to submit material for all departments. Circ: 1,259.

FREELANCE POTENTIAL: 95% written by non-staff writers. Publishes 10 freelance submissions yearly. Receives 1–2 unsolicited mss yearly.

SUBMISSIONS AND PAYMENT: Send complete ms. Accepts hard copy only. SAE/IRC. Responds in 2–4 weeks. All rights. Articles and fiction, 500–750 words. Depts/columns, word lengths vary. No payment. Provides 5 contributor's copies.

Calgary's Child

#723, 105-150 Crowfoot Crescent NW
Calgary, Alberta T3G 3T2
Canada

Editor: Ellen Percival

DESCRIPTION AND INTERESTS: Much of the editorial in *Calgary's Child* is focused on parental empowerment. Targeted toward Calgary-area families, particularly those with children in their early teens and younger, it offers informative articles on family issues, education, children's health, and child care, as well as information about community and recreational events. It is published six times each year. Circ: 70,000.
Website: www.calgaryschild.com

FREELANCE POTENTIAL: 90% written by non-staff writers. Publishes 210 freelance submissions yearly; 20% by authors new to the magazine.

SUBMISSIONS AND PAYMENT: Sample copy available at website. Query with outline. Accepts email queries to calgaryschild@shaw.ca. No simultaneous submissions. Response time varies. Exclusive Calgary rights. Articles, 400–500 words; to $50 Canadian. Payment policy varies. Provides 2 contributor's copies.

Canoe & Kayak Magazine

10526 NE 68th Street, Suite 3
Kirkland, WA 98033

Editor

DESCRIPTION AND INTERESTS: Flatwater and whitewater boating are the subjects of this magazine, published seven times each year. Destination and technique pieces appear along with equipment reviews. Circ: 63,000.
Website: www.canoekayak.com

FREELANCE POTENTIAL: 90% written by non-staff writers. Publishes 25 freelance submissions yearly; 5% by unpublished writers, 25% by authors who are new to the magazine. Receives 240 queries and unsolicited mss yearly.

SUBMISSIONS AND PAYMENT: Sample copy and guidelines, free with 9x12 SASE (7 first-class stamps). Query or send complete ms. Accepts email submissions to editor@canoekayak.com. Responds in 6–8 weeks. All rights. Articles, 400–2,000 words. Depts/columns, 150–750 words. Written material, $.15–$.50 per word. Pays within 30 days of publication. Provides 1 contributor's copy.

Camping Today

126 Hermitage Road
Butler, PA 16001

Editor: DeWayne Johnston

DESCRIPTION AND INTERESTS: Members of the Family Campers and RVers Association turn to *Camping Today* for articles on camping sites, recreational vehicles, and family travel. Each of its 10 yearly issues also features product and campsite reviews. Circ: 10,000.
Website: www.fcrv.org

FREELANCE POTENTIAL: 40% written by nonstaff writers. Publishes 15–20 freelance submissions yearly; 10% by unpublished writers. Receives 240 unsolicited mss yearly.

SUBMISSIONS AND PAYMENT: Writers' guidelines and theme list available. Send complete ms with artwork (JPEG files). Accepts hard copy. SASE. Responds in 2 months. One-time rights. Articles, 1,000–3,000 words. Depts/columns, word lengths vary. Written material, $35–$150. Pays on publication. Provides 1+ contributor's copies.

Catalyst Chicago
Independent Reporting on Urban Schools

332 South Michigan Avenue, Suite 500
Chicago, IL 60604

Editor-in-Chief: Veronica Anderson

DESCRIPTION AND INTERESTS: The mission of *Catalyst Chicago* is to improve the education of children through authoritative journalism covering education reform, policy, programs, and laws, and to encourage a dialogue between parents, educators, and public policy makers. It is published five times each year. Writers may present their credentials to be considered for future assignments. Circ: 9,000.
Website: www.catalyst-chicago.org

FREELANCE POTENTIAL: 20% written by nonstaff writers. Publishes 10–20 freelance submissions yearly; 25% by authors who are new to the magazine. Receives 45 queries yearly.

SUBMISSIONS AND PAYMENT: Sample copy and writers' guidelines, $2. Query or send letter of introduction. Accepts hard copy. SASE. Response time varies. All rights. Articles, to 2,300 words; $1,700. Pays on acceptance. Provides 1 contributor's copy.

Cat Fancy

3 Burroughs
Irvine, CA 92618

Editor: Susan Logan

DESCRIPTION AND INTERESTS: *Cat Fancy* strives to provide education as well as entertainment to cat owners and lovers. Each monthly issue covers feline news, profiles, and advice. Circ: 290,000.
Website: www.catchannel.com

FREELANCE POTENTIAL: 95% written by non-staff writers. Publishes 150 freelance submissions yearly; 10% by unpublished writers, 70% by new authors. Receives 500+ queries yearly.

SUBMISSIONS AND PAYMENT: Guidelines available. Query with clips between January 1 and May 1 only. Accepts email to query@catfancy.com. Availability of artwork improves chance of acceptance. Responds by August. First rights. Articles, 600–1,000 words. Depts/columns, 600 words. 35mm slides; high-resolution digital images with contact sheets. All material, payment rates vary. Pays on publication. Provides 2 contributor's copies.

Charlotte Baby & Child

81519 Alexander
Chapel Hill, NC 27157

Editorial Department

DESCRIPTION AND INTERESTS: This upscale annual targets parents with children up to the age of six who live in Charlotte, North Carolina, and the surrounding area. It features profiles of area moms, families, and businesses, as well as informational articles on health, home, child care, finance, recreation, and local events. Circ: 10,000.
Website: www.charlottebabyandchild.com

FREELANCE POTENTIAL: 50% written by non-staff writers. Publishes 30 freelance submissions each year.

SUBMISSIONS AND PAYMENT: Sample copy available at website. Query with clips or writing samples. Accepts hard copy and email submissions to editor@charlottebabyandchild.com. SASE. Responds in 2 months. Rights vary. Written material, word lengths and payment rates vary. Pays on publication. Provides 2 contributor's copies.

Chickadee

Bayard Press Canada
10 Lower Spadina Avenue, Suite 400
Toronto, Ontario M4V 2V2
Canada

Submissions Editor

DESCRIPTION AND INTERESTS: *Chickadee* is a discovery magazine for beginning readers ages six to nine whose thirst for knowledge and appetite for humor are insatiable. Each of its 10 yearly issues is built around a central theme, and is designed to enhance children's independent reading and problem-solving skills. It does this through articles, stories, puzzles, and experiments. All work is assigned. It does not review unsolicited manuscripts or queries. Circ: 85,000.
Website: www.owlkids.com

FREELANCE POTENTIAL: 5% written by non-staff writers. Publishes 1 freelance submission each year.

SUBMISSIONS AND PAYMENT: Sample copy, $4. Guidelines and theme list available. Send résumé only. Accepts hard copy. No unsolicited mss or queries. All rights. Fiction, 650–700 words; $250. Pays on acceptance. Provides 2 contributor's copies.

Childbirth Magazine

375 Lexington Avenue
New York, NY 10017

Managing Editor: Kate Kelly

DESCRIPTION AND INTERESTS: All aspects related to the third trimester of pregnancy and childbirth are covered in this annual publication. Information can be found on labor, delivery, and recovery, as well as child care and parenting. It welcomes queries on childbirth and recovery topics. All work is done by assignment only; unsolicited manuscripts are not accepted. Circ: 230,000.
Website: www.americanbaby.com

FREELANCE POTENTIAL: 55% written by non-staff writers. Publishes 10 freelance submissions each year.

SUBMISSIONS AND PAYMENT: Query. No unsolicited mss. Accepts hard copy. SASE. Responds in 6 weeks. First serial rights. Articles, 1,000–2,000 words. Depts/columns, word lengths vary. Written material, payment rates vary. Pays on acceptance. Provides 5 contributor's copies.

Children's Advocate

Action Alliance for Children
The Hunt Home
1201 Martin Luther King Jr. Way
Oakland, CA 94612-1217

Editor: Jeanne Tepperman

DESCRIPTION AND INTERESTS: *Children's Advocate* contains news of public policy issues that affect children in California. Published six times each year in English and Spanish, it uses California writers only, and does not buy unsolicited submissions. Writers wishing to join its group of freelancers should send résumés and clips. Circ: 15,000.
Website: www.4children.org

FREELANCE POTENTIAL: 60% written by nonstaff writers. Publishes 24 freelance submissions each year.

SUBMISSIONS AND PAYMENT: Sample copy and writers' guidelines available. All work is assigned. Send résumé and writing samples. Accepts hard copy. SASE. First North American rights. Articles, 500 or 1,000 words. Depts/columns, 500 or 1,000 words. Written material, $.25 per word. Pays on acceptance. Provides 3 contributor's copies.

Chirp

Bayard Press Canada
10 Lower Spadina Avenue, Suite 400
Toronto, Ontario M4V 2V2
Canada

Submissions Editor

DESCRIPTION AND INTERESTS: *Chirp* is called a "See and Do, Laugh and Learn" magazine because that is what it entices its young audience to do. Designed for children ages two through six, the magazine uses games, puzzles, rhymes, and stories to teach the relationship between pictures and words. It is published nine times each year. *Chirp* does not accept unsolicited manuscripts or article queries. All work is done on assignment. Circ: 60,000.
Website: www.owlkids.com

FREELANCE POTENTIAL: 10% written by nonstaff writers. Publishes 1–3 freelance submissions yearly; 1% by unpublished writers.

SUBMISSIONS AND PAYMENT: Sample copy, $3.50. Guidelines available. Send résumé only. No unsolicited mss. Accepts hard copy. All rights. Written material, 300–400 words; payment rates vary. Pays on acceptance. Provides 2 contributor's copies.

Cincinnati Family Magazine

10945 Reed Hartman Highway, Suite 221
Cincinnati, OH 45242

Editor: Sherry Hang

DESCRIPTION AND INTERESTS: This monthly magazine serves as an informational resource for parents living in the Cincinnati region. It contains articles on parenting, education, and childcare issues, as well as a calendar of events and other recreational information. Circ: 55,000.
Website: www.cincinnatifamily.com

FREELANCE POTENTIAL: 50% written by nonstaff writers. Publishes 12–15 freelance submissions yearly; 5% by unpublished writers, 5% by authors who are new to the magazine. Receives 360–420 queries, 180–240 unsolicited mss yearly.

SUBMISSIONS AND PAYMENT: Guidelines and editorial calendar available. Query or send complete ms. Accepts hard copy and email submissions to sherryh@daycommail.com. SASE. Response time varies. First rights. Articles, word lengths vary; $75–$125. Depts/columns, word lengths and payment rates vary. Pays 30 days after publication.

Cincinnati Parent

1071 Celestial Street, Suite 1104
Cincinnati, OH 45202

Editor: Jason Jones

DESCRIPTION AND INTERESTS: Previously published as *All About Kids Parenting Magazine*, this monthly magazine provides a mix of family-related articles and local news for parents in the greater Cincinnati area. Topics cover health, education, and parenting issues. A comprehensive calendar of events as well as local resource directories appear each month. Circ: 120,000.
Website: www.cincinnatiparent.com

FREELANCE POTENTIAL: 25–40% written by nonstaff writers. Publishes 30 freelance submissions yearly.

SUBMISSIONS AND PAYMENT: Guidelines available at website. Send complete ms. Accepts disk submissions (text files only) and email submissions to jason@cincinnatiparent.com. SASE. Response time varies. Rights vary. Articles and depts/columns, word lengths and payment rates vary. Payment policy varies. Provides 1 contributor's copy.

Civilized Revolt

107 Crestview Drive
Morgantown, WV 26505

Editor: Derek Wehrwein

DESCRIPTION AND INTERESTS: This online magazine is written by and for young adults, providing a platform for them to share their political and cultural views. It appeals to a young, politically-minded, conservative audience. Updated every two weeks, it seeks commentary on newsworthy current events. Articles containing bashing, foul language, and baseless accusations are not accepted. Hits per month: 7,100.
Website: www.civilizedrevolt.com

FREELANCE POTENTIAL: 15% written by non-staff writers. Publishes 20 freelance submissions yearly; 75% by unpublished writers, 33% by authors who are new to the magazine. Receives 20 unsolicited mss yearly.

SUBMISSIONS AND PAYMENT: Guidelines available at website. Send complete ms. Accepts email submissions to editor@civilizedrevolt.com. Responds in 2–3 weeks. All rights. Articles, 400–900 words. No payment.

Clubhouse

P.O. Box 15
Berrien Springs, MI 49103

Editor: Elaine Trumbo

DESCRIPTION AND INTERESTS: Though still filled with stories, puzzles, games, and activities for children ages nine through twelve, *Clubhouse* is no longer available in print. It has become exclusively a monthly e-zine, with a distinctive Christian perspective and a positive message, dedicated to creative kids. At this time, it is not accepting new material. Writers may check the website for updates to that policy. Hits per month: Unavailable.
Website: www.yourstoryhour.org/clubhouse

FREELANCE POTENTIAL: 85% written by non-staff writers. Publishes several freelance submissions yearly; 75% by unpublished writers, 95% by authors who are new to the magazine.

SUBMISSIONS AND PAYMENT: Sample copy available at website. Send complete ms. Accepts hard copy. SASE. Response time varies. All rights. Articles and fiction, 1,500 words. B/W line art. All material, payment rates vary.

Classic Toy Trains

21027 Crossroads Circle
Waukesha, WI 53187

Editor: Carl Swanson

DESCRIPTION AND INTERESTS: Toy train enthusiasts read this magazine for its photo-filled articles on collecting and operating toy trains. It is published nine times each year. How-to articles and reviews are needed. Circ: 55,000.
Website: www.classictoytrains.com

FREELANCE POTENTIAL: 60% written by non-staff writers. Publishes 40–50 freelance submissions yearly; 20% by unpublished writers, 20% by authors who are new to the magazine. Receives 96 queries, 60 unsolicited mss yearly.

SUBMISSIONS AND PAYMENT: Sample copy, $4.95 ($3 postage). Prefers query; will accept complete ms. Accepts hard copy, disk submissions (Microsoft Word), and email to editor@ classictoytrains.com. SASE. Responds in 3 months. All rights. Articles, 500–5,000 words; $75 per page. Depts/columns, word lengths and payment rates vary. Pays on acceptance. Provides 1 contributor's copy.

Coastal Family Magazine

340 Eisenhower Drive, Suite 240
Savannah, GA 31406

Managing Editor: Laura Gray

DESCRIPTION AND INTERESTS: Targeting families in the Savannah area and the South Carolina Lowcountry, this monthly magazine addresses topics such as education, health and fitness, family finances, and parenting. Regional travel and local recreation are also covered. The "Ages & Stages" section, which covers parenting issues for infants through teens, offers the best opportunity for freelancers. The magazine also accepts reprints. Circ: 18,000.
Website: www.coastalfamily.com

FREELANCE POTENTIAL: 85% written by non-staff writers. Publishes 12–15 freelance submissions yearly. Receives 100–150 queries yearly.

SUBMISSIONS AND PAYMENT: Guidelines available. Accepts hard copy and email queries to editor@coastalfamily.com. SASE. Response time varies. One-time print and web rights. Written material, word lengths vary; $25–$40. Pays 1–2 months after publication.

Coins

F & W Publications
700 East State Street
Iola, WI 54990

Editor: Robert Van Ryzin

DESCRIPTION AND INTERESTS: The pages of this monthly magazine, geared to coin enthusiasts and collectors, are filled with information on collecting and buying coins, medals, and tokens. Personal experience pieces, profiles, and how-to articles are featured as well as value guides and an events calendar. The magazine usually does not use freelancers unless they are coin experts. Circ: 60,000.
Website: www.coinsmagazine.net

FREELANCE POTENTIAL: 40% written by non-staff writers. Publishes 70 freelance submissions yearly; 5% by authors who are new to the magazine. Receives 36–60 queries yearly.

SUBMISSIONS AND PAYMENT: Sample copy and guidelines, free. Query. Accepts hard copy. SASE. Responds in 1–2 months. All rights. Articles, 1,500–2,500 words; $.04 per word. Work for hire. Pays on publication. Provides contributor's copies upon request.

Community Education Journal

3929 Old Lee Highway, Suite 91A
Fairfax, VA 22030

Editor

DESCRIPTION AND INTERESTS: Serving as a forum for the exchange of ideas and practices in community education, this journal addresses issues with adult education, and after-school and summer programs for kindergarten through high school. Coverage of successful programs is featured. Submissions should follow AP style. *Community Education Journal* is published quarterly by the National Community Education Association. Circ: Unavailable.
Website: www.ncea.com

FREELANCE POTENTIAL: 98% written by non-staff writers. Publishes 24 freelance submissions each year.

SUBMISSIONS AND PAYMENT: Sample copy, guidelines, and theme list, $5. Accepts email submissions to ncea@ncea.com (Microsoft Word or PDF attachments). Responds in 1–2 months. All rights. Articles, 1,500–2,000 words. No payment.

ColumbiaKids

1911 Pacific Avenue
Tacoma, WA 98402

Managing Editor: Stephanie Lile

DESCRIPTION AND INTERESTS: This free online magazine for children ages 4 to 14 focuses on the Pacific Northwest region, including Alaska, Washington, Oregon, and Idaho. It features well-researched articles and creative stories set in the region. *CoumbiaKids* is a sister publication to *Columbia: the Magazine of Northwest History*. It premiered in August 2008 and is published twice each year. Hits per month: 5,000+.
Website: www.columbiakidsmagazine.org

FREELANCE POTENTIAL: 80% written by non-staff writers. Publishes several freelance submissions yearly.

SUBMISSIONS AND PAYMENT: Sample copy available at website. Send complete ms with bibliography. Accepts hard copy SASE. Response time varies. First world and archival rights. Articles, 800–1,200 words; $200. Depts/columns, 200–500 words; $50–$100. Pays on publication.

Cookie

4 Times Square
New York, NY 10036

Acquisitions: Mireille Hyde

DESCRIPTION AND INTERESTS: *Cookie* sets itself apart from other parenting magazines by targeting busy parents who live a sophisticated lifestyle. Published 10 times each year, it focuses on issues relevant to parents of children up to age six, and regularly covers topics such as health and fitness, nutrition, family travel, fashion, and beauty. When submitting queries, writers should include an organization plan for the article. Circ: 400,000.
Website: www.cookiemag.com

FREELANCE POTENTIAL: 50% written by non-staff writers. Publishes 10 freelance submissions yearly. Receives 600–1,200 queries yearly.

SUBMISSIONS AND PAYMENT: Query. Accepts hard copy and email queries to editor@cookiemag.com (include "Freelance Pitch" in the subject line). SASE. Response time varies. Rights vary. Articles, word lengths vary; payment rates vary. Pays on publication.

Craft Bits

P.O. Box 3106
Birkdale, Queensland 4159
Australia

Editor: Shellie Wilson

DESCRIPTION AND INTERESTS: This independent online magazine offers craft ideas and projects for the whole family. It features a special kids craft section with projects for young children and craft activities for small groups. Among its other categories are jewelry making, recycled crafts, soap making recipes, scrapbooking, and holiday crafts. Circ: 60,000.
Website: www.craftbits.com

FREELANCE POTENTIAL: 5% written by non-staff writers. Publishes many freelance submissions yearly.

SUBMISSIONS AND PAYMENT: Query with photographs (JPEG images). Queries for complicated craft ideas should include photos to accompany the step-by-step instructions. Accepts email queries to staff@craftbits.com. Responds in 1–2 months. All rights. Written material, word lengths vary; $45 (USD). Kill fee, $10. Pays on publication.

Curriculum Review

Paperclip Communications
125 Paterson Avenue
Little Falls, NJ 07424

Editor: Laura Betti

DESCRIPTION AND INTERESTS: Providing teachers and school administrators with the latest strategies and resources to succeed in their jobs, *Curriculum Review* contains articles, curriculum ideas, and teaching techniques. The newsletter is published nine times during the school year, from September to May. Writers who have educational backgrounds are invited to submit material. Circ: 5,000.
Website: www.curriculumreview.com

FREELANCE POTENTIAL: 2% written by non-staff writers. Publishes 10 freelance submissions yearly. Receives 24 queries yearly.

SUBMISSIONS AND PAYMENT: Sample copy, free with 9x12 SASE (2 first-class stamps). Query. Accepts hard copy. SASE. Responds in 1 month. One-time rights. Articles, to 4,000 words. Depts/columns, word lengths vary. Written material, payment rates vary. Payment policy varies. Provides contributor's copies.

Creative Child Magazine

2505 Anthem Village Drive, Suite E619
Henderson, NV 89052

Editor: Scott Reichart

DESCRIPTION AND INTERESTS: Parents looking to nurture and stimulate their child's creativity will find helpful tips, timely information, and craft ideas for all age levels in this magazine. Articles also cover such topics as family travel, health, and safety. Book and media reviews also appear. The magazine is published six times each year. Circ: 50,000.
Website: www.creativechild.com

FREELANCE POTENTIAL: 25% written by non-staff writers. Publishes 20 freelance submissions each year.

SUBMISSIONS AND PAYMENT: Sample copy, free with 9x12 SASE (4 first-class stamps). Query. Accepts hard copy and email queries to info@creativechild.com. SASE. Responds in 1–3 months. First print and electronic rights. Articles and depts/columns, word lengths vary; payment rates vary. Pays on publication. Provides contributor's copies.

Dirt Rider Magazine

6420 Wilshire Boulevard, 17th Floor
Los Angeles, CA 90048-5515

Editor

DESCRIPTION AND INTERESTS: A monthly magazine geared toward dirt bike racers of all ages, this publication features rider profiles, riding tips, new product information and reviews, news from the dirt bike racing circuit, and competition results. It welcomes editorial queries from freelancers. Photographs are strongly encouraged. Circ: Unavailable.
Website: www.dirtrider.com

FREELANCE POTENTIAL: 20% written by non-staff writers. Publishes 20 freelance submissions each year.

SUBMISSIONS AND PAYMENT: Query. Accepts hard copy. SASE. Response time varies. Rights vary. Articles, word lengths vary. 5x5 JPEG images at 300 dpi; include captions and photo releases for all persons depicted in photographs. All material, payment rates vary. Pays on publication. Provides 1 contributor's copy.

Discoveries

WordAction Publishing Company
2923 Troost Avenue
Kansas City, MO 64109

Assistant Editor: Kimberly Adams

DESCRIPTION AND INTERESTS: WordAction is a not-for-profit organization that provides curriculum resources that enable people of all ages to discover God's Word. *Discoveries* is its take-home story paper for children that reinforces its Sunday school curriculum. Each issue contains stories that portray believable experiences for kids ages eight to ten; articles that offer advice, personal experience, and family service projects; and crafts and puzzles. Circ: 35,000.
Website: www.wordaction.com

FREELANCE POTENTIAL: 75% written by non-staff writers. Of the freelance submissions published yearly, 20% are by unpublished writers.

SUBMISSIONS AND PAYMENT: Guidelines and theme list available. Query or send complete ms. Accepts hard copy. SASE. Responds in 6–8 weeks. Multiple-use rights. Articles, 150 words; $15. Stories, to 500 words; $25. Pays on acceptance. Provides 1 contributor's copy.

Dollhouse Miniatures

68132 250th Avenue
Kasson, MN 55944

Editor-in-Chief: Terrence Lynch

DESCRIPTION AND INTERESTS: Written for hobbyists of small-scale reproductions, this magazine dedicates its pages to profiles of miniaturists, detailed instructions for projects, and reviews of new products. It is published monthly. Circ: 25,000.
Website: www.dhminiatures.com

FREELANCE POTENTIAL: 75% written by non-staff writers. Publishes 100 freelance submissions yearly; 10% by unpublished writers, 30% by authors who are new to the magazine. Receives 60 queries yearly.

SUBMISSIONS AND PAYMENT: Guidelines available at website. Sample copy, $4.95 with 9x12 SASE ($1.95 postage). Query with outline. Accepts hard copy and email queries to tlynch@madavor.com. SASE. Responds in 2 months. All rights. Articles and depts/columns, word lengths vary; $75 per magazine page. Pays on publication. Provides 1 contributor's copy.

Dog Fancy

BowTie, Inc.
P.O. Box 6050
Mission Viejo, CA 92690-6050

Managing Editor: Hazel Kelly

DESCRIPTION AND INTERESTS: Devoted to the care and enjoyment of all dogs, this monthly magazine is written for breeders, trainers, showers, and pet owners. Thoroughly researched articles are featured in each issue covering breed profiles, canine health, basic care, behavior, and training. Circ: 270,000.
Website: www.dogfancy.com

FREELANCE POTENTIAL: 80% written by non-staff writers. Publishes 20–25 freelance submissions yearly; 25% by authors who are new to the magazine. Receives 1,200 queries yearly.

SUBMISSIONS AND PAYMENT: Guidelines available. Sample copy, $4.50 at newsstands. Query with résumé, outline, and writing samples. Accepts hard copy. SASE. Responds in 2–3 months. First North American serial rights. Articles, 1,200–1,800 words. Depts/columns, 650 words. All material, payment rates vary. Pays on publication. Provides 2 contributor's copies.

Dolls

P.O. Box 5000
Iola, WI 54945

Editor: Carie Ferg

DESCRIPTION AND INTERESTS: Serious doll collectors read this magazine for its complete coverage of dolls and doll collecting. Its articles cover a wide range of dolls, collectors, manufacturers, shows, and the history of dolls. It is not interested in articles that are purely nostalgic or sentimental. It is published 10 times each year. Circ: 65,000.
Website: www.dollsmagazine.com

FREELANCE POTENTIAL: 85% written by non-staff writers. Publishes 50 freelance submissions yearly; 10% by authors who are new to the magazine. Receives 40 unsolicited mss yearly.

SUBMISSIONS AND PAYMENT: Send complete ms. Accepts email submissions to carief@jonespublishing.com. Availability of artwork improves chance of acceptance. SASE. Response time varies. One-time rights. Articles, 1,000–2,000 words; $200. Color prints, slides, or transparencies; JPEG files at 300 dpi. Pays on publication.

Dyslexia Online Magazine

Submissions: Teresa Burns

DESCRIPTION AND INTERESTS: Geared to parents, teachers, and students alike, this online publication focuses on children and adults with dyslexia. Readers will find a mix of self-help, informational, and personal experience pieces on the condition. Book reviews and teaching techniques are also featured. Many of the articles are written in both English and Spanish, and all material is permanently available at the website. *Dyslexia Online Magazine* was founded in 1998. Hits per month: 150,000.
Website: www.dyslexia-parent.com/magazine

FREELANCE POTENTIAL: 50% written by nonstaff writers.

SUBMISSIONS AND PAYMENT: Sample copy and writers' guidelines available at website. Send complete ms. Accepts email submissions to dyslextest@aol.com (no attachments). Responds in 2 weeks. All rights. Articles, word lengths vary. No payment.

Earthwatch Institute Journal

3 Clock Tower Place, Suite 100
Maynard, MA 01754-0075

Editor: Jennifer Goebel

DESCRIPTION AND INTERESTS: As the official member publication of Earthwatch Institute, this journal is committed to educating, inspiring, and empowering its readers in scientific field research. Its pages are filled with news and research from the Institute, as well as member profiles. Distributed around the world, it is published three times each year. Circ: 25,000.
Website: www.earthwatch.org

FREELANCE POTENTIAL: 30% written by nonstaff writers. Publishes 2–3 freelance submissions yearly. Receives 60 queries, 36 unsolicited mss each year.

SUBMISSIONS AND PAYMENT: Sample copy and guidelines available. Query or send complete ms. Prefers email submissions to jgoebel@earthwatch.org; will accept hard copy. SASE. Response time varies. First North American serial rights. Articles, 1,500–2,000 words; $500–$1,000. Pays on publication.

Early Years

3035 Valley Avenue, Suite 103
Winchester, VA 22601

Submissions Editor

DESCRIPTION AND INTERESTS: Parents of children in preschool and kindergarten are the target audience for this newsletter, which is published nine times each year. Short informational and how-to articles and tips are featured to promote school readiness and to help parents participate in their children's educational success. Circ: 60,000.
Website: www.rfeonline.com

FREELANCE POTENTIAL: 100% written by nonstaff writers. Publishes 80 freelance submissions yearly; 28% by unpublished writers. Receives 36 queries yearly.

SUBMISSIONS AND PAYMENT: Sample copy, free with 9x12 SASE (2 first-class stamps). Query with résumé and clips. Accepts hard copy. SASE. Responds in 1 month. All rights. Articles, 225–300 words. Depts/columns, 175–200 words. Written material, $.60 per word. Pays on acceptance. Provides 5 contributor's copies.

The Education Revolution

417 Roslyn Road
Roslyn Heights, NY 11577

Executive Director: Jerry Mintz

DESCRIPTION AND INTERESTS: Targeting educators, administrators, and parents, this quarterly magazine offers the latest information on alternative education. Articles highlight successful public and private alternative schools, and provide homeschooling information. Published by the Alternative Education Resource organization, it also includes job opportunities, conference details, contact information, and international news. Queries are welcome from writers experienced in the education field. Circ: 5,000.
Website: www.educationrevolution.org

FREELANCE POTENTIAL: 20% written by nonstaff writers. Publishes 10 freelance submissions yearly; 40% by authors who are new to the magazine. Receives 180 queries yearly.

SUBMISSIONS AND PAYMENT: Query. Accepts hard copy. SASE. Responds in 1 month. Rights vary. Articles and depts/columns, word lengths vary. No payment.

EFCA Today

418 Fourth Street NE
Charlottesville, VA 22902

Editor: Diane McDougall

DESCRIPTION AND INTERESTS: Evangelical Free Church of America (EFCA) pastors, elders, deacons, Sunday school teachers, and ministry volunteers read this quarterly journal to participate in a national conversation about spirituality and the wide variety of issues facing the church and church members today. Writers who are not affiliated with EFCA are unlikely to fit the magazine's needs. Circ: 30,000.
Website: www.efca.org/today

FREELANCE POTENTIAL: 80% written by non-staff writers. Publishes several freelance submissions yearly.

SUBMISSIONS AND PAYMENT: Sample copy and guidelines, $1 with 9x12 SASE (5 first-class stamps). Query. Accepts email queries to dianemc@journeygroup.com. Response time varies. First rights. Articles, 200–700 words. Cover theme articles, 300–1,000 words. Written material, $.23 per word. Pays on acceptance.

The Elementary School Journal

University of Missouri, College of Education
202 London Hall
Columbia, MO 65211-1150

Managing Editor

DESCRIPTION AND INTERESTS: For more than 100 years, *The Elementary School Journal* has served researchers, teacher educators, and practitioners in elementary and middle school education. Appearing five times each year, it publishes peer-reviewed articles on education theory and research. The magazine prefers to publish original studies that contain data about school and classroom processes. It also publishes the latest research in child development. Circ: 2,200.
Website: www.journals.uchicago.edu/ESJ

FREELANCE POTENTIAL: 100% written by nonstaff writers. Publishes several freelance submissions yearly.

SUBMISSIONS AND PAYMENT: Sample copy, $13.50. Guidelines available at website. Send 4 copies of complete ms with 100- to 150-word abstract. Accepts hard copy and disk submissions. SASE. Response time varies. Rights vary. Articles, word lengths vary. No payment.

Encyclopedia of Youth Studies

130 Essex Street
South Hamilton, MA 01982

Editor: Dean Borgman

DESCRIPTION AND INTERESTS: With an audience of teachers, social workers, and youth ministers, this online encyclopedia strives to improve the lives of children and young adults. More than 200 topics of youth culture are covered, with well-researched articles, training materials, and program ideas. It is actively seeking submissions. Hits per month: Unavailable.
Website: www.centerforyouth.org

FREELANCE POTENTIAL: 20% written by non-staff writers. Publishes 5–10 freelance submissions yearly; 85% by unpublished writers, 85% by authors who are new to the magazine. Receives 48 queries, 12 unsolicited mss yearly.

SUBMISSIONS AND PAYMENT: Sample copy and guidelines available at website. Query or send complete ms. Accepts email submissions to cys@centerforyouth.org. Responds to queries in 1 week, to mss in 1 month. All rights. Articles, 600 words. No payment.

Entertainment Magazine

P.O. Box 3355
Tucson, AZ 85722

Publisher: Robert Zucker

DESCRIPTION AND INTERESTS: Focusing on the greater Tucson area, this online magazine is updated daily with information on community activities that are of interest to families. It looks for submissions pertaining to health and fitness, music, regional news, recreation, college, and careers. Product promotions are not accepted. Hits per month: 1 million.
Website: www.emol.org

FREELANCE POTENTIAL: 90% written by non-staff writers. Publishes 300+ freelance submissions yearly; 75% by unpublished writers, 25% by authors who are new to the magazine. Receives 3,000+ queries yearly.

SUBMISSIONS AND PAYMENT: Sample copy and writers' guidelines available at website. Query. Accepts email queries to publisher@ emol.org. Responds in 1–2 days. Author retains rights. Articles, to 1,000 words. B/W digital images. No payment.

Equestrian

4047 Iron Works Parkway
Lexington, KY 40511

Editor: Brian Sosby

DESCRIPTION AND INTERESTS: Equestrians and horse enthusiasts are the target audience of this publication from the United States Equestrian Federation. Published 10 times each year, it features profiles of top equestrians and horses, information on USEF competitions, and articles on horse care in every issue. Circ: 90,000.
Website: www.usef.org

FREELANCE POTENTIAL: 50% written by non-staff writers. Publishes 60 freelance submissions yearly; 10% by authors who are new to the magazine. Receives 240 queries yearly.

SUBMISSIONS AND PAYMENT: Sample copy and guidelines available. Query with résumé and writing samples. Accepts email queries to bsosby@usef.org. Responds in 1 week. First rights. Articles, 2,000–3,000 words. Depts/columns, 500–1,000 words. Written material, payment rates vary. Kill fee, 50%. Pays on publication. Provides 1 contributor's copy.

Family-Life Magazine

100 Professional Center Drive, Suite 104
Rohnert Park, CA 94928

Publisher/Editor: Sharon Gowan

DESCRIPTION AND INTERESTS: A monthly regional publication, *Family-Life Magazine* offers parents informative articles on family life issues and area entertainment. Education, the arts, child care, health and fitness, and family activities are regularly covered. It is distributed in the Sonoma, Mendocino, and Lake counties of California. Circ: 40,000.
Website: www.family-life.us

FREELANCE POTENTIAL: 40% written by non-staff writers. Publishes 24–36 freelance submissions yearly; 10% by unpublished writers. Receives 120+ unsolicited mss yearly.

SUBMISSIONS AND PAYMENT: Guidelines and editorial calendar available. Send complete ms. Accepts email submissions to sharon@family-life.us (in body of email or as Microsoft Word attachment). Response time varies. One-time rights. Articles, 650–1,150 words; $.08 per word. Pays on publication.

eSchoolNews

7920 Norfolk Avenue, Suite 900
Bethesda, MD 20814

Editor: Greg Downey

DESCRIPTION AND INTERESTS: Read by school administrators and technology educators, this online magazine provides coverage of a range of technology issues affecting kindergarten through college classrooms. Industry news and events as well as the latest products, services, and strategies are featured in each monthly issue. The magazine is also available in a print version. Circ: Unavailable.
Website: www.eschoolnews.com

FREELANCE POTENTIAL: 20% written by non-staff writers. Publishes 6–8 freelance submissions yearly. Receives 100 unsolicited mss yearly.

SUBMISSIONS AND PAYMENT: Sample copy available at website. Prefers query; will accept complete ms. Accepts hard copy and email to gdowney@eschoolnews.com. SASE. Response time varies. Rights vary. Articles and depts/columns, word lengths and payment rates vary. Pays on acceptance.

Family Safety & Health

1121 Spring Lake Drive
Itasca, IL 60143

Editor: Tim Hodson

DESCRIPTION AND INTERESTS: This quarterly is filled with informative articles about safety and health issues in the home, at the job, and on the road. It is published by the National Safety Council. Currently all work is assigned. It is not accepting queries or unsolicited manuscripts, but prospective writers are encouraged to send in writing samples to familiarize the editors with their caliber of work. Circ: 225,000.
Website: www.nsc.org

FREELANCE POTENTIAL: 1% written by non-staff writers. Publishes 5 freelance submissions yearly; 20% by authors new to the magazine.

SUBMISSIONS AND PAYMENT: Sample copy, $4 with 9x12 SASE ($.77 postage). No queries or unsolicited mss; send résumé and clips only. All writing is done on work-for-hire basis. Accepts hard copy. All rights. Articles, 1,200 words. Written material, payment rates vary. Pays on acceptance. Provides 2 contributor's copies.

Farm & Ranch Living

5925 Country Lane
Greendale, WI 53129

Editor

DESCRIPTION AND INTERESTS: This lifestyle magazine fills its pages with interesting and unique stories about present-day farmers and ranchers. It is published six times each year. Circ: 350,000.
Website: www.farmandranchliving.com

FREELANCE POTENTIAL: 90% written by non-staff writers. Publishes 36 freelance submissions yearly; 50% by unpublished writers, 50% by authors who are new to the magazine. Receives 120 queries and unsolicited mss yearly.

SUBMISSIONS AND PAYMENT: Sample copy, $2. Query or send complete ms. Accepts hard copy and email submissions to editors@ farmandranchliving.com. Availability of artwork improves chance of acceptance. SASE. Responds in 6 weeks. One-time rights. Articles, 1,200 words. Depts/columns, 350 words. Color prints. Written material, $10–$150. Pays on publication. Provides 1 contributor's copy.

FatherMag.com

P.O. Box 231891
Houston, TX 77223

Managing Editor: John Gill

DESCRIPTION AND INTERESTS: Based on the belief that fatherhood is a man's most important work, this e-zine provides articles that inform, educate, motivate, and inspire fathers everywhere. Subjects include positive parenting, custody issues, and discipline. *FatherMag.com* posts two editions on the same site: one that focuses on family life, and another that focuses on family strife. It seeks submissions from anyone—father or other—who has ideas to share and stories to tell. Hits per month: 1 million.
Website: www.fathermag.com

FREELANCE POTENTIAL: 95% written by non-staff writers. Publishes 50 freelance submissions yearly; 50% by authors new to the magazine.

SUBMISSIONS AND PAYMENT: Sample copy and guidelines available at website. Query. Accepts email queries to jgill@fathermag.com. Response time varies. One-time rights. Articles and fiction, word lengths vary. No payment.

Fido Friendly

P.O. Box 160
Marsing, ID 83639

Editor

DESCRIPTION AND INTERESTS: Written for people who travel with their dogs, this magazine features travel information for U.S. and Canadian destinations. Published six times each year, it offers reviews of dog-friendly cities, states, hotels, and camping sites. Writers are welcome to submit articles about travel experiences with their dogs. Circ: 44,000.
Website: www.fidofriendly.com

FREELANCE POTENTIAL: 60% written by non-staff writers. Publishes 6–12 freelance submissions yearly; 10% by unpublished writers, 10% by authors who are new to the magazine. Receives 120 queries yearly.

SUBMISSIONS AND PAYMENT: Sample copy and guidelines, $4.95. Query with sample paragraph. Accepts email queries to fieldeditor@ fidofriendly.com. Responds in 1 month. First rights. Articles, 800–1,200 words; $.10 per word. Pays on publication. Provides 1 author's copy.

FineScale Modeler

21027 Crossroads Circle
P.O. Box 1612
Waukesha, WI 53187

Editor: Matthew Usher

DESCRIPTION AND INTERESTS: Both beginner and advanced modelers turn to this magazine for step-by-step how-to feature articles on model building of cars, boats, planes, and military vehicles. It is published 10 times each year. Circ: 60,000.
Website: www.finescale.com

FREELANCE POTENTIAL: 85% written by non-staff writers. Publishes 40 freelance submissions yearly; 20% by authors who are new to the magazine. Receives 200 unsolicited mss yearly.

SUBMISSIONS AND PAYMENT: Sample copy, $4.95 with 9x12 SASE. Query or send complete ms. Accepts hard copy, disk submissions with hard copy, and email submissions to editor@ finescale.com. SASE. Responds in 1–4 months. All rights. Articles, 750–3,000 words. Depts/ columns, word lengths vary. Written material, $60–$75 per page. Pays on acceptance. Provides 1 contributor's copy.

Fort Lauderdale Family Magazine

7045 SW 69th Avenue
South Miami, FL 33143

Publisher: Janet Jupiter

DESCRIPTION AND INTERESTS: Child care, health and fitness, social issues, and pets are all covered in this monthly magazine for Fort Lauderdale-area families. Each issue also includes regional news, movie reviews, and information on local events. Preference is given to writers who either live in the Fort Lauderdale community or have a strong knowledge of the area. Circ: Unavailable.
Website: www.fortlauderdalefamily.com

FREELANCE POTENTIAL: 30% written by nonstaff writers. Publishes 15–20 freelance submissions yearly.

SUBMISSIONS AND PAYMENT: Sample copy available at website. Query. Accepts hard copy and email queries to familymag@bellsouth.net. SASE. Response time varies. One-time rights. Articles and depts/columns, word lengths vary; payment rates vary. Pays on publication. Provides contributor's copies.

Golfer Girl Magazine

P.O. Box 804
Del Mar, CA 92014

Submissions: Libby Hooton

DESCRIPTION AND INTERESTS: Edited in part by teenage girls, *Golfer Girl Magazine* targets girls ages 8 to 17 who love the game of golf. Each issue features inspiring and educational articles about the sport and its female players, as well as other topics of interest to this age group. Published quarterly, it is the first magazine to target junior female golfers. Circ: 10,000.
Website: www.golfergirlmagazine.com

FREELANCE POTENTIAL: 100% written by nonstaff writers. Publishes many freelance submissions yearly; 40% by authors who are new to the magazine.

SUBMISSIONS AND PAYMENT: Send complete ms, preferably with art. Accepts email submissions to libby@golfergirlmagazine.com; copy claude@golfergirlmagazine.com. Response time varies. Rights vary. Written material, word lengths vary. No payment. Provides a 2-year subscription.

Gay Parent Magazine

P.O. Box 750852
Forest Hills, NY 11375-0852

Editor: Angeline Acaln

DESCRIPTION AND INTERESTS: Gay and lesbian parents read this magazine for information on gay-friendly schools, camps, and family vacations. Published six times each year, it also features personal experience pieces about important issues. Circ: 10,000.
Website: www.gayparentmag.com

FREELANCE POTENTIAL: 3% written by non-staff writers. Publishes 6 freelance submissions yearly; 1% by authors who are new to the magazine. Receives 75 unsolicited mss yearly.

SUBMISSIONS AND PAYMENT: Sample copy and guidelines, $3.50. Send complete ms. Accepts email submissions to gayparentmag@gmail.com. Availability of artwork improves chance of acceptance. Response time varies. One-time rights. Articles, 500–1,000 words; $.10 per word. Color prints or digital images. Pays on publication. Provides contributor's copies upon request.

Go! Magazine

2711 South Loop Drive, Suite 4700
Ames, IA 50010

Editor: Michele Regenold

DESCRIPTION AND INTERESTS: Iowa State University's Center for Transportation and Education produces this e-zine to show teenagers and young adults the exciting possibilities for careers in transportation, whether they be related to cars, trucks, trains, planes, or ships. Updated six times each year, it also publishes personal essays. Hits per month: Unavailable.
Website: www.go-explore-trans.org

FREELANCE POTENTIAL: 50–75% written by nonstaff writers. Publishes 2 freelance submissions yearly.

SUBMISSIONS AND PAYMENT: Sample copy, guidelines, and theme list available at website. Query. Accepts email queries to editor@go-explore-trans.com. Response time varies. First world electronic and archival rights. Articles, 1,000–1,500 words. Depts/columns, 750–1,250 words. Written material, $.25 per word. Artwork, $25 per published photo. Payment policy varies.

Good Housekeeping

Hearst Corporation
300 West 57th Street
New York, NY 10019-5288

Executive Editor: Judith Coyne

DESCRIPTION AND INTERESTS: A popular woman's magazine since 1885, *Good House-keeping* focuses on health, fashion, beauty, family life, and careers. It is published monthly. Circ: 25 million.
Website: www.goodhousekeeping.com

FREELANCE POTENTIAL: 80% written by nonstaff writers. Publishes 50+ freelance submissions yearly. Receives 18,000–24,000 queries yearly.

SUBMISSIONS AND PAYMENT: Guidelines available at website. Sample copy, $2.50 at newsstands. Query with résumé and clips for nonfiction; SASE. Send complete ms for fiction; mss not returned. Accepts hard copy. Responds in 4–6 weeks. All rights for nonfiction; first North American serial rights for fiction. Articles, 750–2,500 words; to $2,000. Essays, to 1,000 words; to $750. Fiction, to 3,000 words; payment rates vary. Pays on acceptance. Provides 1 contributor's copy.

Grandparents Magazine

281 Rosedale Avenue
Wayne, PA 19087

Editor: Katrina Hayday Wester

DESCRIPTION AND INTERESTS: This popular online magazine is dedicated to celebrating the joys of being a grandparent. Editorial content includes informational articles, profiles, essays, and interviews. Product reviews of books, toys, and baby items are also featured. A good portion of the material includes ideas and activities for spending time with grandchildren of all ages. It claims to be the number one destination on Google for grandparents. Hits per month: Unavailable.
Website: www.grandparentsmagazine.net

FREELANCE POTENTIAL: Publishes several freelance submissions yearly.

SUBMISSIONS AND PAYMENT: Sample copy and writer's guidelines available at website. Query. Accepts email queries to katrina@grandparentsmagazine.net. Response time varies. Electronic rights. Articles, word lengths vary. No payment.

Grand Rapids Family

549 Ottawa Avenue NW, Suite 201
Grand Rapids, MI 49503

Editor: Carole Valade

DESCRIPTION AND INTERESTS: Parents living in western Michigan use this monthly magazine as a resource for the many different issues they face while raising children. Topics include child care and education, family issues, health, adoption, and travel. Profiles, reviews, and news, all with a local angle, are also featured. It is interested in profiles of outstanding children and family travel articles. Circ: 30,000.
Website: www.grfamily.com

FREELANCE POTENTIAL: 20% written by nonstaff writers. Publishes 15 freelance submissions each year.

SUBMISSIONS AND PAYMENT: Guidelines available with #10 SASE. Query or send complete ms. Accepts hard copy. SASE. Responds to queries in 2 months, to mss in 6 months. All rights. Articles and depts/columns, word lengths vary; payment rates vary. B/W or color prints; $25. Kill fee, $25. Pays on publication.

Harford County Kids

P.O. Box 1666
Bel Air, MD 21014

Publisher: Joan Fernandez

DESCRIPTION AND INTERESTS: Families residing in and around Maryland's Harford County read this magazine for its varied coverage of family-friendly topics. Each monthly issue is filled with informational and how-to articles covering pregnancy, child development, health and fitness, and education. Profiles of area families and coverage of local activities and events are regular features. Circ: 28,500.
Website: www.harfordcountykids.com

FREELANCE POTENTIAL: 100% written by nonstaff writers. Publishes 20 freelance submissions yearly.

SUBMISSIONS AND PAYMENT: Guidelines available. Query. Accepts hard copy and email queries to joanf@aboutdelta.com. SASE. Response time varies. First print and electronic rights. Articles and depts/columns, word lengths vary; payment rates vary. Pays on publication. Provides 1 contributor's copy.

Henry Parents Magazine

3651 Peachtree Pkwy, Suite 325
Suwanee, GA 30024

Editor: Terrie Carter

DESCRIPTION AND INTERESTS: Targeting parents in Henry County, Georgia, this monthly offers informational articles on recreation, sports, local events, family and parenting, finance, and health. It also features profiles of local families and businesses. *Henry Parents* prefers working with writers who are familiar with the county. Circ: Unavailable.
Website: www.henryparents.com

FREELANCE POTENTIAL: 75% written by non-staff writers. Publishes several freelance submissions yearly.

SUBMISSIONS AND PAYMENT: Sample copy available at website. Query or send complete ms. Accepts email submissions to editor@henryparents.com. Responds in 1 month. First and non-exclusive online archival rights. Articles, 500–1,000 words. Depts/columns, word lengths vary. Written material, payment rates vary. Pays on publication. Provides 2 contributor's copies.

Highlights High Five

807 Church Street
Honesdale, PA 18431-1895

Editor: Kathleen Hayes

DESCRIPTION AND INTERESTS: A member of the *Highlights* magazine family, this monthly is written specifically for children ages two through six. *High Five*'s developmentally appropriate fiction and nonfiction general interest stories are designed to be read aloud to the emergent reader, and read independently by children in kindergarten through second grade. Contents include animal stories, humor, folktales, and contemporary fiction. It is also filled with engaging illustrations, puzzles, games, crafts, and recipes that are kid-friendly and fun to make. Currently, all of the editorial content is commissioned or written by staff. Check the website for changes to this policy. Circ: Unavailable.
Website: www.highlights.com

FREELANCE POTENTIAL: Publishes few freelance submissions yearly.

SUBMISSIONS AND PAYMENT: Not seeking freelance submissions at this time.

The High School Journal

Editorial Office, School of Education
University of North Carolina, CB#3500
Chapel Hill, NC 27599

Submissions: George Noblit

DESCRIPTION AND INTERESTS: This quarterly for secondary school teachers and administrators publishes articles dealing with adolescent growth, development, interests, beliefs, and learning as they affect school practice. It also reports on successful teaching techniques and student/teacher/administrator interaction. Circ: 800.
Website: www.uncpress.unc.edu

FREELANCE POTENTIAL: 100% written by nonstaff writers. Publishes 20–30 freelance submissions yearly; 25% by unpublished writers, 85% by authors who are new to the magazine. Receives 324 unsolicited mss yearly.

SUBMISSIONS AND PAYMENT: Sample copy, $7.50 with 9x12 SASE. Send 3 copies of complete ms. Accepts email submissions to gwn@unc.edu. Responds in 3–4 months. All rights. Articles, 1,500–2,500 words. Depts/columns, 300–400 words. No payment. Provides 3 contributor's copies.

High School Writer Senior

Writer Publications
P.O. Box 718
Grand Rapids, MN 55477-0718

Editor: Emily Benes

DESCRIPTION AND INTERESTS: A mix of poems, articles, and stories, all written by senior high school students, is found in this newspaper published six times each year. Circ: 44,000.
Website: www.writerpublications.com

FREELANCE POTENTIAL: 100% written by nonstaff writers. Publishes 300 freelance submissions yearly; 95% by unpublished writers, 75% by authors who are new to the magazine. Receives 100 unsolicited mss yearly.

SUBMISSIONS AND PAYMENT: Sample copy, free. Accepts submissions from senior high school students of subscribing teachers only. Send complete ms. Accepts hard copy, email submissions to writer@mx3.com (ASCII attachments), and simultaneous submissions if identified. SASE. Response time varies. One-time rights. Articles and fiction, to 2,000 words. Poetry, no line limits. No payment. Provides 1 contributor's copy.

High School Years

128 North Royal
Front Royal, VA 22630

Submissions Editor

DESCRIPTION AND INTERESTS: Helping high school students achieve academic success is the goal of this publication, which is distributed directly to parents. Topics include homework, careers, and parenting. Published monthly throughout the school year, it is currently not accepting freelance material. Circ: 300,000.
Website: www.rfeonline.com

FREELANCE POTENTIAL: 100% written by nonstaff writers. Publishes 80 freelance submissions yearly; 25% by unpublished writers. Receives 36 unsolicited mss yearly.

SUBMISSIONS AND PAYMENT: Sample copy, guidelines, and editorial calendar, free with 9x12 SASE. Query with résumé and clips when submission policy changes. SASE. Responds in 1 month. All rights. Articles, 225–300 words. Depts/columns, 175–200 words. Written material, $.60 per word. Pays on acceptance. Provides 5 contributor's copies.

Home & School Connection

128 North Royal Avenue
Front Royal, VA 22630

Submissions: Jennifer Hutchinson

DESCRIPTION AND INTERESTS: Parents of elementary school children read *Home & School Connection* for information and advice on promoting success in school. Published monthly in newsletter format, it offers articles written in an upbeat and conversational tone. Topics covered include parenting skills, careers, special needs, family issues, health, nature, and science. It is currently not reviewing queries or unsolicited manuscripts, but interested writers may submit résumés for future consideration. Circ: Unavailable.
Website: www.rfeonline.com

FREELANCE POTENTIAL: 100% written by nonstaff writers. Publishes 80 freelance submissions yearly; 28% by unpublished writers, 14% by authors who are new to the magazine.

SUBMISSIONS AND PAYMENT: Sample copy, free with 9x12 SASE (2 first-class stamps). Send résumé only. SASE. Response time varies.

Hit Parader

210 Route 4 East, Suite 211
Paramus, NJ 07652

Editor

DESCRIPTION AND INTERESTS: Fans of heavy metal and hard rock find the latest information and trends in this monthly magazine. Band profiles, concert information, and instrument and sound equipment reviews are featured. Circ: 150,000.
Website: www.hitparader.com

FREELANCE POTENTIAL: 10% written by nonstaff writers. Publishes 10 freelance submissions yearly; 50% by authors who are new to the magazine.

SUBMISSIONS AND PAYMENT: Sample copy, $4.99 at newsstands. Query. Accepts hard copy. Availability of artwork improves chance of acceptance. SASE. Responds in 1–3 months. First rights. Articles, 1,000 words; $100–$150. Other material, word lengths and payment rates vary. 3x5 and 5x7 B/W prints and color transparencies. Pays on publication. Provides 1 author's copy.

Home Times Family Newspaper

P.O. Box 22547
West Palm Beach, FL 33416

Editor: Dennis Lombard

DESCRIPTION AND INTERESTS: *Home Times* is a monthly tabloid newspaper that reports on international, national, and local news, people, and issues—all with a biblical worldview. It's looking for local stringers to write personality features on faith, parenting, and education. All articles must be positive in nature. Circ: 8,000.
Website: www.hometimes.org

FREELANCE POTENTIAL: 90% written by nonstaff writers. Publishes 25 freelance submissions yearly; 25% by unpublished writers, 50% by authors who are new to the magazine. Receives 100 unsolicited mss yearly.

SUBMISSIONS AND PAYMENT: Sample copy and guidelines, $3. Send ms. Accepts hard copy and simultaneous submissions. SASE. Responds in 1 month. One-time print and electronic rights. Articles, to 900 words; $5–$50. Depts/columns, word lengths and payment rates vary. Pays on publication. Provides author's copies on request.

Horse Illustrated

P.O. Box 6050
Mission Viejo, CA 92690

Editor: Elizabeth Moyer

DESCRIPTION AND INTERESTS: Written for horse owners and riders, this monthly supports responsible ownership of horses, supplies tips on riding and caring for horses, and covers equine issues. Most work is assigned, but new writers are welcome to send a query with a sample of their writing. Circ: 200,000.
Website: www.horseillustrated.com

FREELANCE POTENTIAL: 80% written by non-staff writers. Publishes 10–20 freelance submissions yearly. Receives 480 queries, 240 mss yearly.

SUBMISSIONS AND PAYMENT: Guidelines available. Prefers complete ms; will accept query with detailed outline, resources, and clips. Accepts hard copy. No simultaneous submissions. SASE. Responds in 1+ months. First North American serial rights. Articles, 1,500–1,800 words. Depts/columns, 1,000–1,400 words. Written material, $50–$400. Pays on publication. Provides 2 contributor's copies.

Hot Rod

6420 Wilshire Boulevard
Los Angeles, CA 90048

Editor: David Freiburger

DESCRIPTION AND INTERESTS: Hot rod enthusiasts turn to this monthly magazine for the latest news and information about the sport. Articles cover racing events, car maintenance and repairs, and collecting. Driver profiles are also included. New writers have the best chance of publication by submitting material for car news or car features. Circ: 680,000.
Website: www.hotrod.com

FREELANCE POTENTIAL: 15% written by non-staff writers. Publishes 24 freelance submissions yearly. Receives 288 queries yearly.

SUBMISSIONS AND PAYMENT: Sample copy, $3.50 at newsstands. Guidelines available. Query. Accepts hard copy. SASE. Response time varies. All rights. Articles, 3,000 characters per page; $250–$300 per page. Depts/columns, word lengths vary; $100 per page. B/W and color prints and 35mm color transparencies; payment rates vary. Pays on publication.

I.D.

Cook Communications Ministries
4050 Lee Vance View
Colorado Springs, CO 80918

Editor: Doug Mauss

DESCRIPTION AND INTERESTS: Bible stories, along with lessons and activities, can be found in this weekly Sunday school journal for high school students. School, health, careers, sports, nature, recreation, and social issues are among the topics covered. Check the website for submission information, as the magazine plans to resume accepting freelance submissions in the near future. Circ: 50,000.
Website: www.davidccook.com

FREELANCE POTENTIAL: 30% written by non-staff writers.

SUBMISSIONS AND PAYMENT: Guidelines available. Send résumé. No queries or unsolicited ms. All articles are assigned. Accepts hard copy. SASE. Responds in 6 months. Rights vary. Articles, 600–1,200 words; $50–$300 depending on experience. B/W and color prints; payment rates vary. Pays on acceptance. Provides 1 contributor's copy.

I Love Cats

16 Meadow Hill Lane
Armonk, NY 10504

Editor: Lisa Allmendinger

DESCRIPTION AND INTERESTS: Cat afficionados of all ages read this magazine for all sorts of feline information. It features profiles, information on cat health and behavior, and tips for owners. No talking cats, stories written from a cat's point of view, or poetry. The availability of digital (JPEG) images improves your chances of acceptance. It is published six times each year. Circ: 25,000.
Website: www.iluvcats.com

FREELANCE POTENTIAL: 90% written by non-staff writers. Publishes 100+ freelance submissions yearly; 60% by unpublished writers, 70% by authors who are new to the magazine. Receives 6,000 unsolicited mss yearly.

SUBMISSIONS AND PAYMENT: Sample copy and guidelines, $5 with 9x12 SASE. Send complete ms. Accepts email to ilovecatseditorial@sbcglobal.net. Responds in 1–2 months. All rights. Articles and fiction, 500–1,000 words, $50–$100. Pays on publication. Provides 1 copy.

Indian Life Newspaper

P.O. Box 3765
Redwood Post Office, Redwood Centre
Winnipeg, Manitoba R2L IL6
Canada

Editor: Jim Uttley

DESCRIPTION AND INTERESTS: Published six times each year, this tabloid is dedicated to covering the social, cultural, and spiritual needs of the North American Indian Church. All submissions should be written from a Christian perspective. Most of its writers are Native, but non-Native writers are published from time to time. Circ: 20,000.
Website: www.indianlife.org

FREELANCE POTENTIAL: 80% written by non-staff writers. Publishes 10 freelance submissions yearly; 2% by unpublished writers, 25% by authors who are new to the magazine. Receives 276 unsolicited mss yearly.

SUBMISSIONS AND PAYMENT: Sample copy, $2.50 with #9 SAE. Prefers query; will accept complete ms. Accepts hard copy and disk submissions. SAE/IRC. Responds in 1 month. First rights. Articles, 250–2,000 words; $25–$100. Pays on publication. Provides 15 author's copies.

Inside Kung-Fu

Action Pursuit Group
265 South Anita Drive, Suite 120
Orange, CA 92868-3310

Editor: Dave Cater

DESCRIPTION AND INTERESTS: Beginner and experienced practitioners read *Inside Kung-Fu* for in-depth information on all aspects of Chinese-style martial arts. The magazine, published monthly, welcomes articles on traditional forms of fighting, weapons, and history, and profiles of well-known practitioners. Circ: 65,000.
Website: www.insidekung-fu.com

FREELANCE POTENTIAL: 80% written by non-staff writers. Publishes 80–100 freelance submissions yearly; 50% by unpublished writers, 50% by authors who are new to the magazine. Receives 504 queries yearly.

SUBMISSIONS AND PAYMENT: Sample copy and guidelines, $2.95 with 9x12 SASE. Query. Accepts hard copy and email queries to dave.cater@apg-media.com. SASE. Responds in 4–6 weeks. First rights. Articles, 1,500 words. Depts/columns, 750 words. Written material, payment rates vary. Pays on publication.

Junior Storyteller

P.O. Box 205
Masonville, CO 80541

Editor: Vivian Dubrovin

DESCRIPTION AND INTERESTS: Geared toward elementary through junior high students with an interest in storytelling, this quarterly provides tips on the art of storytelling as well as a variety of stories to tell. It is particularly interested in stories children can tell. Knowledge of storytelling and kids' activities is more important than writing experience. Circ: 500.
Website: www.storycraft.com

FREELANCE POTENTIAL: 30% written by non-staff writers. Receives several queries yearly.

SUBMISSIONS AND PAYMENT: Guidelines available at website. Sample copy, $4 with 6x9 SASE. Query. Accepts hard copy and email queries to jrstoryteller@storycraft.com. Availability of artwork may improve chance of acceptance. SASE. Response time varies. First rights. Articles and fiction, 500–1,000 words; $50–$125. Pays on acceptance. Provides 10 contributor's copies.

Kahani

P.O. Box 590155
Newton Centre, MA 02459

Editor: Monika Jain

DESCRIPTION AND INTERESTS: This children's literary magazine, inspired by South Asian culture in America, fills its pages with empowering stories, educational activities, fun facts, and art. Published quarterly, it is looking for submissions that complement the upcoming editorial themes, which may be found at the website. Circ: Unavailable.
Website: www.kahani.com

FREELANCE POTENTIAL: 50% written by non-staff writers. Publishes several freelance submissions yearly.

SUBMISSIONS AND PAYMENT: Query with clips for articles; send complete ms for fiction. Accepts email submissions to writers@kahani.com; include "Feature Query" or "Fiction Submission" in the subject line. Responds only if interested for articles; in 1 month for fiction. Rights vary. Articles, 400–600 words. Fiction, to 950 words. No payment. Provides copies.

Kansas 4-H Journal

116 Umberger Hall
KSU
Manhattan, KS 66506-3714

Editor: Rhonda Atkinson

DESCRIPTION AND INTERESTS: A mix of photo-essays, how-to articles, and personal experience pieces fills the pages of this magazine for 4-H members living in Kansas. Published eight times each year, it covers member news, events, programs, and issues. Writers actively involved in 4-H programs are welcome to submit material that captures the essence of 4-H programs and members. Circ: 10,000.

FREELANCE POTENTIAL: 60% written by nonstaff writers. Publishes 100 freelance submissions yearly; 10% by unpublished writers, 20% by authors who are new to the magazine. Receives 696 queries and unsolicited mss yearly.

SUBMISSIONS AND PAYMENT: Sample copy and editorial calendar, $5. Query or send complete ms. Accepts hard copy. SASE. Response time varies. Rights vary. Articles, 500 words; payment rates vary. Payment policy varies.

Keyboard

1111 Bay Hill Drive
San Bruno, CA 94066

Editor-in-Chief: Ernie Rideout

DESCRIPTION AND INTERESTS: Keyboardists of all levels read this magazine for its comprehensive technical coverage. Published monthly, it also includes artist profiles, product guides, and music reviews. Circ: 61,000.
Website: www.keyboardmag.com

FREELANCE POTENTIAL: 25–35% written by nonstaff writers. Publishes 120 freelance submissions yearly; 35% by unpublished writers, 55% by authors who are new to the magazine. Receives 60–120 unsolicited mss yearly.

SUBMISSIONS AND PAYMENT: Sample copy and guidelines available via email request. Send complete ms with résumé. Accepts hard copy and email submissions to keyboard@ musicplayer.com. SASE. Responds in 3 months. All rights. Articles, 500–3,000 words. Depts/ columns, 400–600 words. Written material, payment rates vary. Pays on publication. Provides 5 contributor's copies.

Kids

341 East Lancaster Avenue
Downingtown, PA 19335

Editor: Bob Ludwick

DESCRIPTION AND INTERESTS: Distributed in the elementary and intermediate schools of Pennsylvania's Chester County, *Kids* features informational articles on school programs, events, and news, as well as inspiring profiles of local teachers and students. Published monthly, the tabloid-style magazine is sponsored by the Chester County Intermediate Unit and its public schools. Circ: 45,000.

FREELANCE POTENTIAL: 90% written by nonstaff writers. Publishes 120 freelance submissions yearly; 20% by unpublished writers. Receives several queries yearly.

SUBMISSIONS AND PAYMENT: Sample copy and editorial calendar, free with 9x12 SASE. Query with résumé. Accepts hard copy. SASE. Responds in 1 week. All rights. Articles and depts/columns, to 500 words. No payment. Provides 2 contributor's copies.

Kidsandkaboodle.com

1169 Mount Rushmore Way
Lexington, KY 40515

Editor: Jennifer Anderson

DESCRIPTION AND INTERESTS: Families living in central Kentucky read this e-zine for informative articles on child development, health, and parenting issues. It also offers a special section for expectant parents. Local events and new product information are also covered in each monthly issue. Hits per month: 50,000.
Website: www.kidsandkaboodle.com

FREELANCE POTENTIAL: 20% written by nonstaff writers. Publishes 20 freelance submissions yearly; 50% by unpublished writers, 50% by authors who are new to the magazine. Receives 36–48 unsolicited mss yearly.

SUBMISSIONS AND PAYMENT: Sample copy available at website. Send complete ms. Accepts email submissions to editor@ kidsandkaboodle.com. Response time varies. All rights. Written material, word lengths vary. No payment.

Kids Discover

192 Lexington Avenue, Room 1003
New York, NY 10016-6913

Editor: Stella Sands

DESCRIPTION AND INTERESTS: Each issue of this monthly children's magazine explores an individual theme. Articles are fun, educational, and fact-filled, with artwork and photography to complement. Areas of focus include science, history, ecology, geography, and architecture. Its target audience is children ages six to twelve. The magazine has a group of writers it works with on a regular basis; therefore, it is not accepting unsolicited submissions at this time. Circ: 400,000.
Website: www.kidsdiscover.com

FREELANCE POTENTIAL: 100% written by staff writers. Receives 120 queries yearly.

SUBMISSIONS AND PAYMENT: Sample copy available upon email request to editor@kidsdiscover.com. Not accepting queries or unsolicited mss at this time. Rights vary. Written material, word lengths and payment rates vary. Payment policy varies.

KidSpirit Magazine

77 State Street
Brooklyn, NY 11201

Editor: Elizabeth Dabney Hochman

DESCRIPTION AND INTERESTS: This nonprofit magazine is written by and for 11- to 15-year-old "deep thinkers with big ideas." It tackles life's big questions—including ethics, morals, science, and spirituality. Each quarterly issue offers fiction, essays, and poetry that center around one thought-provoking theme. The goal of *KidSpirit* is to draw readers in and to encourage further exploration of the issue's theme. Circ: 5,000.
Website: www.kidspiritmagazine.com

FREELANCE POTENTIAL: 100% written by nonstaff writers. Publishes 40 freelance submissions yearly; 100% by unpublished writers.

SUBMISSIONS AND PAYMENT: Guidelines and theme list available. Accepts submissions from children ages 11 to 15 only. Query with author bio. Accepts hard copy and email submissions to info@kidspirit.com. SASE. Responds in 1 month. Rights vary. Written material, word lengths vary. No payment. Provides 2 copies.

The Kids' Storytelling Club

P.O. Box 205
Masonville, CO 80541

Editor: Vivian Dubrovin

DESCRIPTION AND INTERESTS: This e-zine is dedicated to children who love storytelling. It offers basic techniques for beginning storytellers, storytelling news, and storytelling activities. It has information for parents and teachers, too. Its editors would like to see more articles on local activities and opportunities for young story-tellers, such as scout badge programs and after-school workshops. Hits per month: 4,000+.
Website: www.storycraft.com

FREELANCE POTENTIAL: 70% written by non-staff writers. Publishes many freelance submissions yearly; many by unpublished writers, most by authors who are new to the magazine.

SUBMISSIONS AND PAYMENT: Guidelines available at website. Query. Accepts hard copy and email queries to jrstoryteller@storycraft.com. SASE. Response time varies. First rights. Articles, 500 words. Fiction, 250–500 words. Written material, $25. Pays on acceptance.

Kiki

214 East 8th Street, 5th Floor
Cincinnati, OH 45202

Editor-in-Chief: Jamie G. Bryant

DESCRIPTION AND INTERESTS: Written "for girls with style and substance," this quarterly magazine has articles that help expand tween girls' creative horizons with topics such as fashion, sewing, art, design, business, travel, careers, and personal grooming. It is not interested in beauty tips or articles on sex, puberty, or boy/girl relationships. Circ: Unavailable.
Website: www.kikimag.com

FREELANCE POTENTIAL: 50% written by non-staff writers. Of the freelance submissions published yearly, 75% are by authors who are new to the magazine.

SUBMISSIONS AND PAYMENT: Sample copy, $7.95. Guidelines available at website. Query. Accepts hard copy. SASE. Responds in 3 months. All rights. Articles, 250–750 words; sidebars, 150 words. Written material, $.50–$1 per word. Pays on publication. Provides 2 contributor's copies.

Kindred

P.O. Box 971
Mullumbimby, New South Wales 2482
Australia

Editor: Kali Wendorf
U.S. Editorial Contact: Lisa Reagan

DESCRIPTION AND INTERESTS: The mission of this parenting magazine is to promote sustainability and natural living within family life. Articles focus on personal growth, natural parenting, social justice, and environmental health. It is published four times each year. Circ: Unavailable.
Website: www.kindredmagazine.com.au

FREELANCE POTENTIAL: 80% written by nonstaff writers. Publishes 12–16 freelance submissions yearly; 70% by unpublished writers, 70% by authors who are new to the magazine. Receives 120 queries, 120 unsolicited mss yearly.

SUBMISSIONS AND PAYMENT: Sample copy and guidelines available. Query or send complete ms with list of sources. Accepts email submissions to kali@kindredmagazine.com.au (Microsoft Word attachments). Response time varies. First rights. Articles, word lengths vary. No payment.

The Learning Edge

Clonlara School
1289 Jewett
Ann Arbor, MI 48104

Editor: Judy Gelner

DESCRIPTION AND INTERESTS: *The Learning Edge* is written by and for students and families enrolled in the Clonlara School Home-Based Education Program. It covers education-related topics as well as issues specific to home schooling. The magazine also publishes education news and opinion pieces. At this time, it will review submissions from Clonlara students and families only. Circ: 1,000.
Website: www.clonlara.org

FREELANCE POTENTIAL: 25% written by nonstaff writers. Of the freelance submissions published yearly, 10% are by unpublished writers, 1% are by authors who are new to the magazine. Receives 12 queries yearly.

SUBMISSIONS AND PAYMENT: Sample copy available upon request. All articles are assigned to Clonlara students. No unsolicited mss or queries. Rights vary. Written material, word lengths vary. No payment.

Kiwibox.com

330 West 38th Street, Suite 1604
New York, NY 10018

Submissions Editor: Kristen Dunleavy

DESCRIPTION AND INTERESTS: A popular Internet forum for high school and college students, *Kiwibox.com* publishes teen-written fiction and nonfiction only. Science, technology, sports, music, movies, dating, social issues, and fashion are some of the topics covered. Genre fiction also appears. The website is designed and run by teens and is updated weekly. Any well-written material that would interest young adults ages 14 to 21 will be considered. Hits per month: 10 million.
Website: www.kiwibox.com

FREELANCE POTENTIAL: 90% written by nonstaff writers. Publishes numerous freelance submissions yearly.

SUBMISSIONS AND PAYMENT: Sample copy available at website. Send complete ms. Accepts email submissions only to kristen@kiwibox.net. Responds in 2 weeks. All rights. Articles, 350 words. Fiction, word lengths vary. No payment.

Learning Through History

P.O. Box 1858
Cranberry Township, PA 16066-1858

Editor: Ron Thompson

DESCRIPTION AND INTERESTS: Each of the six themed issues of this bimonthly magazine, showcases a period in history with interesting articles, crafts, and activities. Writers must request guidelines through website. Circ: 3,500.
Website: www.learningthroughhistory.com

FREELANCE POTENTIAL: 100% written by nonstaff writers. Publishes 100 freelance submissions yearly; 20–30% by unpublished writers, 40% by authors who are new to the magazine.

SUBMISSIONS AND PAYMENT: Accepts query with synopsis and clips from established writers. Accepts complete ms from less experienced writers. Accepts email submissions to submissions@ learningthroughhistory.com (Microsoft Word or text attachments). Responds in 1 month. All rights. Articles, 1,000–1,400 words; $75. Arts and crafts projects, word lengths vary; $50. Kill fee, 50%. Pays on publication. Provides 1 contributor's copy.

Literary Mama

1416 11th Avenue
San Francisco, CA 94122

Senior Editor: Caroline Grant

DESCRIPTION AND INTERESTS: This quarterly online literary magazine features writing by mothers about the complexities and many faces of motherhood. It features poetry, fiction, and creative nonfiction. With the exception of its columns, the vast majority of the content is written by freelancers. Hits per month: 35,000.
Website: www.literarymama.com

FREELANCE POTENTIAL: 75% written by nonstaff writers. Publishes 130 freelance submissions yearly; 75% by authors new to the magazine.

SUBMISSIONS AND PAYMENT: Guidelines available at website. Query for profiles, reviews, and columns. Send complete ms for all other work. Accepts email submissions only. See website for appropriate editor and email address. Responds in 1–4 months. Non-exclusive rights. Articles and fiction, to 6,000 words. Depts/columns, 1,000–1,600 words. Poetry, no line limits. No payment.

Look-Look Magazine

732 North Highland Avenue
Los Angeles, CA 90038

Editor

DESCRIPTION AND INTERESTS: This magazine, launched in 2003 in print format, was recently revamped as an e-zine. It is dedicated to showcasing the work of artists, writers, and poets between the ages of 12 and 20. Updated twice each year, it publishes fiction, creative nonfiction, humor, poetry, and articles about the arts. Hits per month: Unavailable.
Website: www.look-lookmagazine.com

FREELANCE POTENTIAL: 95% written by nonstaff writers. Publishes 120 freelance submissions yearly; 100% by unpublished writers, 100% by authors who are new to the magazine. Receives 8,000 queries yearly.

SUBMISSIONS AND PAYMENT: Sample copy available at website. Query. Accepts queries through website only. Response time varies. All rights. Articles and fiction, to 2,000 words. Poetry, lengths vary. B/W or color prints, transparencies, or digital images; line art. No payment.

Little Rock Family

22 East Second Street
Little Rock, AR 72201

Submissions Editor

DESCRIPTION AND INTERESTS: Published monthly, this regional newsletter is written for busy parents living in central Arkansas. A wide range of topics are covered in its articles, including parenting and family issues, health care, fashion, fitness, child care, education, and recreation. Regional events, resources, and family activities are included in each issue. Writers are welcome to submit informational and educational articles of interest to families. Circ: 20,000.
Website: www.littlerockfamily.com

FREELANCE POTENTIAL: 100% written by nonstaff writers. Publishes many freelance submissions yearly.

SUBMISSIONS AND PAYMENT: Query. Accepts hard copy. SASE. Response time varies. First rights. Articles and depts/columns, word lengths vary; payment rates vary. Payment policy varies.

Lowcountry Parent

1180 Sam Rittenberg Boulevard
Charleston, SC 29407

Submissions Editor: Shannon Brigham

DESCRIPTION AND INTERESTS: Distributed free to families living in Charleston, South Carolina, and its surrounding areas, *Lowcountry Parent* offers valuable and timely information on parenting and family issues, child development, health, and education. It is published monthly. Circ: 38,000.
Website: www.lowcountry.com

FREELANCE POTENTIAL: Publishes many freelance submissions yearly; 10% by authors who are new to the magazine. Receives 1,200 queries each year.

SUBMISSIONS AND PAYMENT: Sample copy, free. Writers' guidelines available. Query with sample pages. Accepts email queries to lowcountryparent@aol.com. Responds in 3 days. One-time rights. Articles and depts/columns, word lengths vary. Written material, $15–$100. Pays on publication. Provides 3 contributor's copies.

The Lutheran Digest

P.O. Box 4250
Hopkins, MN 55343

Editor: David L. Tank

DESCRIPTION AND INTERESTS: The pages of *The Lutheran Digest* are filled with articles and poetry that reflect the Lutheran perspective on life. It seeks to encourage readers to help others embrace the philosophies of the faith through articles on education, family, relationships, and religion. Submissions should be well written and teach without preaching. Circ: 70,000.
Website: www.lutherandigest.com

FREELANCE POTENTIAL: 100% written by nonstaff writers. Publishes 80 freelance submissions yearly; 30% by authors who are new to the magazine. Receives 200 unsolicited mss yearly.

SUBMISSIONS AND PAYMENT: Sample copy, $3.50 with 6x9 SASE. Guidelines available. Send complete ms with biography. Accepts hard copy. SASE. Responds in 2–3 months. One-time and second rights. Articles, word lengths vary; $35. Poetry, 2 stanzas; payment rates vary. Pays on acceptance. Provides 1 contributor's copy.

Mad Magazine

1700 Broadway
New York, NY 10019

Submissions Editor: Amy Vozeolas

DESCRIPTION AND INTERESTS: Young adults and adults alike read this monthly magazine for the humorous parodies and satire that fill its pages. Current events, pop culture, social issues, relationships, and television programs are highlighted through comic strips, short fiction, and articles. Circ: Unavailable.
Website: www.madmag.com

FREELANCE POTENTIAL: 90% written by nonstaff writers. Publishes 25 freelance submissions each year.

SUBMISSIONS AND PAYMENT: Sample copy, $4.99 at newsstands. Guidelines available at website. Query or send complete ms. Prefers email to submissions@madmagazine.com; will accept hard copy if submission includes artwork. SASE. Responds if interested. All rights. Written material, word lengths vary; $500 per printed page. Artwork, payment rates vary. Pays on acceptance. Provides 1 contributor's copy.

Mad Kids

1700 Broadway
New York, NY 10019

Submissions Editor: Charles Kadau

DESCRIPTION AND INTERESTS: Targeting children ages seven to twelve, *Mad Kids* entertains its readers with a blend of comic strips, activities, and humorous fiction and nonfiction. Published quarterly, each issue includes a Q&A with a celebrity. It is the sister publication to *Mad Magazine.* Circ: Unavailable.
Website: www.madmag.com

FREELANCE POTENTIAL: 90% written by nonstaff writers. Publishes 10–15 freelance submissions yearly.

SUBMISSIONS AND PAYMENT: Sample copy, $4.99 at newsstands. Guidelines available at website. Query or send complete ms. Prefers email to submissions@madmagazine.com; will accept hard copy if submission includes artwork. SASE. Responds if interested. All rights. Written material, word lengths vary; $500 per printed page. Artwork, payment rates vary. Pays on acceptance. Provides 1 contributor's copy.

Metro Augusta Parent

700 Broad Street
Augusta, GA 30901

Editor: Amy Christian

DESCRIPTION AND INTERESTS: This regional monthly publication for parents covers everything from recreational activities, dining options, schools, and sports, to book and movie reviews. General family and parenting topics are also featured. A "Just for Parents" column has recently been added. It is particularly interested in material that relates to teens. Circ: Unavailable.
Website: www.augustaparent.com

FREELANCE POTENTIAL: 80% written by nonstaff writers. Publishes 50 freelance submissions yearly; 5% by unpublished writers, 5% by new authors. Receives 240 queries yearly.

SUBMISSIONS AND PAYMENT: Sample copy, free with SASE. Query. Accepts hard copy and email queries to editor@augustaparent.com. SASE. Response time varies. First rights. Articles and depts/columns, word lengths and payment rates vary. Payment policy varies. Provides 1 contributor's copy.

Metro Parent

P.O. Box 13660
Portland, OR 97213

Editor: Marie Sherlock

DESCRIPTION AND INTERESTS: Distributed free of charge in Portland's metropolitan area, *Metro Parent* provides articles of interest to parents. Coverage includes child development, nutrition, education, and travel. Published monthly, it includes calendars of local events and activities. Only work from local writers is considered. Circ: 45,000.
Website: www.metro-parent.com

FREELANCE POTENTIAL: 75% written by non-staff writers. Publishes 50 freelance submissions yearly; 20% by unpublished writers. Receives 240 queries yearly.

SUBMISSIONS AND PAYMENT: Sample copy and theme list, $2. Query with outline. Accepts hard copy, email to editor@metro-parent.com, and simultaneous submissions if identified. SASE. Responds in 1 month. Rights vary. Articles and depts/columns, word lengths and payment rates vary. Pays on publication.

Minnesota Parent

1115 Hennepin Avenue South
Minneapolis, MN 55403

Editor: Tricia Cornell

DESCRIPTION AND INTERESTS: The goal of this magazine is to support and enrich the lives of families living in Minnesota. Each issue's articles run the gamut from travel, popular culture, and new product reviews to social issues and relevant legislation. Coverage of regional family events is also provided. Published monthly, its focus is on parenting children from birth through the school-age years. Circ: 70,000.
Website: www.mnparent.com

FREELANCE POTENTIAL: 50% written by non-staff writers. Publishes 24 freelance submissions each year.

SUBMISSIONS AND PAYMENT: Query. Accepts hard copy and email queries to tcornell@mnpubs.com. SASE. Response time varies. First serial and electronic rights. Articles and depts/columns, word lengths vary; $50–$350. Pays on publication. Provides 2 contributor's copies.

Miami Family Magazine

7045 SW 69th Avenue
South Miami, FL 33143

Publisher: Janet Jupiter

DESCRIPTION AND INTERESTS: This monthly magazine for Miami families provides valuable information on local activities, sporting events, and school programs. Profiles of local residents with interesting stories to tell are featured. Included in each issue are cooking and pets columns, and book and movie reviews. It prefers to work with writers living in or thoroughly familiar with the Miami area. Circ: Unavailable.
Website: www.miamifamilymagazine.com

FREELANCE POTENTIAL: 30% written by non-staff writers. Publishes 15–20 freelance submissions yearly.

SUBMISSIONS AND PAYMENT: Sample copy available at website. Query. Accepts hard copy and email queries to familymag@bellsouth.net. SASE. Response time varies. One-time rights. Articles and depts/columns, word lengths and payment rates vary. Pays on publication. Provides contributor's copies.

Model Airplane News

Air Age Publishing
20 Westport Road
Wilton, CT 06897

Editor-in-Chief: Debra Cleghorn

DESCRIPTION AND INTERESTS: Avid aero-modelers turn to this monthly magazine for its comprehensive how-to articles on building and flying model airplanes. Detailed product reviews are also provided, and pages are filled with color photos, charts, and graphs. Circ: 95,000.
Website: www.modelairplanenews.com

FREELANCE POTENTIAL: 90% written by non-staff writers. Publishes 100+ freelance submissions yearly; 33% by authors who are new to the magazine. Receives 144–288 queries yearly.

SUBMISSIONS AND PAYMENT: Sample copy and guidelines, $3.50 with 9x12 SASE. Query with outline and biography describing model experience. Accepts hard copy. Availability of artwork improves chance of acceptance. SASE. Responds in 6 weeks. All North American serial rights. Articles, 1,700–2,000 words; $175–$600. 35mm color slides. Pays on publication. Provides up to 6 contributor's copies.

Montessori Life

281 Park Avenue South
New York, NY 10010

Co-Editors: Kathy Carey and Carey Jones

DESCRIPTION AND INTERESTS: Parents and Montessori educators turn to this magazine for its up-to-date coverage of the research, issues, trends, and curricula in education. It is published quarterly. Circ: 10,500.
Website: www.amshq.org

FREELANCE POTENTIAL: 90% written by non-staff writers. Publishes 40 freelance submissions yearly; 30% by unpublished writers, 30% by authors who are new to the magazine. Receives 120–240 unsolicited mss yearly.

SUBMISSIONS AND PAYMENT: Sample copy, $5 with 9x12 SASE. Guidelines and editorial calendar available at website. Send complete ms. Accepts email submissions to edmontessorilife@aol.com. Responds in 3 months. All rights. Articles, 1,000–4,000 words. Fiction, 1,000–5,000 words. Depts/columns, 500–1,000 words. Written material, payment rates vary. Pays on publication. Provides 1–5 contributor's copies.

Moo-Cow Fan Club

P.O. Box 165
Peterborough, NH 03458

Editor: Becky Ances

DESCRIPTION AND INTERESTS: The engaging characters and format found in this quarterly e-zine invite children ages eight and up to read and play without being bombarded by pop media and advertisements. Each edition is dedicated to one fun topic, with stories, articles, and activities. At this time it is not reviewing freelance queries or unsolicited manuscripts. Hits per month: Unavailable.
Website: www.moocowfanclub.com

FREELANCE POTENTIAL: 20% written by non-staff writers. Publishes 8 freelance submissions yearly; 60% by unpublished writers, 90% by authors who are new to the magazine. Receives 100+ queries yearly.

SUBMISSIONS AND PAYMENT: Guidelines and theme list available. Query with sample article during open submission period. Check website for updates and changes regarding the submission of freelance material.

Mother Verse

2663 Highway 3
Two Harbors, MN 55616

Editor: Melanie Mayo-Laasko

DESCRIPTION AND INTERESTS: Published quarterly, this "Journal of Contemporary Motherhood" explores global experiences across cultural, geographical, economic, and political lines. It seeks submissions of works of literature, art, interviews, and reviews that explore motherhood in a smart, honest manner. Circ: 1,000.
Website: www.motherverse.com

FREELANCE POTENTIAL: 95% written by non-staff writers. Publishes 80–100 freelance submissions yearly; 10% by unpublished writers, 95% by authors who are new to the magazine. Receives 500 queries and unsolicited mss yearly.

SUBMISSIONS AND PAYMENT: Sample copy and guidelines available at website. Query or send ms. Accepts email to submissions@motherverse.com (RTF attachments or in body of email). Responds in 2–5 months. One-time rights. Written material, word lengths vary. No payment. Provides 1 contributor's copy.

Motivos Bilingual Magazine

P.O. Box 34391
Philadelphia, PA 19101

Publisher: Jenee Chiznick

DESCRIPTION AND INTERESTS: Written by and for young adults interested in Latino culture, *Motivos Bilingual Magazine* publishes socially and culturally relevant articles. College and career information, ethnic material, and life issues of interest to this age group are found in each issue. The magazine appears quarterly. Circ: 45,000.
Website: www.motivosmag.com

FREELANCE POTENTIAL: 50% written by non-staff writers. Publishes 30 freelance submissions yearly; 40% by unpublished writers, 50% by authors who are new to the magazine.

SUBMISSIONS AND PAYMENT: Guidelines available at website. Send complete ms. Accepts hard copy and email submissions to editor@motivosmag.com. SASE. Response time varies. First North American serial rights. Articles, 400–800 words; payment rates vary. Payment policy varies. Provides 2 contributor's copies.

Mysteries Magazine

P.O. Box 490
Walpole, NH 03608

Editor: Kim Guarnaccia

DESCRIPTION AND INTERESTS: Published quarterly, *Mysteries Magazine* offers factual and thought-provoking articles, as well as book reviews, on historical and ancient mysteries, unexplained paranormal events, and bizarre scientific breakthroughs. Circ: 15,000.
Website: www.mysteriesmagazine.com

FREELANCE POTENTIAL: 30% written by non-staff writers. Publishes 12 freelance submissions yearly; 20% by authors who are new to the magazine. Receives 240 queries yearly.

SUBMISSIONS AND PAYMENT: Sample copy, $8. Guidelines available at website. Query. Accepts email queries only to editor@mysteriesmagazine.com. Responds in 1 month. First North American serial rights. Articles, 3,000–5,000 words. Depts/columns, 1,200–1,500 words. Book reviews, 200–500 words. Written material, $.05 per word. Pays on publication. Provides 2 contributor's copies.

National Geographic Explorer

1145 17th Street NW
Washington, DC 20036

Editor

DESCRIPTION AND INTERESTS: Used as classroom material for grades two through six, *National Geographic Explorer* allows students to take a voyage around the world through its pages. Three nonfiction articles on science and social studies topics appear in each issue, accompanied by kid-friendly glossaries. A teacher's guide is also included. The magazine is published seven times during the school year. Circ: Unavailable.
Web: http://magma.nationalgeographic.com/ngexplorer

FREELANCE POTENTIAL: 5% written by non-staff writers. Publishes few freelance submissions each year.

SUBMISSIONS AND PAYMENT: Send résumé only. All material written on assignment. Accepts hard copy. SASE. Responds only if interested. All rights. Articles and fiction, word lengths and payment rates vary. Pays on acceptance.

NASSP Bulletin

Sage Publications
2455 Teller Road
Thousand Oaks, CA 91320

Editor: Len Foster

DESCRIPTION AND INTERESTS: As the official publication of the National Association of Secondary School Principals, this quarterly magazine serves as a professional journal for middle-level and high school principals. It contains articles on strategic planning and decision-making, promoting student learning and motivation, and education reform. Circ: Unavailable.
Website: http://bulletin.sagepub.com

FREELANCE POTENTIAL: 100% written by nonstaff writers. Publishes 20–25 freelance submissions yearly; 15% by unpublished writers, 30% by new authors . Receives 192 mss yearly.

SUBMISSIONS AND PAYMENT: Sample copy, free with 9x12 SASE. Guidelines available at website. Send complete ms (written in APA style) with abstract. Prefers email to lenf@wsu.edu (Microsoft Word attachments); will accept hard copy with CD. SASE. Responds in 4–6 weeks. All rights. No payment. Provides 2 contributor's copies.

Natural Solutions

2995 Wilderness Place, Suite 205
Boulder, CO 80301

Editor: Linda Sparrowe

DESCRIPTION AND INTERESTS: The latest news and information on alternative medicine is the focus of this magazine that is published 10 times each year. It covers topics including health, vaccinations, spiritual and mental health, and nutrition. Most articles are written on assignment. Circ: 225,000.
Website: www.naturalsolutionsmag.com

FREELANCE POTENTIAL: 95% written by non-staff writers. Publishes 40–50 freelance submissions yearly; 25% by authors who are new to the magazine. Receives 30 queries yearly.

SUBMISSIONS AND PAYMENT: Sample copy, $4.95 at newsstands. Guidelines available. Query with clips. Accepts email queries to editor@naturalsolutionsmag.com. Response time varies. All rights. Written material, word lengths and payment rates vary. Pays within 45 days of acceptance. Provides 2 contributor's copies.

Neapolitan Family Magazine

P.O. Box 110656
Naples, FL 34108

Editor: Andrea Breznay

DESCRIPTION AND INTERESTS: A monthly magazine written exclusively for families living in Collier County, Florida, this publication contains articles about local issues that relate to parenting, family, and education. It also contains information on recreational opportunities in the area. Local freelancers are preferred. Circ: 11,000.
Website: www.neafamily.com

FREELANCE POTENTIAL: 50% written by non-staff writers. Publishes 25 freelance submissions yearly; 20% by authors who are new to the magazine. Receives 100 unsolicited mss yearly.

SUBMISSIONS AND PAYMENT: Guidelines and editorial calendar available at website. Sample copy, free with 9x12 SASE. Send complete ms. Prefers email submissions to andrea@ neafamily.com; will accept hard copy. SASE. Responds in 1 month. Rights vary. Articles and depts/columns, word lengths and payment rates vary. Pays on publication.

New Jersey Suburban Parent

Middlesex Publications
850 Route 1
North Brunswick, NJ 08902

Editor: Melodie Dhondt

DESCRIPTION AND INTERESTS: This award-winning tabloid provides New Jersey parents with a valuable guide to fun and educational activities in the state each month. It also features articles and columns covering parenting topics such as family health, child care, education, and summer camps. Circ: 78,000.
Website: www.njparentweb.com

FREELANCE POTENTIAL: 80% written by non-staff writers. Publishes 12 freelance submissions yearly; 20% by unpublished writers, 40% by new authors. Receives 1,440 queries yearly.

SUBMISSIONS AND PAYMENT: Sample copy, guidelines, and editorial calendar, free with 9x12 SASE. Query with writing samples. Accepts hard copy and simultaneous submissions if identified. SASE. Responds in 1–2 months. Rights vary. Articles, 700–1,000 words; $30. B/W and color prints, payment rates vary. Pays on acceptance. Provides 1+ contributor's copies.

New Jersey Family

1122 Route 22 West
Mountainside, NJ 07092

Editor: Farn Dupre

DESCRIPTION AND INTERESTS: With the stated mission of helping New Jersey parents be the best parents they can be, this monthly magazine covers issues facing families today. Its editors would like to see more parenting and education articles that are specific to northern and central New Jersey. Circ: 126,000.
Website: www.njfamily.com

FREELANCE POTENTIAL: 60% written by non-staff writers. Publishes 150 freelance submissions yearly; 33% by unpublished writers, 33% by authors who are new to the magazine. Receives 240 queries yearly.

SUBMISSIONS AND PAYMENT: Guidelines available at website. Query with writing samples. Accepts email to editor@njcountyfamily.com (no attachments). Response time varies. First rights. Articles, 750–1,000 words. Depts/columns, word lengths vary. Written material, payment rates vary. Payment policy varies.

The Northland

P.O. Box 841
Schumacher, Ontario P0N 1G0
Canada

Submissions Editor

DESCRIPTION AND INTERESTS: Appearing quarterly, this magazine features articles on topics of interest to members of the Diocese of Moosonee of the Anglican Church of Canada. It features church news, sermons, prayers, and essays about living a spiritual life. Because it is directed to church members only, all content must be relevant to the parishes that comprise the diocese, such as church celebrations, events, and parishioners' accomplishments. Circ: 500.

FREELANCE POTENTIAL: 100% written by nonstaff writers. Publishes several freelance submissions yearly. Receives few unsolicited mss each year.

SUBMISSIONS AND PAYMENT: Sample copy available. Send complete ms. Accepts hard copy. SASE. Response time varies. Rights vary. Articles and depts/columns, word lengths vary. No payment.

North State Parent

P.O. Box 1602
Mount Shasta, CA 96067

Editorial Department: Lisa Shara

DESCRIPTION AND INTERESTS: Read by parents living in California's Butte, Shasta, Tehama, and southern Siskiyou counties, this regional magazine shares information and experiences that address a variety of lifestyles, cultures, and beliefs. It strives to promote a positive, healthy, and peaceful environment for children of all ages while respecting cultural differences. It is published monthly. Circ: Unavailable.
Website: www.northstateparent.com

FREELANCE POTENTIAL: 90% written by nonstaff writers. Publishes 20 freelance submissions each year.

SUBMISSIONS AND PAYMENT: Guidelines available at website. Send complete ms. Accepts hard copy and email submissions to lisa@northstateparent.com. SASE. Response time varies. First rights. Articles, 700–1,000 words. Depts/columns, 300–500 words. Written material, $35–$75. Pays on publication.

The Olive Branch
The Youth Magazine of Seeds of Peace

P.O. Box 25045
Jerusalem, Israel

Editor: Eric Kapenga

DESCRIPTION AND INTERESTS: This magazine is written and produced by and for young adults from regions of conflict who are part of the Seeds of Peace program. Published quarterly, it offers factual articles on politics, current events, and the environment; and profiles of young people. Circ: 16,000.
Website: www.seedsofpeace.org

FREELANCE POTENTIAL: 100% written by nonstaff writers. Publishes 200 freelance submissions yearly; 95% by unpublished writers, 90% by authors who are new to the magazine.

SUBMISSIONS AND PAYMENT: Sample copy available at website. Accepts submissions from members of Seeds of Peace only. Query or send complete ms. Accepts hard copy and email submissions to eric@seedsofpeace.org. SASE. Response time varies. Rights vary. Written material, word lengths vary. No payment.

The Numismatist

American Numismatic Association
818 North Cascade Avenue
Colorado Springs, CO 80903-3279

Editor-in-Chief: Barbara J. Gregory

DESCRIPTION AND INTERESTS: Written for members of the American Numismatic Association, this monthly magazine provides a forum for collectors of coins, medals, tokens, and paper money. Circ: 30,500.
Website: www.money.org

FREELANCE POTENTIAL: 60% written by nonstaff writers. Publishes 36 freelance submissions yearly; 20% by unpublished writers, 10% by authors who are new to the magazine. Receives 48 unsolicited mss yearly.

SUBMISSIONS AND PAYMENT: Sample copy and guidelines, free with 9x12 SASE ($2.50 postage). Send complete ms with biography. Prefers email submissions to editor@money.org; will accept hard copy and disk submissions. SASE. Responds in 8–10 weeks. Perpetual nonexclusive rights. Articles, to 3,500 words; $.12 per word. Pays on publication. Provides 5 contributor's copies.

Owl

Bayard Press Canada
10 Lower Spadina Avenue, Suite 400
Toronto, Ontario M5V 2Z2
Canada

Submissions Editor

DESCRIPTION AND INTERESTS: This is a discovery magazine for children ages eight through twelve. Appearing nine times each year, *Owl* is dedicated to entertaining kids with contemporary information about the world around them, and encouraging them to enjoy, respect, and conserve their natural environment. All work is assigned; do not send queries or manuscripts. Circ: 104,000.
Website: www.owlkids.com

FREELANCE POTENTIAL: 60% written by nonstaff writers. Publishes 1–3 freelance submissions yearly; 5% by unpublished writers, 10% by authors who are new to the magazine.

SUBMISSIONS AND PAYMENT: Sample copy, $4.28 Canadian. Guidelines available. Send résumé only. No queries or unsolicited mss. Accepts hard copy. All rights. Articles, 500–1,000 words; $200–$500. Pays on acceptance. Provides 1 contributor's copy.

Parenting for High Potential

P.O. Box 21351
Baton Rouge, LA 70894

Editor: Dr. Jennifer Jolly

DESCRIPTION AND INTERESTS: This magazine, published quarterly, is dedicated to helping parents of talented and gifted students develop their child's individual potential. Each issue includes special features, expert advice, reviews of recommended software, and book reviews, all relating to how parents can make a difference in their children's lives. It is seeking articles by writers with extensive knowledge about high-ability children and youth. Circ: Unavailable.
Website: www.nagc.org

FREELANCE POTENTIAL: 100% written by non-staff writers. Publishes 10–12 freelance submissions yearly; 50% by authors who are new to the magazine. Receives 20–30 unsolicited mss yearly.

SUBMISSIONS AND PAYMENT: Guidelines available. Send complete ms. Accepts email submissions to jollyphp@gmail.com. Responds in 6–8 weeks. First rights. Articles and depts/columns, word lengths vary. No payment.

ParentingHumor.com

P.O. Box 2128
Weaverville, NC 28787

Editor

DESCRIPTION AND INTERESTS: Focusing on the funny side of parenting, this e-zine features a range of humorous articles for parents to enjoy. Topics include pregnancy, relationships, health and beauty, and all types of parenting issues. The website is updated weekly. Writers with a humorous parenting story to tell are welcome to submit their material. Hits per month: Unavailable.
Website: www.parentinghumor.com

FREELANCE POTENTIAL: 98% written by non-staff writers. Publishes 350 freelance submissions yearly. Receives 300 queries yearly.

SUBMISSIONS AND PAYMENT: Sample copy, guidelines, and submission form available at website. Query. Accepts email queries to staff@parentinghumor.com. Response time varies. One-time electronic rights. Articles, word lengths vary. No payment. Offers an author's biography and a link to the author's website.

Parents' Choice

Parents' Choice Foundation
201 West Padonia Road, Suite 303
Timonium, MD 21093

Editor: Claire Green

DESCRIPTION AND INTERESTS: This e-zine is dedicated to serving the interests of parents who want to know about products and toys for their children *before* they buy. It features information on products that are safe, age-appropriate, educational, or socially sound. The website is updated monthly, and a newsletter is published twice each year. Reviews of toys and television shows are welcome. Hits per month: 1 million+.
Website: www.parentschoice.org

FREELANCE POTENTIAL: 80% written by non-staff writers. Publishes many freelance submissions yearly.

SUBMISSIONS AND PAYMENT: Sample copy available at website. Query or send complete ms. Accepts hard copy, email queries to submissions@parents-choice.org, and simultaneous submissions if identified. SASE. Response time varies. All rights. Articles, to 1,500 words; payment rates vary. Pays on acceptance.

Parents Express

290 Commerce Drive
Fort Washington, PA 19034

Submissions Editor: Daniel Sean Kaye

DESCRIPTION AND INTERESTS: Serving parents in the Philadelphia area and southern New Jersey, *Parents Express* offers its readers news, information, and advice on family issues, entertainment, and local events. It is published monthly. Circ: 49,000.
Website: www.parents-express.net

FREELANCE POTENTIAL: 30% written by non-staff writers. Publishes 25–35 freelance submissions yearly; 25% by unpublished writers, 75% by authors who are new to the magazine. Receives several queries yearly.

SUBMISSIONS AND PAYMENT: Sample copy, free with 9x12 SASE ($2.14 postage). Query with clips or writing samples. Accepts hard copy and email to dkaye@montgomerynews.com. SASE. Responds in 1 month. One-time rights. Articles, 300–1,000 words; $35–$200. Depts/columns, 600–800 words; payment rates vary. Pays on publication. Provides contributor's copies.

Parents Magazine

375 Lexington Avenue
New York, NY 10017

Editor

DESCRIPTION AND INTERESTS: This monthly magazine for parents and expectant parents provides informational and inspirational articles about all sorts of parenting issues. It covers child health, safety, and development; education and school issues; emotional development; and behavior. It also has information on pregnancy. A number of its articles are personal experience pieces, with writers sharing their trials and triumphs. Circ: Unavailable.
Website: www.parents.com

FREELANCE POTENTIAL: 50% written by non-staff writers. Publishes 50 freelance submissions yearly. Receives 1,200 queries yearly.

SUBMISSIONS AND PAYMENT: Sample copy, $3.50 with 9x12 SASE (4 first-class stamps). Query with clips. Accepts hard copy. SASE. Responds in 6 weeks. Rights vary. Articles and depts/columns, word lengths and payment rates vary. Pays on publication. Provides 2 author's copies.

Plum Magazine

276 Fifth Avenue, Suite 302
New York, NY 10001

Submissions: Mary Jane Horton

DESCRIPTION AND INTERESTS: Part health magazine, part lifestyle magazine, *Plum* is a maternity magazine for women age 35 and older. Twice each year it offers articles on issues that are unique to being pregnant later in life, as well as information on newborn care, postpartum issues, and stories about motherhood. Circ: 500,000.
Website: www.plummagazine.com

FREELANCE POTENTIAL: 90% written by non-staff writers. Publishes 25 freelance submissions yearly; 10% by authors who are new to the magazine. Receives 150 queries and mss yearly.

SUBMISSIONS AND PAYMENT: Guidelines available via email to editor@plummagazine.com. Query or send complete ms. Accepts email submissions to editor@plummagazine.com. Response time varies. All rights. Articles and depts/columns, word lengths vary. No payment. Provides 1 contributor's copy.

The Pink Chameleon

Editor: Dorothy Paula Freda

DESCRIPTION AND INTERESTS: Short fiction, nonfiction, and poetry—all for a family-oriented audience—are found in *The Pink Chameleon*. Hits per month: 100.
**Website: www.geocities.com/
        thepinkchameleon/index.html**

FREELANCE POTENTIAL: 95% written by non-staff writers. Publishes 50–100 freelance submissions yearly; 40% by unpublished writers, 60% by authors who are new to the magazine. Receives 120 unsolicited mss yearly.

SUBMISSIONS AND PAYMENT: Writers' guidelines available at website. Send complete ms with brief author biography from January 1 to April 31 and September 1 through October 31 only. Accepts email submissions to dpfreda@juno.com (no attachments). Response time varies. One-year electronic rights. Fiction and articles, to 2,500 words. Poetry, to 36 lines. No payment.

Prairie Messenger

Box 190, 100 College Drive
Muenster, Saskatchewan S0K 2Y0
Canada

Associate Editor: Maureen Weber

DESCRIPTION AND INTERESTS: This weekly tabloid for Catholics living in Saskatchewan and Manitoba covers religious news as well as current events, the arts, and social issues—all from a Catholic perspective. Opinion pieces are also published. Circ: 6,800.
Website: www.prairiemessenger.ca

FREELANCE POTENTIAL: 60% written by non-staff writers. Publishes 10 freelance submissions yearly. Receives 30 queries and mss yearly.

SUBMISSIONS AND PAYMENT: Sample copy, $1 with 9x12 SAE/IRC. Writer's guidelines available at website. Query or send complete ms. Accepts email submissions to pm.canadian@stpeterspress.ca. Responds in 1 month. First rights. Articles, 700 words; payment rates vary. Depts/columns, 700 words; $55 Canadian. Color prints or transparencies and line art; payment rates vary. Pays at the end of each month. Provides 1 contributor's copy.

Premier Baby and Child

5100 Windance Place
Holly Springs, NC 27540

Editor and Publisher: Robyn Mangrum

DESCRIPTION AND INTERESTS: *Premier Baby and Child* is the Triangle region's upscale magazine for parents with children from birth to age six. It features profiles of local moms and offers a strong focus on local businesses, events, and products. This annual publication is distributed free through hospitals and pediatrician's offices. Circ: 35,000.
Website: www.premierbaby.com

FREELANCE POTENTIAL: 100% written by nonstaff writers. Publishes several freelance submissions yearly; 100% by unpublished writers.

SUBMISSIONS AND PAYMENT: Sample copy and guidelines available. Query with résumé and clips. Accepts hard copy and email queries to publisher@premierbaby.com. SASE. Response time varies. All rights. Articles, 250–500 words. Depts/columns, word lengths vary. Written material, $50–$100. Pays on publication. Provides 2 contributor's copies.

PTO Today

100 Stonewall Boulevard, Suite 3
Wrentham, MA 02093

Editor-in-Chief: Craig Bystrynski

DESCRIPTION AND INTERESTS: Members and leaders of parent-teacher organizations turn to this publication for information on increasing membership, retaining volunteers, leadership issues, fundraising, and organizational management issues. Published six times each year, it is geared toward PTOs in elementary and middle schools. Circ: 83,000.
Website: www.ptotoday.com

FREELANCE POTENTIAL: 65% written by nonstaff writers. Publishes 40–50 freelance submissions yearly; 5% by unpublished writers, 10% by authors who are new to the magazine. Receives 120–180 queries yearly.

SUBMISSIONS AND PAYMENT: Guidelines available. Query. Accepts email queries to editor@ptotoday.com. Responds in 2 months. First and electronic rights. Articles, word lengths vary; payment rates vary. Pays on publication. Provides 1 contributor's copy.

Purple Circle Magazine

14200 FM 1062
Canyon, TX 79015

Editor: Melita Cramblet

DESCRIPTION AND INTERESTS: Published 10 times each year, this magazine targets the parents of 4-H or FFA exhibitors, teachers, county agents, and the producers of all of the items for these projects, including livestock producers and show suppliers. Submissions should promote the Junior Livestock Show Industry. Circ: 3,300.
Website: www.purplecircle.com

FREELANCE POTENTIAL: 50% written by nonstaff writers. Publishes 40 freelance submissions yearly; 99% by unpublished writers, 1% by authors who are new to the magazine.

SUBMISSIONS AND PAYMENT: Sample copy available. Query or send complete ms. Accepts hard copy. SASE. Response time varies. Exclusive rights. Articles and depts/columns, word lengths vary. No payment. Provides 2 contributor's copies.

Racquetball

1685 West Uintah
Colorado Springs, CO 80904

Editor: Jim Hiser

DESCRIPTION AND INTERESTS: Read by members of the United States Racquetball Association as well as recreational racquetball enthusiasts, this magazine covers all angles of the sport. Appearing six times each year, it provides information on techniques and strategies, tournament coverage, and player profiles. Circ: 40,000.
Website: www.usaracquetball.com

FREELANCE POTENTIAL: 50% written by nonstaff writers. Publishes 24–30 freelance submissions yearly; 80% by unpublished writers, 20% by authors who are new to the magazine. Receives 100 queries yearly.

SUBMISSIONS AND PAYMENT: Sample copy and guidelines, $4. Prefers query; will accept complete ms. Prefers email to jhiser@usra.org; will accept hard copy. SASE. Responds in 9 weeks. One-time rights. Articles, 1,500–2,000 words. Depts/columns, 500–1,000 words. Written material, $.03–$.07 per word. Pays on publication.

Radio Control Boat Modeler

Air Age Publishing
20 Westport Road
Wilton, CT 06897

Executive Editor: Matt Higgins

DESCRIPTION AND INTERESTS: Enthusiasts of radio-controlled model boats turn to this quarterly magazine for its comprehensive and full-color articles on this hobby. Building, racing, event coverage, and product reviews are covered in each issue. Circ: 55,000.
Website: www.rcboatmodeler.com

FREELANCE POTENTIAL: 70% written by non-staff writers. Publishes 20–25 freelance submissions yearly; 75% by unpublished writers. Receives 180 queries yearly.

SUBMISSIONS AND PAYMENT: Sample copy and guidelines, free with 9x12 SASE. Query with outline and brief biography. Accepts hard copy and email queries to rcboatmodeler@airage.com. Availability of artwork improves chance of acceptance. B/W prints and 35 mm slides. SASE. Responds in 1–3 months. All rights. Articles, 1,000–2,000 words; $50–$500. Pays on publication. Provides 2 contributor's copies.

Radio Control Car Action

Air Age Publishing
20 Westport Road
Wilton, CT 06897

Executive Editor: Matt Higgins

DESCRIPTION AND INTERESTS: Providing in-depth information on all aspects of radio-controlled gas and electric cars, this popular magazine offers enthusiasts everything from how-to articles and new product reviews to expert advice and racer profiles. Designing, building, painting, detailing, and racing are all featured as is event news and coverage. It is published monthly. Circ: 140,000.
Website: www.rccaraction.com

FREELANCE POTENTIAL: 30% written by non-staff writers. Publishes 50 freelance submissions yearly. Receives 410 unsolicited mss yearly.

SUBMISSIONS AND PAYMENT: Sample copy and guidelines available. Send complete ms with available artwork. Accepts hard copy and disk submissions (ASCII). SASE. Response time varies. All rights. Articles, word lengths and payment rates vary. 35mm color slides. Pays on acceptance. Provides 2 contributor's copies.

Read

Weekly Reader
1 Reader's Digest Road
Pleasantville, NY 10570

Editor: Bryon Cahill

DESCRIPTION AND INTERESTS: Published 16 times during the school year, *Read* is used in classrooms by children ages 12 to 14 as a basis to facilitate reading and discussion. The magazine has recently changed its focus to classic works of plays, short stories, and other material. It rarely uses new material but does maintain a small group of writers for assignments. This Weekly Reader publication does not accept unsolicited submissions. Circ: 160,000.
Website: www.weeklyreader.com

FREELANCE POTENTIAL: 60% written by non-staff writers. Receives 900 queries yearly.

SUBMISSIONS AND PAYMENT: Send letter and résumé to be considered for assignments. No unsolicited mss. Responds only if interested. First North American serial and electronic one-time use rights. Articles, 1,000–2,000 words. Written material, payment rates vary. Pays on acceptance. Provides 5 contributor's copies.

Read, America!

3900 Glenwood Avenue
Golden Valley, MN 55422

Editor & Publisher: Roger Hammer

DESCRIPTION AND INTERESTS: *Read, America!* is a quarterly newsletter for literacy professionals and advocates that contains articles on literacy and literacy-building programs, as well as reading-related news. Children's poetry and stories, which feature a variety of light, fun, interesting, motivating, and thoughtful themes, also appear in each issue. It does not publish fantasy fiction. Circ: 10,000.

FREELANCE POTENTIAL: 50% written by non-staff writers. Publishes 50 freelance submissions yearly; 100% by authors who are new to the magazine. Receives 1,500 unsolicited mss yearly.

SUBMISSIONS AND PAYMENT: Sample copy and guidelines, $7.50. Send complete ms. Accepts hard copy. No simultaneous submissions. SASE. Responds in 2–3 months. All rights. Articles and fiction, to 1,000 words; $50. Pays on acceptance.

Redbook

Hearst Corporation
300 West 57th Street, 22nd Floor
New York, NY 10019

Articles Department

DESCRIPTION AND INTERESTS: This popular consumer magazine for women features articles on a range of topics, including current events, relationships, health, sex, and family life. It is interested in queries for articles about social issues and strengthening a marriage. All queries must include a list of sources. *Redbook* is published monthly. Circ: 2.3 million.
Website: www.redbookmag.com

FREELANCE POTENTIAL: 5% written by non-staff writers. Publishes 10 freelance submissions yearly; 2% by unpublished writers. Receives 9,960+ queries yearly.

SUBMISSIONS AND PAYMENT: Sample copy, $2.99 at newsstands. Guidelines available at website. Query with clips. Accepts hard copy. SASE. Responds in 3–4 months. All rights. Articles, 1,000–3,000 words; $.75–$1 per word. Depts/columns, 1,000–5,000 words; payment rates vary. Pays on acceptance.

Reptiles

P.O. Box 6050
Mission Viejo, CA 92690

Editor: Russ Case

DESCRIPTION AND INTERESTS: Pet owners and veterinarians who care for reptiles and amphibians find the latest health, nutrition, and breeding information in this monthly magazine. Articles submitted must demonstrate knowledge of reptiles and herpetology. Circ: 40,000.
Website: www.reptilechannel.com

FREELANCE POTENTIAL: 60% written by non-staff writers. Publishes 55 freelance submissions yearly; 50% by unpublished writers, 40% by authors who are new to the magazine. Receives 120 queries yearly.

SUBMISSIONS AND PAYMENT: Sample copy, $4.50 at newsstands. Query or send complete ms. Accepts hard copy. No simultaneous submissions. SASE. Responds in 2–3 months. First North American serial rights. Articles and depts/columns, word lengths vary; payment rates vary. Payment policy varies. Provides 2 contributor's copies.

The Rock

Cook Communications Ministries
4050 Lee Vance View
Colorado Springs, CO 80918

Editor: Doug Mauss

DESCRIPTION AND INTERESTS: *The Rock* is a weekly resource for Sunday school teachers and students. Its editorial content is geared toward middle school-aged children, and includes inspirational stories alongside Bible lessons. It also features games, puzzles, and activities, all with a God- and Bible-centered theme. Most work is staff written, and it is not currently reviewing unsolicited manuscripts. Circ: 50,000.
Website: www.cookministries.com

FREELANCE POTENTIAL: 10% written by non-staff writers. Publishes 2–3 freelance submissions yearly; 20% by unpublished writers.

SUBMISSIONS AND PAYMENT: Guidelines available at website. Send résumé or writing samples only. No unsolicited mss or queries. Accepts hard copy. SASE. Response time varies. Rights negotiable. Written material, word lengths and payment rates vary. Pays on acceptance. Provides 1 contributor's copy.

Rugby Magazine

33 Kings Highway
Orangeburg, NY 10962

Editor: Ed Hagerty

DESCRIPTION AND INTERESTS: This monthly magazine provides a wealth of information on the sport of rugby. Information on championships played in the U.S. as well as international matches and other news items are featured regularly. Personality profiles of those involved in the sport are sought. Circ: 10,500.
Website: www.rugbymag.com

FREELANCE POTENTIAL: 50% written by non-staff writers. Publishes 400 freelance submissions yearly; 50% by unpublished writers, 50% by authors who are new to the magazine. Receives 600 queries and unsolicited mss yearly.

SUBMISSIONS AND PAYMENT: Sample copy and guidelines, $4 with 9x12 SASE ($1.70 postage). Query or send complete ms. Accepts hard copy and disk submissions. SASE. Responds in 2 weeks. All rights. Written material, word lengths and payment rates vary. Pays on publication. Provides 3 contributor's copies.

Scholastic News

557 Broadway
New York, NY 10012

Submissions Editor, Editions 1–3: Janis Behrens
Submissions Editor, Editions 4–6: Lee Baier

DESCRIPTION AND INTERESTS: Distributed weekly in classrooms to elementary school students, *Scholastic News* portrays current events around the world through its interactive and age-appropriate articles. There are six editions, one for each grade from first through sixth. A comprehensive website showcases additional articles. News articles and topics of interest to children are sought. Circ: 1 million+.
Website: www.scholastic.com

FREELANCE POTENTIAL: 5% written by non-staff writers. Publishes 10–20 freelance submissions yearly.

SUBMISSIONS AND PAYMENT: Query or send complete ms with résumé. Accepts hard copy and simultaneous submissions if identified. Availability of artwork improves chance of acceptance. SASE. Responds in 1–3 months. All rights. Articles, to 500 words; $75–$500. Pays on publication. Provides 3+ contributor's copies.

Scholastic News English/Español

557 Broadway
New York, NY 10012

Editor: Graciella Vidal

DESCRIPTION AND INTERESTS: Designed to help Spanish-speaking children develop the skills they need to transition to English proficiency, this publication features articles in Spanish that, when flipped over, are also printed in English. A teacher's edition contains reproducibles. The magazine is published monthly for students in grades one through three. Circ: 125,000.
Website: www.scholastic.com

FREELANCE POTENTIAL: 10% written by non-staff writers. Publishes several freelance submissions yearly. Receives many unsolicited mss each year.

SUBMISSIONS AND PAYMENT: Sample copy and editorial calendar available at website. Query or send complete ms with résumé. Accepts hard copy and simultaneous submissions if identified. SASE. Responds in 1–3 months. All rights. Articles, to 500 words; $75–$500. Pays on publication. Provides 3+ copies.

Science World

Scholastic Inc.
557 Broadway
New York, NY 10012-3999

Editor: Patricia Janes

DESCRIPTION AND INTERESTS: Illustrated articles on life, Earth, and physical science are found in this magazine that is distributed to junior high and high school classrooms. It is published 14 times each year by Scholastic. Circ: 40,000.
Website: www.scholastic.com/scienceworld

FREELANCE POTENTIAL: 50% written by non-staff writers. Publishes 2 freelance submissions yearly; 10% by authors who are new to the magazine. Receives 120 queries yearly.

SUBMISSIONS AND PAYMENT: Sample copy and guidelines, free with 9x12 SASE. All articles are assigned. Query with list of publishing credits and clips or writing samples. Accepts hard copy. SASE. Responds in 2 months. All rights. Articles, to 750 words; $200–$650. Depts/columns, 200 words; $100–$125. Written material, $.10 per word. Kill fee, 50%. Pays on publication. Provides 2 contributor's copies.

Scott Stamp Monthly

Scott Publishing Company
P.O. Box 828
Sidney, OH 45365

Editor: Donna Houseman

DESCRIPTION AND INTERESTS: Philatelists of all levels read this important resource for stamp collecting. It contains articles on stamp collecting and building and protecting a collection. Published monthly, it also features a catalogue of the latest stamp releases. Circ: 35,000.
Website: www.scottstampmonthly.com

FREELANCE POTENTIAL: 70% written by non-staff writers. Publishes 100 freelance submissions yearly; 15% by unpublished writers, 15% by new authors. Receives 180 queries and mss yearly.

SUBMISSIONS AND PAYMENT: Sample copy and guidelines, $3.50 with 9x12 SASE ($2.07 postage). Prefers query; will accept complete ms. Accepts hard copy and disk submissions (Microsoft Word). SASE. Responds in 1 month. First rights. Articles, 1,000–2,000 words; $75–$150. Depts/columns, word lengths and payment rates vary. Pays on publication. Provides 1 contributor's copy.

Scuola Calcio Coaching Magazine

P.O. Box 15669
Wilmington, NC 28408

Contributing Writer: Antonio Saviano

DESCRIPTION AND INTERESTS: Read by coaches, this magazine is one of the few soccer publications devoted to teaching readers how to improve the skills of very young players. Appearing nine times each year, it includes articles on safety and goal-setting for players as young as five years old, and features activities for specific age levels. Circ: 350+.
Website: www.soccercoachingmagazine.com

FREELANCE POTENTIAL: 10% written by non-staff writers. Publishes 10 freelance submissions yearly; 5% by unpublished writers, 10% by authors who are new to the magazine.

SUBMISSIONS AND PAYMENT: Guidelines available at website. Query. Accepts hard copy and email to magazine@soccercoachingmagazine.com (Microsoft Word attachments). SASE. Response time varies. Worldwide rights. Articles and depts/columns, word lengths and payment rates vary. Payment policy varies.

Sesame Street Magazine

One Lincoln Plaza
New York, NY 10023

Editor: Rebecca Herman

DESCRIPTION AND INTERESTS: Children ages two through five delight in *Sesame Street Magazine* for its funny stories and short articles on topics including animals, health, and manners. It also offers activities that feature the well-known characters of *Sesame Street.* It is published 11 times each year and does not accept freelance submissions. Prospective writers should send a résumé and published clips only. Circ: 650,000.
Website: www.sesamestreet.com

FREELANCE POTENTIAL: 100% staff-written.

SUBMISSIONS AND PAYMENT: All material is written in-house. Will not accept queries of unsolicited manuscripts from freelance writers. Writers interested in working for this magazine should send a résumé and published clips.

Shameless

360A Bloor Street W
P.O. Box 68548
Toronto, Ontario M5S 1X1
Canada

Editor: Megan Griffith-Greene

DESCRIPTION AND INTERESTS: An alternative to the typical teen magazine, *Shameless* seeks to engage and motivate Canadian girls to make a difference. Articles cover a range of topics, from socio-political issues, current events, activism, and arts and culture to health, sexuality, and sports. Published three times each year, it also regularly features profiles of successful women. Thought-provoking and creative articles are sought. Circ: 3,000.
Website: www.shamelessmag.com

FREELANCE POTENTIAL: 30% written by non-staff writers. Publishes 25 freelance submissions each year.

SUBMISSIONS AND PAYMENT: Guidelines available at website. Query with clips. Prefers email to submit@shamelessmag.com; will accept hard copy. SAE/IRC. Response time varies. First and electronic rights. Articles, 600–2,200 words. Profiles, 300–500 words. No payment.

Sierra

85 Second Street, 2nd Floor
San Francisco, CA 94105

Managing Editor

DESCRIPTION AND INTERESTS: Appearing six times each year, *Sierra* is read by environmentally concerned people who are passionate about nature conservation. It seeks strong, well-researched articles on significant environmental and conservation issues. Circ: 620,000.
Website: www.sierraclub.org/sierra

FREELANCE POTENTIAL: 45% written by non-staff writers. Publishes 50 freelance submissions yearly; 10% by authors who are new to the magazine. Receives 480–720 queries yearly.

SUBMISSIONS AND PAYMENT: Sample copy, $3 with 9x12 SASE ($1.99 postage). Guidelines available at website. Query. Accepts hard copy. SASE. Responds in 6–8 weeks. First North American serial, reproduction, and archival rights. Articles, 1,000–3,000 words. Depts/columns, to 1,500 words. Written material, $100–$3,000. Pays on acceptance. Provides 2 contributor's copies.

Simply You Magazine

P.O. Box 284
Phillips, WI 54555-0284

Editor

DESCRIPTION AND INTERESTS: Targeting teens and young adults, this online magazine focuses on enhancing mind, body, and spirit. It is especially looking for more articles that offer coming-of-age and relationship (such as with parents, siblings, or friends) advice. Portions of the magazine appear in a newsletter six times each year. Hits per month: 10,000.
Website: www.simplyyoumagazine.com

FREELANCE POTENTIAL: 25% written by non-staff writers. Publishes 20–40 freelance submissions yearly; 25% by unpublished writers, 50% by authors who are new to the magazine. Receives 12–36 unsolicited mss yearly.

SUBMISSIONS AND PAYMENT: Sample copy and guidelines, free with #10 SASE. Send complete ms. Accepts e-mail submissions to lynne@simplyyoumagazine.com. Responds in 1–2 months. All rights. Articles, word lengths vary. No payment. Provides 1 contributor's copy.

Small Town Life

1046 Barnett Hill Road
Punxsutawney, PA 15767

Editor: Jennifer Forrest

DESCRIPTION AND INTERESTS: Feel-good articles and personal experience pieces suitable for all ages can be found in *Small Town Life*. Published six times each year, it prides itself on working with new writers. Please note that it does not publish fiction and does not welcome fiction queries or submissions. Circ: 5,000.
Website: www.smalltownlifemagazine.com

FREELANCE POTENTIAL: 80% written by non-staff writers. Publishes 60 freelance submissions yearly; 10% by unpublished writers, 25% by authors who are new to the magazine.

SUBMISSIONS AND PAYMENT: Sample copy, $5. Guidelines available at website. Query with clips; or send complete ms. Accepts disk submissions and email submissions to editor@smalltownlifemagazine.com. SASE. Response time varies. First rights. Articles, 3–4 pages. Depts/columns, 500–700 words. No payment. Provides 2 contributor's copies.

Skiing

5720 Flatiron Parkway
Boulder, CO 80301

Associate Editor: Scott Gornall

DESCRIPTION AND INTERESTS: Ski buffs of all levels read this magazine for its informational and how-to articles on skiing and other winter sports. Published six times each year, it seeks articles about small ski areas. Circ: 300,000.
Website: www.skinet.com

FREELANCE POTENTIAL: 60% written by non-staff writers. Publishes 50 freelance submissions yearly; 2% by unpublished writers, 5% by authors who are new to the magazine. Receives 180 queries yearly.

SUBMISSIONS AND PAYMENT: Sample copy and guidelines, $2.50 with 9x12 SASE ($1 postage). Query with clips or writing samples. No simultaneous submissions. Prefers email to scott.gornall@bonniercorp.com; will accept hard copy. SASE. Responds in 2–4 months. First universal and all media rights. Articles and depts/columns, word lengths vary; $1 per word. Pays on acceptance. Provides contributor's copies.

Soccer Youth

P.O. Box 983
Morehead, KY 40351

Editor

DESCRIPTION AND INTERESTS: Targeted to boys and girls ages 7 through 14 and their parents, this magazine bills itself as "the nation's soccer magazine for kids." It offers articles on all aspects of youth soccer, from skill-building to fundraising. Profiles of professional soccer teams and players are other regular features, as are puzzles, quizzes, and comics. It is published six times each year. Circ: 50,000.
Website: www.socceryouth.com

FREELANCE POTENTIAL: 50% written by non-staff writers. Publishes 10–20 freelance submissions yearly.

SUBMISSIONS AND PAYMENT: Query with word count and availability of artwork. Prefers email submissions to mailbox@socceryouth.com; will accept hard copy. SASE. Response time varies. All rights. Articles and depts/columns, word lengths vary; payment rates vary. Pays on publication.

Socialist Appeal

P.O. Box 4244
St. Paul, MN 55104

Editor: John Peterson

DESCRIPTION AND INTERESTS: *Socialist Appeal* is the official newspaper of the Workers International League. Published eight times each year, its goal is to provide an analysis of U.S. and international politics from a Marxist, working-class perspective. The magazine was created to defend the ideas of revolutionary Marxism and fight for socialism in the U.S. and around the world. It is read by American workers and young adults looking for an alternative to the capitalist system and the status quo. Circ: 700.
Website: www.socialistappeal.org

FREELANCE POTENTIAL: 90% written by non-staff writers. Publishes 80–100 freelance submissions yearly; 25% by new authors.

SUBMISSIONS AND PAYMENT: Sample copy available at website. Query or send complete ms. Accepts hard copy. SASE. Response time varies. Rights vary. Articles, word lengths vary. No payment.

Spirit

Sisters of St. Joseph of Carondelet
1884 Randolph Avenue
St. Paul, MN 55105-1700

Editor: Joan Mitchell

DESCRIPTION AND INTERESTS: *Spirit* is a take-home paper for teens in religious education classes. It is published 28 times each year and features inspirational articles to help steer young adults in the right direction. Most of its material is based on Scripture, highlighting lessons from the Bible that are still practical in today's society. At this time, *Spirit* is not accepting unsolicited material, as most of its writing is done in-house. Check the website for changes to this policy. Circ: 25,000.
Website: www.goodgroundpress.com

FREELANCE POTENTIAL: 50% written by non-staff writers. Publishes 1–2 freelance submissions each year.

SUBMISSIONS AND PAYMENT: Sample copy, free. Not accepting queries or unsolicited material at this time.

Sporting Youth GA

P.O. Box 1137
Watkinsville, GA 30677

Editor: Barbara W. Peterson

DESCRIPTION AND INTERESTS: Targeting children, teens, and parents living in northeast Georgia, this free publication covers school sports and sporting events in the area. Profiles of local and celebrity athletes are featured in every issue, as well as training tips from coaches, college athletes, and professionals. Distributed every other month, its goal is to provide a positive resource that will help young people get active. Circ: 5,000.
Website: www.sportingyouthga.com

FREELANCE POTENTIAL: 90% written by non-staff writers. Publishes 25 freelance submissions each year.

SUBMISSIONS AND PAYMENT: Sample copy available at website. Query or send complete ms. Accepts email submissions to mail@sportingyouthga.com. Responds in 1 month. One-time rights. Articles and depts/columns, word lengths vary. No payment. Provides copies.

Story Station

Editor

DESCRIPTION AND INTERESTS: *Story Station* is an online publication filled with stories for children and young adults. It accepts submissions in all genres and prefers well-plotted stories with upbeat endings. All of its stories feature a protagonist in the 6- to 12-year-old age range. Hits per month: Unavailable.
**Website: www.viatouch.com/Learn/
    StoryStation**

FREELANCE POTENTIAL: 100% written by nonstaff writers. Publishes 50–100 freelance submissions yearly.

SUBMISSIONS AND PAYMENT: Sample copy and guidelines available at website. Send complete ms with brief author bio. Accepts email submissions to storystation@viatouch.com. Responds in 1–2 months. Electronic rights for 120 days. Fiction, 1,500–3,000 words; $.01 per word. Pays on publication.

The Storyteller

2441 Washington Road
Maynard, AR 72444

Editor: Regina Williams

DESCRIPTION AND INTERESTS: *The Storyteller* targets new and emerging writers with advice columns, high-quality fiction, and essays. Circ: 700.
**Website: www.thestorytellermagazine.com
www.freewebs.com/fossilcreek**

FREELANCE POTENTIAL: 90% written by non-staff writers. Publishes 350 freelance submissions yearly; 80% by unpublished writers, 30% by authors who are new to the magazine. Receives 300 unsolicited mss yearly.

SUBMISSIONS AND PAYMENT: Sample copy, $6 with 9x12 SASE ($1.54 postage). Writers' guidelines available. Send complete ms. Accepts hard copy and simultaneous submissions if identified. SASE. Responds in 1–2 weeks. First North American serial rights. Fiction and nonfiction, 1,200–2,000 words; $0.025 per word. Poetry, to 40 lines; $1 per poem. Payment policy varies.

Supertwins

P.O. Box 306
East Islip, NY 11730

Editor: Maureen Boyle

DESCRIPTION AND INTERESTS: The name doesn't leave much room for interpretation—this quarterly magazine is written for parents of twins and multiples up to age five. It is filled with informative articles about issues unique to parents of multiples, as well as inspirational and motivational pieces, crafts, recipes, and first-person essays. It is published by Mothers of Supertwins (MOST). Circ: Unavailable.
Website: www.MOSTonline.org

FREELANCE POTENTIAL: 100% written by nonstaff writers. Publishes 16 freelance submissions yearly. Receives 100 queries yearly.

SUBMISSIONS AND PAYMENT: Sample copy, $5 with 9x12 SASE. Query or send complete ms. Accepts hard copy and email submissions to info@mostonline.org. SASE. Response time varies. Rights vary. Articles and depts/columns, word lengths vary; payment rates vary. Pays on publication. Provides 2 contributor's copies.

Storytelling Magazine

National Storytelling Network
132 Boone Street, Suite 5
Jonesborough, TN 37659

Submissions: Kit Rogers

DESCRIPTION AND INTERESTS: Teachers, librarians, and professional storytellers are the target audience for this magazine that is published by the National Storytelling Network. Read-aloud stories and resources as well as news and applications of oral storytelling traditions can be found in its pages. It appears six times each year. Experienced storytellers with something new to tell are encouraged to submit their queries for review. Circ: 6,000.
Website: www.storynet.org

FREELANCE POTENTIAL: 50% written by non-staff writers. Publishes 100 freelance submissions yearly. Receives 48 unsolicited mss yearly.

SUBMISSIONS AND PAYMENT: Sample copy, $6. Query. Accepts email queries to kit@ storynet.org. Response time varies. First North American serial rights. Articles, 1,000–2,000 words. Depts/columns, 500 words. No payment. Provides 2 contributor's copies.

Surfing

950 Calle Amanecer, Suite C
San Clemente, CA 92673

Editor: Evan Slater

DESCRIPTION AND INTERESTS: Targeting experienced surfers ages 13 to 30, *Surfing* provides informative and highly readable articles on how to improve skills and enjoy this sport's lifestyle. It is published monthly. Circ: 105,000.
Website: www.surfingmagazine.com

FREELANCE POTENTIAL: 20% written by non-staff writers. Publishes 15 freelance submissions yearly; 50% by unpublished writers. Receives 72 unsolicited mss yearly.

SUBMISSIONS AND PAYMENT: Sample copy, $3.99 at newsstands. Guidelines available. Query or send complete ms. Accepts hard copy, disk submissions (QuarkXPress or Microsoft Word), and simultaneous submissions if identified. SASE. Responds in 1 month. One-time rights. Articles, 2,000–3,000 words. Depts/columns, 35–500 words. Written material, $.10–$.25 per word. Pays on publication. Provides 2 contributor's copies.

Synchro Swimming USA

201 South Capitol Avenue, Suite 901
Indianapolis, IN 46225

Editor: Taylor D. Payne

DESCRIPTION AND INTERESTS: Up-to-date information and news on the sport of synchronized swimming can be found in this online magazine. Updated quarterly by United States Synchronized Swimming, it covers the sport's teams, coaches, and judges, as well as competition results, member news, and information on new gear and equipment. Freelance writers are welcome to submit new and interesting material on related topics and event coverage. Hits per month: 7,000.
Website: www.usasynchro.org

FREELANCE POTENTIAL: 50% written by nonstaff writers. Publishes 15 freelance submissions yearly; 50% by unpublished writers, 50% by authors who are new to the magazine.

SUBMISSIONS AND PAYMENT: Query or send complete ms. Accepts hard copy. SASE. Response time varies. All rights. Articles, word lengths vary. No payment.

Teachers Interaction

Concordia Publishing House
3358 South Jefferson Avenue
St. Louis, MO 63118-3698

Editor: Thomas A. Nummela

DESCRIPTION AND INTERESTS: Sunday school teachers of the Lutheran Church–Missouri Synod read this quarterly magazine. Teaching tips and strategies, and articles on child development and theology appear. It is not currently reviewing submissions from new writers. Circ: 12,000.
Website: www.cph.org

FREELANCE POTENTIAL: 95% written by nonstaff writers. Publishes 20 freelance submissions yearly; 10% by unpublished writers. Receives 48 unsolicited mss yearly.

SUBMISSIONS AND PAYMENT: Sample copy, $4.99. Guidelines available. Query or send complete ms; include Social Security number. Prefers email submissions to tom.nummela@cph.org; will accept hard copy. SASE. Responds in 3 months. All rights. Articles, to 1,100 words; $55–$110. Depts/columns, 400 words; $20–$40. Pays on publication. Provides 1 author's copy.

Teachers & Writers

520 Eighth Avenue, Suite 2020
New York, NY 10018

Editor: Susan Karwoska

DESCRIPTION AND INTERESTS: This quarterly provides teachers and other educators with creative ideas and exercises for teaching creative writing to students in kindergarten through college. Circ: 3,000.
Website: www.twc.org

FREELANCE POTENTIAL: 60% written by nonstaff writers. Publishes 8 freelance submissions yearly; 5% by unpublished writers, 50% by authors who are new to the magazine. Receives 50 unsolicited mss yearly.

SUBMISSIONS AND PAYMENT: Guidelines available. Send complete ms. Accepts hard copy and simultaneous submissions if identified. SASE. Response time varies. First serial rights. Articles, 700–5,000 words. Depts/columns, word lengths vary. Written material, $20 per printed column. Pays on publication. Provides 10 contributor's copies.

Teen Strings

255 West End Avenue
San Rafael, CA 94901

Editorial Director: Greg Cahill

DESCRIPTION AND INTERESTS: Teens and young adults who play string instruments turn to this magazine for articles written about and by other young musicians. Published quarterly, it features articles on playing technique, performance, and auditioning. Each issue also contains a personal experience essay written by a teen string player. The magazine is open to working with unpublished writers, particularly those with strong backgrounds in playing or teaching string instruments. Circ: Unavailable.
Website: www.teenstrings.com

FREELANCE POTENTIAL: 64% written by nonstaff writers. Publishes 6 freelance submissions each year.

SUBMISSIONS AND PAYMENT: Sample copy available. Query. Accepts email queries to greg@stringletter.com. Response time varies. Rights vary. Articles, 1,000 words; $300. Pays on publication.

Teen Times

1910 Association Drive
Reston, VA 20191

Communications Coordinator: Leslie Shields

DESCRIPTION AND INTERESTS: Teen members of FCCLA (Family, Career, and Community Leaders of America) receive this quarterly magazine. Articles cover topics including careers, leadership, community service, health and fitness, and family. It accepts submissions from student members only. Circ: 220,000.
Website: www.fcclainc.org

FREELANCE POTENTIAL: 50% written by nonstaff writers. Publishes 12–16 freelance submissions yearly; 100% by unpublished writers, 10% by new authors. Receives 12–24 queries yearly.

SUBMISSIONS AND PAYMENT: Sample copy, free with 9x12 SASE. FCCLA student members only, query or send complete ms. Accepts hard copy. Availability of artwork improves chance of acceptance. SASE. Response time varies. All rights. Articles and depts/columns, word lengths vary. No payment. Provides contributor's copies upon request.

Tidewater Parent

258 Granby Street
Norfolk, VA 23510

Editor: Jennifer O'Donnell

DESCRIPTION AND INTERESTS: Parents living in Virginia's Hampton Roads area value this magazine for its comprehensive coverage of family issues as well as local resources. Each monthly issue features articles on education, health, child development, and parenting. Distributed at schools, doctor's offices, and libraries, it works with area writers only. Circ: 48,000.
Website: www.tidewaterparent.com

FREELANCE POTENTIAL: 90% written by nonstaff writers. Publishes 40 freelance submissions yearly; 10% by unpublished writers, 50% by authors who are new to the magazine. Receives 72 unsolicited mss yearly.

SUBMISSIONS AND PAYMENT: Send complete ms. Will accept previously published mss that can be reprinted. Accepts hard copy. SASE. Response time varies. Rights vary. Articles, 800–1,200 words; $25. Kill fee, 50%. Pays on publication. Provides 1 contributor's copy.

ThisWeek Community Newspapers

7801 N. Central Drive
Lewis Center, OH 43035

Editor: Staci Perkins

DESCRIPTION AND INTERESTS: *ThisWeek Community Newspapers* consists of 22 regional publications serving all of central Ohio. Local news and event coverage, sports, education, and articles on parenting and families are a part of each monthly issue. It prefers to work with local writers or those who have knowledge of these communities that include Bexley, New Albany, Grandview, and Marysville. Circ: Unavailable.
Website: www.thisweeknews.com

FREELANCE POTENTIAL: 100% written by nonstaff writers. Publishes 100 freelance submissions yearly.

SUBMISSIONS AND PAYMENT: Sample copy, free with 9x12 SASE. Query. Accepts email queries to editorial@thisweeknews.com (no attachments). Responds in 1 month. First or reprint rights. Written material, word lengths vary; $.10–$.20 per word. Pays on publication. Provides 2 contributor's copies.

Today's Christian
People of Faith, Stories of Hope

465 Gundersen Drive
Carol Stream, IL 60188

Editorial Coordinator: Cynthia Thomas

DESCRIPTION AND INTERESTS: This magazine has wide appeal, with a varied demographic of Christian readers. It is published six times each year, and publishes articles, interviews, personal essays, and humor. Circ: 85,000.
Website: www.todays-christian.com

FREELANCE POTENTIAL: 80% written by nonstaff writers. Publishes 25–30 freelance submissions yearly; 10% by unpublished writers, 10% by authors who are new to the magazine. Receives 1,000 unsolicited mss yearly.

SUBMISSIONS AND PAYMENT: Sample copy, free with 6x9 SASE (4 first-class stamps). Send complete ms. Accepts hard copy and email submissions to tceditor@todays-christian.com. SASE. Responds in 2 months. First serial rights. Articles, 700–2,800 words. Depts/columns, word lengths vary. Written material, $.15–$.25 per word. Pays on publication. Provides 2 contributor's copies.

Toy Farmer

7496 106th Avenue SE
LaMoure, ND 58458-9404

Editorial Assistant: Cheryl Hegvik

DESCRIPTION AND INTERESTS: Targeting avid collectors of farm toys from around the world, this monthly magazine includes all the latest information and trends related to the hobby. Manufacturer profiles, farm toy histories and nostalgia, and upcoming auctions and shows are featured. Circ: 27,000.
Website: www.toyfarmer.com

FREELANCE POTENTIAL: 100% written by nonstaff writers. Publishes 50 freelance submissions yearly; 20% by unpublished writers, 20% by authors who are new to the magazine. Receives numerous queries yearly.

SUBMISSIONS AND PAYMENT: Sample copy, guidelines, and editorial calendar, $5 with 9x12 SASE. Query with writing samples. Accepts hard copy. SASE. Responds in 1 month. First rights. Articles, 1,500 words. Depts/columns, 800 words. Written material, $.10 per word. Pays on publication. Provides 2 contributor's copies.

Toy Trucker & Contractor

7469 106 Avenue SE
LaMoure, ND 58458-9404

Editorial Assistant: Cheryl Hegvik

DESCRIPTION AND INTERESTS: This monthly hobbyist publication targets collectors of toy trucks and construction equipment. New product information, toy show events, informational articles, and profiles of collectors are also featured in each issue. Circ: 8,000.
Website: www.toytrucker.com

FREELANCE POTENTIAL: 100% written by nonstaff writers. Publishes 60 freelance submissions yearly; 10% by unpublished writers, 20% by authors who are new to the magazine. Receives 12 queries yearly.

SUBMISSIONS AND PAYMENT: Guidelines and editorial calendar available. Query with writing samples. Accepts hard copy. SASE. Responds in 1 month. First North American serial rights. Articles, 1,000–5,000 words. Depts/columns, word lengths vary. Written material, $.10 per word. Pays on publication. Provides 2 contributor's copies.

TransWorld Snowboarding

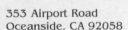

353 Airport Road
Oceanside, CA 92058

Managing Editor: Annie Fast

DESCRIPTION AND INTERESTS: Appearing eight times each year, *TransWorld Snowboarding* brings all the news of the sport—as well as the newest gear, athletes, and best places to snowboard—to avid snowboarders. Most of its readers are teens and young adults. It's looking for snowboarding news and resort reviews from a snowboarder's point of view. Circ: 1.4 million.
Website: www.twsnow.com

FREELANCE POTENTIAL: 40% written by nonstaff writers. Publishes 20 freelance submissions yearly; 10% by authors who are new to the magazine. Receives 120 queries yearly.

SUBMISSIONS AND PAYMENT: Sample copy, $3.99 at newsstands. Guidelines and theme list available. Query. Accepts email queries to annie.fast@transworld.net. Responds in 1 month. Rights vary. Articles, to 1,600 words. Depts/columns, 300 words. Written material, $.35 per word. Pays on publication.

Turtle Trails and Tales

P.O. Box 19623
Reno, NV 89511

Editor: Virginia Castleman

DESCRIPTION AND INTERESTS: A blend of articles, personal experience pieces, and short stories are found in the pages of this multicultural magazine for children. Each of the six issues published each year is themed, with all material promoting tolerance and understanding toward other cultures. Circ: 20,000+.

FREELANCE POTENTIAL: 40% written by nonstaff writers. Publishes 12 freelance submissions yearly; 80% by unpublished writers. Receives 36 unsolicited mss yearly.

SUBMISSIONS AND PAYMENT: Sample copy, guidelines, and theme list available. Send complete ms. Accepts hard copy and email submissions to vcastleman@sbcglobal.net. SASE. Responds in 3 months. First rights. Articles, 750 words; $25. Fiction, 500–800 words; payment rates vary. Pays on publication. Provides 1 contributor's copy.

Twist

270 Sylvan Avenue
Englewood Cliffs, NJ 07632

Associate Editor: Ellen Collis

DESCRIPTION AND INTERESTS: Celebrity news and pop culture are the mainstay of this publication for teens. *Twist* offers entertainment news, fashion and beauty tips, and articles on relationships. Quizzes and celebrity interviews are also always of interest. It is published 10 times each year. Circ: 230,000.
Website: www.twistmagazine.com

FREELANCE POTENTIAL: 5% written by non-staff writers. Publishes 10 freelance submissions yearly; 5% by unpublished writers, 5% by authors who are new to the magazine. Receives 240 queries yearly.

SUBMISSIONS AND PAYMENT: Sample copy, $3.99 with 9x12 SASE. Guidelines available. Query. Accepts hard copy. SASE. Responds in 2–3 weeks. First North American serial rights. Written material, word lengths and payment rates vary. Pays on acceptance. Provides 2 contributor's copies.

Vibrant Life

55 West Oak Ridge Drive
Hagerstown, MD 21740

Editor: Charles Mills

DESCRIPTION AND INTERESTS: This health magazine promotes mental clarity, spiritual balance, and physical health within a vegan lifestyle and from a Christian perspective. It is published six times each year. Circ: 28,500.
Website: www.vibrantlife.com

FREELANCE POTENTIAL: 95% written by non-staff writers. Publishes 25 freelance submissions yearly; 50% by unpublished writers, 25% by authors who are new to the magazine. Receives 400 queries, 400 unsolicited mss yearly.

SUBMISSIONS AND PAYMENT: Sample copy and guidelines, $1 with 9x12 SASE (3 first-class stamps). Prefers complete ms; will accept query. Accepts hard copy and email to vibrantlife@rhpa.org (Microsoft Word attachments). SASE. Responds in 1 month. First world, reprint, and electronic rights. Articles, 450–1,500 words; $75–$300. Pays on acceptance. Provides 3 contributor's copies.

Vegetarian Journal

P.O. Box 1463
Baltimore, MD 21203

Managing Editor: Debra Wasserman

DESCRIPTION AND INTERESTS: Since 1982, this quarterly magazine has been providing vegetarians and vegans with information and support. It features informational, educational articles, and recipes that make it easy for anyone, including children, to be vegan in any situation. It also sponsors an essay contest for children. It is not interested in articles about supplements, or in self-help pieces. Circ: 20,000.
Website: www.vrg.org

FREELANCE POTENTIAL: 60% written by nonstaff writers. Publishes 12 freelance submissions yearly; 5% by authors who are new to the magazine.

SUBMISSIONS AND PAYMENT: Sample copy, $4. Query with author credentials. Accepts hard copy. SASE. Responds in 1 week. Submit seasonal material 1 year in advance. One-time rights. Articles, word lengths vary; $200. Pays on acceptance. Provides 3+ contributor's copies.

The Village Family

501 40th Street S
Fargo, ND 58103

Editor: Laurie Neill

DESCRIPTION AND INTERESTS: Positive, upbeat articles for parents are the focus of this family magazine, written exclusively for residents of the greater Fargo region. Appearing six times each year, it also features articles on education, child care, child health, and other parenting matters. Circ: 25,000.
Website: www.thevillagefamily.org

FREELANCE POTENTIAL: 80% written by non-staff writers. Publishes 30 freelance submissions yearly; 60% by unpublished writers. Receives 1,200+ queries and unsolicited mss yearly.

SUBMISSIONS AND PAYMENT: Guidelines available. Query or send complete ms with author bio. Accepts hard copy and email submissions to magazine@thevillagefamily.org. SASE. Response time varies. First and electronic rights. Articles, to 1,500 words. Depts/columns, word lengths vary. Written material, $.10–$.15 per word. Reprints, $30–$50. Pays on publication.

Volta Voices

Alexander Graham Bell Association
for the Deaf and Hard of Hearing
3417 Volta Place NW
Washington, DC 20007-2778

Editor

DESCRIPTION AND INTERESTS: Focusing on hearing loss and spoken language education, *Volta Voices* features helpful and up-to-date information for the deaf and hearing impaired and their families, as well as for professionals working in the field. Articles cover new teaching methods, technology, early intervention, and issues associated with hearing loss. It is published six times each year. Circ: 5,500.
Website: www.agbell.org

FREELANCE POTENTIAL: 90% written by nonstaff writers. Publishes 6–8 freelance submissions yearly; 50% by unpublished writers. Receives 24 unsolicited mss yearly.

SUBMISSIONS AND PAYMENT: Sample copy available at website. Send complete ms. Accepts email to editor@agbell.org (Microsoft Word attachments). Responds in 1–3 months. All rights. Articles, 500–2,000 words. No payment. Provides 3 contributor's copies.

Wanna Bet?

North American Training Institute
314 West Superior Street, Suite 508
Duluth, MN 55802

Submissions

DESCRIPTION AND INTERESTS: Written by and for middle school-aged kids, this online publication focuses on the concerns of gambling. Age-appropriate articles address all forms of betting and gambling and the problems they cause, whether it is a parent or the child himself who is addicted. Healthy alternatives are also discussed. Updated monthly, it has adult and youth advisors along with a junior editor. *Wanna Bet?* is published by the North American Training Institute. Hits per month: 60,000.
Website: www.nati.org

FREELANCE POTENTIAL: 25% written by nonstaff writers. Publishes many freelance submissions yearly.

SUBMISSIONS AND PAYMENT: Sample copy available at website. Query or send complete ms. Accepts email submissions to info@nati.org. Response time varies. Electronic rights. Articles, word lengths vary. No payment.

The Water Skier

USA Water Ski
1251 Holy Cow Road
Polk City, FL 33868-8200

Editor: Scott Atkinson

DESCRIPTION AND INTERESTS: Water-skiing enthusiasts read this magazine for comprehensive information about technique and training, accompanied by action photography. Also found in its pages are competition results and profiles of teams and athletes. *The Water Skier*, the official publication of the American Water Ski Association, is published nine times each year. Circ: 35,000.
Website: www.usawaterski.org

FREELANCE POTENTIAL: 20% written by nonstaff writers. Publishes 10–12 freelance submissions yearly; 10% by authors who are new to the magazine. Receives 20–30 queries yearly.

SUBMISSIONS AND PAYMENT: Sample copy, $1.25 with 9x12 SASE. Query. Accepts hard copy. SASE. Responds in 1 month. All rights. Articles, 1,000 words. Fiction, 500–1,000 words. Written material, payment rates vary. Provides 1 contributor's copy.

Wild West

Weider History Group
741 Miller Drive SE, Suite D-2
Leesburg, VA 20175

Editor: Greg Lalire

DESCRIPTION AND INTERESTS: Read by history enthusiasts as well as high school teachers, *Wild West* portrays the settlement of the American West through its lively and well-documented articles. Published six times each year, it seeks highly readable and accurate submissions. Circ: 80,000.
Website: www.thehistorynet.com

FREELANCE POTENTIAL: 80% written by nonstaff writers. Publishes 60 freelance submissions yearly; 10% by unpublished writers, 20% by authors who are new to the magazine. Receives 250 queries yearly.

SUBMISSIONS AND PAYMENT: Sample copy and guidelines, $6. Query with résumé, outline, illustration ideas, source lists, and clips or writing samples. Accepts hard copy. SASE. Responds in 4–6 months. All rights. Articles, to 3,500 words; $300. Depts/columns, to 1,200 words; $150. Pays on publication.

Women Today Magazine

Box 300 STN "A"
Vancouver, British Columbia V6C 2X3
Canada

Senior Editor: Claire Colvin

DESCRIPTION AND INTERESTS: Active, modern women are the audience for this e-zine that updates its content monthly. Inspiring women to be the best version of themselves by pursuing excellence in all areas of their lives, it covers health, spirituality, culture, and relationships. Poetry, fiction, and artwork are not used. Hits per month: Unavailable.
Website: www.womentodaymagazine.com

FREELANCE POTENTIAL: 30–50% written by nonstaff writers. Publishes 20–30 freelance submissions yearly; 15% by unpublished writers, 25% by new authors. Receives 450 mss yearly.

SUBMISSIONS AND PAYMENT: Sample copy and writers' guidelines available at website. Send complete ms. Accepts email submissions to info@womentodaymagazine.com. Responds in 4–6 weeks. One-time rights. Articles, 1,000–1,500 words. Depts/columns, 700–1,000 words. No payment.

World Around You

Laurent Clerc National Deaf Education Center
KDES, Suite 3600
800 Florida Avenue NE
Washington, DC 20002

Submissions: Michael Walton

DESCRIPTION AND INTERESTS: People who are hearing-impaired turn to this e-zine for informational and inspirational articles on careers, lifestyles, and achievements of other deaf people. It also features profiles of notable deaf people and first-person essays. It is produced five times each year by Gallaudet University, which is the only college in the world dedicated entirely to deaf students. Hits per month: Unavailable.
**Website: http://clerccenter.gallaudet.edu/
    worldaroundyou/**

FREELANCE POTENTIAL: 10% written by nonstaff writers. Publishes 3–5 freelance submissions yearly. Receives 48 queries yearly.

SUBMISSIONS AND PAYMENT: Sample copy available at website. Query. Accepts hard copy and email queries to michael.walton@gallaudet.edu. SASE. Responds in 1 month. Rights negotiable. Articles, word lengths vary; payment rates vary. Pays on publication.

Youngbucks-outdoors.com

10350 Highway 80 East
Montgomery, AL 36117

Online Editor: Daniel Dye

DESCRIPTION AND INTERESTS: This e-zine, a Buckmasters publication, offers a wholesome presentation of hunters and hunting. Geared for children ages 7 through 13, it presents articles that encourage kids to enjoy nature and outdoor adventure. It features stories and articles about animals, hunting and fishing, and science. Hits per month: 200,000.
Website: www.youngbucksoutdoors.com

FREELANCE POTENTIAL: 60% written by nonstaff writers. Publishes 15–20 freelance submissions yearly.

SUBMISSIONS AND PAYMENT: Guidelines available. Query with detailed photo information. Availability of artwork improves chance of acceptance. Accepts hard copy and email to ddye@buckmasters.com. SASE. Responds in 1 week. First rights. Articles, 400 words. Fiction, to 500 words. Color prints and transparencies. All material, payment rates vary. Pays on publication.

Your Big Backyard

National Wildlife Federation
11100 Wildlife Center Drive
Reston, VA 20190

Editorial Department

DESCRIPTION AND INTERESTS: This monthly from the National Wildlife Federation targets children ages three to seven. A sister publication to *Ranger Rick*, it covers animals and nature, piquing the interest of young children everywhere. It also offers recipes and activities. *Your Big Backyard* does not accept unsolicited material. Circ: 400,000.
Website: www.nwf.org/YourBigBackyard

FREELANCE POTENTIAL: 10% written by nonstaff writers.

SUBMISSIONS AND PAYMENT: Not accepting queries or unsolicited manuscripts. Prospective writers should send their résumés and clips.

Your Child

820 Second Avenue
New York, NY 10017

Editor: Kay E. Pomerantz

DESCRIPTION AND INTERESTS: Raising and educating Jewish children is the focus of this online newsletter, which is updated three times each year. Articles cover parenting, religious, and social issues. Holiday and family project ideas, as well as media reviews, are also included. Articles on positive child-raising techniques with Jewish angles are welcome. Circ: 3,000.
Website: www.uscj.org

FREELANCE POTENTIAL: Of the freelance submissions published yearly, 50% are by unpublished writers.

SUBMISSIONS AND PAYMENT: Sample copy, free with 9x12 SASE ($.55 postage). Query or send complete ms. Accepts hard copy and email queries to kaykp@aol.com. Availability of artwork improves chance of acceptance. SASE. Response time varies. All rights. Articles, word lengths vary. 8x10 B/W transparencies; line art. No payment. Provides 1 contributor's copy.

Youthrunner

P.O. Box 1156
Lake Oswego, OR 97035

Editor: Dan Kesterson

DESCRIPTION AND INTERESTS: The core readership of this monthly magazine includes members of middle school and high school track teams and cross-country teams, as well as their coaches. With articles on training and performance, and tips from the pros, *Youthrunner* always welcomes new ideas and advice from coaches. Circ: Unavailable.
Website: www.youthrunner.com

FREELANCE POTENTIAL: 50% written by non-staff writers. Publishes 10 freelance submissions yearly; 50% by unpublished writers, 50% by authors who are new to the magazine.

SUBMISSIONS AND PAYMENT: Sample copy available. Send complete ms. Accepts email submissions to dank@youthrunner.com. Response time varies. First rights. Articles and depts/columns, word lengths and payment rates vary. Payment policy varies.

Yummy Mummy

13988 Maycrest Way, Unit 140
Richmond, British Columbia V6V 3C3
Canada

Managing Editor: Michelle Froese

DESCRIPTION AND INTERESTS: Available to families throughout the mainland in British Columbia, *Yummy Mummy* is a guide to all things mom. Published eight times yearly, it also devotes two issues per year to dads. It is particularly interested in tips to make a mom's life easier. Circ: 10,000.
Website: www.yummymummymag.com

FREELANCE POTENTIAL: 75% written by non-staff writers. Publishes 40 freelance submissions yearly; 35% by authors who are new to the magazine. Receives 300 queries yearly.

SUBMISSIONS AND PAYMENT: Sample copy available at website. Query. Accepts email queries to editor@yummymummymag.com. Response time varies. One-time print and electronic rights. Articles, 600–800 words. Depts/columns, staff written. Written material, payment rates vary. Pays on publication. Provides contributor's copies upon request.

ZooGoer

National Zoological Park
3001 Connecticut Avenue NW
Washington, DC 20008

Editor: Cindy Han

DESCRIPTION AND INTERESTS: Entertaining and informative articles on natural history, wildlife biology, and conservation are found in this magazine. Published six times each year, it is sent to members of the Friends of the National Zoo. Circ: 30,000.
Website: www.fonz.org/zoogoer.htm

FREELANCE POTENTIAL: 70% written by non-staff writers. Publishes 25 freelance submissions yearly; 15% by unpublished writers, 25% by new authors. Receives 15 queries and mss yearly.

SUBMISSIONS AND PAYMENT: Guidelines available. Query with synopsis and clips; or send complete ms. Accepts hard copy, disk submissions (Microsoft Word), and email to hanc@si.edu. SASE. Responds in 1–2 months. First rights. Articles, 2,500–3,000 words. Depts/columns, 800–1,500 words. Written material, $.50 per word. Pays on publication. Provides 5 contributor's copies.

Contests and Awards

Selected Contests and Awards

Entering a writing contest will provide you with a chance to have your work read by established writers and qualified editors. Winning or placing in a contest or an award program can open the door to publication and recognition of your writing. If you don't win, try to read the winning entry if it is published; doing so will give you some insight into how your work compares with its competition.

For both editors and writers, contests generate excitement. For editors, contests are a source to discover new writers. Entries are more focused because of the contest guidelines, and therefore more closely target an editor's current needs.

For writers, every contest entry is read, often by more than one editor, as opposed to unsolicited submissions that are often relegated to a slush pile.

And you don't have to be the grand-prize winner to benefit—non-winning manuscripts are often purchased by the publication for future issues.

To be considered for the contests and awards that follow, your entry must fulfill all of the requirements mentioned. Most are looking for unpublished article or story manuscripts, while a few require published works. Note special entry requirements, such as whether or not you can submit the material yourself, need to be a member of an organization, or are limited in the number of entries you can send. Also, be sure to submit your article or story in the standard manuscript submission format.

For each listing, we've included the address, the contact, a description, the entry requirements, the deadline, and the prize. In some cases, the 2009 deadlines were not available at press time. We recommend that you write to the addresses provided or visit the websites to request an entry form and the contest guidelines, which usually specify the current deadline.

Appalachian Writers Association Contests

Dr. Christina Walton, Assistant Professor
Morehead State University
401 J Ginger Hall
Morehead, KY 40351

DESCRIPTION: The Appalachian Writers Association sponsors annual contests in the categories of short story, essay, playwriting, and poetry. The competition is open to members of AWA, and accepts previously unpublished material only.
Website: www.moreheadstate.edu

REQUIREMENTS: Entry fee, $5 (for up to 3 entries). Accepts hard copy. Submit 2 copies of entry. Visit the website or send an SASE for complete category guidelines and further information.

PRIZES: A grant of $2,500 will be presented to each of the winners.

DEADLINE: Applications are accepted between September 15 and October 31.

Arizona Literary Contest

Contest Coordinator
6145 West Echo Lane
Glendale, AZ 85302

DESCRIPTION: Sponsored by the Arizona Authors Association, this annual contest is open to all writers. It accepts original submissions in the categories of short story, essay/article, poetry, and novel.
Website: www.azauthors.com/contest.html

REQUIREMENTS: Entry fees vary for each category and range from $10 to $30. Check website for word lengths for each category. Accepts hard copy. Manuscripts will not be returned. Complete guidelines available at the website or with an SASE.

PRIZES: Winners are published in *Arizona Literary Magazine*. Cash prizes are also awarded.

DEADLINE: Entries are accepted between January 1 and July 1.

Isaac Asimov Award

University of South Florida
School of Mass Communications
4202 East Fowler
Tampa, FL 33620

DESCRIPTION: Undergraduate students may submit entries to this annual award, which has a goal of promoting and encouraging the writing of high-quality science fiction and fantasy.
Website: www.asimovs.com

REQUIREMENTS: Open to full-time college students only. Entry fee, $10. Limit 3 entries per competition. Entries must be between 1,000 and 10,000 words. Include a cover sheet with author's name, address, and university. Author's name should not appear on the manuscript.

PRIZES: Winner receives a cash prize of $500 and will be considered for publication in *Asimov's Science Fiction Magazine*.

DEADLINE: December 15.

Baker's Plays High School Playwriting Contest

Attn: Associate Editor
45 West 25th Street
New York, NY 10010

DESCRIPTION: Sponsored by Baker's Plays, this annual contest looks to inspire young adults to keep American theater alive.
Website: www.bakersplays.com

REQUIREMENTS: Open to high school students only. No entry fee or word length limitations. Plays must be accompanied by the signature of a sponsoring high school English teacher. Accepts hard copy. Include an SASE for return of manuscript. Visit the website or send an SASE for complete guidelines and entry form.

PRIZES: First-place winner receives a cash award of $500 with a royalty-earning contract from Baker's Plays. Second- and third-place winners receive cash prizes.

DEADLINE: January 30.

Waldo M. and Grace C. Bonderman Youth Theatre Playwriting Competition

Indiana Repertory Theatre
140 West Washington Street
Indianapolis, IN 46204

DESCRIPTION: The Indiana Repertory Theatre established this annual award to encourage the writing of theatrical scripts for young audiences. It is open to all writers and accepts original scripts or adaptations with proper proof that the original work is in the public domain or that permission has been granted by the copyright holder.
Website: www.indianarep.com/Bonderman

REQUIREMENTS: No entry fee. Entries should have a running time of approximately 45 minutes. Limit one entry per competition. Accepts hard copy.

PRIZES: Awards are presented to the top 10 finalists. Four cash awards of $1,000 are also awarded to the playwrights whose entries are selected for development.

DEADLINE: August 31.

ByLine Magazine Contests

Contests
P.O. Box 111
Albion, NY 14411

DESCRIPTION: ByLine sponsors monthly contests for writers in several different categories, including personal memoir, juvenile short story, poetry, and inspirational article. These contests are open to all writers.
Website: www.bylinemag.com

REQUIREMENTS: Entry fees range from $3 to $5 depending on category. Multiple entries are accepted. Accepts hard copy. Send an SASE or visit the website for complete category information and further guidelines.

PRIZES: Cash prizes ranging from $10 to $70 are presented to the winners. Runners-up also receive cash awards in each category. Winning entries for the Annual Literary Awards are published in ByLine and receive a cash award of $250.

DEADLINE: Varies for each category.

Canadian Writer's Journal Short Fiction Contest

Canadian Writer's Journal
Box 1178
New Liskeard, Ontario P0J 1P0
Canada

DESCRIPTION: Offered semi-annually, this contest is sponsored by Canadian Writer's Journal and accepts original, unpublished stories in any genre.
Website: www.cwj.ca

REQUIREMENTS: Entry fee, $5. Entries should be approximately 1,200 words. Multiple entries are not accepted. Accepts hard copy. Author's name should not appear on manuscript. Include a cover sheet with author's name, address, and title of entry. Manuscripts will not be returned. Visit the website for complete information.

PRIZES: First-place winner receives a cash prize of $100. Second- and third-place winners receive cash prizes of $50 and $25, respectively.

DEADLINE: September 30 and March 31.

CAPA Competition

C/O Dan Uitti
Connecticut Authors and Publishers Association
P.O. Box 715
Avon, CT 06001

DESCRIPTION: Open to residents of Connecticut only, this competition is held annually and is sponsored by the Connecticut Authors and Publishers Association. It accepts original, unpublished entries in the categories of children's story, short story, personal essay, and poetry.
Website: http://aboutcapa.com

REQUIREMENTS: Entry fee, $10 for one story or essay and up to 3 poems. Children's stories, to 2,000 words. Personal essays, to 1,500 words. Poetry, to 30 lines. Multiple entries are accepted. Accepts hard copy. Submit 4 copies of each entry. Manuscripts are not returned.

PRIZES: First-place winners in each category receive a cash prize of $100. Second-place winners receive cash awards of $50.

DEADLINE: May 31.

Children's Writer Contests

Children's Writer
95 Long Ridge Rd
West Redding, CT 06896

DESCRIPTION: *Children's Writer* newsletter sponsors two contests each year with varying themes for previously unpublished, original work. Upcoming themes for 2009 are sports nonfiction and readaloud folktale or fantasy.
Website: www.childrenswriter.com

REQUIREMENTS: No entry fee for subscribers. For non-subscribers, a $13 entry fee includes an 8-month subscription to the newsletter. Word length requirements vary for each category, but range from 200 to 750 words. Multiple entries are accepted. Accepts hard copy. Manuscripts are not returned.

PRIZES: Cash prizes vary for each contest. Winning entries are published in *Children's Writer*.

DEADLINE: February 28 and October 31.

CNW Writing Competition

CNW/FFWA
P.O. Box A
North Stratford, NH 03590

DESCRIPTION: This annual competition offers prizes in several categories, including children's short story, children's nonfiction, novel chapter, nonfiction chapter, and poetry. It is open to all writers.
Website: www.writers-editors.com

REQUIREMENTS: Entry fees and word lengths vary for each category. Multiple entries are accepted. Accepts hard copy. Do not staple entries. Each entry must include an official entry form, available at the website. Visit website for complete list of categories and other details.

PRIZES: First- through third-place winners will be awarded in each category. Winners receive cash prizes ranging from $50 to $100.

DEADLINE: March 15.

Kimberly Colen Grant

Box 20322
Park West Finance Station
New York, NY 10025-1512

DESCRIPTION: This grant is available to new authors and illustrators who have not yet published their first children's book. Grants are offered in early reader/picture book and middle-grade/young adult categories.
Website: www.scbwi.org

REQUIREMENTS: Applicants should send a 1-page letter (approximately 300 words) describing the work they hope will become their first children's book. It must describe a work-in-progress or a complete project. Do not send writing or illustrations with application.

PRIZES: A grant of $2,500 will be presented to each of the winners.

DEADLINE: Applications are accepted between September 15 and October 31.

Crossquarter Annual Short Science Fiction Contest

P.O. Box 23749
Santa Fe, NM 87502

DESCRIPTION: Sponsored by Crossquarter Publishing, this annual contest awards top-quality science fiction short stories that demonstrate the best of the human spirit. Open to all writers, it will not consider entries with elements of horror.
Website: www.crossquarter.com

REQUIREMENTS: Entry fee, $15; $10 for each additional entry. All entries should not exceed 7,500 words. Accepts hard copy. Visit the website or send an SASE for complete competition information.

PRIZES: First-place winner receives $250 and publication. Second- through fourth-place winners receive cash prizes ranging from $50 to $125 and publication of their entries.

DEADLINE: January 15.

Shubert Fendrich Memorial Playwriting Contest

Pioneer Drama Service, Inc.
P.O. Box 4267
Englewood, CO 80155-4267

DESCRIPTION: Held annually, this playwriting contest presents its winners with publication and an advance on royalties. Plays may be on any subject that is appropriate for family viewing.
Website: www.pioneerdrama.com

REQUIREMENTS: No entry fee. Plays should run between 20 and 90 minutes. Accepts hard copy. All entries must include a cover letter with title of entry, synopsis, cast list, proof of production, number of sets, scenes, and, if applicable, musical score on CD. Writers who have previously been published by Pioneer Drama Service are not eligible.

PRIZES: Winner receives a $1,000 advance on royalties in addition to publication.

DEADLINE: March 1.

Foster City International Writing Contest

C/o Foster City Parks & Rec Department
650 Shell Boulevard
Foster City, CA 94404

DESCRIPTION: Held annually, this contest accepts entries in the categories of children's story, fiction, humor, poetry, and personal essay. The competition is open to original, unpublished work only.
**Website: www.geocities.com/
fostercity_writers**

REQUIREMENTS: Entry fee, $12. Multiple entries are accepted. Children's stories and fiction, to 3,000 words. Humor and personal experience pieces, to 500 words. Poetry, to 2,000 words. Accepts hard copy and email submissions to fostercity_writers@yahoo.com (RTF or Word attachments). Visit the website for further details.

PRIZES: First place winners in each category receive a cash prize of $150.

DEADLINE: December 30.

H. E. Francis Award

Department of English
University of Alabama at Huntsville
Huntsville, AL 35899

DESCRIPTION: Sponsored by the Ruth Hindman Foundation and UAH English Department, this contest is held annually. It accepts original, previously unpublished short story entries.
Website: www.uah.edu/hefranciscontest/

REQUIREMENTS: Entry fee, $15. Entries should not exceed 5,000 words. Accepts hard copy. Send an SASE or visit the website for complete guidelines.

PRIZES: First-place winners receive a cash prize of $1,000.

DEADLINE: December 31.

Friends of the Library Contest

Decatur Public Library
130 N. Franklin Street
Decatur, IL 62523

DESCRIPTION: Open to all writers, this contest accepts entries in the categories of essay, fiction, juvenile fiction, rhymed poetry, and unrhymed poetry. It accepts original, unpublished entries only.
Website: www.decatur.lib.il.us

REQUIREMENTS: Entry fee, $3. Limit 5 entries per competition. Accepts hard copy. Essays, to 2,000 words. Fiction and juvenile fiction, to 3,000 words. Poetry, to 40 lines.

PRIZES: First-place winner receives a cash award of $50. Second- and third-place winners receive cash prizes of $30 and $20, respectively.

DEADLINE: September 25.

John Gardner Memorial Prize for Fiction

Harpur Palate
English Dept., Binghamton University
Box 6000
Binghamton, NY 13902-6000

DESCRIPTION: Honoring John Gardner for his dedication to the creative writing program at Binghamton University, this contest is open to submissions of previously unpublished fiction in any genre.
Website: http://harpurpalate.binghamton.edu

REQUIREMENTS: Entry fee, $15. Multiple entries are accepted under separate cover only. Accepts hard copy. Include a cover sheet with author's name, address, phone number, email address, and title. Author's name should not appear on entry. Manuscripts are not returned.

PRIZES: Winner receives a cash award of $500 and publication in *Harpur Palate*.

DEADLINE: March 31.

Paul Gillette Memorial Writing Contest

Pikes Peak Writers
4164 Austin Bluffs Parkway, #246
Colorado Springs, CO 80918

DESCRIPTION: This annual writing contest accepts original entries in children's fiction, young adult fiction, mystery, historical fiction, and creative nonfiction.
Website: www.ppwc.net

REQUIREMENTS: Entry fee, $30 for members; $40 for non-members. Word lengths vary for each category. Manuscript critiques are available for an additional $20. Accepts hard copy and email submissions to pgcontest@gmail.com. All entries must be accompanied by an official entry form, cover letter, and 2 copies of manuscript.

PRIZES: First-place winners in each category receive a cash prize of $100. Second-place winners receive a cash prize of $50.

DEADLINE: November 1.

Highlights for Children Fiction Contest

Fiction Contest
803 Church Street
Honesdale, PA 18431

DESCRIPTION: Sponsored by the popular children's magazine, this annual contest is open to all writers and looks for well-written short stories for ages 2 through 12. Entries should not contain violence, crime, or derogatory humor.
Website: www.highlights.com

REQUIREMENTS: No entry fee. Entries should not exceed 500 words. Multiple entries are accepted. Accepts hard copy. Include an SASE for return of manuscript. Send an SASE or visit the website for complete competition guidelines.

PRIZES: Winner receives a cash award of $1,000 and publication in *Highlights for Children* (requires all rights).

DEADLINE: Entries must be postmarked between January 1 and January 31.

Insight Writing Contest

Insight Magazine
55 West Oak Ridge Drive
Hagerstown, MD 21740-7390

DESCRIPTION: *Insight* magazine sponsors this annual contest that puts value on the mechanics of good writing, particularly with a spiritual message. It accepts entries of short nonfiction and poetry that target young people ages 14 to 22.
Website: www.insightmagazine.org

REQUIREMENTS: No entry fee. Entries should be between 1,500 and 2,000 words (no longer than 7 pages). Accepts hard copy and email submissions to insight@rhpa.org. Include a cover letter with author's name, address, phone number, and Social Security number. Author's name must not appear on entry. Include an SASE for return of manuscript.

PRIZES: Winners receive cash awards ranging from $150 to $250.

DEADLINE: June 2.

Magazine Merit Award

The Society of Children's Book Writers
and Illustrators
8271 Beverly Boulevard
Los Angeles, CA 90048

DESCRIPTION: The Society of Children's Book
Writers and Illustrators sponsors this annual
award to honor outstanding stories and articles
for young people that were published during the
previous calendar year.
Website: www.scbwi.org

REQUIREMENTS: No entry fee. Open to SCBWI
members only. Submit 4 copies of published
work showing proof of publication date. Include
4 cover sheets with author's name, mailing
address, phone number, title, category, name of
publication, and date of issue. Accepts hard
copy. Visit the website for further details.

PRIZES: Winners in each category receive a
plaque. Honor certificates are also awarded.

DEADLINE: Entries are accepted between
January 31 and December 15 of each year.

Milkweed Fiction Prize

Milkweed Editions
1011 Washington Avenue South, Suite 300
Minneapolis, MN 55415

DESCRIPTION: Milkweed Editions presents this
annual prize to reward the best fiction manuscripts it receives during the calendar year with
publication and a cash advance. Manuscripts can
be a collection of short stories or individual stories previously published in magazines or
anthologies.
Website: www.milkweed.org

REQUIREMENTS: No entry fee. Manuscripts
previously submitted to Milkweed Editions are
not eligible. Complete guidelines are available at
the website.

PRIZES: Winner receives publication and a cash
advance of $5,000.

DEADLINE: Ongoing.

National Children's Theatre Competition

Actors' Playhouse/Miracle Theatre
280 Miracle Mile
Coral Gables, FL 33134

DESCRIPTION: Musical scripts for young audiences ages 5 to 12 are the focus of this annual
competition. It is open to all writers and prefers
entries that can be enjoyed by both children and
adults.
Website: www.actorsplayhouse.org

REQUIREMENTS: Entry fee, $10. Multiple
entries are accepted. Entries should have a running time of 45–60 minutes. Entries should have
a maximum of 8 adult actors to play any number
of roles. Accepts hard copy. Visit the website or
send an SASE for complete information.

PRIZES: Winner receives a cash prize of $500
and a production of their musical.

DEADLINE: April 1.

New Millennium Writings Awards

P.O. Box 2463
Room M2
Knoxville, TN 37901

DESCRIPTION: This competition is sponsored
by the literary journal *New Millennium Writings*
and is open to all writers. It accepts entries in
the categories of fiction, nonfiction, short fiction,
and poetry in all subject areas.
Website: www.newmillenniumwritings.com

REQUIREMENTS: Entry fee, $17. Fiction and
nonfiction, to 6,000 words. Short fiction, to
1,000 words. Poetry, to 3 poems. Accepts
hard copy and electronic submissions through
the website. Visit the website or send an SASE
for complete guidelines and further category
information.

PRIZES: First-place winners in each category
receive a cash prize of $1,000.

DEADLINE: November 15.

NWA Nonfiction Contest

National Writers Association
10940 S. Parke Road, #508
Parker, CO 80134

DESCRIPTION: Sponsored by the National Writers Association, this annual award accepts previously unpublished nonfiction entries. The competition is open to all writers and looks to encourage high-quality nonfiction writing.
Website: www.nationalwriters.com

REQUIREMENTS: Entry fee, $18. Multiple entries are accepted under separate cover. Entries should not exceed 5,000 words. Accepts hard copy. All entries must be accompanied by an entry form (available with an SASE or at the website).

PRIZES: First-place winner receives a cash prize of $200. Second- and third-place winners receive $100 and $50, respectively.

DEADLINE: December 31.

NWA Short Story Contest

National Writers Association
10940 S. Parke Road, #508
Parker, CO 80134

DESCRIPTION: This annual contest encourages the development of creative skills and recognizes and rewards outstanding ability in short story writing. The contest is open to all writers and accepts previously unpublished, original work only.
Website: www.nationalwriters.com

REQUIREMENTS: Entry fee, $15. Multiple entries are accepted under separate cover. Entries should not exceed 5,000 words. Accepts hard copy. All entries must be accompanied by an entry form (available with an SASE or at the website).

PRIZES: First-place winner receives a cash prize of $250. Second- and third-place winners receive $100 and $50, respectively.

DEADLINE: July 1.

Pacific Northwest Writers Association Literary Contest

PMB 2717
1420 NW Gilman Blvd, Ste. 2
Issaquah, WA 98027

DESCRIPTION: The Pacific Northwest Writers Association sponsors this annual contest that offers awards in 11 categories including juvenile short story or picture book, nonfiction/memoir, young adult novel, and screenwriting.
Website: www.pnwa.org

REQUIREMENTS: Entry fee, $35 for members; $50 for non-members. Limit one entry per category. Word lengths vary for each category; check website for complete guidelines. Accepts hard copy. Send 2 copies of manuscript. Include a 3x5 card with author's name, address, and title of entry.

PRIZES: Winners in each category receive a cash prize of $1,000 and publication.

DEADLINE: February 22.

Pockets Annual Fiction Contest

Attn: Lynn Gilliam
Box 340004
1908 Grand Avenue
Nashville, TN 37203-0004

DESCRIPTION: Fiction entries are accepted by this contest, sponsored by *Pockets* magazine. It is open to all types of fiction with the exception of historical and biblical fiction. The competition is open to all writers.
Website: www.pockets.org

REQUIREMENTS: No entry fee. Entries should be between 1,000 and 1,600 words. Multiple entries are accepted. Accepts hard copy. Manuscript must list accurate word count on cover sheet. Entries not adhering to the word guideline lengths will be disqualified. Send an SASE for return of manuscript.

PRIZES: Winner receives a cash award of $1,000 and publication in *Pockets*.

DEADLINE: Submissions must be postmarked between March 1 and August 15.

San Antonio Writers Guild Writing Contests

P.O. Box 34775
San Antonio, TX 78265

DESCRIPTION: This annual contest offers prizes in the categories of children's literature, fiction, nonfiction, and poetry. It is open to all writers and accepts previously unpublished, original work only.
Website: www.sawritersguild.org

REQUIREMENTS: Entry fee, $10 for members, $20 for non-members. Submit two copies of each entry. Accepts hard copy. Complete guidelines and category information are available with an SASE or at the website.

PRIZES: First-place winners receive a cash prize of $100. Second- and third-place winners receive cash prizes of $50 and $25, respectively.

DEADLINE: First Thursday of October.

Seventeen Magazine Fiction Contest

Seventeen Magazine
1440 Broadway, 13th Floor
New York, NY 10018

DESCRIPTION: Each year *Seventeen* magazine sponsors this annual fiction contest open to writers between the ages of 13 and 21. It accepts original short stories that exemplify creativity, originality, and writing ability.
Website: www.seventeen.com

REQUIREMENTS: No entry fee. Entries should not exceed 2,000 words. Multiple entries are accepted. Accepts hard copy. Send an SASE or visit the website for further information.

PRIZES: Grand-prize winner receives a cash prize of $2,000 and publication in *Seventeen*. Cash prizes and possible publication are awarded to second- and third-place winners.

DEADLINE: January 2.

Skipping Stones Awards

Skipping Stones
P.O. Box 3939
Eugene, OR 97403

DESCRIPTION: These annual awards look to cultivate awareness of our multicultural world without perpetuating stereotypes or biases. Entries should promote cooperation, non-violence, and an appreciation of nature. Entries may be published magazine articles, books, or educational videos.
Website: www.skippingstones.org

REQUIREMENTS: Entry fee, $3. Send 4 copies of each entry. Only entries published in the year preceding the contest are eligible. Send an SASE, visit the website, or send an email to editor@skippingstones.org for further information.

PRIZES: Cash prizes are awarded to first-through fourth-place winners.

DEADLINE: February.

Southwest Writers Contests

Southwest Writers Workshop
3721 Morris NE
Albuquerque, NM 87110

DESCRIPTION: The Southwest Writers Workshop contests offer awards in several categories including middle-grade short story, children's picture book, genre story, and young adult short story. The competition accepts unpublished material only.
Website: www.southwestwriters.org

REQUIREMENTS: Entry fee, $25 for members; $45 for non-members. Word lengths vary for each category; check website for specific information. Accepts hard copy. Author's name should not appear on manuscript itself. Include a cover letter with author's name, title of entry, and contact information.

PRIZES: Winners receive cash awards ranging from $50 to $150. First-place winners also compete for the $1,000 Storyteller Award.

DEADLINE: May 1.

Sydney Taylor Manuscript Competition

Association of Jewish Libraries
c/o Aileen Grossberg
67 Park Street
Montclair, NJ 07042

DESCRIPTION: The Association of Jewish Libraries sponsors this annual award to encourage aspiring writers of Jewish children's books. It looks for fiction manuscripts of interest to children ages 8 to 11. Entries must have universal appeal and should reveal positive aspects of Jewish life.
Website: www.jewishlibraries.com

REQUIREMENTS: No entry fee. Entries should be between 64 and 200 pages in length. Limit one entry per competition. Submit 3 copies of manuscript. Accepts hard copy. Send an SASE or visit the website for further information.

PRIZES: Winner receives a cash prize of $1,000 and possible publication.

DEADLINE: December 15.

Tennessee Williams One-Act Play Competition

Tennessee Williams New Orleans Literary Festival
938 Lafayette Street, Suite 514
New Orleans, LA 70113

DESCRIPTION: Honoring previously unpublished playwrights, this competition is open to all writers and looks for high-quality one-act play submissions.
Website: www.tennesseewilliams.org

REQUIREMENTS: Entry fee, $20. Plays should be one act only and have a 60-minute running time. Accepts hard copy. All entries must include an entry form, available with an SASE or at the website.

PRIZES: Winner receives a cash prize of $1,500 and a staged reading of the winning play.

DEADLINE: Entries are accepted between September 1 and November 15.

Utah Original Writing Competition

617 E South Temple
Salt Lake City, UT 84102

DESCRIPTION: This competition is held annually to promote and reward excellence from writers living in Utah. Awards are presented in several categories including juvenile essay, short story, juvenile book, and general nonfiction.
Website: http://arts.utah.gov/literature/ comprules.html

REQUIREMENTS: No entry fee. Word lengths vary for each category; check website for specific information. Limit one entry per category. Manuscripts will not be returned. Entrants must reside in Utah. Accepts hard copy.

PRIZES: Winners receive cash prizes ranging from $750 to $1,000.

DEADLINE: June 27.

Paul A. Witty Short Story Award

International Reading Association
P.O. Box 8139
Newark, DE 19714-8139

DESCRIPTION: The International Reading Association sponsors this annual award that looks for high-quality short stories published for the first time during the previous calendar year. Entries should serve as a literary standard and encourage young people to read periodicals.
Website: www.reading.org

REQUIREMENTS: No entry fee. No word length limitations. Accepts hard copy accompanied by a copy of the periodical. Limit three entries per magazine. Authors or publishers may nominate pieces for this award. For additional information, visit the website or send an SASE.

PRIZES: A cash award of $1,000 is presented to the winner.

DEADLINE: December 1.

WOW! Women on Writing Flash Fiction Contests

Contest Coordinator

DESCRIPTION: Every three months, *WOW! Women on Writing,* e-zine holds a contest for flash fiction submissions that target women. Entries should be lighthearted, imaginative, and creative.
Website: www.wow-womenonwriting.com

REQUIREMENTS: Entry fee, $5. Multiple entries are accepted. Entries should be between 250 and 500 words. Accepts email only to contests@wow-womenonwriting.com and electronic submissions through the website.

PRIZES: First-place winner receives a cash award of $200. Second- and third-place winners receive cash awards of $150 and $100, respectively. All winners will be published on the WOW! website.

DEADLINE: February 28; May 31; August 31; and November 30.

Writers' Journal Writing Contests

P.O. Box 394
Perham, MN 56573-0374

DESCRIPTION: *Writers' Journal* sponsors numerous contests each year that are open to writers at all levels of experience. Contest categories include horror/ghost stories, romance, fiction, and poetry.
Website: www.writersjournal.com

REQUIREMENTS: Entry fees range from $3 to $5 depending on contest. Word lengths and complete guidelines vary for each contest. Check the website or send an SASE for further details. Accepts hard copy.

PRIZES: Winners of these contests receive cash prizes and publication in *Writers' Journal.*

DEADLINE: Varies for each category.

The Writing Conference, Inc. Writing Contests

P.O. Box 664
Ottawa, KS 66067-0664

DESCRIPTION: With a goal of encouraging young people to develop a love of writing, these writing contests are open to children and young adults and accept short stories, short nonfiction, and poetry.
Website: www.writingconference.com

REQUIREMENTS: No entry fee. Limit one entry per competition. Accepts hard copy. Visit the website or send an SASE for further information.

PRIZES: Winners in each category receive publication in *The Writers' Slate.*

DEADLINE: January 8.

Writing for Children Competition

Writers' Union of Canada
90 Richmond Street, Suite 200
Toronto, Ontario M5C 1P1
Canada

DESCRIPTION: Promoting new and emerging writers, this annual contest is sponsored by the Writers' Union of Canada and is open to Canadian residents only. It accepts fiction or nonfiction entries targeting children of all ages.
Website: www.writersunion.ca

REQUIREMENTS: Entry fee, $15. Multiple entries are accepted. Entries should not exceed 1,500 words. Accepts hard copy. Send an SASE or visit the website for complete guidelines.

PRIZES: Winner receives a cash prize of $1,500. The Writers' Union of Canada will submit the winning entry to several children's publishers.

DEADLINE: April 24.

Indexes

2009 Market News

New Listings ☆

Action Pursuit Games
Administrator
American Educator
Appalachian Writer's
    Association Contests
Beckett Plushie Pals
Bonbon
Brass
Broomstix for Kids
Camping Today
Charlotte Baby & Child
Cincinnati Parent
ColumbiaKids
Craft Bits
Creative Connections
Crossquarter Annual Short
    Science Fiction Contest
Crow Toes Quarterly
Dance International
Equestrian
Faith Today
Fox Valley Kids
H. E. Francis Award
Friends of the Library
    Contest
Henry Parents Magazine
Imagination-Cafe

Inland Empire Family
James Hubbard's My Family
    Doctor
Junior Shooters
JVibe
Key Club
Kid Magazine Writers
The Kids' Ark
KidSpirit
Kiki Magazine
Lacrosse Magazine
Living for the Whole Family
The Lutheran Digest
MetroKids
My Light Magazine
National Children's Theatre
    Competition
Natural Life
Natural Solutions
N.E.W. Kids
New Millennium Writings
    Awards
The Olive Branch
The Pink Chameleon
Playground Magazine
Premier Baby and Child
Purple Circle Magazine

Rainbow Rumpus
Sierra
Soccer Youth
Spaceports & Spidersilk
Stories for Children
    Magazine
Story Station
The Storyteller
t&T News
TC Magazine
Teach Kids Essentials
This Week Community
    Newspapers
Toy Trucker & Contractor
Transworld Snowboarding
The Vegetarian Journal
Writers' Journal Writing
    Contests
Your Big Backyard
Yummy Mummy

Deletions and Name Changes

Aquila: Unable to contact

Atlanta Sporting Family: Unable to contact

Big Country Peacock Chronicle: Ceased publication

B'nai B'rith Magazine: Removed per editor's request

Caring for Kids: Using reprints from land-grant colleges only

Child Life: Ceased publication

Child Welfare Report: Ceased publication

Church Educator: Ceased publication

College and Junior Tennis: Ceased publication

Cyberteens Zeen: Suspended operations

Daughters: Ceased publication

East of the Web: Unable to contact

Family Magazine/La Familia: Removed per editor's request

FamilyRapp.com: Unable to contact

Fandangle Magazine: Ceased publication

Girlfriend: Ceased publication

InQuest Gamer: Ceased publication

Keynoter: See listing for **Key Club**

Kids Hall of Fame News: Ceased publication

Know Your World Extra: Ceased publication

Knucklebones: Suspended publication

Life Learning Magazine: See listing for **Natural Life Magazine**

Living: See listing for **Living for the Whole Family**

Loud Magazine: Ceased publication

Middle Years: Unable to contact

MomsVoice.com: Ceased publication

Mr. Marquis' Museletter: See listing for **Creative Connections**

No Crime: Ceased publication

Northern Michigan Family Magazine: Suspended publication

Potluck Children's Literary Magazine: Ceased publication

St. Louis Parent: Unable to contact

Social Studies and the Young Learner: Removed at editor's request

Student Leader: On hiatus

Synapse: Ceased publication

Teaching Music: Not using freelance material at this time

Teaching PreK–8: Ceased publication

Technology & Learning: Ceased publication

Teenage Christian: See listing for **TC Magazine**

Teen Trend Magazine: Magazine up for sale; status uncertain

Today's Playground: See listing for **Playground Magazine**

True Girl: Ceased publication

Vegetarian Baby & Child: Ceased publication

Vegetarianteen.com: Ceased publication

Ventura County Parent Magazine: Ceased publication

Vertical Thought: Not using freelance material at this time

Wee Ones: Ceased publication

West Tennessee Parent & Family: Ceased publication

With: Ceased publication

With Kids: Unable to contact

Wondertime: Magazine undergoing major changes at this time

Fifty+ Freelance

You can improve your chances of selling by submitting to magazines that fill their pages with freelance material. Of the 678 markets listed in this directory, we have listed 87 markets that buy at least 50% of their freelance material from writers who are new to the magazine. Of course, there are no guarantees; but if you approach these magazines with well-written manuscripts targeted to their subject, age range, and word-limit requirements, you can increase your publication odds.

AIM Magazine
The ALAN Review
Alateen Talk
Amazing Kids!
American Secondary Education
The Apprentice Writer
Autism Asperger's Digest
bNetS@vvy
Brain, Child
Bread For God's Children
Brilliant Star
Capper's
Cat Fancy
Childhood Education
Children's Ministry
Cicada
The Claremont Review
The Clearing House
Clubhouse
Creative Connections
Creative Kids
Crow Toes Quarterly
The Dabbling Mum
Dimensions of Early Childhood
Dovetail
Educational Horizons
Education Forum
Education Week
Elementary School Writer
Encyclopedia of Youth Studies
Fort Myers & SW Florida Magazine

Gifted Education Press Quarterly
Group
Highlights for Children
The High School Journal
High School Writer (Junior High Edition)
High School Writer (Senior)
The Illuminata
I Love Cats
Indy's Child
Inside Out
Insight
Jack And Jill
Journal of School Health
Justine Magazine
Kaleidoscope
Kansas School Naturalist
Keyboard
Kindred
The Lamp-Post
Leading Edge
Learning and Leading with Technology
Literary Mama
Look-Look Magazine
Momentum
Mom Writer's Literary Magazine
Mothering
Mother Verse Magazine
My Light Magazine
North Star Family Matters
The Olive Branch
Our Children

Parents Express-Pennsylvania
The Pink Chameleon
PresenTense Magazine
Principal
Read, America!
Relate Magazine
Reunions Magazine
SchoolArts
School Library Journal
Seek
Shine Brightly
Sisterhood Agenda
Skipping Stones
South Jersey Mom
Sparkle
Stone Soup
Stories for Children Magazine
Teen Voices
Transitions Abroad
The Universe in the Classroom
Voice of Youth Advocates
What If?
Writers' Journal
Young People's Press
Zamoof! Magazine

Category Index

To help you find the appropriate market for your manuscript or query letter, we have compiled a category and subject index listing magazines according to their primary editorial interests. Pay close attention to the markets that overlap. For example, when searching for a market for your rock-climbing adventure story for 8- to 12-year-old readers, you might look under the categories "Adventure Stories" and "Middle-grade (Fiction)." If you have an idea for an article about blue herons for early readers, look under the categories "Animals/Pets" and "Early Reader (Nonfiction)" to find possible markets. Always check the magazine's listing for explanations of specific needs.

For your convenience, we have listed below all of the categories that are included in this index. If you don't find a category that exactly fits your material, try to find a broader term that covers your topic.

Adventure Stories
Animals (Fiction)
Animals/Pets (Nonfiction)
Audio/Video
Bilingual (Nonfiction)
Biography
Boys' Magazines
Canadian Magazines
Career/College
Child Care
Computers
Contemporary Fiction
Crafts/Hobbies
Current Events
Drama
Early Reader (Fiction)
Early Reader (Nonfiction)
Education/Classroom
Factual/Informational
Family/Parenting
Fantasy
Folktales/Folklore
Games/Puzzles/Activities
Geography
Gifted Education
Girls' Magazines
Health/Fitness

Historical Fiction
History
Horror
How-to
Humor (Fiction)
Humor (Nonfiction)
Inspirational Fiction
Language Arts
Mathematics
Middle-grade (Fiction)
Middle-grade (Nonfiction)
Multicultural/Ethnic (Fiction)
Multicultural/Ethnic (Nonfiction)
Music
Mystery/Suspense
Nature/Environment (Fiction)
Nature/Environment (Nonfiction)
Personal Experience
Photo-Essays
Popular Culture
Preschool (Fiction)
Preschool (Nonfiction)
Profile/Interview

Read-aloud Stories
Real-life/Problem-solving
Rebus
Recreation/Entertainment
Regional (Fiction)
Regional (Nonfiction)
Religious (Fiction)
Religious (Nonfiction)
Reviews
Romance
Science Fiction
Science/Technology
Self-help
Services/Clubs
Social Issues
Special Education
Sports (Fiction)
Sports (Nonfiction)
Travel
Western
Writing
Young Adult (Fiction)
Young Adult (Nonfiction)
Young Author (Fiction)
Young Author (Nonfiction)

Historical Fiction

Mystery/Suspense

Nature/Environment (Fiction)

Nature/Environment (Nonfiction)

Personal Experience

Read–aloud Stories

Real–life/ Problem–solving

Rebus

Recreation/ Entertainment

Religious (Fiction)

Self–help

Services/Clubs

Social Issues

Special Education

Sports (Fiction)

Sports (Nonfiction)

Young Author (Fiction)

Young Author (Nonfiction)

Magazine and Contest Index

The following codes have been used to indicate each publication's readership: **YA**=Young adults, **A**=Adults, **E**=Educators (including librarians, teachers, administrators, student group leaders, and child-care professionals), **F**=Family (general interest), **P**=Parents. We have listed age ranges when specified by the editor.

If you do not find a particular magazine, turn to Market News on page 338.

★ indicates a newly listed magazine